Introduction to
Probability Models
Fourth Edition

Fourth Edition

INTRODUCTION TO PROBABILITY MODELS

Sheldon M. Ross

Department of Industrial Engineering and Operations Research
University of California, Berkeley
Berkeley, California

ACADEMIC PRESS, INC.
Harcourt Brace Jovanovich, Publishers
Boston   San Diego   New York
Berkeley   London   Sydney   Tokyo   Toronto

ACADEMIC PRESS, INC.
1250 Sixth Avenue, San Diego, CA 92101

United Kingdom Edition published by
ACADEMIC PRESS LIMITED
24–28 Oval Road, London NW1 7DX

Table 11.2 is reprinted from Sheldon M. Ross, "Approximations in
Renewel Theory," *Probability in the Engineering and Informational
Sciences,* 1(2):163-175 (1987), published by Cambridge University
Press.

Library of Congress Cataloging-in-Publication Data

Ross, Sheldon M.
Introduction to probability models / Sheldon M. Ross.—4th ed.
    p. cm.
Bibliography: p.
Includes index.
ISBN 0-12-598464-2
1. Probabilities.   I. Title.
QA273.R84   1989   89-6539
519.2—dc20   CIP

PRINTED IN THE UNITED STATES OF AMERICA

90 91 92   9 8 7 6 5 4 3 2

Preface to the Fourth Edition

For the most part, the fourth edition is quite similar in spirit to the third. It differs from the third primarily by the addition of material in the chapters on discrete time Markov chains, continuous time Markov chains, renewal theory, and simulation. For instance, examples concerning communication protocols and waiting times for patterns were added to the Markov chain chapter, as was some additional material on the reverse chain. The chapter on continuous time Markov chains includes additional material on birth and death processes as well as new material on computing the transition probability function. The renewal theory chapter includes, among other new things, a section on computing the renewal function, an application of results from the 2-state continuous time Markov chain to compute the renewal function when the interarrival distribution is the convolution or the mixture of two exponentials, and an application to quality control. There is additional material in the Simulation chapter concerning ways of efficiently generating nonhomogeneous Poisson processes, and the use of "conditioning" to reduce the variance when trying to estimate the average customer delay in a variety of queueing models.

Preface

This text is intended as an introduction to elementary probability theory and stochastic processes. It is particularly well-suited for those wanting to see how probability theory can be applied to the study of phenomena in fields such as engineering, management science, the physical and social sciences, and operations research.

It is generally felt that there are two approaches to the study of probability theory. One approach is heuristic and nonrigorous and attempts to develop in the student an intuitive feel for the subject which enables him or her to "think probabilistically." The other approach attempts a rigorous development of probability by using the tools of measure theory. It is the first approach that is employed in this text. However, as it is extremely important in both understanding and applying probability theory to be able to "think probabilistically," this text should also be useful to students interested primarily in the second approach.

Chapters 1 and 2 deal with basic ideas of probability theory. In Chapter 1 an axiomatic framework is presented, while in Chapter 2 the important concept of a random variable is introduced.

Chapter 3 is concerned with the subject matter of conditional probability and conditional expectation. "Conditioning" is one of the key tools of probability theory, and it is stressed throughout the book. When properly used, conditioning often enables us to easily solve problems that at first glance seem quite difficult. The final section of this chapter, not in the original edition, presents applications to (1) a computer list problem, (2) a random graph, and (3) the Polya urn model and its relation to the Bose-Einstein distribution.

In Chapter 4 we come into contact with our first random, or stochastic, process, known as a Markov chain, which is widely applicable to the study of many real-world phenomena. New applications to genetics and production processes are presented. The concept of time reversibility is introduced and its usefulness illustrated. In the final section we consider a model for optimally making decisions known as a Markovian decision process.

In Chapter 5 we are concerned with a type of stochastic process known as a counting process. In particular, we study a kind of counting process known as a Poisson process. The intimate relationship between this process and the exponential distribution is discussed.

Chapter 6 considers Markov chains in continuous time with an emphasis on birth and death models. Time reversibility is shown to be a useful concept, as it is in the study of discrete-time Markov chains. The final section presents the computationally important technique of uniformization.

Chapter 7, the renewal theory chapter, is concerned with a type of counting process more general than the Poisson. By making use of renewal reward processes, limiting results are obtained and applied to various fields.

Chapter 8 deals with queueing, or waiting line, theory. After some preliminaries dealing with basic cost identities and types of limiting probabilities, we consider exponential queueing models and show how such models can be analyzed. Included in the models we study is the important class known as network of queues. We then study models in which some of the distributions are allowed to be arbitrary.

Chapter 9 is concerned with reliability theory. This chapter will probably be of greatest interest to the engineer and operations researcher.

Ideally, this text would be used in a one-year course in probability models. Other possible courses would be a one-semester course in introductory probability theory (involving Chapters 1–3 and parts of others) or a course in elementary stochastic processes. It is felt that the textbook is flexible enough to be used in a variety of possible courses. For example, I have used Chapters 5 and 8, with smatterings from Chapters 4 and 6, as the basis of an introductory course in queueing theory.

There are many examples worked out throughout the text, and there are also a large number of problems to be worked by students. Answers to selected problems appear in the text, and a separate solutions manual is available to instructors using the text.

Contents

3. Conditional Probability and Conditional Expectation

4. Markov Chains

5. The Exponential Distribution and the Poisson Process

6. Continuous-Time Markov Chains

9. Reliability Theory

10. Brownian Motion and Stationary Processes

11. Simulation

Introduction to
Probability Models
Fourth Edition

Chapter 1

Introduction to Probability Theory

1. Introduction

Any realistic model of a real-world phenomenon must take into account the possibility of randomness. That is, more often than not, the quantities we are interested in will not be predictable in advance but, rather, will exhibit an inherent variation that should be taken into account by the model. This is usually accomplished by allowing the model to be probabilistic in nature. Such a model is, naturally enough, referred to as a probability model.

The majority of the chapters of this book will be concerned with different probability models of natural phenomena. Clearly, in order to master both the "model building" and the subsequent analysis of these models, we must have a certain knowledge of basic probability theory. The remainder of this chapter, as well as the next two chapters, will be concerned with a study of this subject.

2. Sample Space and Events

Suppose that we are about to perform an experiment whose outcome is not predictable in advance. However, while the outcome of the experiment will not be known in advance, let us suppose that the set of all possible outcomes is known. This set of all possible outcomes of an experiment is known as the *sample space* of the experiment and is denoted by S.

Some examples are the following.

1. If the experiment consists of the flipping of a coin, then

$$S = \{H, T\}$$

where H means that the outcome of the toss is a head and T that it is a tail.

2. If the experiment consists of tossing a die, then the sample space is

$$S = \{1, 2, 3, 4, 5, 6\}$$

where the outcome i means that i appeared on the die, $i = 1, 2, 3, 4, 5, 6$.

3. If the experiment consists of flipping two coins then the sample space consists of the following four points

$$S = \{(H, H), (H, T), (T, H), (T, T)\}$$

The outcome will be (H, H) if both coins come up heads; it will be (H, T) if the first coin comes up heads and the second comes up tails; it will be (T, H) if the first comes up tails and the second heads; and it will be (T, T) if both coins come up tails.

4. If the experiment consists of tossing two dice, then the sample space consists of the 36 points

$$S = \begin{cases} (1,1), (1,2), (1,3), (1,4), (1,5), (1,6) \\ (2,1), (2,2), (2,3), (2,4), (2,5), (2,6) \\ (3,1), (3,2), (3,3), (3,4), (3,5), (3,6) \\ (4,1), (4,2), (4,3), (4,4), (4,5), (4,6) \\ (5,1), (5,2), (5,3), (5,4), (5,5), (5,6) \\ (6,1), (6,2), (6,3), (6,4), (6,5), (6,6) \end{cases}$$

where the outcome (i,j) is said to occur if i appears on the first die and j on the second die.

5. If the experiment consists of measuring the lifetime of a car, then the sample space consists of all nonnegative real numbers. That is,

$$S = [0, \infty)^* \quad \Diamond$$

---

*The set (a, b) is defined to consist of all points x such that $a < x < b$. The set $[a, b]$ is defined to consist of all points x such that $a \leq x \leq b$. The sets $(a, b]$ and $[a, b)$ are defined, respectively, to consist of all points x such that $a < x \leq b$ and all points x such that $a \leq x < b$.

Any subset E of the sample space S is known as an *event*. Some examples of events are the following.

1′. In Example (1) above, if $E = \{H\}$, then E is the event that a head appears on the flip of the coin. Similarly, if $E = \{T\}$, then E would be the event that a tail appears.

2′. In Example (2), if $E = \{1\}$, then E is the event that one appears on the toss of the die. If $E = \{2, 4, 6\}$, then E would be the event that an even number appears on the toss.

3′. In example (3), if $E = \{(H, H),(H, T)\}$, then E is the event that a head appears on the first coin.

4′. In Example (4), if $E = \{(1, 6), (2, 5), (3, 4), (4, 3), (5, 2), (6, 1)\}$, then E is the event that the sum of the dice equals seven.

5′. In Example (5), if $E = (2, 6)$, then E is the event that the car lasts between two and six years.   ◇

For any two events E and F of a sample space S we define the new event $E \cup F$ to consist of all points which are either in E or in F or in both E and F. That is, the event $E \cup F$ will occur if *either* E or F occurs. For example, in (1) if $E = \{H\}$ and $F = \{T\}$, then

$$E \cup F = \{H, T\}$$

That is, $E \cup F$ would be the whole sample space S. In (2) if $E = \{1, 3, 5\}$ and $F = \{1, 2, 3\}$, then

$$E \cup F = \{1, 2, 3, 5\}$$

and thus $E \cup F$ would occur if the outcome of the die is either a 1 or 2 or 3 or 5. The event $E \cup F$ is often referred to as the *union* of the event E and the event F.

For any two events E and F, we may also define the new event EF, referred to as the *intersection* of E and F, as follows. EF consists of all points which are *both* in E and in F. That is, the event EF will occur only if E and F occur. For example, in (2) if $E = \{1, 3, 5\}$ and $F = \{1, 2, 3\}$, then

$$EF = \{1, 3\}$$

and thus EF would occur if the outcome of the die is either 1 or 3. In example (1) if $E = \{H\}$ and $F = \{T\}$, then the event EF would not consist of any points and hence could not occur. To give such an event a name we shall refer to its as the null event and denote it by $\varnothing$. (That is, $\varnothing$ refers to the event consisting of no points.) If $EF = \varnothing$, then E and F are said to be *mutually exclusive*.

We also define unions and intersections of more than two events in a similar manner. If $E_1, E_2, \ldots$ are events, then the union of these events, denoted by $\cup_{n=1}^{\infty} E_n$, is defined to be that event which consists of all points that are in E_n for at least one value of $n = 1, 2, \ldots$. Similarly, the intersection of the events E_n, denoted by $\Pi_{n=1}^{\infty} E_n$, is defined to be the event consisting of those points that are in all of the events $E_n, n = 1, 2, \ldots$.

Finally, for any event E we define the new event E^c, referred to as the *complement* of E, to consist of all points in the sample space S which are not in E. That is E^c will occur if and only if E does not occur. In Example (4) if $E = \{(1, 6), (2, 5), (3, 4), (4, 3), (5, 2), (6, 1)\}$, then E^c will occur if the sum of the dice does not equal seven. Also note that since the experiment must result in some outcome, it follows that $S^c = \emptyset$.

3. Probabilities Defined on Events

Consider an experiment whose sample space is S. For each event E of the sample space S, we assume that a number $P(E)$ is defined and satisfies the following three conditions:

(i) $0 \leq P(E) \leq 1$.
(ii) $P(S) = 1$.
(iii) For any sequence of events $E_1, E_2, \ldots$ which are mutually exclusive, that is, events for which $E_n E_m = \emptyset$ when $n \neq m$, then

$$P\left(\bigcup_{n=1}^{\infty} E_n \right) = \sum_{n=1}^{\infty} P(E_n)$$

We refer to $P(E)$ as the probability of the event E.

Example 3a In the coin tossing example, if we assume that a head is equally likely to appear as a tail, then we would have

$$P(\{H\}) = P(\{T\}) = \tfrac{1}{2}$$

On the other hand, if we had a biased coin and felt that a head was twice as likely to appear as a tail, then we would have

$$P(\{H\}) = \tfrac{2}{3}, \qquad P(\{T\}) = \tfrac{1}{3} \quad \diamond$$

Example 3b In the die tossing example, if we supposed that all six numbers were equally likely to appear, then we would have $P(\{1\}) = P(\{2\}) = P(\{3\}) = P(\{4\}) = P(\{5\}) = P(\{6\}) = \tfrac{1}{6}$. From (iii) it would follow that the probability of getting an even number would equal

$$P(\{2, 4, 6\}) = P(\{2\}) + P(\{4\}) + P(\{6\})$$
$$= \tfrac{1}{2} \quad \Diamond$$

Remarks We have chosen to give a rather formal definition of probabilities as being functions defined on the events of a sample space. However, it turns out that these probabilities have a nice intuitive property. Namely, if our experiment is repeated over and over again then (with probability 1) the proportion of time that event E occurs will just be $P(E)$. $\Diamond$

Since the events E and E^c are always mutually exclusive and since $E \cup E^c = S$ we have by (ii) and (iii) that

$$1 = P(S) = P(E \cup E^c) = P(E) + P(E^c)$$

or

$$P(E) + P(E^c) = 1 \qquad (3.1)$$

In words, Equation (3.1) states that the probability that an event does not occur is one minus the probability that it does occur.

We shall now derive a formula for $P(E \cup F)$, the probability of all points either in E or in F. To do so, consider $P(E) + P(F)$, which is the probability of all points in E plus the probability of all points in F. Since any point that is in both E and F will be counted twice in $P(E) + P(F)$ and only once in $P(E \cup F)$, we must have

$$P(E) + P(F) = P(E \cup F) + P(EF)$$

or equivalently

$$P(E \cup F) = P(E) + P(F) - P(EF) \qquad (3.2)$$

Note that in the case that E and F are mutually exclusive (that is, when $EF = \emptyset$), then Equation (3.2) states that

$$P(E \cup F) = P(E) + P(F) - P(\emptyset)$$
$$= P(E) + P(F)$$

a result which also follows from condition (iii). (Why is $P(\emptyset) = 0$?)

Example 3c Suppose that we toss two coins, and suppose that we assume that each of the four points in the sample space $S =$

$\{(H, H), (H, T), (T, H), (T, T)\}$ are equally likely and hence have probability $\frac{1}{4}$. Let

$$E = \{(H, H), (H, T)\} \quad \text{and} \quad F = \{(H, H), (T, H)\}$$

That is, E is the event that the first coin falls heads, and F is the event that the second coin falls heads.

By Equation (3.2) we have that $P(E \cup F)$, the probability that either the first or the second coin falls heads is given by

$$\begin{aligned}
P(E \cup F) &= P(E) + P(F) - P(EF) \\
&= \tfrac{1}{2} + \tfrac{1}{2} - P(\{H, H\}) \\
&= 1 - \tfrac{1}{4} = \tfrac{3}{4}
\end{aligned}$$

This probability could, of course, have been computed directly since

$$P(E \cup F) = P(\{H, H), (H, T), (T, H)\}) = \tfrac{3}{4} \quad \Diamond$$

We may also calculate the probability that any one of the three events E or F or G occurs. This is done as follows

$$P(E \cup F \cup G) = P((E \cup F) \cup G)$$

which by Equation (3.2) equals

$$P(E \cup F) + P(G) - P((E \cup F)G)$$

Now we leave it for the reader to show that the events $(E \cup F)G$ and $EG \cup FG$ are equivalent, and hence the above equals

$$\begin{aligned}
P(E &\cup F \cup G) \\
&= P(E) + P(F) - P(EF) + P(G) - P(EG \cup FG) \\
&= P(E) + P(F) - P(EF) + P(G) - P(EG) - P(FG) + P(EGFG) \\
&= P(E) + P(F) + P(G) - P(EF) - P(EG) - P(FG) + P(EFG) \quad (3.3)
\end{aligned}$$

In fact, it can be shown by induction that for any n events $E_1, E_2, E_3, \ldots, E_n$

$$\begin{aligned}
P(E_1 &\cup E_2 \cup \cdots \cup E_n) \\
&= \sum_i P(E_i) - \sum_{i<j} P(E_iE_j) + \sum_{i<j<k} P(E_iE_jE_k) \\
&\quad - \sum_{i<j<k<l} P(E_iE_jE_kE_l) + \cdots + (-1)^{n+1} P(E_1E_2 \ldots E_n) \quad (3.4)
\end{aligned}$$

In words, Equation (3.4) states that the probability of the union of n

events equals the sum of the probabilities of these events taken one at a time minus the sum of the probabilities of these events taken two at a time plus the sum of the probabilities of these events taken three at a time, and so on.

4. Conditional Probabilities

Suppose that we toss two dice and suppose that each of the 36 possible outcomes are equally likely to occur and hence have probability $\frac{1}{36}$. Suppose that we observe that the first die is a four. Then, given this information, what is the probability that the sum of the two dice equals six? To calculate this probability we reason as follows: Given that the initial die is a four, it follows that there can be at most six possible outcomes of our experiment, namely, (4, 1), (4, 2), (4, 3), (4, 4), (4, 5), and (4, 6). Since each of these outcomes originally had the same probability of occurring, they should still have equal probabilities. That is, given that the first die is a four, then the (conditional) probability of each of the outcomes (4, 1), (4, 2), (4, 3), (4, 4), (4, 5), (4, 6) is $\frac{1}{6}$ while the (conditional) probability of the other 30 points in the sample space is 0. Hence, the desired probability will be $\frac{1}{6}$.

If we let E and F denote respectively the event that the sum of the dice is six and the event that the first die is a four, then the probability just obtained is called the conditional probability that E occurs given that F has occurred and is denoted by

$$P(E|F)$$

A general formula for $P(E|F)$ which is valid for all events E and F is derived in the same manner as above. Namely, if the event F occurs, then in order for E to occur it is necessary that the actual occurrence be a point in both E and in F, that is it must be in EF. Now, as we know that F has occurred, it follows that F becomes our new sample space and hence the probability that the event EF occurs will equal the probability of EF relative to the probability of F. That is

$$P(E|F) = \frac{P(EF)}{P(F)} \tag{4.1}$$

Note that Equation (4.1) is only well defined when $P(F) > 0$ and hence $P(E|F)$ is only defined when $P(F) > 0$.

Example 4a Suppose cards numbered one through ten are placed in a hat, mixed up, and then one of the cards is drawn. If we are told that the number on the drawn card is at least five, then what is the conditional probability that it is ten?

Solution: Let E denote the event that the number of the drawn card is ten, and let F be the event that it is at least five. The desired probability is $P(E|F)$. Now, from Equation (4.1)

$$P(E|F) = \frac{P(EF)}{P(F)}$$

However, $EF = E$ since the number of the card will be both ten and at least five if and only if it is number ten. Hence,

$$P(E|F) = \frac{\frac{1}{10}}{\frac{6}{10}} = \frac{1}{6}   \quad \Diamond$$

Example 4b   A family has two children. What is the conditional probability that both are boys given that at least one of them is a boy? Assume that the sample space S is given by $S = \{(b, b), (b, g), (g, b), (g, g)\}$, and all outcomes are equally likely. ((b, g) means for instance that the older child is a boy and the younger child a girl.)

Solution: Letting E denote the event that both children are boys, and F the event that at least one of them is a boy, then the desired probability is given by

$$P(E|F) = \frac{P(EF)}{P(F)}$$

$$= \frac{P(\{(b, b)\})}{P(\{(b, b), (b, g), (g, b)\})} = \frac{\frac{1}{4}}{\frac{3}{4}} = \frac{1}{3}   \quad \Diamond$$

Example 4c   Bev can either take a course in computers or in chemistry. If Bev takes the computer course, then she will receive an A grade with probability $\frac{1}{2}$, while if she takes the chemistry course then she will receive an A grade with probability $\frac{1}{3}$. Bev decides to base her decision on the flip of a fair coin. What is the probability that Bev will get an A in chemistry?

Solution: If we let F be the event that Bev takes chemistry and E denote the event that she receives an A in whatever course she takes, then the desired probability is $P(EF)$. This is calculated by using Equation (4.1) as follows:

$$P(EF) = P(F)P(E|F)$$
$$= \tfrac{1}{2}\tfrac{1}{3} = \tfrac{1}{6} \quad \Diamond$$

Example 4d Suppose an urn contains seven black balls and five white balls. We draw two balls from the urn without replacement. Assuming that each ball in the urn is equally likely to be drawn, what is the probability that both drawn balls are black?

Solution: Let F and E denote respectively the events that the first and second ball drawn is black. Now, given that the first ball selected is black, there are six remaining black balls and five white balls, and so $P(E|F) = \tfrac{6}{11}$. As $P(F)$ is clearly $\tfrac{7}{12}$, our desired probability is

$$P(EF) = P(F)P(E|F)$$
$$= \tfrac{7}{12}\tfrac{6}{11} = \tfrac{42}{132} \quad \Diamond$$

Example 4e Suppose that each of three men at a party throws his hat into the center of the room. The hats are first mixed up and then each man randomly selects a hat. What is the probability that none of the three men winds up with his own hat?

Solution: We shall solve the above by first calculating the complementary probability that at least one man winds up with his own hat. Let us denote by E_i, $i = 1, 2, 3$, the event that the ith man winds up with his own hat. In order to calculate the probability $P(E_1 \cup E_2 \cup E_3)$, we first note that

$$P(E_i) = \tfrac{1}{3}, \quad i = 1, 2, 3$$
$$P(E_iE_j) = \tfrac{1}{6}, \quad i \neq j$$
$$P(E_1E_2E_3) = \tfrac{1}{6} \tag{4.2}$$

To see why Equation (4.2) is correct, consider first $P(E_iE_j) = P(E_i)P(E_j|E_i)$. Now $P(E_i)$, the probability that the ith man selects his own hat, is clearly $\tfrac{1}{3}$ since he is equally likely to select any of the three hats. On the other hand, given that the ith man has selected his own hat, then there remain two hats that the jth man may select from, and as one of these two is his own hat, it follows that with probability $\frac{1}{2}$ he will select it. That is, $P(E_j|E_i) = \frac{1}{2}$ and so

$$P(E_iE_j) = P(E_i)P(E_j|E_i) = \frac{1}{3}\frac{1}{2} = \frac{1}{6}$$

To calculate $P(E_1E_2E_3)$ we write

$$P(E_1E_2E_3) = P(E_1E_2)P(E_3|E_1E_2)$$
$$= \frac{1}{6}P(E_3|E_1E_2)$$

However, given that the first two men get their own hats it follows that the third man must also get his own hat (since there are no other hats left). That is, $P(E_3|E_1E_2) = 1$ and so

$$P(E_1E_2E_3) = \frac{1}{6}$$

Now, from Equation (3.4) we have that

$$P(E_1 \cup E_2 \cup E_3) = P(E_1) + P(E_2) + P(E_3) - P(E_1E_2)$$
$$- P(E_1E_3) - P(E_2E_3) + P(E_1E_2E_3)$$
$$= 1 - \frac{1}{2} + \frac{1}{6}$$
$$= \frac{2}{3}$$

Hence, the probability that none of the men winds up with his own hat is $1 - \frac{2}{3} = \frac{1}{3}$. ◊

5. Independent Events

Two events E and F are said to be *independent* if

$$P(EF) = P(E)P(F)$$

By Equation (4.1) this implies that E and F are independent if $P(E|F) = P(E)$ (which also implies that $P(F|E) = P(F)$). That is, E and F are independent if knowledge that F has occurred does not affect the probability that E occurs. That is, the occurrence of E is independent of whether or not F occurs.

Two events E and F which are not independent are said to be *dependent*.

Example 5a Suppose we toss two fair dice. Let E_1 denote the event that the sum of the dice is six and F denote the event that the first die equals four. Then

$$P(E_1F) = P(\{4, 2\}) = \frac{1}{36}$$

while

$$P(E_1)P(F) = \tfrac{5}{36}\tfrac{1}{6} = \tfrac{5}{216}$$

and hence E_1 and F are not independent. Intuitively, the reason for this is clear for if we are interested in the possibility of throwing a six (with two dice), then we will be quite happy if the first die lands four (or any of the numbers 1, 2, 3, 4, 5) for then we still have a possibility of getting a total of six. On the other hand, if the first die landed six, then we would be unhappy as we would no longer have a chance of getting a total of six. In other words, our chance of getting a total of six depends upon the outcome of the first die and hence E_1 and F cannot be independent.

Let E_2 be the event that the sum of the dice equals seven. Is E_2 independent of F? The answer is yes since

$$P(E_2F) = P(\{(4, 3)\}) = \tfrac{1}{36}$$

while

$$P(E_2)P(F) = \tfrac{1}{6}\tfrac{1}{6} = \tfrac{1}{36}$$

We leave it for the reader to present the intuitive argument why the event that the sum of the dice equals seven is independent of the outcome on the first die. ◇

The definition of independence can be extended to more than two events. The events $E_1, E_2, \ldots, E_n$ are said to be independent if for every subset $E_{1'}, E_{2'}, \ldots, E_{r'}, r \leq n$, of these events

$$P(E_{1'}E_{2'} \cdots E_{r'}) = P(E_{1'})P(E_{2'}) \cdots P(E_{r'})$$

Intuitively, the events $E_1, E_2 \ldots, E_n$ are independent if knowledge of the occurrence of any of these events has no effect on the probability of any other event.

Example 5b (Pairwise Independent Events That Are Not Independent): Let a ball be drawn from an urn containing four balls, numbered 1, 2, 3, 4. Let $E = \{1, 2\}$, $F = \{1, 3\}$, $G = \{1, 4\}$. If all four outcomes are assumed equally likely, then

$$P(EF) = P(E)P(F) = \tfrac{1}{4}$$
$$P(EG) = P(E)P(G) = \tfrac{1}{4}$$
$$P(FG) = P(F)P(G) = \tfrac{1}{4}$$

However,

$$\tfrac{1}{4} = P(EFG) \neq P(E)P(F)P(G)$$

Hence, even though the events, E, F, G are pairwise independent, they are not jointly independent.   ◊

Suppose that a sequence of experiments, each of which results in either a "success" or a "failure," are to be performed. Let E_i, $i \geq 1$, denote the event that the ith experiment results in a success. If for all $i_1, i_2, \ldots, i_n$

$$P(E_{i_1}E_{i_2} \cdots E_{i_n}) = \prod_{j=1}^{n} P(E_{i_j})$$

we say that the sequence of experiments consists of *independent trials*.

Example 5c   The successive flips of a coin consists of independent trials if we assume (as is usually done) that the outcome on any flip is not influenced by the outcomes on earlier flips. A "success" might consist of the outcome heads and a "failure" tails, or possibly the reverse.   ◊

6. Bayes' Formula

Let E and F be events. We may express E as

$$E = EF \cup EF^c$$

for in order for a point to be in E, it must either be in both E and F, or it must be in E and not in F. Since EF and EF^c are obviously mutually exclusive, we have that

$$\begin{aligned} P(E) &= P(EF) + P(EF^c) \\ &= P(E|F)P(F) + P(E|F^c)P(F^c) \\ &= P(E|F)P(F) + P(E|F^c)(1 - P(F)) \end{aligned} \tag{6.1}$$

Equation (6.1) states that the probability of the event E is a weighted average of the conditional probability of E given that F has occurred and the conditional probability of E given that F has not occurred, each conditional probability being given as much weight as the event it is conditioned on has of occurring.

Example 6a   Consider two urns. The first containing two white and seven black balls, and the second containing five white and six black balls. We flip a fair coin and then draw a ball from the first urn or the second urn depending upon whether the outcome was heads or tails. What is the conditional probability that the outcome of the toss was heads given that a white ball was selected?

Solution: Let W be the event that a white ball is drawn, and let H be the event that the coin comes up heads. The desired probability $P(H|W)$ may be calculated as follows.

$$
\begin{aligned}
P(H|W) &= \frac{P(HW)}{P(W)} \\
&= \frac{P(W|H)P(H)}{P(W)} \\
&= \frac{P(W|H)P(H)}{P(W|H)P(H) + P(W|H^c)P(H^c)} \\
&= \frac{\frac{2}{9}\frac{1}{2}}{\frac{2}{9}\frac{1}{2} + \frac{5}{11}\frac{1}{2}} = \frac{22}{67} \quad \diamondsuit
\end{aligned}
$$

Example 6b In answering a question on a multiple choice test a student either knows the answer or he guesses. Let p be the probability that he knows the answer and $1 - p$ the probability that he guesses. Assume that a student who guesses at the answer will be correct with probability $1/m$, where m is the number of multiple-choice alternatives. What is the conditional probability that a student knew the answer to a question given that he answered it correctly?

Solution: Let C and K denote respectively the event that the student answers the question correctly and the event that he actually knows the answer. Now

$$
\begin{aligned}
P(K|C) &= \frac{P(KC)}{P(C)} \\
&= \frac{P(C|K)P(K)}{P(C|K)P(K) + P(C|K^c)P(K^c)} \\
&= \frac{p}{p + (1/m)(1 - p)} \\
&= \frac{mp}{1 + (m - 1)p}
\end{aligned}
$$

Thus, for example, if $m = 5$, $p = \frac{1}{2}$, then the probability that a student knew the answer to a question he correctly answered is $\frac{5}{6}$. ◊

Example 6c A laboratory blood test is 95 percent effective in detecting a certain disease when it is, in fact, present. However, the test also yields a "false positive" result for 1 percent of the healthy persons tested. (That is, if a healthy person is tested, then, with probability .01, the test result will imply he has the disease.) If .5 percent of the population actually has the disease, what is the probability a person has the disease given that his test result is positive?

Solution: Let D be the event that the tested person has the disease, and E the event that his test result is positive. The desired probability $P(D|E)$ is obtained by

$$P(D|E) = \frac{P(DE)}{P(E)}$$

$$= \frac{P(E|D)P(D)}{P(E|D)P(D) + P(E|D^c)P(D^c)}$$

$$= \frac{(.95)(.005)}{(.95)(.005) + (.01)(.995)}$$

$$= \frac{95}{294} \approx .323$$

Thus, only 32 percent of those persons whose test results are positive actually have the disease.   ◊

Equation (6.1) may be generalized in the following manner. Suppose that $F_1, F_2, \ldots , F_n$ are mutually exclusive events such that $\cup_{i=1}^n F_i = S$. In other words, exactly one of the events $F_1, F_2, \ldots , F_n$ will occur. By writing

$$E = \bigcup_{i=1}^n EF_i$$

and using the fact that the events EF_i, $i = 1, \ldots , n$, are mutually exclusive, we obtain that

$$P(E) = \sum_{i=1}^{n} P(EF_i)$$

$$= \sum_{i=1}^{n} P(E|F_i)P(F_i) \tag{6.2}$$

Thus, Equation (6.2) shows how, for given events $F_1, F_2, \ldots, F_n$ of which one and only one must occur, we can compute $P(E)$ by first "conditioning" upon which one of the F_i occurs. That is, it states that $P(E)$ is equal to a weighted average of $P(E|F_i)$, each term being weighted by the probability of the event on which it is conditioned.

Suppose now that E has occurred and we are interested in determining which one of the F_j also occurred. By Equation (6.2) we have that

$$P(F_j|E) = \frac{P(EF_j)}{P(E)}$$

$$= \frac{P(E|F_j)P(F_j)}{\sum_{i=1}^{n} P(E|F_i)P(F_i)} \tag{6.3}$$

Equation (6.3) is known as Bayes' formula.

Example 6d You know that a certain letter is equally likely to be in any one of three different folders. Let α_i be the probability that you will find your letter upon making a quick examination of folder i if the letter is, in fact, in folder i, $i = 1, 2, 3$. (We may have $\alpha_i < 1$.) Suppose you look in folder 1 and do not find the letter. What is the probability that the letter is in folder 1?

Solution: Let F_i, $i = 1, 2, 3$, be the event that the letter is in folder i; and let E be the event that a search of folder 1 does not come up with the letter. We desire $P(F_1|E)$. From Bayes' formula we obtain

$$P(F_1|E) = \frac{P(E|F_1)P(F_1)}{\sum_{i=1}^{3} P(E|F_i)P(F_i)}$$

$$= \frac{(1 - \alpha_1)\frac{1}{3}}{(1 - \alpha_1)\frac{1}{3} + \frac{1}{3} + \frac{1}{3}} = \frac{1 - \alpha_1}{3 - \alpha_1} \quad \diamond$$

Problems

1. A box contains three marbles: one red, one green, and one blue. Consider an experiment that consists of taking one marble from the box then replacing it in the box and drawing a second marble from the box. What is the sample space? If, at all times, each marble in the box is equally likely to be selected, what is the probability of each point in the sample space?

2. Repeat 1 when the second marble is drawn without replacing the first marble.

3. A coin is to be tossed until a head appears twice in a row. What is the sample space for this experiment? If the coin is fair, then what is the probability that it will be tossed exactly four times?

4. Let E, F, G be three events. Find expressions for the events that of E, F, G
   (a)   only F occurs,
   (b)   both E and F but not G occurs,
   (c)   at least one event occurs,
   (d)   at least two events occur,
   (e)   all three events occur,
   (f)   none occurs,
   (g)   at most one occurs,
   (h)   at most two occur.

5. An individual uses the following gambling system at Las Vegas. He bets $1 that the roulette wheel will come up red. If he wins, he quits. If he loses then he makes the same bet a second time only that this time he bets $2; and then regardless of the outcome, quits. Assuming that he has a probability of $\frac{1}{2}$ of winning each bet, what is the probability that he goes home a winner? Why is this system not used by everyone?

6. Show that $E(F \cup G) = EF \cup EG$.

7. Show that $(E \cup F)^c = E^c F^c$.

8. If $P(E) = .9$ and $P(F) = .8$, show that $P(EF) \geq .7$. In general, show that

$$P(EF) \geq P(E) + P(F) - 1$$

This is known as Bonferroni's inequality.

9. We say that $E \subset F$ is every point in E is also in F. Show that if $E \subset F$, then

$$P(F) = P(E) + P(FE^c) \geq P(E)$$

10. Show that

$$P\left(\bigcup_{i=1}^{n} E_i\right) \leq \sum_{i=1}^{n} P(E_i)$$

This is known as Boole's inequality.

Hint: Either use Equation (3.2) and mathematical induction, or else show that $\bigcup_{i=1}^{n} E_i = \bigcup_{i=1}^{n} F_i$, where $F_1 = E_1$, $F_i = E_i \prod_{j=1}^{i-1} E_j^c$, and use property (iii) of a probability.

11. If two fair dice are tossed, what is the probability that the sum is i, $i = 2, 3, \ldots, 12$?

12. Let E and F be mutually exclusive events in the sample space of an experiment. Suppose that the experiment is repeated until either event E or event F occurs. What does the sample space of this new super experiment look like? Show that the probability that event E occurs before event F is $P(E)/[P(E) + P(F)]$.

Hint: Argue that the probability that the original experiment is performed n times and E appears on the nth time is $P(E) \times (1 - p)^{n-1}$, $n = 1, 2, \ldots$, where $p = P(E) + P(F)$. Add these probabilities to get the desired answer.

13. The dice game craps is played as follows. The player throws two dice, and if the sum is seven or eleven, then he wins. If the sum is two, three, or twelve, then he loses. If the sum is anything else, then he continues throwing until he either throws that number again (in which case he wins) or he throws a seven (in which case he loses). Calculate the probability that the player wins.

14. The probability of winning on a single toss of the dice is p. A starts, and if he fails, he passes the dice to B, who then attempts to win on his toss. They continue tossing the dice back and forth until one of them wins. What are their respective probabilities of winning?

15. Argue that $F = EF \cup EF^c$, $E \cup F = E \cup FE^c$.

16. Use Problem 15 to show that $P(E \cup F) = P(E) + P(F) - P(EF)$.

17. Suppose each of three persons tosses a coin. If the outcome of one of the tosses differs from the other outcomes, then the game ends. If not, then the persons start over and retoss their coins. Assuming fair coins, what is the probability that the game will end with the first round

of tosses? If all three coins are biased and have a probability $\frac{1}{4}$ of landing heads, then what is the probability that the game will end at the first round?

18. Assume that each child that is born is equally likely to be a boy or a girl. If a family has two children, what is the probability that they both be girls given that (a) the eldest is a girl, (b) at least one is a girl?

19. Two dice are rolled. What is the probability that at least one is a six? If the two faces are different, what is the probability that at least one is a six?

20. Three dice are thrown. What is the probability the same number appears on exactly two of the three dice?

21. Suppose that 5 percent of men and 0.25 percent of women are color-blind. A color-blind person is chosen at random. What is the probability of this person being male? Assume that there are an equal number of males and females.

22. What is the conditional probability that the first die is six given that the sum of the dice is seven?

23. Suppose all n men at a party throw their hats in the center of the room. Each man then randomly selects a hat. Show that the probability that none of the n men winds up with his own hat is $\frac{1}{2!} - \frac{1}{3!} + \frac{1}{4!} - + \cdots (-1)^n/n!$. Note that as $n \to \infty$ this converges to e^{-1}. Is this surprising? (Would you have incorrectly thought that this probability would go to 1 as $n \to \infty$?)

24. In a class there are four freshman boys, six freshman girls, and six sophomore boys. How many sophomore girls must be present if sex and class are to be independent when a student is selected at random?

25. Mr. Jones has devised a gambling system for winning at roulette. When he bets, he bets on red, and places a bet only when the ten previous spins of the roulette have landed on a black number. He reasons that his chance of winning is quite large since the probability of eleven consecutive spins resulting in black is quite small. What do you think of this system?

26. Consider two boxes, one containing one black and one white marble, the other, two black and one white marble. A box is selected at random and a marble is drawn at random from the selected box. What is the probability that the marble is black?

27. In Problem 26, what is the probability that the first box was the one selected given that the marble is white?

28. Urn 1 contains two white balls and one black ball, while urn 2 contains one white ball and five black balls. One ball is drawn at random from urn 1 and placed in urn 2. A ball is then drawn from urn 2. It happens to be white. What is the probability that the transferred ball was white?

29. Stores *A, B,* and *C* have 50, 75, 100 employees, and respectively 50, 60, and 70 percent of these are women. Resignations are equally likely among all employees, regardless of sex. One employee resigns and this is a woman. What is the probability that she works in store *C*?

30. (a) A gambler has in his pocket a fair coin and a two-headed coin. He selects one of the coins at random, and when he flips it, it shows heads. What is the probability that it is the fair coin? (b) Suppose that he flips the same coin a second time and again it shows heads. What is now the probability that it is the fair coin? (c) Suppose that he flips the same coin a third time and it shows tails. What is now the probability that it is the fair coin?

31. There are three coins in a box. One is a two-headed coin, another is a fair coin, and the third is a biased coin which comes up heads 75 percent of the time. When one of the three coins is selected at random and flipped, it shows heads. What is the probability that it was the two-headed coin?

32. Suppose we have ten coins which are such that if the *i*th one is flipped then heads will appear with probability $i/10$, $i = 1, 2, \ldots, 10$. When one of the coins is randomly selected and flipped, it shows heads. What is the conditional probability that it was the fifth coin?

33. Urn 1 has five white and seven black balls. Urn 2 has three white and twelve black balls. We flip a fair coin. If the outcome is heads, then a ball from urn 1 is selected, while if the outcome is tails, then a ball from urn 2 is selected. Suppose that a white ball is selected. What is the probability that the coin landed tails?

34. An urn contains *b* black balls and *r* red balls. One of the balls is drawn at random, but when it is put back in the urn *c* additional balls of the same color are put in with it. Now suppose that we draw another ball. Show that the probability that the first ball drawn was black given that the second ball drawn was red is $b/(b + r + c)$.

35. Three prisoners are informed by their jailer that one of them has been chosen at random to be executed, and the other two are to be freed. Prisoner A asks the jailer to tell him privately which of his fellow prisoners will be set free, claiming that there would be no harm in divulging this information, since he already knows that at least one will go free. The jailer refuses to answer this question, pointing out that if A knew which of his fellows were to be set free, then his own probability of being executed would rise from $\frac{1}{3}$ to $\frac{1}{2}$, since he would then be one of two prisoners. What do you think of the jailer's reasoning?

References

Reference [2] provides a colorful introduction to some of the earliest developments in probability theory. References [3], [4], and [7] are all excellent introductory texts in modern probability theory. Reference [5] is the definitive work which established the axiomatic foundation of modern mathematical probability theory. Reference [6] is a nonmathematical introduction to probability theory and its applications, written by one of the greatest mathematicians of the eighteenth century.

1. L. Breiman, "Probability," Addison-Wesley, Reading, Massachusetts, 1968.

2. F. N. David, "Games, Gods, and Gambling," Hafner, New York, 1962.

3. W. Feller, "An Introduction to Probability Theory and Its Applications," Vol. I., John Wiley, New York, 1957.

4. B. V. Gnedenko, "Theory of Probability," Chelsea, New York, 1962.

5. A. N. Komogorov, "Foundations of the Theory of Probability," Chelsea, New York, 1956.

6. Marquis de Laplace, "A Philosophical Essay on Probabilities," 1825 (English Translation), Dover, New York, 1951.

7. S. Ross, "A First Course in Probability," Second Edition, Macmillan, New York, 1984.

Chapter 2

Random Variables

1. Random Variables

It frequently occurs that in performing an experiment we are mainly interested in some function of the outcome as opposed to the actual outcome itself. For instance, in tossing dice we are often interested in the sum of the two dice and are not really concerned about the actual outcome. That is, we may be interested in knowing that the sum is seven and not be concerned over whether the actual outcome was $(1, 6)$ or $(2, 5)$ or $(3, 4)$ or $(4, 3)$ or $(5, 2)$ or $(6, 1)$. These quantities of interest, or more formally, these real-valued functions defined on the sample space, are known as *random variables*.

Since the value of a random variable is determined by the outcome of the experiment, we may assign probabilities to the possible values of the random variable.

Example 1a Letting X denote the random variable which is defined as the sum of two fair dice, then

$$P\{X = 2\} = P\{(1, 1)\} = \tfrac{1}{36}$$

$$P\{X = 3\} = P\{(1, 2), (2, 1)\} = \tfrac{2}{36}$$

$$P\{X = 4\} = P\{(1, 3), (2, 2), (3, 1)\} = \tfrac{3}{36}$$

$$P\{X = 5\} = P\{(1, 4), (2, 3), (3, 2), (4, 1)\} = \tfrac{4}{36}$$

$P\{X = 6\} = P\{(1, 5), (2, 4), (3, 3), (4, 2), (5, 1)\} = \frac{5}{36}$

$P\{X = 7\} = P\{(1, 6), (2, 5), (3, 4), (4, 3), (5, 2), (6, 1)\} = \frac{6}{36}$

$P\{X = 8\} = P\{(2, 6), (3, 5), (4, 4), (5, 3), (6, 2)\} = \frac{5}{36}$

$P\{X = 9\} = P\{(3, 6), (4, 5), (5, 4), (6, 3)\} = \frac{4}{36}$

$P\{X = 10\} = P\{(4, 6), (5, 5), (6, 4)\} = \frac{3}{36}$

$P\{X = 11\} = P\{(5, 6), (6, 5)\} = \frac{2}{36}$

$P\{X = 12\} = P\{(6, 6)\} = \frac{1}{36}$ (1.1)

In other words, the random variable X can take on any integral value between two and twelve, and the probability that it takes on each value is given by Equation (1.1). Since X must take on one of the values two through twelve, we must have that

$$1 = P\left\{\bigcup_{i=2}^{12} \{X = n\}\right\} = \sum_{n=2}^{12} P\{X = n\}$$

which may be checked from Equation (1.1). $\Diamond$

Example 1b   For a second example, suppose that our experiment consists of tossing two fair coins. Letting Y denote the number of heads appearing, then Y is a random variable taking on one of the values 0, 1, 2 with respective probabilities

$$P\{Y = 0\} = P\{(T, T)\} = \tfrac{1}{4}$$

$$P\{Y = 1\} = P\{(T, H),(H, T)\} = \tfrac{2}{4}$$

$$P\{Y = 2\} = P\{(H, H)\} = \tfrac{1}{4}$$

Of course, $P\{Y = 0\} + P\{Y = 1\} + P\{Y = 2\} = 1$ $\Diamond$

Example 1c   Suppose that we toss a coin having a probability p of coming up heads, until the first head appears. Letting N denote the number of flips required, then, assuming that the outcome of successive flips are independent, N is a random variable taking on one of the values 1, 2, 3, . . . , with respective probabilities

$P\{N = 1\} = P\{H\} = p$

$P\{N = 2\} = P\{(T, H)\} = (1 - p)p$

$P\{N = 3\} = P\{(T, T, H)\} = (1 - p)^2 p$

$\qquad\vdots$

$P\{N = n\} = P\{(\underbrace{T, T, . . . ,T}_{n-1}, H)\} = (1 - p)^{n-1}p, \quad n \geq 1$

As a check, note that

$$P\left\{\bigcup_{n=1}^{\infty} \{N = n\}\right\} = \sum_{n=1}^{\infty} P\{N = n\}$$

$$= p \sum_{n=1}^{\infty} (1 - p)^{n-1}$$

$$= \frac{p}{1 - (1 - p)}$$

$$= 1 \quad \diamond$$

Example 1d Suppose that our experiment consists in seeing how long a battery can operate before wearing down. Suppose also that we are not primarily interested in the actual lifetime of the battery but are only concerned about whether or not the battery lasts at least two years. In this case, we may define the random variable I by

$$I = \begin{cases} 1, & \text{if the lifetime of the battery is two or more years} \\ 0, & \text{otherwise.} \end{cases}$$

If E denotes the event that the battery lasts two or more years, then the random variable I is known as the *indicator* random variable for the event E. (Note that I equals 1 or 0 depending upon whether or not E occurs.) $\quad \diamond$

Example 1e Suppose that independent trials, each of which results in any of m possible outcomes with respective probabilities $p_1, \ldots,$ p_m, $\sum_{i=1}^{m} p_i = 1$, are continually performed. Let X denote the number of trials needed until each outcome has occurred at least once.

Rather than directly considering $P\{X = n\}$ we will first determine $P\{X > n\}$, the probability that at least one of the outcomes has not yet occurred after n trials. Letting A_i denote the event that outcome i has not yet occurred after the first n trials, $i = 1, \ldots,$ m, then

$$P\{X > n\} = P\left(\bigcup_{j=1}^{m} A_i\right)$$

$$= \sum_{i=1}^{m} P(A_i) - \sum_{i < j} \sum P(A_i A_j)$$

$$+ \sum_{i < j < k} \sum \sum P(A_i A_j A_k) - \ldots + (-1)^{m+1} P(A_1 \ldots A_m)$$

Now, $P(A_i)$ is the probability that each of the first n trials all result in a non-i outcome, and so by independence

$$P(A_i) = (1 - p_i)^n$$

Similarly, $P(A_iA_j)$ is the probability that the first n trials all result in a non-i and non-j outcome, and so

$$P(A_iA_j) = (1 - p_i - p_j)^n$$

As all of the other probabilities are similar, we see that

$$P\{X > n\} = \sum_{i=1}^{m}(1 - p_i)^n - \sum_{i < j}\sum(1 - p_i - p_j)^n$$
$$+ \sum_{i < j < k}\sum\sum(1 - p_i - p_j - p_k)^n - \cdots$$

Since $P\{X = n\} = P\{X > n - 1\} - P\{X > n\}$, we see upon using the algebraic identity $(1 - a)^{n+1} - (1 - a)^n = a(1 - a)^{n-1}$, that

$$P\{X = n\} = \sum_{i=1}^{m}p_i(1 - p_i)^{n-1} - \sum_{i < j}\sum(p_i + p_j)(1 - p_i - p_j)^{n-1}$$
$$+ \sum_{i < j < k}\sum\sum(p_i + p_j + p_k)(1 - p_i - p_j - p_k)^{n-1} - \cdots \quad \diamond$$

In all of the above examples, the random variables of interest took on either a finite or a countable number of possible values. Such random variables are called *discrete*. However, there also exist random variables that take on a continuum of possible values. These are known as *continuous* random variables. One example is the random variable denoting the lifetime of a car, when the car's lifetime is assumed to take on any value in some interval (a, b).

The *cumulative distribution function* (cdf) (or more simply the *distribution function*) $F(\cdot)$ of the random variable X is defined for any real number b, $-\infty < b < \infty$, by

$$F(b) = P\{X \le b\}$$

In words, $F(b)$ denotes the probability that the random variable X takes on a value which will be less than or equal to b. Some properties of the cdf F are

(i) $F(b)$ is a nondecreasing function of b,
(ii) $\lim_{b\to\infty}F(b) = F(\infty) = 1$,
(iii) $\lim_{b\to-\infty}F(b) = F(-\infty) = 0$.

Property (i) follows since for $a < b$ the event $\{X \le a\}$ is contained in the event $\{X \le b\}$, and so it must have a smaller probability. Properties (ii) and (iii) follow since X must take on some finite value.

All probability questions about X can be answered in terms of the cdf $F(\cdot)$. For example,

$$P\{a < X \le b\} = F(b) - F(a) \quad \text{for all} \quad a < b$$

This follows since we may calculate $P\{a < X \le b\}$ by first computing the probability that $X \le b$ (that is, $F(b)$) and then subtracting from this the probability that $X \le a$ (that is, $F(a)$).

If we desire the probability that X is strictly smaller than b, we may calculate this probability by

$$P\{X < b\} = \lim_{h \to 0^+} P\{X \le b - h\}$$
$$= \lim_{h \to 0^+} F(b - h)$$

where $\lim_{h \to 0^+}$ means that we are taking the limit as h decreases to 0. Note that $P\{X < b\}$ does not necessarily equal $F(b)$ since $F(b)$ also includes the probability that X includes b.

2. Discrete Random Variables

As was previously mentioned, a random variable that can take on at most a countable number of possible values is said to be *discrete*. For a discrete random variable X, we define the *probability mass function* $p(a)$ of X by

$$p(a) = P\{X = a\}$$

The probability mass function $p(a)$ is positive for at most a countable number of values of a. That is, if X must assume one of the values x_1, $x_2, \ldots$, then

$$p(x_i) > 0, \quad i = 1, 2, \ldots$$
$$p(x) = 0, \quad \text{all other values of } x$$

Since X must take on one of the values x_i, we have

$$\sum_{i=1}^{\infty} p(x_i) = 1$$

The cumulative distribution function F can be expressed in terms of $p(a)$ by

$$F(a) = \sum_{\text{all } x_i \leq a} p(x_i)$$

For instance, suppose X has a probability mass function given by

$$p(1) = \tfrac{1}{2}, \qquad p(2) = \tfrac{1}{3}, \qquad p(3) = \tfrac{1}{6}$$

then, the cumulative distribution function F of X is given by

$$F(a) = \begin{cases} 0, & a < 1 \\ \tfrac{1}{2}, & 1 \leq a < 2 \\ \tfrac{5}{6}, & 2 \leq a < 3 \\ 1, & 3 \leq a \end{cases}$$

This is graphically presented in Figure 2.1

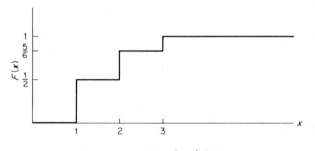

Figure 2.1   Graph of $F(x)$.

Discrete random variables are often classified according to their probability mass function. We now consider some of these random variables.

2.1. The Bernoulli Random Variable

Suppose that a trial, or an experiment, whose outcome can be classified as either a "success" or as a "failure" is performed. If we let X equal 1 if the outcome is a success and 0 if it is a failure, then the probability mass function of X is given by

$$p(0) = P\{X = 0\} = 1 - p$$
$$p(1) = P\{X = 1\} = p \qquad \qquad (2.1)$$

where p, $0 \leq p \leq 1$, is the probability that the trial is a "success."

A random variable X is said to be a *Bernoulli* random variable if its probability mass function is given by Equation (2.1) for some $p \in (0, 1)$.

2.2. The Binomial Random Variable

Suppose that n independent trials, each of which results in a "success" with probability p and in a "failure" with probability $1 - p$, are to be performed. If X represents the number of successes that occur in the n trials, then X is said to be a *binomial* random variable with parameters (n, p).

The probability mass function of a binomial random variable having parameters (n, p) is given by

$$p(i) = \binom{n}{i} p^i (1 - p)^{n-i}, \qquad i = 0, 1, \ldots, n \qquad (2.2)$$

where $\binom{n}{i} = n!/(n - i)!i!$ equals the number of different groups of i objects that can be chosen from a set of n objects. The validity of Equation (2.2) may be verified by first noting that the probability of any particular sequence of the n outcomes containing i successes and $n - i$ failures is, by the assumed independence of trails, $p^i (1 - p)^{n-i}$. Equation (2.2) then follows since there are $\binom{n}{i}$ different sequences of the n outcomes leading to i successes and $n - i$ failures. For instance, if $n = 3$, $i = 2$, then there are $\binom{3}{2} = 3$ ways in which the three trials can result in two successes. Namely, any one of the three outcomes (s, s, f), (s, f, s), (f, s, s), where the outcome (s, s, f) means that the first two trials are successes and the third a failure. Since each of the three outcomes (s, s, f), (s, f, s), (f, s, s) has a probability $p^2 (1 - p)$ of occurring the desired probability is thus $\binom{3}{2} p^2 (1 - p)$.

Note that, by the binomial theorem, the probabilities sum to one, that is,

$$\sum_{i=0}^{\infty} p(i) = \sum_{i=0}^{n} \binom{n}{i} p^i (1 - p)^{n-i} = (p + (1 - p))^n = 1$$

Example 2a Four fair coins are flipped. If the outcomes are assumed independent, what is the probability that two heads and two tails are obtained?

Solution: Letting X equal the number of heads ("successes") that appear, then X is a binomial random variable with parameters $(n = 4, p = \frac{1}{2})$. Hence, by Equation (2.2),

$$P\{X = 2\} = \binom{4}{2} \left(\frac{1}{2}\right)^2 \left(\frac{1}{2}\right)^2 = \frac{3}{8} \quad \diamond$$

Example 2b   It is known that all items produced by a certain machine will be defective with probability .1, independently of each other. What is the probability that in a sample of three items, at most one will be defective?

Solution: If X is the number of defective items in the sample, then X is a binomial random variable with parameters $(3, .1)$. Hence, the desired probability is given by

$$P\{X = 0\} + P\{X = 1\} = \tbinom{3}{0}(.1)^0(.9)^3 + \tbinom{3}{1}(.1)^1(.9)^2 = .972 \quad \Diamond$$

Example 2c   Suppose that an airplane engine will fail, when in flight, with probability $1 - p$ independently from engine to engine; suppose that the airplane will make a successful flight if at least 50 percent of its engines remain operative. For what values of p is a four-engine plane preferable to a two-engine plane?

Solution: As each engine is assumed to fail or function independently of what happens with the other engines, it follows that the number of engines remaining operative is a binomial random variable. Hence, the probability that a four-engine plane makes a successful flight is

$$\tbinom{4}{2}p^2(1 - p)^2 + \tbinom{4}{3}p^3(1 - p) + \tbinom{4}{4}p^4(1 - p)^0$$
$$= 6p^2(1 - p)^2 + 4p^3(1 - p) + p^4$$

whereas the corresponding probability for a two-engine plane is

$$\tbinom{2}{1}p(1 - p) + \tbinom{2}{2}p^2 = 2p(1 - p) + p^2$$

Hence the four-engine plane is safer if

$$6p^2(1 - p)^2 + 4p^3(1 - p) + p^4 \geq 2p(1 - p) + p^2$$

or equivalently if

$$6p(1 - p)^2 + 4p^2(1 - p) + p^3 \geq 2 - p$$

which simplifies to

$$3p^3 - 8p^2 + 7p - 2 \geq 0 \quad \text{or} \quad (p - 1)^2(3p - 2) \geq 0$$

which is equivalent to

$$3p - 2 \geq 0 \quad \text{or} \quad p \geq \tfrac{2}{3}$$

Hence, the four-engine plane is safer when the engine success probability is at least as large as $\frac{2}{3}$, whereas the two-engine plane is safer when this probability falls below $\frac{2}{3}$.   ◊

Example 2d Suppose that a particular trait of a person (such as eye color or left handedness) is classified on the basis of one pair of genes and suppose that *d* represents a dominant gene and *r* a recessive gene. Thus a person with *dd* genes is pure dominance, one with *rr* is pure recessive, and one with *rd* is hybrid. The pure dominance and the hybrid are alike in appearance. Children receive one gene from each parent. If, with respect to a particular trait, two hybrid parents have a total of four children, what is the probability that exactly three of the four children have the outward appearance of the dominant gene?

Solution: If we assume that each child is equally likely to inherit either of two genes from each parent, the probabilities that the child of two hybrid parents will have *dd*, *rr*, or *rd* pairs of genes are, respectively, $\frac{1}{4}, \frac{1}{4}, \frac{1}{2}$. Hence, as an offspring will have the outward appearance of the dominant gene if its gene pair is either *dd* or *rd*, it follows that the number of such children is binomially distributed with parameters $(4, \frac{3}{4})$. Thus the desired probability is

$$\binom{4}{3}\left(\frac{3}{4}\right)^3\left(\frac{1}{4}\right)^1 = \frac{27}{64} \quad ◊$$

Remark on Terminology

If X is a binomial random variable with parameters (n, p), then we say that X has a binomial distribution with parameters (n, p).

2.3. The Geometric Random Variable

Suppose that independent trials, each having a probability p of being a success, are performed until a success occurs. If we let X be the number of trials required until the first success, then X is said to be a *geometric* random variable with parameter p. Its probability mass function is given by

$$p(n) = P\{X = n\} = (1 - p)^{n-1}p, \quad n = 1, 2, \ldots \tag{2.3}$$

Equation (2.3) follows since in order for X to equal n it is necessary and sufficient that the first $n - 1$ trials are failures and the nth trial is a success. Equation (2.3) follows since the outcome of the successive trials are assumed to be independent.

To check that $p(n)$ is a probability mass function, we note that

$$\sum_{n=1}^{\infty} p(n) = p \sum_{1}^{\infty} (1 - p)^{n-1} = 1$$

2.4. The Poisson Random Variable

A random variable X, taking on one of the values $0, 1, 2, \ldots$, is said to be a *Poisson* random variable with parameter λ, if for some $\lambda > 0$,

$$p(i) = P\{X = i\} = e^{-\lambda} \frac{\lambda^i}{i!}, \quad i = 0, 1, \ldots \tag{2.4}$$

Equation (2.4) defines a probability mass function since

$$\sum_{i=0}^{\infty} p(i) = e^{-\lambda} \sum_{i=0}^{\infty} \frac{\lambda^i}{i!} = e^{-\lambda} e^{\lambda} = 1$$

The Poisson random variable has a wide range of applications in a diverse number of areas, as will be seen in Chapter 5.

An important property of the Poisson random variable is that it may be used to approximate a binomial random variable when the binomial parameter n is large and p is small. To see this, suppose that X is a binomial random variable with parameters (n, p), and let $\lambda = np$. Then

$$P\{X = i\} = \frac{n!}{(n-i)!i!} p^i (1 - p)^{n-i}$$

$$= \frac{n!}{(n-i)!i!} \left(\frac{\lambda}{n}\right)^i \left(1 - \frac{\lambda}{n}\right)^{n-i}$$

$$= \frac{n(n-1) \cdots (n-i+1)}{n^i} \frac{\lambda^i}{i!} \frac{(1 - \lambda/n)^n}{(1 - \lambda/n)^i}$$

Now, for n large and p small

$$\left(1 - \frac{\lambda}{n}\right)^n \approx e^{-\lambda}, \qquad \frac{n(n-1) \cdots (n-i+1)}{n^i} \approx 1, \qquad \left(1 - \frac{\lambda}{n}\right)^i \approx 1$$

Hence, for n large and p small,

$$P\{X = i\} \approx e^{-\lambda} \frac{\lambda^i}{i!}$$

Example 2e   Suppose that the number of typographical errors on a single page of this book has a Poisson distribution with parameter

$\lambda = 1$. Calculate the probability that there is at least one error on this page.

Solution:

$$P\{X \geq 1\} = 1 - P\{X = 0\} = 1 - e^{-1} \approx .633 \quad \Diamond$$

Example 2f If the number of accidents occurring on a highway each day is a Poisson random variable with parameter $\lambda = 3$, what is the probability that no accidents occur today?

Solution:

$$P\{X = 0\} = e^{-3} \approx .05 \quad \Diamond$$

Example 2g Consider an experiment that consists of counting the number of α-particles given off in a one-second interval by one gram of radioactive material. If we know from past experience that, on the average, 3.2 such α-particles are given off, what is a good approximation to the probability that no more than 2 α-particles will appear?

Solution: If we think of the gram of radioactive material as consisting of a large number n of atoms each of which has probability $3.2/n$ of disintegrating and sending off an α-particle during the second considered, then we see that, to a very close approximation, the number of α-particles given off will be a poisson random variable with parameter $\lambda = 3.2$. Hence the desired probability is

$$P\{X \leq 2\} = e^{-3.2} + 3.2e^{-3.2} + \frac{(3.2)^2}{2} e^{-3.2} \approx .382$$

3. Continuous Random Variables

In this section, we shall concern ourselves with random variables whose set of possible values is uncountable. Let X be such a random variable. We say that X is a *continuous* random variable is there exists a non-negative function $f(x)$, defined for all real $x \in (-\infty, \infty)$, having the property that for any set B of real numbers

$$P\{X \in B\} = \int_B f(x)dx \qquad (3.1)$$

The function $f(x)$ is called the *probability density function* of the random variable X.

In words, Equation (3.1) states that the probability that X will be in B may be obtained by integrating the probability density function over the set B. Since X must assume some value, $f(x)$ must satisfy

$$1 = P\{X \in (-\infty,\infty)\} = \int_{-\infty}^{\infty} f(x)dx$$

All probability statements about X can be answered in terms of $f(x)$. For instance. letting $B = [a, b]$, we obtain from Equation (3.1) that

$$P\{a \leq X \leq b\} = \int_{a}^{b} f(x)dx \qquad (3.2)$$

If we let $a = b$ in the above, then

$$P\{X = a\} = \int_{a}^{a} f(x)dx = 0$$

In words, this equation states that the probability that a continuous random variable will assume any *particular* value is zero.

The relationship between the cumulative distribution $F(\cdot)$ and the probability density $f(\cdot)$ is expressed by

$$F(a) = P\{X \in (-\infty, a]\} = \int_{-\infty}^{a} f(x)dx$$

Differentiating both sides of the above yields

$$\frac{d}{da} F(a) = f(a)$$

That is, the density is the derivative of the cumulative distribution function. A somewhat more intuitive interpretation of the density function may be obtained from Equation (3.2) as follows:

$$P\left\{a - \frac{\epsilon}{2} \leq X \leq a + \frac{\epsilon}{2}\right\} = \int_{a-\epsilon/2}^{a+\epsilon/2} f(x)dx \approx \epsilon f(a)$$

when ϵ is small. In other words, the probability that X will be contained in an interval of length ϵ around the point a is approximately $\epsilon f(a)$. From this, we see that $f(a)$ is a measure of how likely it is that the random variable will be near a.

There are several important continuous random variables that appear frequently in probability theory. The remainder of this section is devoted to a study of certain of these random variables.

3.1. The Uniform Random Variable

A random variable is said to be uniformly distributed over the interval $(0, 1)$ if its probability density function is given by

$$f(x) = \begin{cases} 1, & 0 < x < 1 \\ 0, & \text{otherwise} \end{cases}$$

Note that the above is a density function since $f(x) \geq 0$ and $\int_{-\infty}^{\infty} f(x)dx = \int_0^1 dx = 1$. Since $f(x) > 0$ only when $x \in (0, 1)$, it follows that X must assume a value in $(0, 1)$. Also, since $f(x)$ is constant for $x \in (0, 1)$, X is just as likely to be "near" any value in $(0, 1)$ as any other value. To check this, note that for any $0 < a < b < 1$,

$$P\{a \leq X \leq b\} = \int_a^b f(x)dx = b - a$$

In other words, the probability that X is in any particular subinterval of $(0, 1)$ equals the length of that subinterval.

In general, we say that X is a uniform random variable on the interval (α, β) if its probability density function is given by

$$f(x) = \begin{cases} \dfrac{1}{\beta - \alpha}, & \text{if } \alpha < x < \beta \\ 0, & \text{otherwise} \end{cases} \tag{3.3}$$

Example 3a Calculate the cumulative distribution function of a random variable uniformly distributed over (α, β).

Solution: Since $F(a) = \int_{-\infty}^a f(x)dx$, we obtain from Equation (3.3) that

$$F(a) = \begin{cases} 0, & a \leq \alpha \\ \dfrac{a - \alpha}{\beta - \alpha}, & \alpha < a < \beta \\ 1, & a \geq \beta \end{cases} \quad \Diamond$$

Example 3b If X is uniformly distributed over $(0, 10)$, calculate the probability that (i) $X < 3$, (ii) $X > 7$, (iii) $1 < X < 6$.

Solution:

$$P\{X < 3\} = \frac{\int_0^3 dx}{10} = \frac{3}{10}$$

$$P\{X > 7\} = \frac{\int_7^{10} dx}{10} = \frac{3}{10}$$

$$P\{1 < X < 6\} = \frac{\int_1^6 dx}{10} = \frac{1}{2} \quad \Diamond$$

3.2. Exponential Random Variables

A continuous random variable whose probability density function is given, for some $\lambda > 0$, by

$$f(x) = \begin{cases} \lambda e^{-\lambda x}, & \text{if } x \geq 0 \\ 0, & \text{if } x < 0 \end{cases}$$

is said to be an exponential random variable with parameter λ. These random variables will be extensively studied in Chapter 5, and so we will content ourselves here with just calculating the cumulative distribution function F:

$$F(a) = \int_0^a \lambda e^{-\lambda x} = 1 - e^{-\lambda a}, \quad a \geq 0$$

Note that $F(\infty) = \int_0^\infty \lambda e^{-\lambda x} dx = 1$, as, of course, it must.

3.3. Gamma Random Variables

A continuous random variable whose density is given by

$$f(x) = \begin{cases} \dfrac{\lambda e^{-\lambda x}(\lambda x)^{\alpha-1}}{\Gamma(\alpha)}, & \text{if } x \geq 0 \\ 0, & \text{if } x < 0 \end{cases}$$

for some $\lambda > 0$, $\alpha > 0$ is said to be a gamma random variable with parameters α, λ. The quantity $\Gamma(\alpha)$ is called the gamma function and is defined by

$$\Gamma(\alpha) = \int_0^\infty e^{-x}x^{\alpha-1}dx$$

It is easy to show by induction that for integral α, say $\alpha = n$,

$$\Gamma(n) = (n-1)!$$

3.4. Normal Random Variables

We say that X is a normal random variable (or simply that X is normally distributed) with parameters μ and σ^2 if the density of X is given by

$$f(x) = \frac{1}{\sqrt{2\pi}\sigma}\, e^{-(x-\mu)^2/2\sigma^2}, \quad -\infty < x < \infty$$

This density function is a bell-shaped curve that is symmetric around μ (see Figure 2.2).

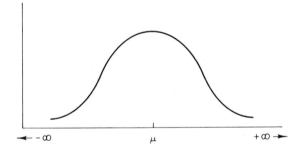

Figure 2.2. Normal density function.

An important fact about normal random variables is that if X is normally distributed with parameters μ and σ^2 then $Y = \alpha X + \beta$ is normally distributed with parameters $\alpha\mu + \beta$ and $\alpha^2\sigma^2$. To prove this suppose first that $\alpha > 0$ and note that $F_Y(\cdot)^*$ the cumulative distribution function of the random variable Y is given by

---

*When there is more than one random variable under consideration, we shall denote the cumulative distribution function of a random variable Z by $F_Z(\cdot)$. Similarly, we shall denote the density of Z by $f_Z(\cdot)$.

$$F_Y(a) = P\{Y \le a\}$$

$$= P\{\alpha X + \beta \le a\}$$

$$= P\left\{ X \le \frac{a - \beta}{\alpha} \right\}$$

$$= F_x\left(\frac{a - \beta}{\alpha} \right)$$

$$= \int_{-\infty}^{(a-\beta)/\alpha} \frac{1}{\sqrt{2\pi}\sigma} e^{-(x-\mu)^2/2\sigma^2} dx$$

$$= \int_{-\infty}^{a} \frac{1}{\sqrt{2\pi}\alpha\sigma} \exp\left\{ \frac{-(v - (\alpha\mu + \beta))^2}{2\alpha^2\sigma^2} \right\} dv \qquad (3.4)$$

where the last equality is obtained by the change in variables $v = \alpha x + \beta$. However, since $F_Y(a) = \int_{-\infty}^{a} f_Y(v)dv$, it follows from Equation (3.4) that the probability density function $f_Y(\cdot)$ is given by

$$f_Y(v) = \frac{1}{\sqrt{2\pi}\alpha\sigma} \exp\left\{ \frac{-(v - (\alpha\mu + \beta))^2}{2(\alpha\sigma)^2} \right\}, \quad -\infty < v < \infty$$

Hence, Y is normally distributed with parameters $\alpha\mu + \beta$ and $(\alpha\sigma)^2$. A similar result is also true when $\alpha < 0$.

One implication of the above result is that if X is normally distributed with parameters μ and σ^2 then $Y = (X - \mu)/\sigma$ is normally distributed with parameters 0 and 1. Such a random variable Y is said to have the *standard* or *unit* normal distribution.

4. Expectation of a Random Variable

4.1. The Discrete Case

If X is a discrete random variable having a probability mass function $p(x)$, then the *expected value* of X is defined by

$$E[X] = \sum_{x:p(x)>0} xp(x)$$

In other words, the expected value of X is a weighted average of the possible values that X can take on, each value being weighted by the probability that X assumes that value. For example, if the probability mass function of X is given by

$$p(1) = \tfrac{1}{2} = p(2)$$

then

$$E[X] = 1(\tfrac{1}{2}) + 2(\tfrac{1}{2}) = \tfrac{3}{2}$$

is just an ordinary average of the two possible values 1 and 2 that X can assume. On the other hand, if

$$p(1) = \tfrac{1}{3}, \qquad p(2) = \tfrac{2}{3}$$

then

$$E[X] = 1(\tfrac{1}{3}) + 2(\tfrac{2}{3}) = \tfrac{5}{3}$$

is a weighted average of the two possible values 1 and 2 where the value 2 is given twice as much weight as the value 1 since $p(2) = 2p(1)$.

Example 4a Find $E[X]$ where X is the outcome when we roll a fair die.

Solution: Since $p(1) = p(2) = p(3) = p(4) = p(5) = p(6) = \tfrac{1}{6}$, we obtain that

$$E[X] = 1(\tfrac{1}{6}) + 2(\tfrac{1}{6}) + 3(\tfrac{1}{6}) + 4(\tfrac{1}{6}) + 5(\tfrac{1}{6}) + 6(\tfrac{1}{6}) = \tfrac{7}{2} \quad \Diamond$$

Example 4b (Expectation of a Bernoulli Random Variable): Calculate $E[X]$ when X is a Bernoulli random variable with parameter p.

Solution: Since $p(0) = 1 - p$, $p(1) = p$, we have

$$E[X] = 0(1 - p) + 1(p) = p$$

Thus, the expected number of successes in a single trial is just the probability that the trial will be a success.

Example 4c (Expectation of a Binomial Random Variable): Calculate $E[X]$ when X is binomially distributed with parameters n and p.

Solution:

$$E[X] = \sum_{i=0}^{n} ip(i)$$

$$= \sum_{i=0}^{n} i\binom{n}{i}p^{i}(1 - p)^{n-i}$$

$$= \sum_{i=1}^{n} \frac{in!}{(n-i)!i!} p^i(1-p)^{n-i}$$

$$= \sum_{i=1}^{n} \frac{n!}{(n-i)!(i-1)!} p^i(1-p)^{n-i}$$

$$= np \sum_{i=1}^{n} \frac{(n-1)!}{(n-i)!(i-1)!} p^{i-1}(1-p)^{n-i}$$

$$= np \sum_{k=0}^{n-1} \binom{n-1}{k} p^k(1-p)^{n-1-k}$$

$$= np[p + (1-p)]^{n-1}$$

$$= np$$

where the second from the last equality follows by letting $k = i - 1$. Thus, the expected number of successes in n independent trials is n multiplied by the probability that a trial results in a success.   ◊

Example 4d   (Expectation of a Geometric Random Variable): Calculate the expectation of a geometric random variable having parameter p.

Solution: By Equation (2.3), we have

$$E[X] = \sum_{n=1}^{\infty} np(1-p)^{n-1}$$

$$= p \sum_{n=1}^{\infty} nq^{n-1}$$

where $q = 1 - p$,

$$E[X] = p \sum_{n=1}^{\infty} \frac{d}{dq}(q^n)$$

$$= p \frac{d}{dq}\left(\sum_{n=1}^{\infty} q^n\right)$$

$$= p \frac{d}{dq}\left(\frac{q}{1-q}\right)$$

$$= \frac{p}{(1 - q)^2}$$
$$= \frac{1}{p}$$

In words, the expected number of independent trials we need perform until we attain our first success equals the reciprocal of the probability that any one trial results in a success. ◊

Example 4e (Expectation of a Poisson Random Variable): Calculate $E[X]$ if X is a Poisson random variable with parameter λ.

Solution: From Equation (2.4), we have

$$E[X] = \sum_{i=0}^{\infty} \frac{ie^{-\lambda}\lambda^i}{i!}$$
$$= \sum_{i=1}^{\infty} \frac{e^{-\lambda}\lambda^i}{(i - 1)!}$$
$$= \lambda e^{-\lambda} \sum_{i=1}^{\infty} \frac{\lambda^{i-1}}{(i - 1)!}$$
$$= \lambda e^{-\lambda} \sum_{k=0}^{\infty} \frac{\lambda^k}{k!}$$
$$= \lambda e^{-\lambda} e^{\lambda}$$
$$= \lambda$$

where we have used the identity $\sum_{k=0}^{\infty} \lambda^k / k! = e^{\lambda}$. ◊

4.2. The Continuous Case

We may also define the expected value of a continuous random variable. This is done as follows. If X is a continuous random variable having a probability density function $f(x)$, then the expected value of X is defined by

$$E[X] = \int_{-\infty}^{\infty} xf(x)dx$$

Example 4f (Expectation of a Uniform Random Variable): Calculate the expectation of a random variable uniformly distributed over (α, β).

Solution: From Equation (3.3) we have

$$E[X] = \int_\alpha^\beta \frac{x}{\beta - \alpha} \, dx$$

$$= \frac{\beta^2 - \alpha^2}{2(\beta - \alpha)}$$

$$= \frac{\beta + \alpha}{2}$$

In other words, the expected value of a random variable uniformly distributed over the interval (α, β) is just the midpoint of the interval.   ◇

Example 4g   (Expectation of an Exponential Random Variable): Let X be exponentially distributed with parameter λ. Calculate $E[X]$.

Solution:

$$E[X] = \int_0^\infty x\lambda e^{-\lambda x} dx$$

Integrating by parts yields

$$E[X] = -xe^{-\lambda x} \Big|_0^\infty + \int_0^\infty e^{-\lambda x} dx$$

$$= 0 - \frac{e^{-\lambda x}}{\lambda} \Big|_0^\infty$$

$$= \frac{1}{\lambda}   \diamond$$

Example 4h   (Expectation of a Normal Random Variable): Calculate $E[X]$ when X is normally distributed with parameters μ and σ^2.

Solution:

$$E[X] = \frac{1}{\sqrt{2\pi}\sigma} \int_{-\infty}^\infty xe^{-(x-\mu)^2/2\sigma^2} dx$$

Writing x as $(x - \mu) + \mu$ yields

$$E[X] = \frac{1}{\sqrt{2\pi}\sigma} \int_{-\infty}^\infty (x - \mu)e^{-(x-\mu)^2/2\sigma^2} dx + \mu \frac{1}{\sqrt{2\pi}\sigma} \int_{-\infty}^\infty e^{-(x-\mu)^2/2\sigma^2} dx$$

Letting $y = x - \mu$ leads to

$$E[X] = \frac{1}{\sqrt{2\pi}\sigma} \int_{-\infty}^{\infty} y e^{-y^2/2\sigma^2} dy + \mu \int_{-\infty}^{\infty} f(x) dx$$

where $f(x)$ is the normal density. By symmetry, the first integral must be 0, and so

$$E[X] = \mu \int_{-\infty}^{\infty} f(y) dy = \mu \quad \diamond$$

4.3. Expectation of a Function of a Random Variable

Suppose now that we are given a random variable X and its probability distribution (that is, its probability mass function in the discrete case or its probability density function in the continuous case). Suppose also that we are interested in calculating, not the expected value of X, but the expected value of some function of X, say $g(X)$. How do we go about doing this? One way is as follows. Since $g(X)$ is itself a random variable, it must have a probability distribution, which should be computable from a knowledge of the distribution of X. Once we have obtained the distribution of $g(X)$, we can then compute $E[g(X)]$ by the definition of the expectation.

Example 4i Suppose X has the following probability mass function

$$p(0) = .2, \quad p(1) = .5, \quad p(2) = .3$$

Calculate $E[X^2]$.

Solution: Letting $Y = X^2$, we have that Y is a random variable that can take on one of the values 0^2, 1^2, 2^2 with respective probabilities

$$p_Y(0) = P\{Y = 0^2\} = .2$$
$$p_Y(1) = P\{Y = 1^2\} = .5$$
$$p_Y(4) = P\{Y = 2^2\} = .3$$

Hence,

$$E[X^2] = E[Y] = 0(.2) + 1(.5) + 4(.3) = 1.7$$

Note that

---

*When there is more than one random variable involved in a discussion, we shall use the notation, $p_Y(\cdot)$ to stand for the probability mass function of Y, $F_X(\cdot)$ to stand for the distribution function of X, etc.

$$1.7 = E[X^2] \neq (E[X])^2 = 1.21 \quad \diamond$$

Example 4j Let X be uniformly distributed over $(0, 1)$. Calculate $E[X^3]$.

Solution: Letting $Y = X^3$, we calculate the distribution of Y as follows. For $0 \leq a \leq 1$,

$$\begin{aligned}
F_Y(a) &= P\{Y \leq a\} \\
&= P\{X^3 \leq a\} \\
&= P\{X \leq a^{1/3}\} \\
&= a^{1/3}
\end{aligned}$$

where the last equality follows since X is uniformly distributed over $(0, 1)$. By differentiating $F_Y(a)$, we obtain the density of Y, namely,

$$f_Y(a) = \tfrac{1}{3} a^{-2/3}, \qquad 0 \leq a < 1$$

Hence,

$$\begin{aligned}
E[X^3] = E[Y] &= \int_{-\infty}^{\infty} a f_Y(a) da \\
&= \int_0^1 a \tfrac{1}{3} a^{-2/3} da \\
&= \tfrac{1}{3} \int_0^1 a^{1/3} da \\
&= \tfrac{1}{3} \tfrac{3}{4} a^{4/3} \Big|_0^1 \\
&= \tfrac{1}{4} \quad \diamond
\end{aligned}$$

While the above procedure will, in theory, always enable us to compute the expectation of any function of X from a knowledge of the distribution of X, there is, fortunately, an easier way of doing this. The following proposition, known as the "law of the unconscious statistician," shows how we can calculate the expectation of $g(X)$ without first determining the distribution of $g(X)$.

Proposition 4.1. (Law of the Unconscious Statistician*)

---

*This law got its name from "unconscious" statisticians who have used it as if it were the definition of $E[g(X)]$.

(a) If X is a discrete random variable with probability mass function $p(x)$, then for any real-valued function g,

$$E[g(X)] = \sum_{x:p(x)>0} g(x)p(x)$$

(b) If X is a continuous random variable with probability density function $f(x)$, then for any real-valued function g,

$$E[g(X)] = \int_{-\infty}^{\infty} g(x)f(x)dx \quad \Diamond$$

Example 4k Applying the above proposition to Example 4i yields

$$E[X^2] = 0^2(0.2) + (1^2)(0.5) + (2^2)(0.3) = 1.7$$

which, of course, checks with the result derived in Example 4i. $\quad \Diamond$

Example 4l Applying the proposition to Example 4j yields

$$E[X^3] = \int_0^1 x^3 dx \quad (\text{since} \quad f(x) = 1, \quad 0 < x < 1)$$

$$= \tfrac{1}{4} \quad \Diamond$$

A simple corollary of the law of the unconscious statistician is the following.

Corollary 4.1. If a and b are constants, then

$$E[aX + b] = aE[X] + b$$

Proof: In the discrete case,

$$E[aX + b] = \sum_{x:p(x)>0} (ax + b)p(x)$$

$$= a \sum_{x:p(x)>0} xp(x) + b \sum_{x:p(x)>0} p(x)$$

$$= aE[X] + b$$

In the continuous case,

$$E[aX + b] = \int_{-\infty}^{\infty} (ax + b)f(x)dx$$

$$= a \int_{-\infty}^{\infty} xf(x)dx + b \int_{-\infty}^{\infty} f(x)dx$$

$$= aE[X] + b \quad \Diamond$$

The expected value of a random variable X, $E[X]$, is also referred to as the *mean* or the first *moment* of X. The quantity $E[X^n]$, $n \geq 1$, is called the *n*th moment of X. By the above proposition, we note that

$$E[X^n] = \begin{cases} \displaystyle\sum_{x:p(x)>0} x^n p(x), & \text{if } X \text{ is discrete} \\ \displaystyle\int_{-\infty}^{\infty} x^n f(x)dx, & \text{if } X \text{ is continuous} \end{cases}$$

Another quantity of interest is the variance of a random variable X, denoted by $\text{Var}(X)$, which is defined by

$$\text{Var}(X) = E[(X - E[X])^2]$$

Thus, the variance of X measures the expected square of the deviation of X from its expected value.

Example 4m (Variance of the Normal Random Variable): Let X be normally distributed with parameters μ and σ^2. Find $\text{Var}(X)$.

Solution: Recalling (see Example 4h) that $E[X] = \mu$, we have that

$$\text{Var}(X) = E[(X - \mu)^2]$$

$$= \frac{1}{\sqrt{2\pi}\sigma} \int_{-\infty}^{\infty} (x - \mu)^2 e^{-(x-\mu)^2/2\sigma^2} dx$$

Substituting $y = (x - \mu)/\sigma$ yields

$$\text{Var}(X) = \frac{\sigma^2}{\sqrt{2\pi}} \int_{-\infty}^{\infty} y^2 e^{-y^2/2} dy$$

We now employ an identity that can be found in many integration tables, namely $\int_{-\infty}^{\infty} y^2 e^{-y^2/2} dy = \sqrt{2\pi}$. Hence,

$$\text{Var}(X) = \sigma^2$$

Another derivation of $\text{Var}(X)$ will be given in Example 6d. $\Diamond$

Suppose that X is continuous with density f, and let $E[X] = \mu$. Then,

$$\begin{aligned}
\text{Var}(X) &= E[(X - \mu)^2] \\
&= E[X^2 - 2\mu X + \mu^2] \\
&= \int_{-\infty}^{\infty} (x^2 - 2\mu x + \mu^2) f(x) dx \\
&= \int_{-\infty}^{\infty} x^2 f(x) dx - 2\mu \int_{-\infty}^{\infty} x f(x) dx + \mu^2 \int_{-\infty}^{\infty} f(x) dx \\
&= E[X^2] - 2\mu\mu + \mu^2 \\
&= E[X^2] - \mu^2
\end{aligned}$$

A similar proof holds in the discrete case, and so we obtain the useful identity

$$\text{Var}(X) = E[X^2] - (E[X])^2$$

Example 4n Calculate $\text{Var}(X)$ when X represents the outcome when a fair die is rolled.

Solution: As previously noted in Example 4a, $E[X] = \frac{7}{2}$. Also,

$$E[X^2] = 1(\tfrac{1}{6}) + 2^2(\tfrac{1}{6}) + 3^2(\tfrac{1}{6}) + 4^2(\tfrac{1}{6}) + 5^2(\tfrac{1}{6}) + 6^2(\tfrac{1}{6}) = (\tfrac{1}{6})(91)$$

Hence,

$$\text{Var}(X) = \tfrac{91}{6} - (\tfrac{7}{2})^2 = \tfrac{35}{12} \quad \diamond$$

5. Jointly Distributed Random Variables

5.1. Joint Distribution Functions

Thus far, we have concerned ourselves with the probability distribution of a single random variable. However, we are often interested in probability statements concerning two or more random variables. In order to deal with such probabilities we define, for any two random variables X and Y, the *joint cumulative probability distribution function of X and Y* by

$$F(a, b) = P\{X \le a, Y \le b\}, \qquad -\infty < a, b < \infty$$

The distribution of X can be obtained from the joint distribution of X and Y as follows:

$$\begin{aligned}
F_X(a) &= P\{X \le a\} \\
&= P\{X \le a, Y \le \infty\} \\
&= F(a, \infty)
\end{aligned}$$

Similarly, the cumulative distribution function of Y is given by

$$F_Y(b) = P\{Y \leq b\} = F(\infty, b)$$

In the case where X and Y are both discrete random variables, it is convenient to define the *joint probability mass function* of X and Y by

$$p(x, y) = P\{X = x, Y = y\}$$

The probability mass function of X may be obtained from $p(x, y)$ by

$$p_X(x) = \sum_{y:p(x,y)>0} p(x, y)$$

Similarly,

$$p_Y(y) = \sum_{x:p(x,y)>0} p(x, y)$$

We say that X and Y are *jointly continuous* if there exists a function $f(x, y)$, defined for all real x and y, having the property that for all sets A and B of real numbers

$$P\{X \in A, Y \in B\} = \int_B \int_A f(x, y)dxdy$$

The function $f(x, y)$ is called the *joint probability density function* of X and Y. The probability density of X can be obtained from a knowledge of $f(x, y)$ by the following reasoning:

$$P\{X \in A\} = P\{X \in A, Y \in (-\infty, \infty)\}$$

$$= \int_{-\infty}^{\infty} \int_A f(x, y)dxdy$$

$$= \int_A f_X(x)dx$$

where

$$f_X(x) = \int_{-\infty}^{\infty} f(x, y)dy$$

is thus the probability density function of X. Similarly, the probability density function of Y is given by

$$f_Y(y) = \int_{-\infty}^{\infty} f(x, y)dx$$

A variation of the law of the unconscious statistician states that if X and Y are random variables and g is a function of two variables, then

$$E[g(X, Y)] = \sum_y \sum_x g(x, y)p(x, y) \qquad \text{in the discrete case}$$

$$= \int_{-\infty}^{\infty} \int_{-\infty}^{\infty} g(x, y)f(x, y)dxdy \quad \text{in the continuous case}$$

For example, if $g(X, Y) = X + Y$, then, in the continuous case,

$$E[X + Y] = \int_{-\infty}^{\infty} \int_{-\infty}^{\infty} (x + y)f(x, y)dxdy$$

$$= \int_{-\infty}^{\infty} \int_{-\infty}^{\infty} xf(x, y)dxdy + \int_{-\infty}^{\infty} \int_{-\infty}^{\infty} yf(x, y)dxdy$$

$$= \int_{-\infty}^{\infty} x \left(\int_{-\infty}^{\infty} f(x, y)dy \right) dx + \int_{-\infty}^{\infty} y \left(\int_{-\infty}^{\infty} f(x, y)dx \right) dy$$

$$= \int_{-\infty}^{\infty} xf_X(x)dx + \int_{-\infty}^{\infty} yf_Y(y)dy$$

$$= E[X] + E[Y]$$

The same result holds in the discrete case and, combined with the corollary in Section 4.3, yields that for any constants a, b

$$E[aX + bY] = aE[X] + bE[Y] \tag{5.1}$$

Joint probability distributions may also be defined for n random variables. The details are exactly the same as when $n = 2$ and are left as an exercise. The corresponding result to Equation (5.1) states that if $X_1, X_2, \ldots, X_n$ are n random variables, then for any n constants $a_1, a_2, \ldots, a_n$,

$$E[a_1X_1 + a_2X_2 + \cdots + a_nX_n]$$
$$= a_1E[X_1] + a_2E[X_2] + \cdots + a_nE[X_n] \tag{5.2}$$

Example 5a Calculate the expected sum obtained when three fair dice are rolled.

Solution: Let X denote the sum obtained. Then $X = X_1 + X_2 + X_3$ where X_i represents the value of the ith die. Thus,

$$E[X] = E[X_1] + E[X_2] + E[X_3] = 3(\tfrac{7}{2}) = \tfrac{21}{2} \quad \lozenge$$

Example 5b As another example of the usefulness of Equation (5.2), let us use it to obtain the expectation of a binomial random variable having parameters n and p. Recalling that such a random variable X represents the number of successes in n trials when each trial has probability p of being a success, we have that

$$X = X_1 + X_2 + \cdots + X_n$$

where

$$X_i = \begin{cases} 1, & \text{if the } i\text{th trial is a success} \\ 0, & \text{if the } i\text{th trial is a failure} \end{cases}$$

Hence, X_i is a Bernoulli random variable having expectation $E[X_i] = 1(p) + 0(1 - p) = p$. Thus

$$E[X] = E[X_1] + E[X_2] + \cdots + E[X_n] = np$$

This derivation should be compared with the one presented in Example 4c.   ◇

Example 5c At a party N men throw their hats into the center of a room. The hats are mixed up and each man randomly selects one. Find the expected number of men that select their own hats.

Solution: Letting X denote the number of men that select their own hats, we can best compute $E[X]$ by noting that

$$X = X_1 + X_2 + \cdots + X_N$$

where

$$X_i = \begin{cases} 1, & \text{if the } i\text{th man selects his own hat} \\ 0, & \text{otherwise} \end{cases}$$

Now, as the ith man is equally likely to select any of the N hats, it follows that

$$P\{X_i = 1\} = P\{i\text{th man selects his own hat}\} = \frac{1}{N}$$

and so

$$E[X_i] = 1\, P\{X_i = 1\} + 0\, P\{X_i = 0\} = \frac{1}{N}$$

Hence, from Equation (5.2) we obtain that

$$E[X] = E[X_1] + \cdots + E[X_N] = \left(\frac{1}{N}\right) N = 1$$

Hence, no matter how many people are at the party, on the average exactly one of the men will select his own hat. ◇

Example 5d Suppose there are 25 different types of coupons and suppose that each time one obtains a coupon, it is equally likely to be any one of the 25 types. Compute the expected number of different types that are contained in a set of 10 coupons.

Solution: Let X denote the number of different types in the set of 10 coupons. We compute $E[X]$ by using the representation

$$X = X_1 + \cdot \cdot \cdot + X_{25}$$

where

$$X_i = \begin{cases} 1, & \text{if at least one type } i \text{ coupon is in the set of 10} \\ 0, & \text{otherwise} \end{cases}$$

Now,

$$E[X_i] = P\{X_i = 1\}$$
$$= P\{\text{at least one type } i \text{ coupon is in the set of 10}\}$$
$$= 1 - P\{\text{no type } i \text{ coupons are in the set of 10 }\}$$
$$= 1 - (\tfrac{24}{25})^{10}$$

when the last equality follows since each of the 10 coupons will (independently) not be a type i with probability $\tfrac{24}{25}$. Hence,

$$E[X] = E[X_1] + \cdot \cdot \cdot + E[X_{25}] = 25[1 - (\tfrac{24}{25})^{10}] ◇$$

5.2. Independent Random Variables

The random variables X and Y are said to be *independent* if, for all a, b,

$$P\{X \leq a, Y \leq b\} = P\{X \leq a\}P\{Y \leq b\} \tag{5.3}$$

In other words, X and Y are independent if, for all a and b, the events $E_a = \{X \leq a\}$ and $F_b = \{Y \leq b\}$ are independent.

In terms of the joint distribution function F of X and Y, we have that X and Y are independent if

$$F(a, b) = F_X(a)F_Y(b) \quad \text{for all} \quad a, b$$

When X and Y are discrete, the condition of independence reduces to

$$p(x, y) = p_X(x)p_Y(y) \qquad (5.4)$$

while if X and Y are jointly continuous, independence reduces to

$$f(x, y) = f_X(x)f_Y(y) \qquad (5.5)$$

To prove this statement, consider first the discrete version, and suppose that the joint probability mass function $p(x, y)$ satisfies Equation (5.4). Then

$$
\begin{aligned}
P\{X \le a, Y \le b\} &= \sum_{y \le b} \sum_{x \le a} p(x, y) \\
&= \sum_{y \le b} \sum_{x \le a} p_X(x)p_Y(y) \\
&= \sum_{y \le b} p_Y(y) \sum_{x \le a} p_X(x) \\
&= P\{Y \le b\}P\{X \le a\}
\end{aligned}
$$

and so X and Y are independent. That Equation (5.5) implies independence in the continuous case is proven in the same manner, and is left as an exercise.

An important result concerning independence is the following.

Proposition 5.1.   If X and Y are independent, then for any functions h and g

$$E[g(X)h(Y)] = E[g(X)]E[h(Y)]$$

Proof: Suppose that X and Y are jointly continuous. Then

$$
\begin{aligned}
E[g(X)h(Y)] &= \int_{-\infty}^{\infty} \int_{-\infty}^{\infty} g(x)h(y)f(x, y)dxdy \\
&= \int_{-\infty}^{\infty} \int_{-\infty}^{\infty} g(x)h(y)f_X(x)f_Y(y)dxdy \\
&= \int_{-\infty}^{\infty} h(y)f_Y(y)dy \int_{-\infty}^{\infty} g(x)f_X(x)dx \\
&= E[h(Y)]E(g(X))
\end{aligned}
$$

The proof in the discrete case is similar. $\diamond$

The covariance of any two random variables X and Y, denoted by $\text{Cov}(X, Y)$, is defined by

$$\text{Cov}(X,Y) = E[(X - E[X])(Y - E[Y])]$$
$$= E[XY - YE[X] - XE[Y] + E[X]E[Y]]$$
$$= E[XY] - E[Y]E[X] - E[X]E[Y] + E[X]E[Y]$$
$$= E[XY] - E[X]E[Y]$$

Note that if X and Y are independent, then by Proposition 5.1 it follows that Cov $(X, Y) = 0$.

Let us consider now the special case where X and Y are indicator variables for whether or not the events A and B occur. That is, for events A and B, define

$$X = \begin{cases} 1, & \text{if } A \text{ occurs} \\ 0, & \text{otherwise} \end{cases}, \qquad Y = \begin{cases} 1, & \text{if } B \text{ occurs} \\ 0, & \text{otherwise} \end{cases}$$

Then,

$$\text{Cov}(X, Y) = E[XY] - E[X]E[Y]$$

and, as XY will equal 1 or 0 depending upon whether or not both X and Y equal 1, we see that

$$\text{Cov}(X, Y) = P\{X = 1, Y = 1\} - P\{X = 1\}P\{Y = 1\}$$

From this we see that

$$\text{Cov}(X, Y) > 0 \Leftrightarrow P\{X = 1, Y = 1\} > P\{X = 1\}P\{Y = 1\}$$
$$\Leftrightarrow \frac{P\{X = 1, Y = 1\}}{P\{X = 1\}} > P\{Y = 1\}$$
$$\Leftrightarrow P\{Y = 1 | X = 1\} > P\{Y = 1\}$$

That is, the covariance of X and Y is positive if the outcome $X = 1$ makes it more likely that $Y = 1$ (which, as is easily seen by symmetry, also implies the reverse).

In general it can be shown that a positive value of $\text{Cov}(X, Y)$ is an indication that Y tends to increase as X does, whereas a negative value indicates that Y tends to decrease as X increases.

A useful expression for the variance of the random variable $X + Y$ may be obtained in terms of the covariance as follows:

$$\text{Var}(X + Y)$$
$$= E[(X + Y - E[X + Y])^2]$$
$$= E[(X + Y - EX - EY)^2]$$
$$= E[((X - EX) + (Y - EY))^2]$$
$$= E[(X - EX)^2 + (Y - EY)^2 + 2(X - EX)(Y - EY)]$$

$$= E[(X - EX)^2] + E[(Y - EY)^2] + 2E[(X - EX)(Y - EY)]$$
$$= \text{Var}(X) + \text{Var}(Y) + 2\text{Cov}(X, Y) \tag{5.6}$$

If X and Y are independent, then Equation (5.6) reduces to

$$\text{Var}(X + Y) = \text{Var}(X) + \text{Var}(Y) \tag{5.7}$$

Example 5e (Variance of a Binomial Random Variable): Compute the variance of a binomial random variable X with parameters n and p.

Solution: Since such a random variable represents the number of successes in n independent trials when each trial has a common probability p of being a success, we may write

$$X = X_1 + \cdots + X_n$$

where the X_i are independent Bernoulli random variables such that

$$X_i = \begin{cases} 1, & \text{if the } i\text{th trial is a success} \\ 0, & \text{otherwise} \end{cases}$$

Hence, from the obvious generalization of (5.7) we obtain

$$\text{Var}(X) = \text{Var}(X_1) + \cdots + \text{Var}(X_n)$$

But

$$\text{Var}(X_i) = E[X_i^2] - (E[X_i])^2$$
$$= E[X_i] - (E[X_i])^2 \quad \text{since} \quad X_i^2 = X_i$$
$$= p - p^2$$

and thus

$$\text{Var}(X) = np(1 - p) \quad \Diamond$$

The generalization of (5.6) to the case of more than two random variables is

$$\text{Var}\left(\sum_1^n X_i\right) = \sum_1^n \text{Var}(X_i) + 2\sum\sum_{i<j} \text{Cov}(X_i, X_j)$$

(See Problem 55 for a proof.) This is often a useful formula for computing variances.

Example 5f (Sampling from a Finite Population: The Hypergeometric): Consider a population of N individuals, some of whom are in favor of a certain proposition. In particular suppose that Np of them are in favor and $N - Np$ are opposed, where p is assumed to be unknown. We are interested in estimating p, the fraction of the population that is for the proposition, by randomly choosing and then determining the positions of n members of the population.

In such situations as described in the preceding, it is common to use the fraction of the sampled population that are in favor of the proposition as an estimator of p. Hence, if we let

$$X_i = \begin{cases} 1, & \text{if the } i\text{th person chosen was in favor} \\ 0, & \text{otherwise} \end{cases}$$

then the usual estimator of p is $\sum_{i=1}^{n} X_i/n$. Let us now compute its mean and variance. Now

$$E\left[\sum_{i=1}^{n} X_i\right] = \sum_{1}^{n} E[X_i]$$

$$= np$$

where the final equality follows since the ith person chosen is equally likely to be any of the N individuals in the population and so has probability Np/N of being in favor.

$$\text{Var}\left(\sum_{1}^{n} X_i\right) = \sum_{1}^{n} \text{Var}(X_i) + 2 \sum\sum_{i<j} \text{Cov}(X_i, X_j)$$

Now, since X_i is a Bernoulli random variable with mean p, it follows that

$$\text{Var}(X_i) = p(1 - p)$$

Also, for $i \neq j$,

$$\begin{aligned} \text{Cov}(X_i, X_j) &= E[X_i X_j] - E[X_i]E[X_j] \\ &= P\{X_i = 1, X_j = 1\} - p^2 \\ &= P\{X_i = 1\}P\{X_j = 1 | X_i = 1\} - p^2 \\ &= \frac{Np}{N} \frac{(Np - 1)}{N - 1} - p^2 \end{aligned}$$

where the last equality follows since if the ith person to be chosen is in favor, then the jth person chosen is equally likely to be any of the other $N - 1$ of which $Np - 1$ are in favor. Thus we see that

$$\text{Var}\left(\sum_{1}^{n} X_i\right) = np(1 - p) + 2\binom{n}{2}\left[\frac{p(Np - 1)}{N - 1} - p^2\right]$$

$$= np(1 - p) - \frac{n(n - 1)p(1 - p)}{N - 1}$$

and so the mean and variance of our estimator are given by

$$E\left[\sum_{1}^{n} \frac{X_i}{n}\right] = p$$

$$\text{Var}\left[\sum_{1}^{n} \frac{X_i}{n}\right] = \frac{p(1 - p)}{n} - \frac{(n - 1)p(1 - p)}{n(N - 1)}$$

Some remarks are in order: As the mean of the estimator is the unknown value p, we would like its variance to be as small as possible (why is this?), and we see by the above that, as a function of the population size N, the variance increases as N increases. The limiting value, as $N \to \infty$, of the variance is $p(1 - p)/n$, which is not surprising since for N large each of the X_i will be (approximately) independent random variables, and thus $\sum_{1}^{n} X_i$ will have an (approximately) binomial distribution with parameters n and p.

The random variable $\sum_{1}^{n} X_i$ can be thought of as representing the number of white balls obtained when n balls are randomly selected from a population consisting of Np white and $N - Np$ black balls. (Identify a person who favors the proposition with a white ball and one against with a black ball.) Such a random variable is called hypergeometric and has a probability mass function given by

$$P\left\{\sum_{1}^{n} X_i = k\right\} = \frac{\binom{Np}{k}\binom{N - Np}{n - k}}{\binom{N}{n}} \qquad \diamond$$

It is often important to be able to calculate the distribution of $X + Y$ from the distributions of X and Y when X and Y are independent. Suppose first that X and Y are independent, X having a probability density f and Y a probability density g. Then, letting $F_{X+Y}(a)$ be the cumulative distribution function of $X + Y$, we have

$$F_{X+Y}(a) = P\{X + Y \le a\}$$

$$= \iint_{x+y \le a} f(x)g(y)\,dx\,dy$$

$$= \int_{-\infty}^{\infty} \int_{-\infty}^{a-y} f(x)g(y)dxdy$$

$$= \int_{-\infty}^{\infty} \left(\int_{-\infty}^{a-y} f(x)dx \right) g(y)dy$$

$$= \int_{-\infty}^{\infty} F_X(a - y)g(y)dy \qquad (5.8)$$

The cumulative distribution function F_{X+Y} is called the *convolution* of the distributions F_X and F_Y (the cumulative distribution functions of X and Y, respectively).

By differentiating Equation (5.8), we obtain that the probability density function $f_{X+Y}(a)$ of $X + Y$ is given by

$$f_{X+Y}(a) = \frac{d}{da} \int_{-\infty}^{\infty} F_X(a - y)g(y)dy$$

$$= \int_{-\infty}^{\infty} \frac{d}{da} (F_X(a - y))g(y)dy$$

$$= \int_{-\infty}^{\infty} f(a - y)g(y)dy \qquad (5.9)$$

Example 5g (Sum of Two Independent Uniform Random Variables): If X and Y are independent random variables both uniformly distributed on $(0,1)$, then calculate the probability density of $X + Y$.

Solution: From Equation (5.9) since

$$f(a) = g(a) = \begin{cases} 1, & 0 < a < 1 \\ 0, & \text{otherwise} \end{cases}$$

we obtain

$$f_{X+Y}(a) = \int_0^1 f(a - y)dy$$

For $0 \leq a \leq 1$, this yields

$$f_{X+Y}(a) = \int_0^a dy = a$$

For $1 < a < 2$, we get

$$f_{X+Y}(a) = \int_{a-1}^{1} dy = 2 - a$$

Hence,

$$f_{X+Y}(a) = \begin{cases} a, & 0 \le a \le 1 \\ 2 - a, & 1 < a < 2 \\ 0, & \text{otherwise} \end{cases} \quad \diamond$$

Rather than deriving a general expression for the distribution of $X + Y$ in the discrete case, we shall consider an example.

Example 5h (Sums of Independent Poisson Random Variables): Let X and Y be independent Poisson random variables with respective means λ_1 and λ_2. Calculate the distribution of $X + Y$.

Solution: Since the event $\{X + Y = n\}$ may be written as the union of the disjoint events $\{X = k, Y = n - k\}$, $0 \le k \le n$, we have

$$\begin{aligned} P\{X + Y = n\} &= \sum_{k=0}^{n} P\{X = k, Y = n - k\} \\ &= \sum_{k=0}^{n} P\{X = k\}P\{Y = n - k\} \\ &= \sum_{k=0}^{n} e^{-\lambda_1}\frac{\lambda_1^k}{k!} e^{-\lambda_2}\frac{\lambda_2^{n-k}}{(n - k)!} \\ &= e^{-(\lambda_1+\lambda_2)} \sum_{k=0}^{n} \frac{\lambda_1^k \lambda_2^{n-k}}{k!(n - k)!} \\ &= \frac{e^{-(\lambda_1+\lambda_2)}}{n!} \sum_{k=0}^{n} \frac{n!}{k!(n - k)!} \lambda_1^k \lambda_2^{n-k} \\ &= \frac{e^{-(\lambda_1+\lambda_2)}}{n!} (\lambda_1 + \lambda_2)^n \end{aligned}$$

In words, $X_1 + X_2$ has a Poisson distribution with mean $\lambda_1 + \lambda_2$. $\diamond$

The concept of independence may, of course, be defined for more than two random variables. In general, the n random variables $X_1, X_2, \ldots, X_n$ are said to be independent if for all values $a_1, a_2, \ldots, a_n$

$$P\{X_1 \le a_1, X_2 \le a_2, \ldots, X_n \le a_n\}$$
$$= P\{X_1 \le a_1\}P\{X_2 \le a_2\} \cdots P\{X_n \le a_n\}$$

5.3 Joint Probability Distribution of Functions of Random Variables

Let X_1 and X_2 be jointly continuous random variables with joint probability density function $f(x_1, x_2)$. It is sometimes necessary to obtain the joint distribution of the random variables Y_1 and Y_2 which arise as functions of X_1 and X_2. Specifically, suppose that $Y_1 = g_1(X_1, X_2)$ and $Y_2 = g_2(X_1, X_2)$ for some functions g_1 and g_2.

Assume that the functions g_1 and g_2 satisfy the following conditions:

1. The equations $y_1 = g_1(x_1, x_2)$ and $y_2 = g_2(x_1, x_2)$ can be uniquely solved for x_1 and x_2 in terms of y_1 and y_2 with solutions given by, say, $x_1 = h_1(y_1, y_2)$, $x_2 = h_2(y_1, y_2)$.
2. The functions g_1 and g_2 have continuous partial derivatives at all points (x_1, x_2) and are such that the following 2×2 determinant

$$J(x_1, x_2) = \begin{vmatrix} \dfrac{\partial g_1}{\partial x_1} & \dfrac{\partial g_1}{\partial x_2} \\[2ex] \dfrac{\partial g_2}{\partial x_1} & \dfrac{\partial g_2}{\partial x_2} \end{vmatrix} \equiv \frac{\partial g_1}{\partial x_1}\frac{\partial g_2}{\partial x_2} - \frac{\partial g_1}{\partial x_2}\frac{\partial g_2}{\partial x_1} \ne 0$$

at all points (x_1, x_2).

Under these two conditions it can be shown that the random variables Y_1 and Y_2 are jointly continuous with joint density function given by

$$f_{Y_1, Y_2}(y_1, y_2) = f_{X_1, X_2}(x_1, x_2)|J(x_1, x_2)|^{-1} \tag{5.10}$$

where $x_1 = h_1(y_1, y_2)$, $x_2 = h_2(y_1, y_2)$.

A proof of Equation (5.10) would proceed along the following lines:

$$P\{Y_1 \le y_1, Y_2 \le y_2\} = \iint\limits_{\substack{(x_1, x_2): \\ g_1(x_1, x_2) \le y_1 \\ g_2(x_1, x_2) \le y_2}} f_{X_1, X_2}(x_1, x_2)dx_1 dx_2 \tag{5.11}$$

The joint density function can now be obtained by differentiating Equation (5.11) with respect to y_1 and y_2. That the result of this differentiation will be equal to the right-hand side of Equation (5.10) is an exercise in advanced calculus whose proof will not be presented in the present text.

Example 5i　If X and Y are independent gamma random variables with parameters (α, λ) and (β, λ), respectively, compute the joint density of $U = X + Y$ and $V = X/(X + Y)$.

Solution:　The joint density of X and Y is given by

$$f_{X,Y}(x, y) = \frac{\lambda e^{-\lambda x}(\lambda x)^{\alpha-1}}{\Gamma(\alpha)} \frac{\lambda e^{-\lambda y}(\lambda y)^{\beta-1}}{\Gamma(\beta)}$$

$$= \frac{\lambda^{\alpha+\beta}}{\Gamma(\alpha)\Gamma(\beta)} e^{-\lambda(x+y)} x^{\alpha-1} y^{\beta-1}$$

Now, if $g_1(x, y) = x + y$, $g_2(x, y) = x/(x + y)$, then

$$\frac{\partial g_1}{\partial x} = \frac{\partial g_1}{\partial y} = 1 \qquad \frac{\partial g_2}{\partial x} = \frac{y}{(x + y)^2} \qquad \frac{\partial g_2}{\partial y} = -\frac{x}{(x + y)^2}$$

and so

$$J(x, y) = \begin{vmatrix} 1 & 1 \\ \dfrac{y}{(x + y)^2} & \dfrac{-x}{(x + y)^2} \end{vmatrix} = -\frac{1}{x + y}$$

Finally, as the equations $u = x + y$, $v = x/(x + y)$ have as their solutions $x = uv$, $y = u(1 - v)$, we see that

$$f_{U,V}(u, v) = f_{X,Y}[uv, u(1 - v)]u$$

$$= \frac{\lambda e^{-\lambda u}(\lambda u)^{\alpha+\beta-1}}{\Gamma(\alpha + \beta)} \frac{v^{\alpha-1}(1 - v)^{\beta-1}\Gamma(\alpha + \beta)}{\Gamma(\alpha)\Gamma(\beta)}$$

Hence $X + Y$ and $X/(X + Y)$ are independent, with $X + Y$ having a gamma distribution with parameters $(\alpha + \beta, \lambda)$ and $X/(X + Y)$ having density function

$$f_V(v) = \frac{\Gamma(\alpha + \beta)}{\Gamma(\alpha)\Gamma(\beta)} v^{\alpha-1}(1 - v)^{\beta-1}, \quad 0 < v < 1$$

This is called the beta density with parameters (α, β).

The above result is quite interesting. For suppose there are $n + m$ jobs to be performed, with each (independently) taking an exponential amount of time with rate λ for performance, and suppose that we have two workers to perform these jobs. Worker I will do jobs $1, 2, \ldots, n$, and worker II will do the remaining m jobs. If we let X and Y denote the total working times of workers I and II, respectively, then upon using the above result it follows that X and Y will be independent gamma random variables having parameters (n, λ) and (m, λ), respectively. Then the above result yields that independently of the working time needed to complete all $n + m$ jobs (that is, of $X + Y$), the proportion of this work that will be performed by worker I has a beta distribution with parameters (n, m). $\diamond$

When the joint density function of the n random variables $X_1, X_2, \ldots, X_n$ is given and we want to compute the joint density function of $Y_1, Y_2, \ldots, Y_n$, where

$$Y_1 = g_1(X_1, \ldots, X_n) \qquad Y_2 = g_2(X_1, \ldots, X_n), \ldots$$
$$Y_n = g_n(X_1, \ldots, X_n)$$

the approach is the same. Namely, we assume that the functions g_i have continuous partial derivatives and that the Jacobian determinant $J(x_1, \ldots, x_n) \neq 0$ at all points $(x_1, \ldots, x_n)$, where

$$J(x_1, \ldots, x_n) = \begin{vmatrix} \dfrac{\partial g_1}{\partial x_1} & \dfrac{\partial g_1}{\partial x_2} & \cdots & \dfrac{\partial g_1}{\partial x_n} \\[2mm] \dfrac{\partial g_2}{\partial x_1} & \dfrac{\partial g_2}{\partial x_2} & \cdots & \dfrac{\partial g_2}{\partial x_n} \\[2mm] \dfrac{\partial g_n}{\partial x_1} & \dfrac{\partial g_n}{\partial x_2} & \cdots & \dfrac{\partial g_n}{\partial x_n} \end{vmatrix}$$

Furthermore, we suppose that the equations $y_1 = g_1(x_1, \ldots, x_n)$, $y_2 = g_2(x_1, \ldots, x_n), \ldots, y_n = g_n(x_1, \ldots, x_n)$ have a unique solution, say, $x_1 = h_1(y_1, \ldots, y_n), \ldots, x_n = h_n(y_1, \ldots, y_n)$. Under these assumptions the joint density function of the random variables Y_i is given by

$$f_{Y_1, \ldots, Y_n}(y_1, \ldots, y_n) = f_{X_1, \ldots, X_n}(x_1, \ldots, x_n)|J(x_1, \ldots, x_n)|^{-1}$$

where $x_i = h_i(y_1, \ldots, y_n)$, $i = 1, 2, \ldots, n$.

6. Moment Generating Functions

The moment generating function $\phi(t)$ of the random variable X is de-fined for all values t by

$$\phi(t) = E[e^{tX}]$$

$$= \begin{cases} \sum_x e^{tx} p(x) & \text{if } X \text{ is discrete} \\ \int_{-\infty}^{\infty} e^{tx} f(x) dx & \text{if } X \text{ is continuous} \end{cases}$$

We call $\phi(t)$ the moment generating function because all of the moments of X can be obtained by successively differentiating $\phi(t)$. For example,

$$\phi'(t) = \frac{d}{dt} E[e^{tX}]$$

$$= E\left[\frac{d}{dt}(e^{tX})\right]$$

$$= E[Xe^{tX}]$$

Hence,

$$\phi'(0) = E[X]$$

Similarly,

$$\phi''(t) = \frac{d}{dt} \phi'(t)$$

$$= \frac{d}{dt} E[Xe^{tX}]$$

$$= E\left[\frac{d}{dt}(Xe^{tX})\right]$$

$$= E[X^2 e^{tX}]$$

and so

$$\phi''(0) = E[X^2]$$

In general, the nth derivative of $\phi(t)$ evaluated at $t = 0$ equals $E[X^n]$, that is,

$$\phi^n(0) = E[X^n], \qquad n \geq 1$$

We now compute $\phi(t)$ for some common distributions.

Example 6a (The Binomial Distribution with Parameters n and p):

$$\phi(t) = E[e^{tX}]$$

$$= \sum_{k=0}^{n} e^{tk} \binom{n}{k} p^k (1-p)^{n-k}$$

$$= \sum_{k=0}^{n} \binom{n}{k} (pe^t)^k (1-p)^{n-k}$$

$$= (pe^t + 1 - p)^n$$

Hence,

$$\phi'(t) = n(pe^t + 1 - p)^{n-1} pe^t$$

and so

$$E[X] = \phi'(0) = np$$

which checks with the result obtained in Example 4c. Differentiating a second time yields

$$\phi''(t) = n(n-1)(pe^t + 1 - p)^{n-2}(pe^t)^2 + n(pe^t + 1 - p)^{n-1}pe^t$$

and so

$$E[X^2] = \phi''(0) = n(n-1)p^2 + np$$

Thus the variance of X is given

$$\begin{aligned} \text{Var}(X) &= E[X^2] - (E[X])^2 \\ &= n(n-1)p^2 + np - n^2p^2 \\ &= np(1-p) \quad \lozenge \end{aligned}$$

Example 6b (The Poisson Distribution with Mean λ):

$$\phi(t) = E[e^{tX}]$$

$$= \sum_{n=0}^{\infty} \frac{e^{tn}e^{-\lambda}\lambda^n}{n!}$$

$$= e^{-\lambda} \sum_{n=0}^{\infty} \frac{(\lambda e^t)^n}{n!}$$

$$= e^{-\lambda}e^{\lambda et}$$

$$= \exp\{\lambda(e^t - 1)\}$$

Differentiation yields

$$\phi'(t) = \lambda e^t \exp\{\lambda(e^t - 1)\}$$
$$\phi''(t) = (\lambda e^t)^2 \exp\{\lambda(e^t - 1)\} + \lambda e^t \exp\{\lambda(e^t - 1)\}$$

and so

$$E[X] = \phi'(0) = \lambda$$
$$E[X^2] = \phi''(0) = \lambda^2 + \lambda$$
$$\mathrm{Var}(X) = E[X^2] - (E[X])^2$$
$$= \lambda$$

Thus both the mean and the variance of the Poisson equal λ.   ◇

Example 6c   (The Exponential Distribution with Parameter λ):

$$\phi(t) = E[e^{tX}]$$

$$= \int_0^\infty e^{tx} \lambda e^{-\lambda x} dx$$

$$= \lambda \int_0^\infty e^{-(\lambda - t)x} dx$$

$$= \frac{\lambda}{\lambda - t} \quad \text{for} \quad t < \lambda$$

We note by the above derivation that, for the exponential distribution, $\phi(t)$ is only defined for values of t less than λ. Differentiation of $\phi(t)$ yields

$$\phi'(t) = \frac{\lambda}{(\lambda - t)^2}, \qquad \phi''(t) = \frac{2\lambda}{(\lambda - t)^3}$$

Hence,

$$E[X] = \phi'(0) = \frac{1}{\lambda}, \qquad E[X^2] = \phi''(0) = \frac{2}{\lambda^2}$$

The variance of X is thus given by

$$\mathrm{Var}(X) = E[X^2] - (E[X])^2$$

$$= \frac{1}{\lambda^2} \quad ◇$$

Example 6d   (The Normal Distribution with Parameters μ and σ^2):

$$\phi(t) = E[e^{tX}]$$

$$= \frac{1}{\sqrt{2\pi}\sigma} \int_{-\infty}^{\infty} e^{tx} e^{-(x-\mu)^2/2\sigma^2} dx$$

$$= \frac{1}{\sqrt{2\pi}\sigma} \int_{-\infty}^{\infty} \exp\left\{\frac{-(x^2 - 2\mu x + \mu^2 - 2\sigma^2 tx)}{2\sigma^2}\right\} dx$$

Now writing

$$x^2 - 2\mu x + \mu^2 - 2\sigma^2 tx = x^2 - 2(\mu + \sigma^2 t)x + \mu^2$$
$$= (x - (\mu + \sigma^2 t))^2 - (\mu + \sigma^2 t)^2 + \mu^2$$
$$= (x - (\mu + \sigma^2 t))^2 - \sigma^4 t^2 - 2\mu\sigma^2 t$$

we have

$$\phi(t) = \frac{1}{\sqrt{2\pi}\sigma} \exp\left\{\frac{\sigma^4 t^2 + 2\mu\sigma^2 t}{2\sigma^2}\right\} \int_{-\infty}^{\infty} \exp\left\{\frac{-(x - (\mu + \sigma^2 t))^2}{2\sigma^2}\right\} dx$$

$$= \exp\left\{\frac{\sigma^2 t^2}{2} + \mu t\right\} \frac{1}{\sqrt{2\pi}\sigma} \int_{-\infty}^{\infty} \exp\left\{\frac{-(x - (\mu + \sigma^2 t))^2}{2\sigma^2}\right\} dx$$

However,

$$\frac{1}{\sqrt{2\pi}\sigma} \int_{-\infty}^{\infty} \exp\left\{\frac{-(x - (\mu + \sigma^2 t))^2}{2\sigma^2}\right\} dx = P\{-\infty < \bar{X} < \infty\} = 1$$

where $\bar{X}$ is a normally distributed random variable having parameters $\bar{\mu} = \mu + \sigma^2 t$ and $\bar{\sigma}^2 = \sigma^2$. Thus, we have shown that

$$\phi(t) = \exp\left\{\frac{\sigma^2 t^2}{2} + \mu t\right\}$$

By differentiating we obtain

$$\phi'(t) = (\mu + t\sigma^2) \exp\left\{\frac{\sigma^2 t^2}{2} + \mu t\right\}$$

$$\phi''(t) = (\mu + t\sigma^2)^2 \exp\left\{\frac{\sigma^2 t^2}{2} + \mu t\right\} + \sigma^2 \exp\left\{\frac{\sigma^2 t^2}{2} + \mu t\right\}$$

and so

$$E[X] = \phi'(0) = \mu$$
$$E[X^2] = \phi''(0) = \mu^2 + \sigma^2$$

implying that

$$\text{Var}(X) = E[X^2] - E([X])^2$$
$$= \sigma^2 \quad \diamond$$

Tables 6.1 and 6.2 give the moment generating function for some common distributions.

An important property of moment generating functions is that the *moment generating function of the sum of independent random variables is just the product of the individual moment generating functions.* To see this, suppose that X and Y are independent and have moment generating functions $\phi_X(t)$ and $\phi_Y(t)$, respectively. Then $\phi_{X+Y}(t)$, the moment generating function of $X + Y$, is given by

$$\phi_{X+Y}(t) = E[e^{t(X+Y)}]$$
$$= E[e^{tX}e^{tY}]$$
$$= E[e^{tX}]E[e^{tY}]$$
$$= \phi_X(t)\phi_Y(t)$$

where the next to the last equality follows from Proposition 5.1 since X and Y are independent.

Another important result is that the *moment generating function uniquely determines the distribution.* That is, there exists a one-to-one correspondence between the moment generating function and the distribution function of a random variable.

Example 6e Suppose the moment generating function of a random variable X is given by $\phi(t) = e^{3(e^t-1)}$. What is $P\{X = 0\}$?

Table 6.1

Discrete probability distribution	Probability mass function, $p(x)$	Moment generating function, $\phi(t)$	Mean	Variance
Binomial with parameters n, p $0 \le p \le 1$	$\binom{n}{x}p^x(1-p)^{n-x}$, $x = 0, 1, \ldots, n$	$(pe^t + (1-p))^n$	np	$np(1-p)$
Poisson with parameter $\lambda > 0$	$e^{-\lambda}\lambda^x \over x!$, $x = 0, 1, 2, \ldots$	$\exp\{\lambda(e^t - 1)\}$	λ	λ
Geometric with parameter $0 \le p \le 1$	$p(1-p)^{x-1}$, $x = 1, 2, \ldots$	$\dfrac{pe^t}{1-(1-p)e^t}$	$\dfrac{1}{p}$	$\dfrac{1-p}{p^2}$

Table 6.2

Continuous probability distribution	Probability density function, $f(x)$	Moment generating function, $\phi(t)$	Mean	Variance
Uniform over (a, b)	$f(x) = \begin{cases} \dfrac{1}{b-a}, & a < x < b \\ 0, & \text{otherwise} \end{cases}$	$\dfrac{e^{tb} - e^{ta}}{t(b-a)}$	$\dfrac{a+b}{2}$	$\dfrac{(b-a)^2}{12}$
Exponential with parameter $\lambda > 0$	$f(x) = \begin{cases} \lambda e^{-\lambda x}, & x \ge 0 \\ 0, & x < 0 \end{cases}$	$\dfrac{\lambda}{\lambda - t}$	$\dfrac{1}{\lambda}$	$\dfrac{1}{\lambda^2}$
Gamma with parameters (n,λ) $\lambda > 0$	$f(x) = \begin{cases} \dfrac{\lambda e^{-\lambda x}(\lambda x)^{n-1}}{(n-1)!}, & x \ge 0 \\ 0, & x < 0 \end{cases}$	$\left(\dfrac{\lambda}{\lambda - t}\right)^n$	$\dfrac{n}{\lambda}$	$\dfrac{n}{\lambda^2}$
Normal with parameters (μ,σ^2)	$f(x) = \dfrac{1}{\sqrt{2\pi}\sigma}e^{-(x-\mu)^2/2\sigma^2},$ $-\infty < x < \infty$	$\exp\left\{\mu t + \dfrac{\sigma^2 t^2}{2}\right\}$	μ	σ^2

Solution: We see from Table 6.1 that $\phi(t) = e^{3(e^t - 1)}$ is the moment generating function of a Poisson random variable with mean 3. Hence, by the one-to-one correspondence between moment generating functions and distribution functions, it follows that X must be a Poisson random variable with mean 3. Thus, $P\{X = 0\} = e^{-3}$. ◊

Example 6f (Sums of Independent Binomial Random Variables): If X and Y are independent binomial random variables with parameters (n, p) and (m, p), respectively, then what is the distribution of $X + Y$?

Solution: The moment generating function of $X + Y$ is given by

$$\phi_{X+Y}(t) = \phi_X(t)\phi_Y(t) = (pe^t + 1 - p)^n (pe^t + 1 - p)^m$$
$$= (pe^t + 1 - p)^{m+n}$$

But $(pe^t + (1 - p))^{m+n}$ is just the moment generating function of a binomial random variable having parameters $m + n$ and p. Thus, this must be the distribution of $X + Y$. ◊

Example 6g   (Sums of Independent Poisson Random Variables): Calculate the distribution of $X + Y$ when X and Y are independent Poisson random variables with means λ_1 and λ_2, respectively.

Solution:

$$\phi_{X+Y}(t) = \phi_X(t)\phi_Y(t)$$
$$= e^{\lambda_1(e^t - 1)}e^{\lambda_2(e^t - 1)}$$
$$= e^{(\lambda_1 + \lambda_2)(e^t - 1)}$$

Hence, $X + Y$ is Poisson distributed with mean $\lambda_1 + \lambda_2$, verifying the result given in Example 5h.   ◇

Example 6h   (Sums of Independent Normal Random Variables): Show that if X and Y are independent normal random variables with parameters (μ_1, σ_1^2) and (μ_2, σ_2^2), respectively, then $X + Y$ is normal with mean $\mu_1 + \mu_2$ and variance $\sigma_1^2 + \sigma_2^2$.

Solution:

$$\phi_{X+Y}(t) = \phi_X(t)\phi_Y(t)$$
$$= \exp\left\{\frac{\sigma_1^2 t^2}{2} + \mu_1 t\right\} \exp\left\{\frac{\sigma_2^2 t^2}{2} + \mu_2 t\right\}$$
$$= \exp\left\{\frac{(\sigma_1^2 + \sigma_2^2)t^2}{2} + (\mu_1 + \mu_2)t\right\}$$

which is the moment generating function of a normal random variable with mean $\mu_1 + \mu_2$ and variance $\sigma_1^2 + \sigma_2^2$. Hence, the result follows since the moment generating function uniquely determines the distribution.   ◇

It is also possible to define the joint moment generating function of two or more random variables. This is done as follows. For any n random variables $X_1, \ldots, X_n$, the joint moment generating function, $\phi(t_1, \ldots, t_n)$, is defined for all real values of $t_1, \ldots, t_n$ by

$$\phi(t_1, \ldots, t_n) = E[e^{(t_1 X_1 + \cdots + t_n X_n)}]$$

It can be shown that $\phi(t_1, \ldots, t_n)$ uniquely determines the joint distribution of $X_1, \ldots, X_n$.

Example 6i   (The Multivariate Normal Distribution): Let $Z_1, \ldots, Z_n$ be a set of n independent unit normal random variables. If, for some constants a_{ij}, $1 \leq i \leq m$, $1 \leq j \leq n$, and μ_i, $1 \leq i \leq m$,

$$X_1 = a_{11}Z_1 + \cdots + a_{1n}Z_n + \mu_1$$
$$X_2 = a_{21}Z_1 + \cdots + a_{2n}Z_n + \mu_2$$
$$\vdots$$
$$X_i = a_{i1}Z_1 + \cdots + a_{in}Z_n + \mu_i$$
$$\vdots$$
$$X_m = a_{m1}Z_1 + \cdots + a_{mn}Z_n + \mu_m$$

then the random variables $X_1, \ldots, X_m$ are said to have a multivariate normal distribution.

It follows from the fact that the sum of independent normal random variables is itself a normal random variable that each X_i is a normal random variable with mean and variance given by

$$E[X_i] = \mu_i$$
$$\text{Var}(X_i) = \sum_{j=1}^{n} a_{ij}^2$$

The covariance of X_i and X_j is given by

$$\text{cov}(X_i, X_j) = \text{cov}\left(\mu_i + \sum_{k=1}^{n} a_{ik}Z_k, \mu_j + \sum_{l=1}^{n} a_{jl}Z_l\right)$$
$$= \text{cov}\left(\sum_{k=1}^{n} a_{ik}Z_k, \sum_{l=1}^{n} a_{jl}Z_l\right)$$
$$= \sum_{k,l} a_{ik}a_{jl} \, \text{cov}(Z_k, Z_l)$$
$$= \sum_{k=1}^{n} a_{ik}a_{jk}$$

since

$$\text{cov}(Z_k, Z_l) = \begin{cases} 1 & \text{if } k = l \\ 0 & \text{if } k \neq l \end{cases}$$

The joint moment generating function is given by

$$\phi(t_1, \ldots, t_m) = E[e^{(t_1X_1 + \cdots + t_mX_m)}]$$

Now,

$$t_1X_1 + \cdots + t_mX_m = (a_{11}t_1 + a_{21}t_2 + \cdots + a_{m1}t_m)Z_1$$

$$+ (a_{12}t_1 + a_{22}t_2 + \cdot \cdot \cdot + a_{m2}t_m)Z_2$$

$$+$$

$$\vdots$$

$$+ (a_{1n}t_1 + a_{2n}t_2 + \cdot \cdot \cdot + a_{mn}t_m)Z_n$$

$$+ \mu_1 t_1 + \mu_2 t_2 + \cdot \cdot \cdot + \mu_m t_m$$

and thus

$$\sum_{i=1}^{m} t_i X_i$$

has a normal distribution with mean

$$E\left[\sum_{i=1}^{m} t_i X_i\right] = \sum_{i=1}^{m} t_i \mu_i$$

and variance

$$\text{Var}\left(\sum_{i=1}^{m} t_i X_i\right) = \sum_{k=1}^{n}\left(\sum_{i=1}^{m} a_{ik}t_i\right)^2$$

Hence, by using the fact that if Y is a normal random variable with mean μ and variance σ^2, then

$$E[e^Y] = \phi_Y(t)|_{t=1} = e^{\mu - \sigma^2/2}$$

we obtain that

$$\phi(t_1, \ldots, t_m) = E\left[\exp\left\{\sum_{1}^{m} t_i X_i\right\}\right]$$

$$= \exp\left\{\sum_{i=1}^{m} t_i \mu_i - \frac{1}{2}\sum_{k=1}^{n}\left(\sum_{i=1}^{m} a_{ik}t_i\right)^2\right\}$$

Now,

$$\sum_{k=1}^{n}\left(\sum_{i=1}^{m} a_{ik}t_i\right)^2 = \sum_{k=1}^{n}\sum_{i=1}^{m} a_{ik}t_i \sum_{j=1}^{m} a_{jk}t_j$$

$$= \sum_{j=1}^{m}\sum_{i=1}^{m} t_i t_j \sum_{k=1}^{n} a_{ik}a_{jk}$$

$$= \sum_{j=1}^{m}\sum_{i=1}^{m} t_i t_j \, \text{cov}(X_i, X_j)$$

and thus $\phi(t_1, \ldots, t_m)$ can be expressed as

$$\phi(t_1, \ldots, t_m) = \exp\left\{\sum_{i=1}^{m} t_i\mu_i - \frac{1}{2}\sum_{j=1}^{m}\sum_{i=1}^{m} t_it_j \operatorname{cov}(X_i, X_j)\right\}$$

which shows that the joint distribution of $X_1, \ldots, X_m$ is completely determined from a knowledge of the values of $\mu_i = E[X_i]$ and cov (X_i, X_j), $i, j = 1, \ldots, m$. $\diamond$

7. Limit Theorems

We start this section by proving a result known as Markov's inequality.

Proposition 7.1. (Markov's Inequality) If X is a random variable which takes only nonnegative values, then for any value $a > 0$

$$P\{X \geq a\} \leq \frac{E[X]}{a}$$

Proof: We give a proof for the case where X is continuous with density f.

$$
\begin{aligned}
E[X] &= \int_0^\infty xf(x)dx \\
&= \int_0^a xf(x)dx + \int_a^\infty xf(x)dx \\
&\geq \int_a^\infty xf(x)dx \\
&\geq \int_a^\infty af(x)dx \\
&= a\int_a^\infty f(x)dx \\
&= aP\{X \geq a\}
\end{aligned}
$$

and the result is proven. $\diamond$

As a corollary, we obtain

Proposition 7.2. (Chebyshev's Inequality) If X is a random variable with mean μ and variance σ^2, then for any value $k > 0$

$$P\{|X - \mu| \geq k\} \leq \frac{\sigma^2}{k^2}$$

Proof: Since $(X - \mu)^2$ is a nonnegative random variable, we can apply Markov's inequality (with $a = k^2$) to obtain

$$P\{(X - \mu)^2 \geq k^2\} \leq \frac{E[(X - \mu)^2]}{k^2}$$

But since $(X - \mu)^2 \geq k^2$ if and only if $|X - \mu| > k$, the above is equivalent to

$$P\{|X - \mu| > k\} \leq \frac{E[(X - \mu)^2]}{k^2} = \frac{\sigma^2}{k^2}$$

and the proof is complete.   ◊

The importance of Markov's and Chebyshev's inequalities is that they enable us to derive bounds on probabilities when only the mean, or both the mean and the variance, of the probability distribution are known. Of course, if the actual distribution were known, then the desired probabilities could be exactly computed, and we would not need to resort to bounds.

Example 7a Suppose that it is known that the number of items produced in a factory during a week is a random variable with mean 500.

(a) What can be said about the probability that this week's production will exceed 1000?

(b) If the variance of a week's production is known to equal 100, then what can be said about the probability that this week's production will be between 400 and 600?

Solution: Let X be the number of items that will be produced in a week.

(a) By Markov's inequality

$$P\{X > 1000\} \leq \frac{E[X]}{1000} = \frac{500}{1000} = \frac{1}{2}$$

(b) By Chebyshev's inequality

$$P\{|X - 500| \geq 100\} \leq \frac{\sigma^2}{(100)^2} = \frac{1}{100}$$

Hence,

$$P\{|X - 500| < 100\} \geq 1 - \frac{1}{100} = \frac{99}{100}$$

and so the probability that this week's production will be between 400 and 600, is at least .99. ◊

The following theorem, known as the *strong law of large numbers* is probably the most well-known result in probability theory. It states that the average of a sequence of independent random variables having the same distribution will, with probability 1, converge to the mean of that distribution.

Theorem 7.1. (Strong Law of Large Numbers) Let $X_1, X_2, \ldots$ be a sequence of independent random variables having a common distribution, and let $E[X_i] = \mu$. Then, with probability 1,

$$\frac{X_1 + X_2 + \cdots + X_n}{n} \to \mu \quad \text{as} \quad n \to \infty$$

As an example of the above, suppose that a sequence of independent trials are performed. Let E be a fixed event and denote by $P\{E\}$ the probability that E occurs on any particular trial. Letting

$$X_i = \begin{cases} 1, & \text{if } E \text{ occurs on the } i\text{th trial} \\ 0, & \text{if } E \text{ does not occur on the } i\text{th trial} \end{cases}$$

we have by the strong law of large numbers that, with probability 1,

$$\frac{X_1 + \cdots + X_n}{n} \to E[X] = P\{E\} \tag{7.1}$$

Since $X_1 + \cdots + X_n$ represents the number of times that the event E occurs in the first n trials, we may interpret Equation (7.1) as stating that, with probability 1, the limiting proportion of time that the event E occurs is just $P\{E\}$.

Running neck and neck with the strong law of large numbers for the honor of being probability theory's number one result is the *central limit theorem*. Besides its theoretical interest and importance, this theorem provides a simple method for computing approximate probabilities for sums of independent random variables. It also explains the remarkable fact that the empirical frequencies of so many natural "populations" exhibits a bell-shaped (that is, normal) curve. ◊

Theorem 7.2. (Central Limit Theorem) Let $X_1, X_2, \ldots$ be a sequence of independent, identically distributed random variables each with mean μ and variance σ^2. Then the distribution of

$$\frac{X_1 + X_2 + \cdots + X_n - n\mu}{\sigma\sqrt{n}}$$

tends to the standard normal as $n \to \infty$. That is,

$$P\left\{\frac{X_1 + X_2 + \cdots + X_n - n\mu}{\sigma\sqrt{n}} \le a\right\} \to \frac{1}{\sqrt{2\pi}} \int_{-\infty}^{a} e^{-x^2/2}dx$$

as $n \to \infty$.

Note that like the other results of this section, this theorem holds for *any* distribution of the X_i's; herein lies its power.

If X is binomially distributed with parameters n and p then, as shown in Example 6f, X has the same distribution as the sum of n independent Bernoulli random variables each with parameter p. (Recall that the Bernoulli random variable is just a binomial random variable whose parameter n equals 1.) Hence, the distribution of

$$\frac{X - E[X]}{\sqrt{\text{Var}(X)}} = \frac{X - np}{\sqrt{np(1-p)}}$$

approaches the standard normal distribution as n approaches ∞. The normal approximation will, in general, be quite good for values of n satisfying $np(1-p) \ge 10$.   ◊

Example 7b   (Normal Approximation to the Binomial): Let X be the number of times that a fair coin, flipped 40 times, lands heads. Find the probability that $X = 20$. Use the normal approximation and then compare it to the exact solution.

Solution: Since the binomial is a discrete random variable, and the normal a continuous random variable, it leads to a better approximation to write the desired probability as

$$P\{X = 20\} = P\{19.5 < X < 20.5\}$$

$$= P\left\{\frac{19.5 - 20}{\sqrt{10}} < \frac{X - 20}{\sqrt{10}} < \frac{20.5 - 20}{\sqrt{10}}\right\}$$

$$= P\left\{-.16 < \frac{X - 20}{\sqrt{10}} < .16\right\}$$

$$\approx \Phi(.16) - \Phi(-.16)$$

where $\Phi(x)$, the probability that the standard normal is less than x is given by

$$\Phi(x) = \frac{1}{\sqrt{2\pi}} \int_{-\infty}^{x} e^{-y^2/2} dy$$

By the symmetry of the standard normal distribution

$$\Phi(-.16) = P\{N(0, 1) > .16\} = 1 - \Phi(.16)$$

where $N(0, 1)$ is a standard normal random variable. Hence, the desired probability is approximated by

$$P(X = 20\} \approx 2\Phi(.16) - 1$$

Using Table 7.1, we obtain that

$$P\{X = 20\} \approx .1272$$

The exact result is

$$P\{X = 20\} = \binom{40}{20}(\tfrac{1}{2})^{40}$$

which, after much calculation, can be shown to equal .1268. ◇

Example 7c Let X_i, $i = 1, 2, \ldots, 10$ be independent random variables, each being uniformly distributed over $(0, 1)$. Calculate $P\{\Sigma_1^{10} X_i > 7\}$.

Solution: Since $E[X_i] = \tfrac{1}{2}$, $\mathrm{Var}(X_i) = \tfrac{1}{12}$ we have by the central limit theorem that

$$P\left\{\sum_1^{10} X_i > 7\right\} = P\left\{\frac{\Sigma_1^{10} X_i - 5}{\sqrt{10(\tfrac{1}{12})}} > \frac{7 - 5}{\sqrt{10(\tfrac{1}{12})}}\right\}$$

$$\approx 1 - \Phi(2.2)$$

$$= .0139 \quad \diamond$$

8. Stochastic Processes

A *stochastic process* $\{X(t), t \in T\}$ is a collection of random variables. That is, for each $t \in T$, $X(t)$ is a random variable. The index t is often interpreted as time and, as a result, we refer to $X(t)$ as the *state* of the process at time t. For example, $X(t)$ might equal the total number of customers that have entered a supermarket by time t; or the number of customers in the supermarket at time t; or the total amount of sales that have been recorded in the market by time t, etc.

Table 7.1   **Area Φ(x) under the Standard Normal Curve to the Left of x**

x	.00	.01	.02	.03	.04	.05	.06	.07	.08	.09
.0	.5000	.5040	.5080	.5120	.5160	.5199	.5239	.5279	.5319	.5359
.1	.5398	.5438	.5478	.5517	.5557	.5596	.5636	.5675	.5714	.5753
.2	.5793	.5832	.5871	.5910	.5948	.5987	.6026	.6064	.6103	.6141
.3	.6179	.6217	.6255	.6293	.6331	.6368	.6406	.6443	.6480	.6517
.4	.6554	.6591	.6628	.6664	.6700	.6736	.6772	.6808	.6844	.6879
.5	.6915	.6950	.6985	.7019	.7054	.7088	.7123	.7157	.7190	.7224
.6	.7257	.7291	.7324	.7357	.7389	.7422	.7454	.7486	.7517	.7549
.7	.7580	.7611	.7642	.7673	.7704	.7734	.7764	.7794	.7823	.7852
.8	.7881	.7910	.7939	.7967	.7995	.8023	.8051	.8078	.8106	.8133
.9	.8159	.8186	.8212	.8238	.8264	.8289	.8315	.8340	.8365	.8389
1.0	.8413	.8438	.8461	.8485	.8508	.8531	.8554	.8557	.8599	.8621
1.1	.8643	.8665	.8686	.8708	.8729	.8749	.8770	.8790	.8810	.8830
1.2	.8849	.8869	.8888	.8907	.8925	.8944	.8962	.8980	.8997	.9015
1.3	.9032	.9049	.9066	.9082	.9099	.9115	.9131	.9147	.9162	.9177
1.4	.9192	.9207	.9222	.9236	.9251	.9265	.9279	.9292	.9306	.9319
1.5	.9332	.9345	.9357	.9370	.9382	.9394	.9406	.9418	.9429	.9441
1.6	.9452	.9463	.9474	.9484	.9495	.9505	.9515	.9525	.9535	.9545
1.7	.9554	.9564	.9573	.9582	.9591	.9599	.9608	.9616	.9625	.9633
1.8	.9641	.9649	.9656	.9664	.9671	.9678	.9686	.9693	.9699	.9706
1.9	.9713	.9719	.9726	.9732	.9738	.9744	.9750	.9756	.9761	.9767
2.0	.9772	.9778	.9783	.9788	.9793	.9798	.9803	.9808	.9812	.9817
2.1	.9821	.9826	.9830	.9834	.9838	.9842	.9846	.9850	.9854	.9857
2.2	.9861	.9864	.9868	.9871	.9875	.9878	.9881	.9884	.9887	.9890
2.3	.9893	.9896	.9898	.9901	.9904	.9906	.9909	.9911	.9913	.9916
2.4	.9918	.9920	.9922	.9925	.9927	.9929	.9931	.9932	.9934	.9936
2.5	.9938	.9940	.9941	.9943	.9945	.9946	.9948	.9949	.9951	.9952
2.6	.9953	.9955	.9956	.9957	.9959	.9960	.9961	.9962	.9963	.9964
2.7	.9965	.9966	.9967	.9968	.9969	.9970	.9971	.9972	.9973	.9974
2.8	.9974	.9975	.9976	.9977	.9977	.9978	.9979	.9979	.9980	.9981
2.9	.9981	.9982	.9982	.9983	.9984	.9984	.9985	.9985	.9986	.9986
3.0	.9987	.9987	.9987	.9988	.9988	.9989	.9989	.9989	.9990	.9990
3.1	.9990	.9991	.9991	.9991	.9992	.9992	.9992	.9992	.9993	.9993
3.2	.9993	.9993	.9994	.9994	.9994	.9994	.9994	.9995	.9995	.9995
3.3	.9995	.9995	.9995	.9996	.9996	.9996	.9996	.9996	.9996	.9997
3.4	.9997	.9997	.9997	.9997	.9997	.9997	.9997	.9997	.9997	.9998

The set T is called the *index* set of the process. When T is a countable set the stochastic process is said to be *discrete-time* process. If T is an interval of the real line, the stochastic process is said to be a *continuous-time* process. For instance, $(X_n, n = 0, 1, \ldots)$ is a discrete-time stochastic process indexed by the nonnegative integers; while $\{X(t), t \geq 0\}$ is a continuous-time stochastic process indexed by the nonnegative real numbers.

The *state space* of a stochastic process is defined as the set of all possible values that the random variables $X(t)$ can assume.

Thus, a stochastic process is a family of random variables that describes the evolution through time of some (physical) process. We shall see much of stochastic processes in the following chapters of this text.

Problems

1. An urn contains five red, three orange, and two blue balls. Two balls are randomly selected. What is the sample space of this experiment? Let X represent the number of orange balls selected. What are the possible values of X? Calculate $P\{X = 0\}$.

2. Let X represent the difference between the number of heads and the number of tails obtained when a coin is tossed n times. What are the possible values of X?

3. In Problem 2, if the coin is assumed fair, then, for $n = 2$, what are the probabilities associated with the values that X can take on?

4. Suppose a die is rolled twice. What are the possible values that the following random variables can take on?
   (i) The maximum value to appear in the two rolls.
   (ii) The minimum value to appear in the two rolls.
   (iii) The sum of the two rolls.
   (iv) The value of the first roll minus the value of the second roll.

5. If the die in Problem 4 is assumed fair, calculate the probabilities associated with the random variables in (i)–(iv).

6. Suppose five fair coins are tossed. Let E be the event that all coins land heads. Define the random variable I_E

$$I_E = \begin{cases} 1, & \text{if } E \text{ occurs} \\ 0, & \text{if } E^c \text{ occurs} \end{cases}$$

For what outcomes in the original sample space does I_E equal 1? What is $P\{I_E = 1\}$?

7. Suppose a coin having probability .7 of coming up heads is tossed three times. Let X denote the number of heads that appear in the three tosses. Determine the probability mass function of X.

8. Suppose the distribution function of X is given by

$$F(b) = \begin{cases} 0, & b < 0 \\ \frac{1}{2}, & 0 \le b < 1 \\ 1, & 1 \le b < \infty \end{cases}$$

What is the probability mass function of X?

9. If the distribution function of F is given by

$$F(b) = \begin{cases} 0, & b < 0 \\ \frac{1}{2}, & 0 \le b < 1 \\ \frac{3}{5}, & 1 \le b < 2 \\ \frac{4}{5}, & 2 \le b < 3 \\ \frac{9}{10}, & 3 \le b < 3.5 \\ 1, & b \ge 3.5 \end{cases}$$

calculate the probability mass function of X.

10. Suppose three fair dice are rolled. What is the probability that at most one six appears?

11. A ball is drawn from an urn containing three white and three black balls. After the ball is drawn, it is then replaced and another ball is drawn. This goes on indefinitely. What is the probability that of the first four balls drawn, two are white?

12. On a multiple choice exam with three possible answers for each of the five questions, what is the probability that a student would get four or more correct answers just by guessing?

13. An individual claims to have extrasensory perception (ESP). As a test, a fair coin is flipped ten times, and he is asked to predict in advance the outcome. Our individual gets seven out of ten correct. What is the probability he would have done this well if he had no ESP? (Explain why the relevant probability is $P\{X \ge 7\}$ and not $P\{X = 7\}$.)

14. Suppose X has a binomial distribution with parameters 6 and $\frac{1}{2}$. Show that $X = 3$ is the most likely outcome.

15. Let X be binomially distributed with parameters n and p. Show that as k goes from 0 to n, $P(X = k)$ increases monotonically, then decreases monotonically reaching its largest value.
   (a) in the case that $(n + 1)p$ is an integer, when k equals either $(n + 1)p - 1$ or $(n + 1)p$,
   (b) in the case that $(n + 1)p$ is not an integer, when k satisfies $(n + 1)p - 1 < k < (n + 1)p$.

Hint: Consider $P\{X = k\}/P\{X = k - 1\}$ and see for what values of k it is greater or less than 1.

16. An airline knows that five percent of the people making reservations on a certain flight will not show up. Consequently, their policy is to sell 52 tickets for a flight that can only hold 50 passengers. What is the probability that there will be a seat available for every passenger that shows up?

17. Suppose that an experiment can result in one of r possible outcomes, the ith outcome having probability p_i, $i = 1, \ldots, r$, $\Sigma_{i=1}^r p_i = 1$. If n of these experiments are performed, and if the outcome of any one of the n does not affect the outcome of the other $n - 1$ experiments, then show that the probability that the first outcome appears x_1 times, the second x_2 times, and the rth x_r time is

$$\frac{n!}{x_1!x_2! \cdots x_r!} p_1^{x_1} p_2^{x_2} \cdots p_r^{x_r} \quad \text{when} \quad x_1 + x_2 + \cdots + x_r = n$$

This is known as the *multinomial* distribution.

18. Show that when $r = 2$ the multinomial reduces to the binomial.

19. In Problem 17, let X_i denote the number of times the ith outcome appears, $i = 1, \ldots, r$. What is the probability mass function of $X_1 + X_2 + \cdots + X_k$?

20. A television store owner figures that 50 percent of the customers entering his store will purchase an ordinary television set, 20 percent will purchase a color television set, and 30 percent will just be browsing. If five customers enter his store on a certain day, what is the probability that two customers purchase color sets, one customer purchases an ordinary set, and two customers purchase nothing?

21. In Problem 20, what is the probability that our store owner sells three or more televisions on that day?

22. If a fair coin is successively flipped, find the probability that a head first appears on the fifth trial.

23. A coin having a probability p of coming up heads is successively flipped until the rth head appears. Argue that X, the number of flips required, will be n, $n \geq r$, with probability

$$P\{X = n\} = \binom{n-1}{r-1} p^r (1 - p)^{n-r}, \quad n \geq r$$

This is known as the *negative binomial* distribution.

Hint: How many successes must there be in the first $n - 1$ trials?

24. The probability mass function of X is given by

$$p(k) = \binom{r + k - 1}{r - 1} p^r (1 - p)^k, \qquad k = 0, 1, \ldots$$

Give a possible interpretation of the random variable X.
   Hint: See Problem 23.

25. Let X be a Poisson random variable with parameter λ. Show that $P\{X = i\}$ increases monotonically and then decreases monotonically as i increases, reaching its maximum when i is the largest integer not exceeding λ.
   Hint: Consider $P\{X = i\}/P\{X = i - 1\}$.

26. Compare the Poisson approximation with the correct binomial probability for the following cases:
   (i) $P\{X = 2\}$ when $n = 8$, $p = .1$.
   (ii) $P\{X = 9\}$ when $n = 10$, $p = .95$.
   (iii) $P\{X = 0\}$ when $n = 10$, $p = .1$.
   (iv) $P\{X = 4\}$ when $n = 9$, $p = .2$.

27. If you buy a lottery ticket in 50 lotteries, in each of which your chance of winning a prize is $\frac{1}{100}$, what is the (approximate) probability that you will win a prize (a) at least once, (b) exactly once, (c) at least twice?

28. Let X be a random variable with probability density

$$f(x) = \begin{cases} c(1 - x^2), & -1 < x < 1 \\ 0, & \text{otherwise} \end{cases}$$

   (a) What is the value of c?
   (b) What is the cumulative distribution function of X?

29. Let the probability density of X be given by

$$f(x) = \begin{cases} c(4x - 2x^2), & 0 < x < 2 \\ 0, & \text{otherwise} \end{cases}$$

   (a) What is the value of c?
   (b) $P\{\frac{1}{2} < X < \frac{3}{2}\} = ?$

30. The density of X is given by

$$f(x) = \begin{cases} 10/x^2, & \text{for} \quad x > 10 \\ 0, & \text{for} \quad x \leq 10 \end{cases}$$

What is the distribution of X? Find $P\{X > 20\}$.

31. Let $X_1, X_2, \ldots, X_n$ be independent random variables, each having a uniform distribution over $(0,1)$. Let $M = \text{maximum } (X_1, X_2, \ldots, X_n)$. Show that the distribution function of M, $F_M(\cdot)$, is given by

$$F_M(x) = x^n, \qquad 0 \leq x \leq 1$$

What is the probability density function of M?

32. If the density function of X equals

$$f(x) = \begin{cases} ce^{-2x}, & 0 < x < \infty \\ 0, & x < 0 \end{cases}$$

find c. What is $P\{X > 2\}$?

33. The random variable X has the following probability mass function

$$p(1) = \tfrac{1}{2}, \qquad p(2) = \tfrac{1}{3}, \qquad p(24) = \tfrac{1}{6}$$

Calculate $E[X]$.

34. If X is uniformly distributed over $(0, 1)$, calculate $E[X^2]$.

35. Prove that $E[X^2] \geq (E[X])^2$. When do we have equality?

36. Let c be a constant. Show that
   (i) $\text{Var}(cX) = c^2 \, \text{Var}(X)$.
   (ii) $\text{Var}(c + X) = \text{Var}(X)$.

37. A coin, having probability p of landing heads, is flipped until the head appears for the rth time. Let N denote the number of flips required. Calculate $E[N]$.
   Hint: There is an easy way of doing this. It involves writing N as the sum of r geometric random variables.

38. Calculate the variance of the Bernoulli random variable.

39. (a) Calculate $E[X]$ for the maximum random variable of Problem 31. (b) Calculate $E(X)$ for X as in Problem 28. (c) Calculate $E[X]$ for X as in Problem 29.

40. If X is uniform over $(0, 1)$, calculate $E[X^n]$ and $\text{Var}(X^n)$.

41. Calculate, without using moment generating functions, the variance of a binomial random variable with parameters n and p.

42. Suppose that X and Y are independent binomial random vari-

ables with parameters (n, p) and (m, p). Argue probabilistically (no computations necessary) that $X + Y$ is binomial with parameters $(n + m, p)$.

43. Suppose that X and Y are independent continuous random variables. Show that

$$P\{X \leq Y\} = \int_{-\infty}^{\infty} F_X(y) f_Y(y) dy$$

44. Calculate the moment generating function of the uniform distribution on $(0, 1)$. Obtain $E[X]$ and $\text{Var}[X]$ by differentiating.

45. Suppose that X takes on each of the values 1, 2, 3 with probability $\frac{1}{3}$. What is the moment generating function? Derive $E[X]$, $E[X^2]$, and $E[X^3]$ by differentiating the moment generating function and then compare the obtained result with a direct derivation of these moments.

46. Suppose the density of X is given by

$$f(x) = \begin{cases} \frac{1}{4}xe^{-x/2}, & x > 0 \\ 0, & \text{otherwise} \end{cases}$$

Calculate the moment generating function, $E[X]$, and $\text{Var}(X)$.

47. Calculate the moment generating function of a geometric random variable.

48. Show that the sum of independent identically distributed exponential random variables has a gamma distribution.

49. Use Chebyshev's inequality to prove the *weak law of large numbers*. Namely, if $X_1, X_2, \ldots$ are independent and identically distributed with mean μ and variance σ^2 then, for any $\varepsilon > 0$,

$$P\left\{ \left| \frac{X_1 + X_2 + \cdots + X_n}{n} - \mu \right| > \varepsilon \right\} \rightarrow 0 \quad \text{as} \quad n \rightarrow \infty$$

50. Suppose that X is a random variable with mean 10 and variance 15. What can we say about $P\{5 < X < 15\}$?

51. Let $X_1, X_2, \ldots, X_{10}$ be independent Poisson random variables with mean 1.
   (i) Use the Markov inequality to get a bound on $P\{X_1 + \cdots + X_{10} > 15\}$.
   (ii) Use the central limit theorem to approximate $P\{X_1 + \cdots + X_{10} > 15\}$.

52. If X is normally distributed with mean 1 and variance 4, use the tables to find $P\{2 < X < 3\}$.

53. Show that

$$\lim_{n \to \infty} e^{-n} \sum_{k=0}^{n} \frac{n^k}{k!} = \frac{1}{2}$$

Hint: Let X_n be Poisson with mean n. Use the central limit theorem to show that $P\{X_n \le n\} \to \frac{1}{2}$.

54. Let X denote the number of white balls selected when k balls are chosen at random from an urn containing n white and m black balls.
   (i) Compute $P(X = i)$.
   (ii) Let, for $i = 1, 2, \ldots, k$; $j = 1, 2, \ldots, n$

   $$X_i = \begin{cases} 1, & \text{if the ith ball selected is white} \\ 0, & \text{otherwise} \end{cases}$$

   $$Y_j = \begin{cases} 1, & \text{if the jth white ball is selected} \\ 0, & \text{otherwise} \end{cases}$$

Compute $E[X]$ in two ways by expressing X first as a function of the X_i's and then of the Y_j's.

55. Show that

$$\text{Var}\left[\sum_{1}^{n} X_i\right] = \sum_{i=1}^{n} \text{Var}(X_i) + 2 \sum\sum_{i<j} \text{Cov}(X_i, X_j)$$

and then use this to show that

$$\text{Var}(Y) = 1$$

when Y is the number of men that select their own hats in Example 5c.

56. For the multinomial distribution (Problem 17), let N_i denote the number of times outcome i occurs. Find
   (i) $E[N_i]$.
   (ii) $\text{Var}(N_i)$.
   (iii) $\text{Cov}(N_i, N_j)$.
   (iv) Compute the expected number of outcomes which do not occur.

57. Let $X_1, X_2, \ldots$ be a sequence of independent identically distributed continuous random variables. We say that a record occurs at time n if $X_n > \max(X_1, \ldots, X_{n-1})$. That is, X_n is a record if it is larger than each of $X_1, \ldots, X_{n-1}$. Show

(i) $P\{$a record occurs at time $n\} = 1/n$

(ii) $E[$number of records by time $n] = \Sigma_{i=1}^{n} 1/i$

(iii) $\text{Var}($number of records by time $n) = \Sigma_{i=1}^{n}(i - 1)/i^2$

(iv) Let $N = \min\{n:n > 1$ and a record occurs at time $n\}$. Show $E[N] = \infty$.

Hint: for (ii) and (iii) represent the number of records as the sum of indicator (that is, Bernoulli) random variables.

58. Let $a_1 < a_2 < \cdots < a_n$ denote a set of n numbers, and consider any permutation of these numbers. We say that there is an inversion of a_i and a_j in the permutation if $i < j$ and a_j precedes a_i. For instance the permutation 4, 2, 1, 5, 3 has 5 inversions—(4, 2), (4, 1), (4, 3), (2, 1), (5, 3). Consider now a random permutation of $a_1, a_2, \ldots, a_n$—in the sense that each of the $n!$ permutations is equally likely to be chosen—and let N denote the number of inversions in this permutation. Also, let

N_i = number of $k:k < i$, a_i precedes a_k in the permutation

and note that $N = \Sigma_{i=1}^{n} N_i$

(i) Show that $N_1, \ldots, N_n$ are independent random variables.

(ii) What is the distribution of N_i?

(iii) Compute $E[N]$ and $\text{Var}(N)$.

59. Let X and Y be independent random variables with means μ_x and μ_y and variances σ_x^2 and σ_y^2. Show that

$$\text{Var}(XY) = \sigma_x^2\sigma_y^2 + \mu_y^2\sigma_x^2 + \mu_x^2\sigma_y^2$$

60. Let X and Y be independent normal random variables each having parameters μ and σ^2. Show that $X + Y$ is independent of $X - Y$.

61. Let $\phi(t_1, \ldots, t_n)$ denote the joint moment generating function of $X_1, \ldots, X_n$.

(a) Explain how the moment generating function of X_i, $\phi_{X_i}(t_i)$, can be obtained from $\phi(t_1, \ldots, t_n)$.

(b) Show that $X_1, \ldots, X_n$ are independent if and only if

$$\phi(t_1, \ldots, t_n) = \phi_{X_1}(t_1) \ldots \phi_{X_n}(t_n)$$

62. If $Z_1, \ldots, Z_n$ are independent unit normal random variables, then $X = \Sigma_{i=1}^{n} Z_i^2$ is said to be a chi-square random variable with n degrees of freedom. Compute its moment generating function.

References

1. W. Feller, "An Introduction to Probability Theory and Its Applications," Vol. I., John Wiley, New York, 1957.

2. M. Fisz, "Probability Theory and Mathematical Statistics," John Wiley, New York, 1963.

3. E. Parzen, "Modern Probability Theory and Its Applications," John Wiley, New York, 1960.

4. S. Ross, "A First Course in Probability," Second Edition, Macmillan, New York, 1984.

Chapter 3

Conditional Probability and Conditional Expectation

1. Introduction

One of the most useful concepts in probability theory is that of conditional probability and conditional expectation. The reason is twofold. First, in practice, we are often interested in calculating probabilities and expectations when some partial information is available; hence, the desired probabilities and expectations are conditional ones. Secondly, in calculating a desired probability or expectation it is often extremely useful to first "condition" on some appropriate random variable.

2. The Discrete Case

Recall that for any two events E and F, the conditional probability of E given F is defined, as long as $P(F) > 0$, by

$$P(E|F) = \frac{P(EF)}{P(F)}$$

Hence, if X and Y are discrete random variables, then it is natural to define the *conditional probability mass function* of X given that $Y = y$, by

$$p_{X|Y}(x|y) = P\{X = x|Y = y\}$$

$$= \frac{P\{X = x, Y = y\}}{P\{Y = y\}}$$

$$= \frac{p(x, y)}{p_Y(y)}$$

for all values of y such that $P\{Y = y\} > 0$. Similarly, the conditional probability distribution function of X given that $Y = y$ is defined, for all y such that $P\{Y = y\} > 0$, by

$$F_{X|Y}(x|y) = P\{X \le x | Y = y\}$$

$$= \sum_{a \le x} p_{X|Y}(a|y)$$

Finally, the conditional expectation of X given that $Y = y$ is defined by

$$E[X|Y = y] = \sum_x xP\{X = x | Y = y\}$$

$$= \sum_x x p_{X|Y}(x|y)$$

In other words, the definitions are exactly as before with the exception that everything is now conditional on the event that $Y = y$. If X is independent of Y, then the conditional mass function, distribution, and expectation are the same as the unconditional ones. This follows, since if X is independent of Y, then

$$p_{X|Y}(x|y) = P\{X = x | Y = y\}$$

$$= \frac{P\{X = x, Y = y\}}{P\{Y = y\}}$$

$$= \frac{P\{X = x\}P\{Y = y\}}{P\{Y = y\}}$$

$$= P\{X = x\}$$

Example 2a Suppose that $p(x, y)$, the joint probability mass function of X and Y, is given by

$$p(1, 1) = .5, \qquad p(1, 2) = .1, \qquad p(2, 1) = .1, \qquad p(2, 2) = .3$$

Calculate the probability mass function of X given that $Y = 1$.

Solution: We first note that

$$p_Y(1) = \sum_x p(x, 1) = p(1, 1) + p(2, 1) = .6$$

Hence,

$$p_{X|Y}(1|1) = P\{X = 1|Y = 1\}$$
$$= \frac{P\{X = 1, Y = 1\}}{P\{Y = 1\}}$$
$$= \frac{p(1, 1)}{p_Y(1)}$$
$$= \frac{5}{6}$$

Similarly,

$$p_{X|Y}(2|1) = \frac{p(2, 1)}{p_Y(1)} = \frac{1}{6} \quad \Diamond$$

Example 2b If X and Y are independent Poisson random variables with respective means λ_1 and λ_2, then calculate the conditional expected value of X given that $X + Y = n$.

Solution: Let us first calculate the conditional probability mass function of X given that $X + Y = n$. We obtain

$$P\{X = k|X + Y = n\} = \frac{P\{X = k, X + Y = n\}}{P\{X + Y = n\}}$$
$$= \frac{P\{X = k, Y = n - k\}}{P\{X + Y = n\}}$$
$$= \frac{P\{X = k\}P\{Y = n - k\}}{P\{X + Y = n\}}$$

where the last equality follows from the assumed independence of X and Y. Recalling (see Example 5h of Chapter 2) that $X + Y$ has a Poisson distribution with mean $\lambda_1 + \lambda_2$, the preceding equation equals

$$P\{X = k|X + Y = n\} = \frac{e^{-\lambda_1}\lambda_1^k}{k!} \frac{e^{-\lambda_2}\lambda_2^{n-k}}{(n - k)!} \left[\frac{e^{-(\lambda_1 + \lambda_2)}(\lambda_1 + \lambda_2)^n}{n!} \right]^{-1}$$
$$= \frac{n!}{(n - k)!k!} \frac{\lambda_1^k \lambda_2^{n-k}}{(\lambda_1 + \lambda_2)^n}$$
$$= \binom{n}{k} \left(\frac{\lambda_1}{\lambda_1 + \lambda_2} \right)^k \left(\frac{\lambda_2}{\lambda_1 + \lambda_2} \right)^{n-k}$$

In other words, the conditional distribution of X given that

$X + Y = n$, is the binomial distribution with parameters n and $\lambda_1/(\lambda_1 + \lambda_2)$. Hence,

$$E\{X|X + Y = n\} = n \frac{\lambda_1}{\lambda_1 + \lambda_2} \quad \Diamond$$

Example 2c If X and Y are independent binomial random variables with identical parameters n and p, calculate the conditional probability mass function of X given that $X + Y = m$.

Solution: For $k \le \min(n, m)$,

$$
\begin{aligned}
P\{X = k|X + Y = m\} &= \frac{P\{X = k, X + Y = m\}}{P\{X + Y = m\}} \\
&= \frac{P\{X = k, Y = m - k\}}{P\{X + Y = m\}} \\
&= \frac{P\{X = k\}P\{Y = m - k\}}{P\{X + Y = m\}} \\
&= \frac{\binom{n}{k} p^k (1 - p)^{n-k} \binom{n}{m-k} p^{m-k}(1 - p)^{n-m+k}}{\binom{2n}{m}p^m (1 - p)^{2n-m}}
\end{aligned}
$$

where we have used the fact (see Example 6f of Chapter 2) that $X + Y$ is binomially distributed with parameters $(2n, p)$. Hence, the conditional probability mass function of X given that $X + Y = m$ is given by

$$P\{X = k|X + Y = m\} = \frac{\binom{n}{k}\binom{n}{m-k}}{\binom{2n}{m}}, \quad k = 0, 1, \ldots, \min(m, n) \quad (2.1)$$

The distribution given in Equation (2.1) is known as the *hypergeometric* distribution. It arises as the distribution of the number of black balls that are chosen when a sample of m balls is randomly selected from an urn containing n white and n black balls. $\Diamond$

Example 2d Consider an experiment which results in one of three possible outcomes. Outcome i occurring with probability p_i, $i = 1, 2, 3$, $\Sigma_{i=1}^3 p_i = 1$. Suppose that n independent replications of this experiment are performed and let X_i, $i = 1, 2, 3$, denote the number of times outcome i appears. Determine the conditional distribution of X_1 given that $X_2 = m$.

Solution: For $k \le n - m$,

$$P\{X_1 = k | X_2 = m\} = \frac{P\{X_1 = k, X_2 = m\}}{P\{X_2 = m\}}$$

Now if $X_1 = k$ and $X_2 = m$, then it follows that $X_3 = n - k - m$. However,

$$P\{X_1 = k, X_2 = m, X_3 = n - k - m\}$$

$$= \frac{n!}{k!m!(n - k - m)!} \, p_1^k p_2^m p_3^{(n-k-m)} \tag{2.2}$$

This follows since any particular sequence of the n experiments having outcome 1 appearing k times, outcome 2 m times, and outcome 3 $(n - k - m)$ times has probability $p_1^k p_2^m p_3^{(n-k-m)}$ of occurring. Since there are $n!/[k!m!(n - k - m)!]$ such sequences, Equation (2.2) follows.

Therefore, we have

$$P\{X_1 = k | X_2 = m\} = \frac{\dfrac{n!}{k!m!(n - k - m)!} \, p_1^k p_2^m p_3^{(n-k-m)}}{\dfrac{n!}{m!(n - m)!} \, p_2^m (1 - p_2)^{n-m}}$$

where we have used the fact that X_2 has a binomial distribution with parameters n and p_2. Hence,

$$P\{X_1 = k | X_2 = m\} = \frac{(n - m)!}{k!(n - m - k)!} \left(\frac{p_1}{1 - p_2} \right)^k \left(\frac{p_3}{1 - p_2} \right)^{n-m-k}$$

or equivalently, writing $p_3 = 1 - p_1 - p_2$,

$$P\{X_1 = k | X_2 = m\} = \binom{n - m}{k} \left(\frac{p_1}{1 - p_2} \right)^k \left(1 - \frac{p_1}{1 - p_2} \right)^{n-m-k}$$

In other words, the conditional distribution of X_1, given that $X_2 = m$, is binomial with parameters $n - m$ and $p_1/(1 - p_2)$.   ◇

Remarks

(i) The desired conditional probability in Example 2d could also have been computed in the following manner. Consider the $n - m$ experiments which did not result in outcome 2. For each of these experiments, the probability that outcome 1 was obtained is given by

$$P\{\text{outcome } 1 | \text{not outcome } 2\} = \frac{P\{\text{outcome } 1, \text{not outcome } 2\}}{P\{\text{not outcome } 2\}}$$

$$= \frac{p_1}{1 - p_2}$$

It therefore follows that, given $X_2 = m$, the number of times outcome 1 occurs is binomially distributed with parameters $n - m$ and $p_1/(1 - p_2)$.

(ii) Conditional expectations possess all of the properties of ordinary expectations. For instance, such identities as

$$E\left[\sum_{i=1}^{n} X_i | Y = y \right] = \sum_{i=1}^{n} E[X_i | Y = y]$$

remain valid. ◇

Example 2e Consider $n + m$ independent trials, each of which results in a success with probability p. Compute the expected number of successes in the first n trials given that there are k successes in all.

Solution: Letting Y denote the total number of successes, and

$$X_i = \begin{cases} 1, & \text{if the } i\text{th trial is a success} \\ 0, & \text{otherwise} \end{cases}$$

the desired expectation is $E[\sum_{i=1}^{n} X_i | Y = k]$ which is obtained as

$$E\left[\sum_{1}^{n} X_i | Y = k \right] = \sum_{1}^{n} E[X_i | Y = k]$$

$$= n \frac{k}{n + m}$$

where the last equality follows since if there are a total of k successes, then any individual trial will be a success with probability $k/(n + m)$. That is,

$$E[X_i | Y = k] = P\{X_i = 1 | Y = k\}$$

$$= \frac{k}{n + m} \quad ◇$$

3. The Continuous Case

If X and Y have a joint probability density function $f(x,y)$, then the *conditional probability density function* of X, given that $Y = y$, is defined for all values of y such that $f_Y(y) > 0$, by

$$f_{X|Y}(x|y) = \frac{f(x,y)}{f_Y(y)}$$

To motivate this definition, multiply the left side by dx and the right side by $(dxdy)/dy$ to get

$$f_{X|Y}(x|y)dx = \frac{f(x, y)dxdy}{f_Y(y)dy}$$

$$\approx \frac{P\{x \le X \le x + dx, y \le Y \le y + dy\}}{P\{y \le Y \le y + dy\}}$$

$$= P\{x \le X \le x + dx | y \le Y \le y + dy\}$$

In other words, for small values dx and dy, $f_{X|Y}(x|y)dx$ represents the conditional probability that X is between x and $x + dx$ given that Y is between y and $y + dy$.

The *conditional expectation* of X, given that $Y = y$, is defined for all values of y such that $f_Y(y) > 0$, by

$$E[X|Y = y] = \int_{-\infty}^{\infty} x f_{X|Y}(x|y)dx$$

Example 3a   Suppose the joint density of X and Y is given by

$$f(x, y) = \begin{cases} 6xy(2 - x - y), & 0 < x < 1, \quad 0 < y < 1 \\ 0, & \text{otherwise} \end{cases}$$

Compute the conditional expectation of X given that $Y = y$, where $0 < y < 1$.

Solution: We first compute the conditional density

$$f_{X|Y}(x|y) = \frac{f(x, y)}{f_Y(y)}$$

$$= \frac{6xy(2 - x - y)}{\int_0^1 6xy(2 - x - y)dx}$$

$$= \frac{6xy(2 - x - y)}{y(4 - 3y)}$$

$$= \frac{6x(2 - x - y)}{4 - 3y}$$

Hence,

$$E[X|Y = y] = \int_0^1 \frac{6x^2(2 - x - y)dx}{4 - 3y}$$

$$= \frac{(2 - y)2 - \frac{6}{4}}{4 - 3y}$$

$$= \frac{5 - 4y}{8 - 6y} \quad \Diamond$$

Example 3b Suppose the joint density of X and Y is given by

$$f(x, y) = \begin{cases} 4y(x - y)e^{-(x+y)}, & 0 < x < \infty, \quad 0 \le y \le x \\ 0, & \text{otherwise} \end{cases}$$

Compute $E[X|Y = y]$.

Solution: The conditional density of X, given that $Y = y$, is given by

$$f_{X|Y}(x|y) = \frac{f(x, y)}{f_Y(y)}$$

$$= \frac{4y(x - y)e^{-(x+y)}}{\int_y^\infty 4y(x - y)e^{-(x+y)}dx}, \quad x > y$$

$$= \frac{(x - y)e^{-(x+y)}}{\int_y^\infty (x - y)e^{-(x+y)}dx}$$

$$= \frac{(x - y)e^{-x}}{\int_y^\infty (x - y)e^{-x}dx}$$

Integrating by parts shows that the above gives

$$f_{X|Y}(x|y) = \frac{(x-y)e^{-x}}{e^{-y}}$$

$$= (x-y)e^{-(x-y)}$$

Therefore,

$$E[X|Y=y] = \int_{-\infty}^{\infty} x f_{X|Y}(x|y)\,dx$$

$$= \int_{y}^{\infty} x(x-y)e^{-(x-y)}\,dx$$

Integration by parts yields

$$E[X|Y=y] = -x(x-y)e^{-(x-y)}\Big|_{y}^{\infty} + \int_{y}^{\infty} (2x-y)e^{-(x-y)}\,dx$$

$$= \int_{y}^{\infty} (2x-y)e^{-(x-y)}\,dx$$

$$= -(2x-y)e^{-(x-y)}\Big|_{y}^{\infty} + 2\int_{y}^{\infty} e^{-(x-y)}\,dx$$

$$= y + 2 \quad \Diamond$$

Example 3c   The joint density of X and Y is given by

$$f(x, y) = \begin{cases} \frac{1}{2}ye^{-xy}, & 0 < x < \infty, 0 < y < 2 \\ 0, & \text{otherwise} \end{cases}$$

What is $E[e^{X/2}|Y=1]$?

Solution: The conditional density of X, given that $Y = 1$, is given by

$$f_{X|Y}(x|1) = \frac{f(x, 1)}{f_Y(1)}$$

$$= \frac{\frac{1}{2}e^{-x}}{\int_{0}^{\infty} \frac{1}{2}e^{-x}\,dx}$$

Hence, by the law of the unconscious statistician,

$$E[e^{X/2}|Y = 1] = \int_0^\infty e^{x/2} f_{X|Y}(x|1) dx$$

$$= \int_0^\infty e^{x/2} e^{-x} dx$$

$$= 2 \quad \lozenge$$

4. Computing Expectations by Conditioning

Let us denote by $E[X|Y]$ that function of the random variable Y whose value at $Y = y$ is $E[X|Y = y]$. Note that $E[X|Y]$ is itself a random variable. An extremely important property of conditional expectation is that for all random variables X and Y

$$E[X] = E[E[X|Y]] \tag{4.1}$$

If Y is a discrete random variable, then Equation (4.1) states that

$$E[X] = \sum_y E[X|Y = y]P\{Y = y\} \tag{4.1a}$$

while if Y is continuous with density $f_Y(y)$, then Equation (4.1) says that

$$E[X] = \int_{-\infty}^\infty E[X|Y = y] f_Y(y) dy \tag{4.1b}$$

We now give a proof of Equation (4.1) in the case where X and Y are both discrete random variables.

Proof of Equation (4.1) When X and Y Are Discrete
We must show that

$$E[X] = \sum_y E[X|Y = y]P\{Y = y\} \tag{4.2}$$

Now, the right side of the above can be written

$$\sum_y E[X|Y = y]P\{Y = y\} = \sum_y \sum_x xP\{X = x|Y = y\}P\{Y = y\}$$

$$= \sum_y \sum_x x \frac{P\{X = x, Y = y\}}{P\{Y = y\}} P\{Y = y\}$$

$$= \sum_y \sum_x x P\{X = x, Y = y\}$$

$$= \sum_x x \sum_y P\{X = x, Y = y\}$$

$$= \sum_x x P\{X = x\}$$

$$= E[X]$$

and the result is obtained.   ◇

One way to understand Equation (4.2) is to interpret it as follows. It states that to calculate $E[X]$ we may take a weighted average of the conditional expected value of X given that $Y = y$, each of the terms $E[X|Y = y]$ being weighted by the probability of the event on which it is conditioned.

The following examples will indicate the usefulness of Equation (4.1).

Example 4a   (The Mean of a Geometric Distribution): A coin, having probability p of coming up heads, is to be successively flipped until the first head appears. What is the expected number of flips required?

Solution: Let N be the number of flips required, and let

$$Y = \begin{cases} 1, & \text{if the first flip results in a head} \\ 0, & \text{if the first flip results in a tail} \end{cases}$$

Now

$$E[N] = E[N|Y = 1]P\{Y = 1\} + E[N|Y = 0]P\{Y = 0\}$$
$$= pE[N|Y = 1] + (1 - p)E[N|Y = 0] \tag{4.3}$$

However,

$$E[N|Y = 1] = 1, \qquad E[N|Y = 0] = 1 + E[N] \tag{4.4}$$

To see why Equation (4.4) is true consider $E[N|Y = 1]$. Since $Y = 1$, we know that the first flip resulted in heads and so, clearly, the expected number of flips required is 1. On the other hand if $Y = 0$, then the first flip resulted in tails. However, since the successive flips are assumed independent, it follows that, after the first tail, the expected additional number of flips until the first head is just $E[N]$. Hence $E[N|Y = 0] = 1 + E[N]$. Substituting Equation (4.4) into Equation (4.3) yields

$$E[N] = p + (1 - p)(1 + E[N])$$

or

$$E[N] = 1/p \quad \Diamond$$

As the random variable N is a geometric random variable with probability mass function $p(n) = p(1 - p)^{n-1}$, its expectation could easily have been computed from $E[N] = \Sigma_1^\infty np(n)$ without recourse to conditional expectation. However, if the reader attempts to obtain the solution to our second example without using conditional expectation, he will quickly learn what a useful technique "conditioning" can be.

Example 4b A miner is trapped in a mine containing three doors. The first door leads to a tunnel which takes him to safety after two-hour's travel. The second door leads to a tunnel which returns him to the mine after three-hour's travel. The third door leads to a tunnel which returns him to his mine after five hours. Assuming that the miner is at all times equally likely to choose any one of the doors, what is the expected length of time until the miner reaches safety?

Solution: Let X denote the time until the miner reaches safety, and let Y denote the door he initially chooses. Now

$E[X]$
$= E[X|Y = 1]P\{Y = 1\} + E[X|Y = 2]P\{Y = 2\} + E[X|Y = 3]P\{Y = 3\}$
$= \frac{1}{3}(E[X|Y = 1] + E[X|Y = 2] + E[X|Y = 3])$

However,

$$E[X|Y = 1] = 2$$
$$E[X|Y = 2] = 3 + E[X]$$
$$E[X|Y = 3] = 5 + E[X] \quad (4.5)$$

To understand why the above is correct consider, for instance, $E[X|Y = 2]$, and reason as follows. If the miner chooses the second door, then he spends three hours in the tunnel and then returns to his cell. But once he returns to his cell the problem is as before, and hence his expected additional time until safety is just $E[X]$. Hence $E[X|Y = 2] = 3 + E[Y]$. The argument behind the other equalities in Equation (4.5) is similar. Hence,

$$E[X] = \tfrac{1}{3}(2 + 3 + E[X] + 5 + E[X]) \quad \text{or} \quad E[X] = 10 \quad \diamond$$

Example 4c Sam will read either one chapter of his probability book or one chapter of his history book. If the number of misprints in a chapter of his probability book is Poisson distributed with mean 2 and if the number of misprints in his history chapter is Poisson distributed with mean 5, then assuming Sam is equally likely to choose either book, what is the expected number of misprints that Sam will come across?

Solution: Letting X denote the number of misprints and letting

$$Y = \begin{cases} 1, & \text{if Sam chooses his history book} \\ 2, & \text{if Sam chooses his probability book} \end{cases}$$

then

$$\begin{aligned} E[X] &= E[X|Y = 1]P\{Y = 1\} + E[X|Y = 2]P\{Y = 2\} \\ &= 5(\tfrac{1}{2}) + 2(\tfrac{1}{2}) \\ &= \tfrac{7}{2} \quad \diamond \end{aligned}$$

Example 4d   (The Expectation of a Random Number of Random Variables): Suppose that the expected number of accidents per week at an industrial plant is four. Suppose also that the numbers of workers injured in each accident are independent random variables with a common mean of 2. Assume also that the number of workers injured in each accident is independent of the number of accidents that occur. What is the expected number of injuries during a week?

Solution: Letting N denote the number of accidents and X_i the number injured in the ith accident, $i = 1, 2, \ldots$, then the total number of injuries can be expressed as $\sum_{i=1}^{N} X_i$. Now

$$E\left[\sum_{1}^{N} X_i\right] = E\left[E\left[\sum_{1}^{N} X_i|N\right]\right]$$

But

$$E\left[\sum_{1}^{N} X_i|N = n\right] = E\left[\sum_{1}^{n} X_i|N = n\right]$$

$$= E\left[\sum_1^n X_i\right] \quad \text{by the independence of } X_i \text{ and } N$$

$$= nE[X]$$

which yields that

$$E\left[\sum_{i=1}^N X_i | N\right] = NE[X]$$

and thus

$$E\left[\sum_{i=1}^N X_i\right] = E[NE[X]] = E[N]E[X]$$

Therefore, in our example, the expected number of injuries during a week equals $4 \times 2 = 8$. ◊

Example 4e Independent trials, each of which is a success with probability p, are performed until there are k consecutive successes. What is the mean number of necessary trials?

Solution: Let N_k denote the number of necessary trials to obtain k consecutive successes, and let M_k denote its mean. We will obtain a recursive equation for M_k by conditioning on N_{k-1}, the number of trials needed for $k - 1$ consecutive successes. This yields

$$M_k = E[N_k] = E[E[N_k \mid N_{k-1}]]$$

Now,

$$E[N_k \mid N_{k-1}] = N_{k-1} + 1 + (1 - p)E[N_k]$$

where the above follows since if it takes N_{k-1} trials to obtain $k - 1$ consecutive successes, then either the next trial is a success and we have our k in a row or it is a failure and we must begin anew. Taking expectations of both sides of the above yields

$$M_k = M_{k-1} + 1 + (1 - p)M_k$$

or

$$M_k = \frac{1}{p} + \frac{M_{k-1}}{p}$$

Since N_1, the time of the first success, is geometric with parameter p we see that

$$M_1 = \frac{1}{p}$$

and, recursively,

$$M_2 = \frac{1}{p} + \frac{1}{p^2}$$

$$M_3 = \frac{1}{p} + \frac{1}{p^2} + \frac{1}{p^3}$$

and, in general,

$$M_k = \frac{1}{p} + \frac{1}{p^2} + \cdots + \frac{1}{p^k} \quad \diamond$$

Example 4f (Analyzing the Quick-Sort Algorithm): Suppose we are given a set of n distinct values—$x_1, \ldots, x_n$—and we desire to put these values in increasing order, or as it is commonly called, to *sort* them. An efficient procedure for accomplishing this is the quick-sort algorithm which is defined recursively as follows: When $n = 2$ the algorithm compares the 2 values and puts them in the appropriate order. When $n > 2$ it starts by choosing at random one of the n values—say x_i—and then compares each of the other $n - 1$ values with x_i, noting which are smaller and which are larger than x_i. Letting S_i denote the set of elements smaller than x_i, and $\bar{S}_i$ the set of elements greater than x_i, the algorithm now sorts the set S_i and the set $\bar{S}_i$. The final ordering, therefore, consists of the ordered set of the elements in S_i, then x_i, and then the ordered set of the elements in $\bar{S}_i$. For instance, suppose that the set of elements is 10, 5, 8, 2, 1, 4, 7. We start by choosing one of these values at random (that is, each of the 7 values has probability of $\frac{1}{7}$ of being chosen). Suppose, for instance, that the value 4 is chosen. We then compare 4 with each of the other 6 values to obtain

$$\{2, 1\}, 4, \{10, 5, 8, 7\}$$

We now sort the set $\{2, 1\}$ to obtain

$$1, 2, 4, \{10, 5, 8, 7\}$$

Next we choose a value at random from $\{10, 5, 8, 7\}$—say 7 is chosen—and compare each of the other 3 values with 7 to obtain

$$1, 2, 4, 5, 7, \{10, 8\}$$

Finally we sort $\{10, 8\}$ to end up with

$$1, 2, 4, 5, 7, 8, 10$$

One measure of the effectiveness of this algorithm is the expected number of comparisons that it makes. Let us denote by M_n the expected number of comparisons needed by the quick-sort algorithm to sort a set of n distinct values. To obtain a recursion for M_n we condition on the rank of the initial value selected to obtain:

$$M_n = \sum_{j=1}^{n} E[\text{number of comparisons}|\text{value selected is } j\text{th smallest}] \frac{1}{n}$$

Now if the initial value selected is the jth smallest, then the set of values smaller than it is of size $j - 1$, and the set of values greater than it is of size $n - j$. Hence, as $n - 1$ comparisons with the initial value chosen must be made, we see that

$$M_n = \sum_{j=1}^{n} (n - 1 + M_{j-1} + M_{n-j}) \frac{1}{n}$$

$$= n - 1 + \frac{2}{n} \sum_{k=1}^{n-1} M_k \qquad (\text{since } M_0 = 0)$$

or, equivalently,

$$nM_n = n(n - 1) + 2 \sum_{k=1}^{n-1} M_k$$

To solve the above, note that upon replacing n by $n + 1$ we obtain

$$(n + 1)M_{n+1} = (n + 1)n + 2 \sum_{k=1}^{n} M_k$$

Hence, upon subtraction,

$$(n + 1)M_{n+1} - nM_n = 2n + 2M_n$$

or

$$(n + 1)M_{n+1} = (n + 2)M_n + 2n$$

Therefore,

$$\frac{M_{n+1}}{n + 2} = \frac{2n}{(n + 1)(n + 2)} + \frac{M_n}{n + 1}$$

Iterating this gives

$$\frac{M_{n+1}}{n+2} = \frac{2n}{(n+1)(n+2)} + \frac{2(n-1)}{n(n+1)} + \frac{M_{n-1}}{n}$$

$$=$$

$$\vdots$$

$$= 2 \sum_{k=0}^{n-1} \frac{n-k}{(n+1-k)(n+2-k)} \qquad \text{since } M_1 = 0$$

Hence,

$$M_{n+1} = 2(n+2) \sum_{k=0}^{n-1} \frac{n-k}{(n+1-k)(n+2-k)}$$

$$= 2(n+2) \sum_{i=1}^{n} \frac{i}{(i+1)(i+2)}, \qquad n \geq 1$$

Using the identity $i/(i+1)(i+2) = 2/(i+2) - 1/(i+1)$, we can approximate M_{n+1} for large n as follows:

$$M_{n+1} = 2(n+2) \left[\sum_{i=1}^{n} \frac{2}{i+2} - \sum_{i=1}^{n} \frac{1}{i+1} \right]$$

$$\sim 2(n+2) \left[\int_{3}^{n+2} \frac{2}{x} dx - \int_{2}^{n+1} \frac{1}{x} dx \right]$$

$$= 2(n+2) \left[2\log(n+2) - \log(n+1) + \log 2 - 2\log 3 \right]$$

$$= 2(n+2) \left[\log(n+2) + \log \frac{n+2}{n+1} + \log 2 - 2\log 3 \right]$$

$$\sim 2(n+2) \log(n+2) \qquad \diamond$$

The conditional expectation is often useful in computing the variance of a random variable. In particular, we have that

$$\text{Var}(X) = E[X^2] - (E[X])^2$$

$$= E[E[X^2|Y]] - (E[E[X|Y]])^2$$

Example 4g   (The Variance of a Random Number of Random Variables): In Example 4d we showed that if $X_1, X_2, \ldots$ are independent and identically distributed, and if N is a nonnegative integer valued random variable independent of the X's, then

$$E\left[\sum_{i=1}^{N} X_i\right] = E[N]E[X]$$

What can we say about $\text{Var}(\Sigma_{i=1}^{N} X_i)$?

Solution:

$$\text{Var}\left(\sum_{i=1}^{N} X_i\right) = E\left[\left(\sum_{i=1}^{N} X_i\right)^2\right] - \left(E\left[\sum_{i=1}^{N} X_i\right]\right)^2 \qquad (4.6)$$

To compute each of the individual terms, we condition on N

$$E\left[\left(\sum_{i=1}^{N} X_i\right)^2\right] = E\left[E\left[\left(\sum_{i=1}^{N} X_i\right)^2 \middle| N\right]\right]$$

Now, given that $N = n$, $(\Sigma_{i=1}^{N} X_i)^2$ is distributed as the square of the sum of n independent and identically distributed random variables. Hence, using the identity $E[Z^2] = \text{Var}(Z) + (E[Z])^2$, we have that

$$E\left[\left(\sum_{i=1}^{N} X_i\right)^2 \middle| N = n\right] = \text{Var}\left(\sum_{i=1}^{n} X_i\right) + \left(E\left[\sum_{i=1}^{n} X_i\right]\right)^2$$
$$= n\text{Var}(X) + (nE[X])^2$$

Therefore,

$$E\left[\left(\sum_{i=1}^{N} X_i\right)^2 \middle| N\right] = N\,\text{Var}(X) + N^2(E[X])^2$$

Taking expectations of both sides of the above equation yields that

$$E\left[\left(\sum_{i=1}^{N} X_i\right)^2\right] = E[N]\,\text{Var}(X) + E[N^2](E[X])^2$$

Hence, from Equation (4.6) we obtain

$$\text{Var}\left(\sum_{i=1}^{N} X_i\right) = E[N]\,\text{Var}(X) + E[N^2](E[X])^2 - \left(E\left[\sum_{i=1}^{N} X_i\right]\right)^2$$
$$= E[N]\,\text{Var}(X) + E[N^2](E[X])^2 - (E[N]E[X])^2$$
$$= E[N]\,\text{Var}(X) + (E[X])^2(E[N^2] - (E[N])^2)$$
$$= E[N]\,\text{Var}(X) + (E[X])^2\,\text{Var}(N) \qquad \Diamond$$

Example 4h (Variance of the Geometric Distribution): Independent trials, each resulting in a success with probability p, are successively performed. Let N be the time of the first success. Find $\text{Var}(N)$.

Solution: Let $Y = 1$ if the first trial results in a success, and $Y = 0$ otherwise.

$$\text{Var}(N) = E(N^2) - (E[N])^2$$

To calculate $E[N^2]$ and $E[N]$ we condition on Y. For instance,

$$E[N^2] = E[E[N^2|Y]]$$

However,

$$E[N^2|Y = 1] = 1$$
$$E[N^2|Y = 0] = E[(1 + N)^2]$$

These two equations are true since if the first trial results in a success, then clearly $N = 1$ and so $N^2 = 1$. On the other hand, if the first trial results in a failure, then the total number of trials necessary for the first success will equal one (the first trial that results in failure) plus the necessary number of additional trials. Since this latter quantity has the same distribution as N, we get that $E[N^2|Y = 0] = E[(1 + N)^2]$. Hence, we see that

$$E[N^2] = E[N^2|Y = 1]P\{Y = 1\} + E[N^2|Y = 0]P\{Y = 0\}$$
$$= p + E[(1 + N)^2](1 - p)$$
$$= 1 + (1 - p)E[2N + N^2]$$

Since, as was shown in Example 4a, $E[N] = 1/p$, this yields

$$E[N^2] = 1 + \frac{2(1 - p)}{p} + (1 - p)E[N^2]$$

or

$$E[N^2] = \frac{2 - p}{p^2}$$

Therefore,

$$\text{Var}(N) = E[N^2] - (E[N])^2$$
$$= \frac{2 - p}{p^2} - \left(\frac{1}{p}\right)^2$$
$$= \frac{1 - p}{p^2} \quad \Diamond$$

5. Computing Probabilities by Conditioning

Not only can we obtain expectations by first conditioning upon an appropriate random variable, but we may also use this approach to compute probabilities. To see this, let E denote an arbitrary event and define the indicator random variable X by

$$X = \begin{cases} 1, & \text{if } E \text{ occurs} \\ 0, & \text{if } E \text{ does not occur} \end{cases}$$

It follows from the definition of X that

$$E[X] = P(E)$$
$$E[X|Y = y] = P(E|Y = y), \quad \text{for any random variable } Y$$

Therefore, from Equations (4.1a) and (4.1b) we obtain that

$$P(E) = \sum_y P(E|Y = y)P(Y = y), \quad \text{if } Y \text{ is discrete}$$

$$= \int_{-\infty}^{\infty} P(E|Y = y)f_Y(y)dy, \quad \text{if } Y \text{ is continuous}$$

Example 5a Suppose that X and Y are independent continuous random variables having densities f_X and f_Y, respectively. Compute $P\{X < Y\}$.

Solution: Conditioning on the value of Y yields

$$P\{X < Y\} = \int_{-\infty}^{\infty} P\{X < Y|Y = y\}f_Y(y)dy$$

$$= \int_{-\infty}^{\infty} P\{X < y|Y = y\}f_Y(y)dy$$

$$= \int_{-\infty}^{\infty} P\{X < y\}f_Y(y)dy$$

$$= \int_{-\infty}^{\infty} F_X(y)f_Y(y)dy$$

where

$$F_X(y) = \int_{-\infty}^{y} f_X(x)dx \quad \Diamond$$

Example 5b   Suppose that X and Y are independent continuous random variables. Find the distribution of $X + Y$.

Solution: By conditioning on the value of Y we obtain

$$P\{X + Y < a\} = \int_{-\infty}^{\infty} P\{X + Y < a | Y = y\} f_Y(y) dy$$

$$= \int_{-\infty}^{\infty} P\{X + y < a | Y = y\} f_Y(y) dy$$

$$= \int_{-\infty}^{\infty} P\{X < a - y\} f_Y(y) dy$$

$$= \int_{-\infty}^{\infty} F_X(a - y) f_Y(y) dy \quad \Diamond$$

Example 5c   Each customer who enters Larry's clothing store will purchase a suit with probability p. If the number of customers entering the store is Poisson distributed with mean λ, what is the probability that Larry does not sell any suits?

Solution: Let X be the number of suits that Larry sells, and let N denote the number of customers who enter the store. By conditioning on N we see that

$$P\{X = 0\} = \sum_{n=0}^{\infty} P\{X = 0 | N = n\} P\{N = n\}$$

$$= \sum_{n=0}^{\infty} P\{X = 0 | N = n\} e^{-\lambda} \lambda^n / n!$$

Now, given that n customers enter the store, the probability that Larry does not sell any suits is just $(1 - p)^n$. That is, $P\{X = 0 | N = n\} = (1 - p)^n$. Therefore,

$$P\{X = 0\} = \sum_{n=0}^{\infty} \frac{(1 - p)^n e^{-\lambda} \lambda^n}{n!}$$

$$= e^{-\lambda} \sum_{n=0}^{\infty} \frac{(\lambda(1 - p))^n}{n!}$$

$$= e^{-\lambda} e^{\lambda(1-p)}$$

$$= e^{-\lambda p} \quad \Diamond$$

Example 5c (continued) What is the probability that Larry sells k suits?

Solution:

$$P\{X = k\} = \sum_{n=0}^{\infty} \frac{P\{X = k|N = n\}e^{-\lambda}\lambda^n}{n!}$$

Now, given that $N = n$, X has a binomial distribution with parameters n and p. Hence,

$$P\{X = k|N = n\} = \begin{cases} \binom{n}{k}p^k(1-p)^{n-k}, & n \geq k \\ 0, & n < k \end{cases}$$

so that

$$P\{X = k\} = \sum_{n=k}^{\infty} \binom{n}{k} \frac{p^k(1-p)^{n-k}e^{-\lambda}\lambda^n}{n!}$$

$$= \sum_{n=k}^{\infty} \frac{n!}{(n-k)!k!} \frac{(\lambda p)^k(\lambda(1-p))^{n-k}e^{-\lambda}}{n!}$$

$$= \frac{e^{-\lambda}(\lambda p)^k}{k!} \sum_{n=k}^{\infty} \frac{(\lambda(1-p))^{n-k}}{(n-k)!}$$

$$= e^{-\lambda} \frac{(\lambda p)^k}{k!} \sum_{i=0}^{\infty} \frac{(\lambda(1-p))^i}{i!}$$

$$= e^{-\lambda} \frac{(\lambda p)^k}{k!} e^{\lambda(1-p)}$$

$$= e^{-\lambda p} \frac{(\lambda p)^k}{k!}$$

In other words, X has a Poisson distribution with mean λp. $\quad\Diamond$

Example 5d At a party n men take off their hats. The hats are then mixed up and each man randomly selects one. We say that a match occurs if a man selects his own hat. What is the probability of no matches? What is the probability of exactly k matches?

Solution: Let E denote the event that no matches occur, and to make explicit the dependence on n, write $P_n = P(E)$. We start by conditioning on whether or not the first man selects his own hat—call these events M and M^c. Then

$$P_n = P(E) = P(E|M)P(M) + P(E|M^c)P(M^c).$$

Clearly, $P(E|M) = 0$, and so

$$P_n = P(E|M^c)\frac{n-1}{n} \tag{5.1}$$

Now, $P(E|M^c)$ is the probability of no matches when $n - 1$ men select from a set of $n - 1$ hats that does not contain the hat of one of these men. This can happen in either of two mutually exclusive ways. Either there are no matches and the extra man does not select the extra hat (this being the hat of the man that chose first), or there are no matches and the extra man does select the extra hat. The probability of the first of these events is just P_{n-1}, which is seen by regarding the extra hat as "belonging" to the extra man. As the second event has probability $[1/(n-1)]P_{n-2}$, we have

$$P(E|M^c) = P_{n-1} + \frac{1}{n-1}P_{n-2}$$

and thus, from Equation (5.1),

$$P_n = \frac{n-1}{n}P_{n-1} + \frac{1}{n}P_{n-2}$$

or, equivalently,

$$P_n - P_{n-1} = -\frac{1}{n}(P_{n-1} - P_{n-2}) \tag{5.2}$$

However, as P_n is the probability of no matches when n men select among their own hats, we have

$$P_1 = 0, \qquad P_2 = \tfrac{1}{2}$$

and so, from Equation (5.2),

$$P_3 - P_2 = -\frac{(P_2 - P_1)}{3} = -\frac{1}{3!} \quad \text{or} \quad P_3 = \frac{1}{2!} - \frac{1}{3!}$$

$$P_4 - P_3 = -\frac{(P_3 - P_2)}{4} = \frac{1}{4!} \quad \text{or} \quad P_4 = \frac{1}{2!} - \frac{1}{3!} + \frac{1}{4!}$$

and, in general, we see that

$$P_n = \frac{1}{2!} - \frac{1}{3!} + \frac{1}{4!} - \cdots + \frac{(-1)^n}{n!}$$

To obtain the probability of exactly k matches, we consider any fixed group of k men. The probability that they, and only they, select their own hats is

$$\frac{1}{n}\frac{1}{n-1}\cdots\frac{1}{n-(k-1)}P_{n-k}=\frac{(n-k)!}{n!}P_{n-k}$$

where P_{n-k} is the conditional probability that the other $n-k$ men, selecting among their own hats, have no matches. As there are $\binom{n}{k}$ choices of a set of k men, the desired probability of exactly k matches is

$$\frac{P_{n-k}}{k!}=\frac{\dfrac{1}{2!}-\dfrac{1}{3!}+\cdots+\dfrac{(-1)^{n-k}}{(n-k)!}}{k!}$$

which, for n large, is approximately equal to $e^{-1}/k!$.   ◊

Example 5e (The Ballot Problem): In an election, candidate A receives n votes, and candidate B receives m votes where $n > m$. Assuming that all orderings are equally likely, show that the probability that A is always ahead in the count of votes is $(n - m)/(n + m)$.

Solution: Let $P_{n,m}$ denote the desired probability. By conditioning on which candidate receives the last vote counted we have

$$P_{n,m} = P\{A \text{ always ahead}|A \text{ receives last vote}\}\frac{n}{n+m}$$

$$+ P\{A \text{ always ahead}|B \text{ receives last vote}\}\frac{m}{n+m}$$

Now given that A receives the last vote, one can see that the probability that A is always ahead is the same as if A had received a total of $n - 1$ and B a total of m votes. As a similar result is true when we are given that B receives the last vote, we see from the above that

$$P_{n,m} = \frac{n}{n+m}P_{n-1,m} + \frac{m}{m+n}P_{n,m-1} \tag{5.3}$$

We can now prove that $P_{n,m} = (n - m)/(n + m)$ by induction on $n + m$. As it is true when $n + m = 1$, i.e. $P_{1,0} = 1$, assume it whenever $n + m = k$. Then when $n + m = k + 1$, we have by Equation (5.3) and the induction hypothesis that

$$P_{n,m} = \frac{n}{n+m}\frac{n-1-m}{n-1+m} + \frac{m}{m+n}\frac{n-m+1}{n+m-1}$$

$$= \frac{n-m}{n+m}$$

and the result is proven. ◊

The ballot problem has some interesting applications. For example, consider successive flips of a coin which always lands on "heads" with probability p, and let us determine the probability distribution of the first time, after beginning, that the total number of heads is equal to the total number of tails. The probability that the first time this occurs is at time $2n$ can be obtained by first conditioning on the total number of heads in the first $2n$ trials. This yields

$P\{\text{first time equal} = 2n\}$

$= P\{\text{first time equal} = 2n | n \text{ heads in first } 2n\} \binom{2n}{n} p^n (1-p)^n$

Now given a total of n heads in the first $2n$ flips, one can see that all possible orderings of the n heads and n tails are equally likely, and thus the above conditional probability is equivalent to the probability that in an election, in which each candidate receives n votes, one of the candidates is always ahead in the counting until the last vote (which ties them). But by conditioning on whomever receives the last vote, we see that this is just the probability in the ballot problem when $m = n - 1$. Hence,

$$P\{\text{first time equal} = 2n\} = P_{n,n-1}\binom{2n}{n}p^n(1-p)^n$$

$$= \frac{\binom{2n}{n}p^n(1-p)^n}{2n-1}$$

6. Some Applications

6.1. A List Model

Consider n elements—$e_1, e_2, \ldots, e_n$—which are initially arranged in some ordered list. At each unit of time a request is made for one of these elements—e_i being requested, independently of the past, with probability P_i. After being requested the element is then moved to the front of the list. That is, for instance, if the present ordering is

e_1, e_2, e_3, e_4 and if e_3 is requested, then the next ordering is e_3, e_1, e_2, e_4.

We are interested in determining the expected position of the element requested after this process has been in operation for a long time. However, before computing this probability, let us note two possible applications of this model. In the first we have a stack of reference books. At each unit of time a book is randomly selected and is then returned to the top of the stack. In the second application we have a computer receiving requests for elements stored in its memory. The request probabilities for the elements may not be known and so to reduce the average time it takes the computer to locate the element requested (which is proportional to the position of the requested element since the computer locates the element by starting at the beginning and then going down the list), the computer is programmed to replace the requested element at the beginning of the list.

To compute the expected position of the element requested, we start by conditioning on which element is selected. This yields

E[Position of element requested]

$$= \sum_{i=1}^{n} E[\text{Position}|e_i \text{ is selected}]P_i$$

$$= \sum_{i=1}^{n} E[\text{Position of } e_i|e_i \text{ is selected}]P_i \qquad (6.1)$$

$$= \sum_{i=1}^{n} E[\text{Position of } e_i]P_i$$

Now

$$\text{Position of } e_i = 1 + \sum_{j \neq i} I_j$$

where

$$I_j = \begin{cases} 1, & \text{if } e_j \text{ precedes } e_i \\ 0, & \text{otherwise} \end{cases}$$

and so,

$$E[\text{Position of } e_i] = 1 + \sum_{j \neq i} E[I_j]$$

$$= 1 + \sum_{j \neq i} P\{e_j \text{ precedes } e_i\} \qquad (6.2)$$

To compute $P\{e_j$ precedes $e_i\}$, note that e_j will precede e_i if the most recent request for either of them was for e_j. But given that a request is for either e_i or e_j, the probability that it is for e_j is

$$P\{e_j|e_i \text{ or } e_j\} = \frac{P_j}{P_i + P_j}$$

and, thus,

$$P\{e_j \text{ precedes } e_i\} = \frac{P_j}{P_i + P_j}$$

Hence from Equations (6.1) and (6.2) we see that

$$E[\text{Position of element requested}] = 1 + \sum_{i=1}^{n} P_i \sum_{j \neq i} \frac{P_j}{P_i + P_j}$$

This list model will be further analyzed in Section 7 of Chapter 4, where we will assume a different reordering rule—namely that the element requested is moved one closer to the front of the list as opposed to being moved to the front of the list as assumed here. We will show there that the average position of the requested element is less under the one-closer rule than it is under the front-of-the-line rule.

6.2 A Random Graph

A graph consists of a set V of elements called nodes and a set A of pairs of elements of V called arcs. A graph can be represented graphically by drawing circles for nodes and drawing lines between nodes i and j whenever (i, j) is an arc. For instance if $V = \{1, 2, 3, 4\}$ and $A = \{(1, 2), (1, 4), (2, 3), (1, 2), (3, 3)\}$, then we can represent this graph as shown in Figure 3.1. Note that the arcs have no direction (a graph in which the arcs are ordered pairs of nodes is called a directed graph); and that in the above there are multiple arcs connecting nodes 1 and

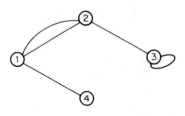

Figure 3.1.   A graph.

2, and a self arc (called a self loop) from node 3 to itself.

We say that there exists a path from node i to node j, $i \neq j$, if there exists a sequence of nodes $i, i_1, \ldots, i_k, j$ such that $(i, i_1), (i_1, i_2), \ldots, (i_k, j)$ are all arcs. If there is a path between each of the $\binom{n}{2}$ distinct pair of nodes we say that the graph is *connected*. The graph in Figure 3.1 is connected but the graph in Figure 3.2 is not. Consider now the following graph where $V = \{1, 2, \ldots, n\}$ and $A = \{(i, X(i)), i = 1, \ldots, n\}$ where the $X(i)$ are independent random variables such that

$$P\{X(i) = j\} = \frac{1}{n}, \quad j = 1, 2, \ldots, n$$

In other words from each node i we select at random one of the n nodes (including possibly the node i itself) and then join node i and the selected node with an arc. Such a graph is commonly referred to as a *random graph*.

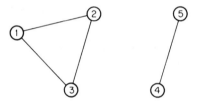

Figure 3.2. A disconnected graph.

We are interested in determining the probability that the random graph so obtained is connected. As a prelude, starting at some node— say node 1—let us follow the sequence of nodes, $1, X(1), X^2(1), \ldots,$ where $X^n(1) = X(X^{n-1}(1))$; and define N to equal the first k such that $X^k(1)$ is not a new node. In other words,

$$N = \text{1st } k \text{ such that } X^k(1) \in [1, X(1), \ldots, X^{k-1}(1)]$$

We can represent this as shown in Figure 3.3 where the arc from $X^{N-1}(1)$ goes back to a node previously visited.

Figure 3.3.

To obtain the probability that the graph is connected we first condition on N to obtain

$$P\{\text{Graph is connected}\} = \sum_{k=1}^{n} P\{\text{Connected}|N = k\}P\{N = k\} \quad (6.3)$$

Now given that $N = k$, the k nodes $1, X(1), \ldots, X^{k-1}(1)$ are connected to each other, and there are no other arcs emanating out of these nodes. In other words, if we regard these k nodes as being one supernode, the situation is the same as if we had one supernode and $n - k$ ordinary nodes with arcs only emanating from the ordinary nodes—each arc going into the supernode with probability k/n. The solution in this situation is obtained from Lemma 6.1 by taking $r = n - k$.

Lemma 6.1.   Given a random graph consisting of nodes $0, 1, \ldots, r$ and r arcs—namely (i, Y_i), $i = 1, \ldots, r$, where

$$Y_i = \begin{cases} j & \text{with probability } \dfrac{1}{r + k}, \quad j = 1, \ldots, r \\[2ex] 0 & \text{with probability } \dfrac{k}{r + k} \end{cases}$$

then

$$P\{\text{Graph is connected}\} = \frac{k}{r + k}$$

(In other words, for the above graph there are $r + 1$ nodes—r ordinary nodes and one supernode. Out of each ordinary node an arc is chosen. The arc goes to the supernode with probability $k/(r + k)$ and to each of the ordinary ones with probability $1/(r + k)$. There is no arc emanating out of the supernode.)

Proof: The proof is by induction on r. As it is true when $r = 1$ for any k, assume it true for all values less than r. Now in the case under consideration, let us first condition on the number of arcs (j, Y_j) for which $Y_j = 0$. This yields

$P\{\text{Connected}\}$

$$= \sum_{i=0}^{r} P\{\text{Connected}|i \text{ of the } Y_j = 0\}\binom{r}{i}\left(\frac{k}{r + k}\right)^i \left(\frac{r}{r + k}\right)^{r-i} \quad (6.4)$$

Now given that exactly i of the arcs are into the supernode, the situation for the remaining $r - i$ arcs which do not go into the supernode is the same as if we had $r - i$ ordinary nodes and one supernode with an arc going out of each of the ordinary nodes— into the supernode with probability i/r and into each ordinary node with probability $1/r$. But by the induction hypothesis the probability that this would lead to a connected graph is i/r.

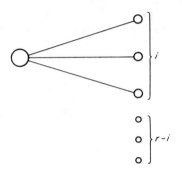

Figure 3.4. The situation given i of the r arcs are into the supernode.

Hence,

$$P\{\text{Connected}|i \text{ of the } Y_j = 0\} = \frac{i}{r}$$

and from Equation (6.4)

$$P\{\text{Connected}\} = \sum_{i=0}^{r} \frac{i}{r} \binom{r}{i} \left(\frac{k}{r+k}\right)^i \left(\frac{r}{r+k}\right)^{r-i}$$

$$= \frac{1}{r} E\left[\text{Binomial}\left(r, \frac{k}{r+k}\right)\right]$$

$$= \frac{k}{r+k}$$

which completes the proof of the lemma. ◇

Hence as the situation given $N = k$ is exactly as described by Lemma 6.1 when $r = n - k$, we see that, for the original graph,

$$P\{\text{Graph is connected}|N = k\} = \frac{k}{n}$$

and from Equation (6.3)

$$P\{\text{Graph is connected}\} = \frac{E(N)}{n} \tag{6.5}$$

To compute $E(N)$ we use the identity

$$E(N) = \sum_{i=1}^{\infty} P\{N \geq i\}$$

which can be proved by defining indicator variables I_i, $i \geq 1$, by

$$I_i = \begin{cases} 1, & \text{if } i \leq N \\ 0, & \text{if } i > N \end{cases}$$

Hence,

$$N = \sum_{i=1}^{\infty} I_i$$

and so

$$E(N) = E\left[\sum_{i=1}^{\infty} I_i\right]$$

$$= \sum_{i=1}^{\infty} E[I_i]$$

$$= \sum_{i=1}^{\infty} P\{N \geq i\} \tag{6.6}$$

Now the event $\{N \geq i\}$ occurs if the nodes $1, X(1), \ldots, X^{i-1}(1)$ are all distinct. Hence,

$$P\{N \geq i\} = \frac{(n-1)}{n} \frac{(n-2)}{n} \cdots \frac{(n-i+1)}{n}$$

$$= \frac{(n-1)!}{(n-i)!n^{i-1}}$$

and so from Equation (6.5) and (6.6),

$$P\{\text{Graph is connected}\} = (n-1)! \sum_{i=1}^{n} \frac{1}{(n-i)!n^i}$$

$$= \frac{(n-1)!}{n^n} \sum_{j=0}^{n-1} \frac{n^j}{j!} \quad (\text{by } j = n - i) \tag{6.7}$$

We can also use Equation (6.7) to obtain a simple approximate expression for the probability that the graph is connected when n is large. To do so, we first note that if X is a Poisson random variable with mean n, then

$$P\{X < n\} = e^{-n} \sum_{j=0}^{n-1} \frac{n^j}{j!}$$

Since Poisson random variable with mean n can be regarded as being the sum of n independent Poisson random variables each with mean 1, it follows by the Central Limit Theorem that for n large such a random variable has approximately a normal distribution and as such has probability $\frac{1}{2}$ of being less than its mean. That is, for n large

$$P\{X < n\} \approx \tfrac{1}{2}$$

and so for n large,

$$\sum_{j=0}^{n-1} \frac{n^j}{j!} \approx \frac{e^n}{2}$$

Hence from Equation (6.7), for n large,

$$P\{\text{Graph is connected}\} \approx \frac{e^n(n-1)!}{2n^n}$$

By employing an approximation due to Stirling which states that for n large

$$n! \approx n^{n+1/2} e^{-n} \sqrt{2\pi}$$

we see that, for n large,

$$P\{\text{Graph is connected}\} \approx \sqrt{\frac{\pi}{2(n-1)}} \, e \left(\frac{n-1}{n} \right)^n$$

and as

$$\lim_{n \to \infty} \left(\frac{n-1}{n} \right)^n = \lim_{n \to \infty} \left(1 - \frac{1}{n} \right)^n = e^{-1}$$

We see that, for n large,

$$P\{\text{Graph is connected}\} \approx \sqrt{\frac{\pi}{2(n-1)}}$$

Now a graph is said to consist of r connected components if its nodes can be partitioned into r subsets so that each of the subsets are connected and there are no arcs between nodes in different subsets. For

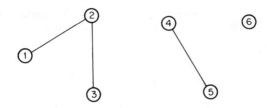

Figure 3.5. A graph having three connected components.

instance, the graph in Figure 3.5 consists of three connected components—namely $\{1, 2, 3\}$, $\{4, 5\}$, and $\{6\}$. Let C denote the number of connected components of our random graph and let

$$P_n(i) = P\{C = i\}$$

where we use the notation $P_n(i)$ to make explicit the dependence on n, the number of nodes. Since a connected graph is by definition a graph consisting of exactly one component, from Equation (6.7) we have

$$P_n(1) = P\{C = 1\}$$

$$= \frac{(n-1)!}{n^n} \sum_{j=0}^{n-1} \frac{n^j}{j!} \tag{6.8}$$

To obtain $P_n(2)$, the probability of exactly two components, let us first fix attention on some particular node—say node 1. In order that a given set of $k - 1$ other nodes—say nodes 2, . . . , k—will along with node 1 constitute one connected component and the remaining $n - k$ a second connected component, we must have

(i) $X(i) \in \{1, 2, . . . , k\}$, for all $i = 1, . . . , k$
(ii) $X(i) \in \{k + 1, . . . , n\}$, for all $i = k + 1, . . . , n$
(iii) The nodes $1, 2, . . . , k$ form a connected subgraph
(iv) The nodes $k + 1, . . . , n$ form a connected subgraph.

The probability of the above occurring is clearly

$$\left(\frac{k}{n}\right)^k \left(\frac{n-k}{n}\right)^{n-k} P_k(1) P_{n-k}(1)$$

and as there are $\binom{n-1}{k-1}$ ways of choosing a set of $k - 1$ nodes from the nodes 2 through n, we have that

$$P_n(2) = \sum_{k=1}^{n-1} \binom{n-1}{k-1} \left(\frac{k}{n}\right)^k \left(\frac{n-k}{n}\right)^{n-k} P_k(1) P_{n-k}(1)$$

and so $P_n(2)$ can be computed from Equation (6.8). In general, the recursive formula for $P_n(i)$ is given by

$$P_n(i) = \sum_{k=1}^{n-i+1} \binom{n-1}{k-1} \left(\frac{k}{n}\right)^k \left(\frac{n-k}{n}\right)^{n-k} P_k(1) P_{n-k}(i-1)$$

To compute $E[C]$, the expected number of connected components, first note that every connected component of our random graph must contain exactly one cycle (a cycle is a set of arcs of the form (i, i_1), $(i_1, i_2), \ldots, (i_{k-1}, i_k), (i_k, i)$ for distinct nodes $i, i_1, \ldots, i_k$). For example, Figure 3.6 depicts a cycle.

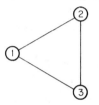

Figure 3.6. A cycle.

The fact that every connected component of our random graph must contain exactly one cycle is most easily proved by noting that if the connected component consists of r nodes, then it must also have r arcs and, hence, must contain exactly one cycle (why?). Thus, we see that

$$E[C] = E[\text{Number of cycles}]$$

$$= E\left[\sum_S I(S)\right]$$

$$= \sum_S E[I(S)]$$

where the sum is over all subsets $S \subset \{1, 2, \ldots, n\}$ and

$$I(S) = \begin{cases} 1, & \text{if the nodes in } S \text{ are all the nodes of a cycle} \\ 0, & \text{otherwise} \end{cases}$$

Now if S consists of k nodes, say $1, \ldots, k$, then

$$E[I(S)] = P\{1, X(1), \ldots, X^{k-1}(1) \text{ are all distinct and}$$
$$\text{contained in } 1, \ldots, k \text{ and } X^k(1) = 1\}$$

$$= \frac{k-1}{n} \frac{k-2}{n} \cdot \cdot \cdot \frac{1}{n} \frac{1}{n} = \frac{(k-1)!}{n^k}$$

Hence, as there are $\binom{n}{k}$ subsets of size k we see that

$$E[C] = \sum_{k=1}^{n} \binom{n}{k} \frac{(k-1)!}{n^k}$$

6.3. Uniform Priors, Polya's Urn Model, and Bose-Einstein Statistics

Suppose that n independent trials, each of which is a success with probability p are performed. If we let X denote the total number of successes, then X is a binomial random variable such that

$$P\{X = k|p\} = \binom{n}{k} p^k (1 - p)^{n-k}, \qquad k = 0, 1, \ldots, n$$

However, let us now suppose that whereas the trials all have the same success probability p, its value is not predetermined but is chosen according to a uniform distribution on $(0, 1)$. (For instance, a coin may be chosen at random from a huge bin of coins representing a uniform spread over all possible values of p, the coin's probability of coming up heads. The chosen coin is then flipped n times.) In this case, by conditioning on the actual value of p, we have that

$$P\{X = k\} = \int_0^1 P\{X = k|p\}f(p)dp$$

$$= \int_0^1 \binom{n}{k} p^k(1 - p)^{n-k} \, dp$$

Now it can be shown that

$$\int_0^1 p^k(1 - p)^{n-k}dp = \frac{k!(n - k)!}{(n + 1)!} \tag{6.9}$$

and thus

$$P\{X = k\} = \binom{n}{k} \frac{k!(n - k)!}{(n + 1)!}$$

$$= \frac{1}{n + 1}, \qquad k = 0, 1, \ldots, n \tag{6.10}$$

In other words, each of the $n + 1$ possible values of X is equally likely.

As an alternate way of describing the above experiment, let us compute the conditional probability that the $(r + 1)$st trial will result in a success given a total of k successes (and $r - k$ failures) in the first r trials.

$$P\{(r + 1)\text{st trial is a success}|k \text{ successes in first } r\}$$

$$= \frac{P\{(r + 1)\text{st is a success, } k \text{ successes in first } r \text{ trials}\}}{P\{k \text{ successes in first } r \text{ trials}\}}$$

$$= \frac{\int_0^1 P\{(r + 1)\text{st is a success, } k \text{ in first } r|p\}dp}{1/(r + 1)}$$

$$= (r + 1)\int_0^1 \binom{r}{k}p^{k+1}(1 - p)^{r-k}dp$$

$$= (r + 1)\binom{r}{k}\frac{(k + 1)!(r - k)!}{(r + 2)!} \quad \text{by Equation (6.9)}$$

$$= \frac{k + 1}{r + 2} \tag{6.11}$$

That is, if the first r trials result in k successes, then the next trial will be a success with probability $(k + 1)/(r + 2)$.

It follows from Equation (6.11) that an alternative description of the stochastic process of the successive outcomes of the trials can be described as follows: There is an urn which initially contains 1 white and 1 black ball. At each stage a ball is randomly drawn and is then replaced along with another ball of the same color. Thus, for instance, if of the first r balls drawn k were white, then the urn at the time of the $(r + 1)$th draw would consist of $k + 1$ white and $r - k + 1$ black, and thus the next ball would be white with probability $(k + 1)/(r + 2)$. If we identify the drawing of a white ball with a successful trial, then we see that this yields an alternate description of the original model. This latter urn model is called *Polya's Urn Model*.

Remarks

(i) In the special case when $k = r$ formula Equation (6.11) is sometimes called Laplace's rule of succession, after the French mathematician Pierre de Laplace. In Laplace's era, this "rule" provoked much controversy, for people attempted to employ it in diverse situations where its validity was questionable. For instance, it was used to justify such propositions as "If you have dined twice at a restaurant and both meals were good, then the next meal also will be good with probability $\frac{3}{4}$," and "Since the sun has risen the past 1,826,213 days, so will it rise tomorrow with probability 1,826,214/1,826,215." The trouble with such claims resides in the fact that it is not at all clear the situation they are describing can be modeled as consisting of independent trials having a common probability of success which is itself uniformly chosen.

(ii) In the original description of the experiment, we referred to the successive trials as being independent, and in fact they are independent when the success probability is known. However, when p is regarded as a random variable, the successive outcomes are no longer independent since knowing whether an outcome is a success or not gives us some information about p, which in turn yields information about the other outcomes. ◊

The preceding can be generalized to situations in which each trial has more than two possible outcomes. Suppose that n independent trials, each resulting in one of m possible outcomes $1, \ldots, m$ with respective probabilities $p_1, \ldots, p_m$ are performed. If we let X_i denote the number of type i outcomes that result in the n trials, $i = 1, \ldots, m$, then the vector $X_1, \ldots, X_m$ will have the multinomial distribution given by

$$P\{X_1 = x_1, X_2 = x_2, \ldots, X_m = x_m | \mathbf{p}\} = \frac{n!}{x_1! \cdots x_m!} p_1^{x_1} p_2^{x_2} \cdots p_m^{x_m}$$

where $x_1, \ldots, x_m$ is any vector of nonnegative integers which sum to n. Now let us suppose that the vector $\mathbf{p} = (p_1, \ldots, p_m)$ is not specified, but instead is chosen by a "uniform" distribution. Such a distribution would be of the form

$$f(p_1, \ldots, p_m) = \begin{cases} c, & 0 \le p_i \le 1, \quad i = 1, \ldots, m, \quad \sum_1^m p_i = 1 \\ 0, & \text{otherwise} \end{cases}$$

The above multivariate distribution is a special case of what is known as the *Dirichlet distribution*, and it is not difficult to show, using the fact that the distribution must integrate to 1, that $c = (m - 1)!$.

The unconditional distribution of the vector $\mathbf{X}$ is given by

$$P\{X_1 = x_1, \ldots, X_m = x_m\}$$
$$= \int\int \cdots \int P\{X_1 = x_1, \ldots, X_m = x_m | p_1, \ldots, p_m\}$$

$$\times f(p_1, \ldots, p_m) dp_1 \cdots dp_m$$

$$= \frac{(m-1)!n!}{x_1! \cdots x_m!} \int\int_{\substack{0 \le p_i \le 1 \\ \Sigma_1^m p_i = 1}} \cdots \int p_1^{x_1} \cdots p_m^{x_m} dp_1 \cdots dp_m$$

Now it can be shown that

$$\int\int_{\substack{0 \le p_i \le 1 \\ \Sigma_1^m p_i = 1}} \cdots \int p_1^{x_1} \cdots p_m^{x_m} dp_1 \cdots dp_m = \frac{x_1! \cdots x_m!}{\left(\sum_1^m x_i + m - 1\right)!} \qquad (6.12)$$

and thus, using the fact that $\Sigma_1^m x_i = n$, we see that

$$P\{X_1 = x_1, \ldots, X_m = x_m\} = \frac{n!(m-1)!}{(n+m-1)!}$$

$$= \binom{n+m-1}{m-1}^{-1} \qquad (6.13)$$

Hence, all of the $\binom{n+m-1}{m-1}$ possible outcomes (there are $\binom{n+m-1}{m-1}$ possible nonnegative integer valued solutions of $x_1 + \cdots + x_m = n$) of the vector $(X_1, \ldots, X_m)$ are equally likely. The probability distribution given by Equation (6.13) is sometimes called the *Bose-Einstein distribution*.

To obtain an alternative description of the foregoing, let us compute the conditional probability that the $(n + 1)$st outcome is of type j if the first n trials have resulted in x_i type i outcomes, $i = 1, \ldots, m$, $\Sigma_1^m x_i = n$. This is given by

$$P\{(n+1)\text{st is } j | x_i \text{ type } i \text{ in first } n, \quad i = 1, \ldots, m\}$$

$$= \frac{P\{(n+1)\text{st is } j, \quad x_i \text{ type } i \text{ in first } n, \quad i = 1, \ldots, m\}}{P\{x_i \text{ type } i \text{ in first } n, \quad i = 1, \ldots, m\}}$$

$$= \frac{\dfrac{n!(m-1)!}{x_1! \cdots x_m!} \int\int \cdots \int p_1^{x_1} \cdots p_j^{x_j+1} \cdots p_m^{x_m} dp_1 \cdots dp_m}{\binom{n+m-1}{m-1}^{-1}}$$

where the numerator is obtained by conditioning on the **p** vector and the denominator is obtained by using Equation (6.13). By Equation (6.12) we have that

$$P\{(n + 1)\text{st is } j | x_i \text{ type } i \text{ in first } n, \quad i = 1, \ldots, m\}$$

$$= \frac{\dfrac{(x_j + 1)n!(m - 1)!}{(n + m)!}}{\dfrac{(m - 1)!n!}{(n + m - 1)!}}$$

$$= \frac{x_j + 1}{n + m} \tag{6.14}$$

Using the Equation (6.14), we can now present an urn model description of the stochastic process of successive outcomes. Namely, consider an urn which initially contains one of each of m types of balls. Balls are then randomly drawn and are replaced along with another of the same type. Hence, if in the first n drawings there have been a total of x_j type j balls drawn, then the urn immediately before the $(n + 1)$st draw will contain $x_j + 1$ type j balls out of a total of $m + n$, and so the probability of a type j on the $(n + 1)$st draw will be as given by Equation (6.14).

Remarks Consider a situation where n particles are to be distributed at random among m possible regions; and suppose that each of the regions appear, at least before the experiment, to have the same physical characteristics. It would thus seem that the most likely distribution for the numbers of particles that fall into each of the regions is the multinomial distribution with $p_i \equiv 1/m$. (This, of course, would correspond to each particle, independent of the others, being equally likely to fall in any of the m regions.) Physicists studying how particles distribute themselves observed the behavior of such particles as photons and atoms containing an even number of elementary particles. However, when they studied the resulting data, they were amazed to discover that the observed frequencies did not follow the multinomial distribution but rather seemed to follow the Bose-Einstein distribution. They were amazed because they could not imagine a physical model for the distribution of particles which would result in all possible outcomes being equally likely. (For instance, if 10 particles are to distribute themselves between 2 regions, it hardly seems reasonable that it is just as likely that both regions will contain 5 particles as it is that all 10 will fall in region 1 or that all 10 will fall in region 2.)

However, from the results of this section we now have a better understanding of the cause of the physicists dilemma. In fact, two possible hypotheses present themselves. First, it may be that the data gathered by the physicists were actually obtained under a variety of different situations, each having its own characteristic **p** vector which gave rise to a uniform spread over all possible **p** vectors. A second possibility (suggested by the urn model interpretation) is that the particles select their regions sequentially and a given particle's probability of falling in a region is roughly proportional to the fraction of the landed particles that are in that region. (In other words, the particles presently in a region provide an "attractive" force on elements which have not yet landed.)   ◊

6.4 In Normal Sampling $\overline{X}$ and S^2 are Independent

Let $X_1, \ldots, X_n$ be independent normal random variables each having mean μ and variance σ^2. Their joint density is

$$f_{X_1, \ldots, X_n}(x_1, \ldots, x_n) = \frac{1}{(2\pi)^{n/2}\sigma^n} \exp\left\{ -\sum_{i=1}^{n} (x_i - \mu)^2/2\sigma^2 \right\} \quad (6.15)$$

Let

$$\overline{X} = \sum_{i=1}^{n} X_i/n$$

$$S^2 = \sum_{i=1}^{n} (X_i - \overline{X})^2 \quad (6.16)$$

One of the most important results in normal sampling theory is that $\overline{X}$, the sample mean, and $S^2/(n-1)$, the sample variance, are independent random variables with S^2/σ^2 having a chi-square distribution with $n-1$ degrees of freedom.

Before proving the above, note the following algebraic identity: If for number $x_1, \ldots, x_n$ we define

$$\bar{x} = \sum_{i=1}^{n} x_i/n, \qquad s^2 = \sum_{i=1}^{n} (x_i - \bar{x})^2$$

then

$$\sum_{i=1}^{n} (x_i - \mu)^2 = \sum_{i=1}^{n} (x_i - \bar{x} + \bar{x} - \mu)^2$$

$$= \sum_{i=1}^{n} (x_i - \bar{x})^2 + n(\bar{x} - \mu)^2$$

or, since $x_1 = n\bar{x} - x_2 - \ldots - x_n$,

$$(n\bar{x} - x_2 - \ldots - x_n - \mu)^2 + \sum_{i=2}^{n} (x_i - \mu)^2 = s^2 + n(\bar{x} - \mu)^2 \quad (6.17)$$

Proposition 6.2. If $X_1, \ldots, X_n$ are independent normal random variables each having mean μ and variance σ^2, then $\bar{X}$ and S^2, as given by (6.16), are independent random variables with $\bar{X}$ being normal with mean μ and variance σ^2/n and S^2/σ^2 having a chi-square distribution with $n - 1$ degrees of freedom.

Proof: The joint density function of $X_1, \ldots, X_n$ is given by (6.15). Make the change of variables

$$\bar{X} = \frac{X_1 + \cdots + X_n}{n}$$

$$X_2 = X_2$$

$$\vdots$$

$$X_n = X_n$$

As the above transformation has Jacobian $1/n$ and as $X_1 = n\bar{X} - X_2 - \cdots - X_n$ it follows that

$$f_{\bar{X}, X_2, \ldots, X_n}(\bar{x}, x_2, \ldots, x_n)$$

$$= n f_{X_1, \ldots, X_n}(n\bar{x} - x_2 - \cdots - x_n, x_2, \ldots, x_n)$$

$$= \frac{n}{(2\pi)^{n/2}\sigma^n} \exp\left\{-\frac{1}{2\sigma^2}\left[(n\bar{x} - x_2 - \cdots - x_n - \mu)^2 + \sum_{i=2}^{n} (x_i - \mu)^2\right]\right\}$$

$$= \frac{n}{(2\pi)^{n/2}\sigma^n} \exp\left\{-\frac{1}{2\sigma^2}[s^2 + n(\bar{x} - \mu)^2]\right\} \quad \text{from (6.17)}$$

where

$$s^2 = \sum_{i=1}^{n} (x_i - \bar{x})^2, \qquad x_1 = n\bar{x} - x_2 - \cdots - x_n \quad (6.18)$$

As $\bar{X}$ is normally distributed with mean μ and variance σ^2/n, we thus obtain that the conditional joint density function of $X_2, \ldots, X_n$ given $\bar{X}$ is as follows:

$$f_{X_2, \ldots, X_n|\bar{X}}(x_2, \ldots, x_n|\bar{x}) = f_{\bar{X}, X_2, \ldots, X_n}(\bar{x}, x_2, \ldots, x_n)/f_{\bar{X}}(\bar{x})$$

$$= \frac{\sqrt{n}}{(\sqrt{2\pi}\,\sigma)^{n-1}} \exp\{-s^2/2\sigma^2\} \quad (6.19)$$

As the above is a joint density function, it must integrate to 1. That is, for all σ^2 and $\bar{x}$

$$\int_{-\infty}^{\infty}\int_{-\infty}^{\infty}\cdots\int_{-\infty}^{\infty} e^{-s^2/2\sigma^2}dx_2 \cdots dx_n = \frac{(\sqrt{2\pi}\,\sigma)^{n-1}}{\sqrt{n}} \qquad (6.20)$$

where s^2 is given by (6.18). From (6.19) we obtain that the conditional moment generating function of S^2/σ^2 given that $\overline{X} = \bar{x}$ is

$$E[e^{tS^2/\sigma^2}|\overline{X} = \bar{x}]$$

$$= \frac{\sqrt{n}}{(\sqrt{2\pi}\,\sigma)^{n-1}}\int\int\cdots\int e^{ts^2/\sigma^2}e^{-s^2/2\sigma^2}dx_2 \cdots dx_n$$

$$= \frac{\sqrt{n}}{(\sqrt{2\pi}\,\sigma)^{n-1}}\int\int\cdots\int e^{-s^2/2\bar{\sigma}^2}dx_2 \cdots dx_n \quad \text{with} \quad \bar{\sigma}^2 = \frac{\sigma^2}{1-2t}$$

$$= \frac{\sqrt{n}}{(\sqrt{2\pi}\,\sigma)^{n-1}}\left(\frac{\sqrt{2\pi}\,\sigma}{\sqrt{1-2t}}\right)^{n-1}\frac{1}{\sqrt{n}} \quad \text{from (6.20)}$$

$$= (1-2t)^{-(n-1)/2}, \qquad t < \tfrac{1}{2} \qquad (6.21)$$

As the conditional moment generating function of S^2 given that $\overline{X} = \bar{x}$ does not depend on $\bar{x}$, we can thus conclude that $\overline{X}$ and S^2 are independent. To show that S^2/σ^2 has a chi-square distribution, let Z denote a unit normal random variable. Then

$$E[e^{tZ^2}] = \frac{1}{\sqrt{2\pi}}\int e^{tx^2}e^{-x^2/2}dx$$

$$= \frac{1}{\sqrt{2\pi}}\int e^{-x^2/2\bar{\sigma}^2}dx \quad \text{where} \quad \bar{\sigma}^2 = (1-2t)^{-1}$$

$$= (1-2t)^{-1/2}$$

Hence, as a chi-square random variable with k degrees of freedom is defined as the sum of the squares of k independent unit normals, it follows that such a random variable would have a moment generating function equal to $(1-2t)^{-k/2}$. Hence, from Equation (6.21), and the uniqueness of moment generating functions, we can conclude that S^2 has a chi-square distribution with $n-1$ degrees of freedom.* ◊

---

*This proof is taken from Shuster, "A Simple Method of Teaching the Independence of $\overline{X}$ and S^2," *The American Statistician*, Feb. 1973.

If Z is a unit normal random variable and χ^2, independent of Z, is a chi-square random variable with n degrees of freedom, then the random variable

$$T_n \equiv \frac{Z}{\sqrt{\chi^2/n}}$$

is said to have a t distribution with n degrees of freedom. The distribution function of T_n has been tabulated and for each $\alpha \in (0, 1)$ the number $t_{n,\alpha}$ such that

$$P\{T_n \geq t_{n,\alpha}\} = \alpha$$

has been determined.

As an immediate corollary of Proposition 6.2 we have

Corollary 6.3. Suppose $X_1, \ldots, X_n$ are independent normal random variables each having mean μ and variance σ^2 and let $\overline{X}$ and S_v^2 denote, respectively, the sample mean and sample variance—that is,

$$\overline{X} = \frac{\sum\limits_{i=1}^{n} X_i}{n}, \qquad S_v^2 = \frac{\sum\limits_{i=1}^{n} (X_i - \overline{X})^2}{n - 1}$$

Then

$$T_{n-1} \equiv \frac{\overline{X} - \mu}{\sigma/\sqrt{n}} \left(\frac{S_v^2}{\sigma^2}\right)^{-1/2} = \frac{\sqrt{n}(\overline{X} - \mu)}{S_v}$$

has a t distribution with $n - 1$ degrees of freedom.

The above corollary is very important in statistical applications. For instance, suppose $X_1, \ldots, X_n$ are independent normal variables having unknown mean μ and known variance σ^2, and suppose we want to use the observed values of the X_i to estimate μ. Since $\sqrt{n}[(\overline{X} - \mu)/\sigma]$ will have a unit normal distribution, it follows that

$$P\left\{-1.96 < \sqrt{n}\,\frac{(\overline{X} - \mu)}{\sigma} < 1.96\right\} = .95$$

or, equivalently

$$P\left\{\overline{X} - 1.96\,\frac{\sigma}{\sqrt{n}} < \mu < \overline{X} + 1.96\,\frac{\sigma}{\sqrt{n}}\right\} = .95$$

That is, 95 percent of the time μ will lie within $1.96 \ (\sigma/\sqrt{n})$ units of the sample average. If we now observe the sample and it turns out that $\overline{X} = \bar{x}$, then we say that "with 95 percent confidence"

$$\bar{x} - 1.96 \frac{\sigma}{\sqrt{n}} < \mu < \bar{x} + 1.96 \frac{\sigma}{\sqrt{n}}$$

That is, "with 95 percent confidence" we assert that the true mean lies within $1.96 \ \sigma/\sqrt{n}$ of the observed sample mean. The interval $(\bar{x} - 1.96 \ \sigma/\sqrt{n}, \bar{x} + 1.96 \ \sigma/\sqrt{n})$ is called a confidence interval for μ.

Let us now suppose that the population variance σ^2 is not known. By letting $S_v^2 = \Sigma_{i=1}^{n} (X_i - \overline{X})^2/(n-1)$ denote the sample variance, then from Corollary 6.3 $\sqrt{n} \ (\overline{X} - \mu)/S_v$ has a t distribution with $n - 1$ degrees of freedom.

Hence, for any $\alpha \ \epsilon \ (0, \frac{1}{2})$

$$P\left\{t_{n-1, \ 1-\alpha/2} \leq \sqrt{n} \ \frac{(\overline{X} - \mu)}{S_v} \leq t_{n-1, \ \alpha/2}\right\} = 1 - \alpha$$

or, using that $t_{n-1, \ 1-\alpha/2} = -t_{n-1, \ \alpha/2}$

$$P\left\{\overline{X} - t_{n-1, \ \alpha/2} \ \frac{S_v}{\sqrt{n}} < \mu < \overline{X} + t_{n-1, \ \alpha/2} \ \frac{S_v}{\sqrt{n}}\right\} = 1 - \alpha$$

Thus, if it is observed that $\overline{X} = \bar{x}$ and $S_v = s_v$, then we can say that "with $100(1-\alpha)$ percent confidence"

$$\mu \epsilon \left(\bar{x} - t_{n-1, \ \alpha/2} \ \frac{S_v}{\sqrt{n}}, \ \bar{x} + t_{n-1, \ \alpha/2} \ \frac{S_v}{\sqrt{n}}\right)$$

Remarks The interpretation of "a $100(1-\alpha)$ percent confidence interval" can be confusing. It should be noted that we are *not* asserting (in the case of σ^2 known) that the probability that $\mu \ \epsilon \ [\bar{x} - 1.96 \ (\sigma/\sqrt{n}), \bar{x} + 1.96 \ (\sigma/\sqrt{n})]$ is .95, for there are no random variables involved in this assertion and thus nothing is random. What we are asserting is that the technique utilized to obtain this interval is such that 95 percent of the time it is employed it will result in an interval in which μ lies. In other words, before the data are observed, we can assert with probability .95 that the interval which will be obtained will contain μ; whereas after the data are obtained, we can only assert that the resultant interval indeed contains μ "with confidence .95."

Problems

1. If X and Y are both discrete, show that $\sum_x p_{X|Y}(x|y) = 1$ for all y such that $p_Y(y) > 0$.

2. The joint probability mass function of X and Y, $p(x, y)$, is given by

$$p(1, 1) = \tfrac{1}{9}, \qquad p(2, 1) = \tfrac{1}{3}, \qquad p(3, 1) = \tfrac{1}{9}$$
$$p(1, 2) = \tfrac{1}{9}, \qquad p(2, 2) = 0, \qquad p(3, 2) = \tfrac{1}{18}$$
$$p(1, 3) = 0, \qquad p(2, 3) = \tfrac{1}{6}, \qquad p(3, 3) = \tfrac{1}{9}$$

Compute $E[X|Y = i]$ for $i = 1, 2, 3$.

3. In Problem 2, are the random variables X and Y independent?

4. An urn contains three white, six red, and five black balls. Six of these balls are randomly selected from the urn. Let X and Y denote respectively the number of white and black balls selected. Compute the conditional probability mass function of X given that $Y = 3$. Also compute $E[X|Y = 1]$.

5. Repeat Problem 4 but under the assumption that when a ball is selected its color is noted, and it is then replaced in the urn before the next selection is made.

6. Suppose $p(x, y, z)$ the joint probability mass function of the random variables X, Y, and Z, is given by

$$p(1, 1, 1) = \tfrac{1}{8}, \qquad p(2, 1, 1) = \tfrac{1}{4},$$
$$p(1, 1, 2) = \tfrac{1}{8}, \qquad p(2, 1, 2) = \tfrac{3}{16},$$
$$p(1, 2, 1) = \tfrac{1}{16}, \qquad p(2, 2, 1) = 0,$$
$$p(1, 2, 2) = 0, \qquad p(2, 2, 2) = \tfrac{1}{4}$$

What is $E[X|Y = 2]$? What is $E[X|Y = 2, Z = 1]$?

7. An unbiased die is successively rolled. Let X and Y denote respectively the number of rolls necessary to obtain a six and a five. Find (a) $E[X]$, (b) $E[X|Y = 1]$, (c) $E[X|Y = 5]$.

8. Show in the discrete case that if X and Y are independent, then

$$E[X|Y = y] = E[X] \quad \text{for all} \quad y$$

9. Suppose X and Y are independent continuous random variables. Show that

$$E[X|Y = y] = E[X] \quad \text{for all} \quad y$$

10. The joint density of X and Y is

$$f(x, y) = \frac{(y^2 - x^2)}{8} e^{-y}, \qquad 0 < y < \infty, \qquad -y \le x \le y$$

Show that $E[X|Y = y] = 0$.

11. The joint density of X and Y is given by

$$f(x, y) = \frac{e^{-x/y} e^{-y}}{y}, \qquad 0 < x < \infty, \qquad 0 < y < \infty$$

Show $E[X|Y = y] = y$

12. Let X be exponential with mean $1/\lambda$; that is,

$$f_X(x) = \lambda e^{-\lambda x}, \qquad 0 < x < \infty$$

Find $E[X|X > 1]$.

13. Let X be uniform over $(0, 1)$. Find $E[X|X < \frac{1}{2}]$.

14. The joint density of X and Y is given by

$$f(x, y) = \frac{e^{-y}}{y}, \qquad 0 < x < y, \qquad 0 < y < \infty$$

Compute $E[X^2|Y = y]$.

15. The random variables X and Y are said to have a bivariate normal distribution if their joint density function is given by

$$f(x, y) = \frac{1}{2\pi\sigma_x\sigma_y\sqrt{1 - \rho^2}} \exp\left\{ -\frac{1}{2(1 - \rho^2)} \right.$$
$$\left. \times \left[\left(\frac{x - \mu_x}{\sigma_x}\right)^2 - \frac{2\rho(x - \mu_x)(y - \mu_y)}{\sigma_x\sigma_y} + \left(\frac{y - \mu_y}{\sigma_y}\right)^2 \right] \right\}$$

for $-\infty < x < \infty$, $-\infty < y < \infty$, where σ_x, σ_y, μ_x, μ_y, and ρ are constants such that $-1 < \rho < 1$, $\sigma_x > 0$, $\sigma_y > 0$, $-\infty < \mu_x < \infty$, $-\infty < \mu_y < \infty$.

(a) Show that X is normally distributed with mean μ_x and variance σ_x^2, and Y is normally distributed with mean μ_y and variance σ_y^2.
(b) Show that the conditional density of X given that $Y = y$ is normal with mean $\mu_x + (\rho\sigma_x/\sigma_y)(y - \mu_y)$ and variance $\sigma_x^2(1 - \rho^2)$.

The quantity ρ is called the correlation between X and Y. It can be shown that

$$\rho = \frac{E[(X - \mu_x)(Y - \mu_y)]}{\sigma_x \sigma_y}$$

$$= \frac{\text{Cov}(X, Y)}{\sigma_x \sigma_y}$$

16. Prove that if X and Y are jointly continuous, then

$$E[X] = \int_{-\infty}^{\infty} E[X|Y = y] f_Y(y) dy$$

17. A prisoner is trapped in a cell containing three doors. The first door leads to a tunnel which returns him to his cell after two-day's travel. The second leads to a tunnel which returns him to his cell after three-day's travel. The third door leads immediately to freedom.

(a) Assuming that the prisoner will always select doors, 1, 2, and 3 with probabilities .5, .3, .2, what is the expected number of days until he reaches freedom?

(b) Assuming that the prisoner is always equally likely to choose among those doors that he has not used, what is the expected number of days until he reaches freedom? (In this version, for instance, if the prisoner initially tries door 1, then when he returns to the cell, he will now select only from doors 2 and 3.)

(c) For parts (a) and (b) find the variance of the number of days until our prisoner reaches freedom.

18. A rat is trapped in a maze. Initially he has to choose one of two directions. If he goes to the right, then he will wander around in the maze for three minutes and will then return to his initial position. If he goes to the left, then with probability $\frac{1}{3}$ he will depart the maze after two minutes of traveling, and with probability $\frac{2}{3}$ he will return to his initial position after five minutes of traveling. Assuming that the rat is at all times equally likely to go to the left or the right, what is the expected number of minutes that he will be trapped in the maze?

19. Find the variance of the amount of time the rat spends in the maze in Problem 18.

20. The number of claims received at an insurance company during a week is a random variable with mean μ_1 and variance σ_1^2 The amount paid in each claim is a random variable with mean μ_2 and variance σ_2^2. Find the mean and variance of the amount of money paid by

the insurance company each week. What independence assumptions are you making? Are these assumptions reasonable?

21. The number of customers entering a store on a given day is Poisson distributed with mean $\lambda = 10$. The amount of money spent by a customer is uniformly distributed over $(0, 100)$. Find the mean and variance of the amount of money that the store takes in on a given day.

22. The conditional variance of X, given the random variable Y, is defined by

$$\text{Var}(X|Y) = E[[X - E(X|Y)]^2|Y]$$

Show that

$$\text{Var}(X) = E[\text{Var}(X|Y)] + \text{Var}(E[X|Y])$$

23. Use Problem 22 to give another proof of the fact that

$$\text{Var}\left(\sum_{i=1}^{N} X_i\right) = E[N]\,\text{Var}(X) + (E[X])^2\,\text{Var}(N)$$

24. Give another proof of Problem 23 by computing the moment generating function of $\sum_{i=1}^{N} X_i$ and then differentiating to obtain its moments.

Hint: Let

$$\phi(t) = E\left[\exp\left(t\sum_{i=1}^{N} X_i\right)\right]$$

$$= E\left[E\left[\exp\left(t\sum_{i=1}^{N} X_i\right)\bigg| N\right]\right]$$

Now,

$$E\left[\exp\left(t\sum_{i=1}^{N} X_i\right)\bigg| N = n\right] = E\left[\exp\left(t\sum_{i=1}^{n} X_i\right)\right] = (\phi_X(t))^n$$

since N is independent of the X's where $\phi_X(t) = E[e^{tX}]$ is the moment generating function for the X's. Therefore,

$$\phi(t) = E[(\phi_X(t))^N]$$

Differentiation yields

$$\phi'(t) = E[N(\phi_X(t))^{N-1}\phi'_X(t)]$$
$$\phi''(t) = E[N(N-1)(\phi_X(t))^{N-2}(\phi'_X(t))^2 + N(\phi_X(t))^{N-1}\phi''_X(t)]$$

Evaluate at $t = 0$ to get the desired result.

25. The number of fish that Steve catches in a day is a Poisson random variable with mean 30. However, on the average, Steve tosses back two out of every three fish he catches. What is the probability that, on a given day, Steve takes home *n* fish. What is the mean and variance of (a) the number of fish he catches, (b) the number of fish he takes home? (What independence assumptions have you made?)

26. There are three coins in a barrel. These coins, when flipped, will come up heads with respective probabilities .3, .5, .7. A coin is randomly selected from among these three and is then flipped ten times. Let N be the number of heads obtained on the ten flips. Find
  (a) $P\{N = 0\}$.
  (b) $P\{N = n\}$, $n = 0, 1, \ldots, 10$.
  (c) Does N have a binomial distribution?
  (d) If you win $1 each time a head appears and you lose $1 each time a tail appears, is this a fair game? Explain.

27. Do Problem 26 under the assumption that each time a coin is flipped, it is then put back in the barrel and another coin is randomly selected. Does N have a binomial distribution now?

28. Explain the relationship between the general formula

$$P(E) = \sum_y P(E|Y = y)P(Y = y)$$

and Bayes' formula.

29. Suppose X is a Poisson random variable with mean λ. The parameter λ is itself a random variable whose distribution is exponential with mean 1. Show that $P\{X = n\} = (\frac{1}{2})^{n+1}$.

30. A coin is randomly selected from a group of ten coins, the *n*th coin having a probability $n/10$ of coming up heads. The coin is then repeatedly flipped until a head appears. Let N denote the number of flips necessary. What is the probability distribution of N? Is N a geometric random variable? When would N be a geometric random variable; that is, what would have to be done differently?

31. Show that

  (a) $E[XY|Y = y] = yE[X|Y = y]$
  (b) $E[g(X, Y)|Y = y] = E[g(X, y)|Y = y]$
  (c) $E[XY] = E[YE[X|Y]]$

32. In the ballot problem (Example 5e of this chapter), compute $P\{A$ is never behind$\}$.

33. An urn contains n white and m black balls which are removed one at a time. If $n > m$, show that the probability that there are always more white than black balls in the urn (until, of course, the urn is empty) equals $(n - m)/(n + m)$. Explain why this probability is equal to the probability that the set of withdrawn balls always contains more white than black balls. (This latter probability is $(n - m)/(n + m)$ by the ballot problem.)

34. Let U_i, $i \geq 1$, denote independent uniform $(0,1)$ random variables. For $0 < a \leq 1$, define N by

$$N = \min\{n : U_1 + \cdots + U_n > a\}$$

(a) Show by induction that

$$P\{N > n\} = \frac{a^n}{n!}$$

(b) Prove that $E[N] = e^a$, and conclude that the expected number of random numbers needed until their sum exceeds 1 is equal to e.

35. Let X_i, $i \geq 1$, be independent uniform $(0,1)$ random variables, and define N by

$$N = \min\{n : X_n < X_{n-1}\}$$

where $X_0 = x$. Let $f(x) = E[N]$.
(a) Derive an integral equation for $f(x)$ by conditioning on X_1.
(b) Differentiate both sides of the equation derived in (a).
(c) Solve the resulting equation obtained in (b).
(d) For a second approach to determining $f(x)$ argue that

$$P\{N \geq k\} = \frac{(1 - x)^{k-1}}{(k - 1)!}$$

(e) Use (d) to obtain $f(x)$.

36. In the list example of Section 6 suppose that the initial ordering at time $t = 0$ is determined completely at random; that is, initially all $n!$ permutations are equally likely. Following the front of the line rule, compute the expected position of the element requested at time t.

Hint: To compute $P\{e_j$ precedes e_i at time $t\}$ condition on whether or not either e_i or e_j have ever been requested prior to t.

37. In the list problem, when the P_i are known, show that the best ordering (best in the sense of minimizing the expected position of the

element requested) is to place the elements in decreasing order of their probabilities. That is, if $P_1 > P_2 > \cdot \cdot \cdot > P_n$, show that $1, 2, \ldots, n$ is the best ordering.

38. Consider the random graph of Section 6 when $n = 5$. Compute the probability distribution of the number of components and verify your solution by using it to compute $E[C]$ and then comparing your solution with

$$E[C] = \sum_{k=1}^{5} \binom{5}{k} \frac{(k-1)!}{5^k}$$

39. (i) From the results of Section 6.3 we can conclude that there are $\binom{n+m-1}{m-1}$ nonnegative integer valued solutions of the equation $x_1 + \ldots + x_m = n$. Prove this directly.
    (ii) How many positive integer valued solutions of $x_1 + \ldots + x_m = n$ are there? *Hint:* Let $y_i = x_i - 1$.
    (iii) For the Bose-Einstein distribution, compute the probability that exactly k of the X_i are equal to 0.

Chapter 4

Markov Chains

1. Introduction

In this chapter, we consider a stochastic process $\{X_n, n = 0, 1, 2, \ldots\}$ that takes on a finite or countable number of possible values. Unless otherwise mentioned, this set of possible values of the process will be denoted by the set of nonnegative integers $\{0, 1, 2, \ldots\}$. If $X_n = i$, then the process is said to be in state i at time n. We suppose that whenever the process is in state i, there is a fixed probability P_{ij} that it will next be in state j. That is, we suppose that

$$P\{X_{n+1} = j | X_n = i, \quad X_{n-1} = i_{n-1}, \ldots, X_1 = i_1, \quad X_0 = i_0\} = P_{ij} \quad (1.1)$$

for all states $i_0, i_1, \ldots, i_{n-1}, i, j$ and all $n \geq 0$. Such a stochastic process is known as a *Markov chain*. Equation (1.1) may be interpreted as stating that, for a Markov chain, the conditional distribution of any future state X_{n+1} given the past states $X_0, X_1, \ldots, X_{n-1}$ and the present state X_n, is independent of the past states and depends only on the present state.

The value P_{ij} represents the probability that the process will, when in state i, next make a transition into state j. Since probabilities are nonnegative and since the process must make a transition into some state, we have that

$$P_{ij} \geq 0, \quad i, j \geq 0; \quad \sum_{j=0}^{\infty} P_{ij} = 1, \quad i = 0, 1, \ldots.$$

Let **P** denote the matrix of one-step transition probabilities P_{ij}, so that

$$\mathbf{P} = \left\| \begin{array}{cccc} P_{00} & P_{01} & P_{02} & \cdots \\ P_{10} & P_{11} & P_{12} & \cdots \\ \vdots & & & \\ P_{i0} & P_{i1} & P_{i2} & \cdots \\ \vdots & \vdots & & \vdots \end{array} \right\|$$

Example 1a (Forecasting the Weather): Suppose that the chance of rain tomorrow depends on previous weather conditions only through whether or not it is raining today and not on past weather conditions. Suppose also that if it rains today, then it will rain tomorrow with probability α; and if it does not rain today, then it will rain tomorrow with probability β.

If we say that the process is in state 0 when it rains and state 1 when it does not rain, then the above is a two-state Markov chain whose transition probabilities are given by

$$\mathbf{P} = \left\| \begin{array}{cc} \alpha & 1 - \alpha \\ \beta & 1 - \beta \end{array} \right\| \qquad \diamondsuit$$

Example 1b (A Communications System): Consider a communications system which transmits the digits 0 and 1. Each digit transmitted must pass through several stages, at each of which there is a probability p that the digit entered will be unchanged when it leaves. Letting X_n denote the digit entering the nth stage, then $\{X_n, n = 0, 1, \ldots\}$ is a two-state Markov chain having a transition probability matrix

$$\mathbf{P} = \left\| \begin{array}{cc} p & 1 - p \\ 1 - p & p \end{array} \right\| \qquad \diamondsuit$$

Example 1c On any given day Gary is either cheerful (C), so-so (S), or glum (G). If he is cheerful today, then he will be C, S, or G tomorrow with respective probabilities .5, .4, .1. If he is feeling so-so today, then he will be C, S, or G tomorrow with probabilities .3, .4, .3. If he is glum today, then he will be C, S, or G tomorrow with probabilities .2, .3, .5.

Letting X_n denote Gary's mood on the nth day, then $\{X_n, n \geq 0\}$ is a three-state Markov chain (state 0 = C, state 1 = S, state 2 = G) with transition probability matrix

$$\mathbf{P} = \left\| \begin{array}{ccc} .5 & .4 & .1 \\ .3 & .4 & .3 \\ .2 & .3 & .5 \end{array} \right\| \quad \Diamond$$

Example 1d (Transforming a Process into a Markov Chain): Suppose that whether or not it rains today depends on previous weather conditions through the last two days. That is, suppose that if it has rained for the past two days, then it will rain tomorrow with probability .7; if it rained today but not yesterday, then it will rain tomorrow with probability .5; if it rained yesterday but not today, then it will rain tomorrow with probability .4; if it has not rained in the past two days, then it will rain tomorrow with probability .2.

If we let the state at time n depend only on whether or not it is raining at time n, then the above model is not a Markov chain (why not?). However, we can transform the above model into a Markov chain by saying that the state at any time is determined by the weather conditions during both that day and the previous day. In other words, we can say that the process is in

state 0   if it rained both today and yesterday,
state 1   if it rained today but not yesterday,
state 2   if it rained yesterday but not today,
state 3   if it did not rain either yesterday or today.

The above would then represent a four-state Markov chain having a transition probability matrix

$$\mathbf{P} = \left\| \begin{array}{cccc} .7 & 0 & .3 & 0 \\ .5 & 0 & .5 & 0 \\ 0 & .4 & 0 & .6 \\ 0 & .2 & 0 & .8 \end{array} \right\|$$

The reader should carefully check the matrix $\mathbf{P}$, and make sure he or she understands how it was obtained. $\Diamond$

Example 1e (A Random Walk Model): A Markov chain whose state space is given by the integers $i = 0, \pm 1, \pm 2, \ldots$ is said to be a random walk if, for some number $0 < p < 1$,

$$P_{i,\,i+1} = p = 1 - P_{i,\,i-1}, \quad i = 0, +1, \ldots.$$

The above Markov chain is called a *random walk* for we may think of it as being a model for an individual walking on a straight line who at each point of time either takes one step to the right with probability p or one step to the left with probability $1 - p$. ◇

Example 1f (A Gambling Model): Consider a gambler who, at each play of the game, either wins \$1 with probability p or loses \$1 with probability $1 - p$. If we suppose that our gambler quits playing either when he goes broke or he attains a fortune of \$N, then the gambler's fortune is a Markov chain having transition probabilities

$$P_{i, i+1} = p = 1 - P_{i, i-1}, \quad i = 1, 2, \ldots, N - 1$$
$$P_{00} = P_{NN} = 1$$

States 0 and N are called *absorbing* states since once entered they are never left. Note that the above is a finite state random walk with absorbing barriers (states 0 and N). ◇

2. Chapman-Kolmogorov Equations

We have already defined the one-step transition probabilities P_{ij}. We now define the n-step transition probabilities P_{ij}^n to be the probability that a process in state i will be in state j after n additional transitions. That is,

$$P_{ij}^n = P\{X_{n+m} = j | X_m = i\}, \quad n \geq 0, \quad i, j \geq 0$$

Of course $P_{ij}^1 = P_{ij}$. The *Chapman-Kolmogorov equations* provide a method for computing these n-step transition probabilities. These equations are

$$P_{ij}^{n+m} = \sum_{k=0}^{\infty} P_{ik}^n P_{kj}^m \quad \text{for all} \quad n, m \geq 0, \quad \text{all } i, j \tag{2.1}$$

and are most easily understood by noting that $P_{ik}^n P_{kj}^m$ represents the probability that starting in i the process will go to state j in $n + m$ transitions through a path which takes it into state k at the nth transition. Hence, summing over all intermediate states k yields the probability that the process will be in state j after $n + m$ transitions. Formally, we have

$$P_{ij}^{n+m} = P\{X_{n+m} = j | X_0 = i\}$$
$$= \sum_{k=0}^{\infty} P\{X_{n+m} = j, X_n = k | X_0 = i\}$$

$$= \sum_{k=0}^{\infty} P\{X_{n+m} = j | X_n = k, X_0 = i\} P\{X_n = k | X_0 = i\}$$

$$= \sum_{k=0}^{\infty} P_{kj}^m P_{ik}^n$$

If we let $\mathbf{P}^{(n)}$ denote the matrix of n-step transition probabilities P_{ij}^n, then Equation (2.1) asserts that

$$\mathbf{P}^{(n+m)} = \mathbf{P}^{(n)} \cdot \mathbf{P}^{(m)}$$

where the dot represents matrix multiplication.* Hence, in particular,

$$\mathbf{P}^{(2)} = \mathbf{P}^{(1+1)} = \mathbf{P} \cdot \mathbf{P} = \mathbf{P}^2$$

and by induction

$$\mathbf{P}^{(n)} = \mathbf{P}^{(n-1+1)} = \mathbf{P}^{n-1} \cdot \mathbf{P} = \mathbf{P}^n$$

That is, the n-step transition matrix may be obtained by multiplying the matrix $\mathbf{P}$ by itself n times.

Example 2a   Consider Example 1a in which the weather is considered as a two-state Markov chain. If $\alpha = .7$ and $\beta = .4$, then calculate the probability that it will rain four days from today given that it is raining today.

Solution: The one-step transition probability matrix is given by

$$\mathbf{P} = \left\| \begin{matrix} .7 & .3 \\ .4 & .6 \end{matrix} \right\|$$

Hence,

$$\mathbf{P}^{(2)} = \mathbf{P}^2 = \left\| \begin{matrix} .7 & .3 \\ .4 & .6 \end{matrix} \right\| \cdot \left\| \begin{matrix} .7 & .3 \\ .4 & .6 \end{matrix} \right\|$$

$$= \left\| \begin{matrix} .61 & .39 \\ .52 & .48 \end{matrix} \right\|$$

---

*If $\mathbf{A}$ is an $N \times M$ matrix whose element in the ith row and jth column is a_{ij} and $\mathbf{B}$ is a $M \times K$ matrix whose element in the ith row and jth column is b_{ij}, then $\mathbf{A} \cdot \mathbf{B}$ is defined to be the $N \times K$ matrix whose element in the ith row and jth column is $\Sigma_{k=1}^{M} a_{ik} b_{kj}$.

$$\mathbf{P}^{(4)} = (\mathbf{P}^2)^2 = \left\|\begin{matrix}.61 & .39 \\ .52 & .48\end{matrix}\right\| \cdot \left\|\begin{matrix}.61 & .39 \\ .52 & .48\end{matrix}\right\|$$

$$= \left\|\begin{matrix}.5749 & .4251 \\ .5668 & .4332\end{matrix}\right\|$$

and the desired probability P_{00}^4 equals .5749. $\diamond$

Example 2b   Consider Example 1d. Given that it rained on Monday and Tuesday, what is the probability that it will rain on Thursday?

Solution: The two-step transition matrix is given by

$$\mathbf{P}^{(2)} = \mathbf{P}^2 = \left\|\begin{matrix}.7 & 0 & .3 & 0 \\ .5 & 0 & .5 & 0 \\ 0 & .4 & 0 & .6 \\ 0 & .2 & 0 & .8\end{matrix}\right\| \cdot \left\|\begin{matrix}.7 & 0 & .3 & 0 \\ .5 & 0 & .5 & 0 \\ 0 & .4 & 0 & .6 \\ 0 & .2 & 0 & .8\end{matrix}\right\|$$

$$= \left\|\begin{matrix}.49 & .12 & .21 & .18 \\ .35 & .20 & .15 & .30 \\ .20 & .12 & .20 & .48 \\ .10 & .16 & .10 & .64\end{matrix}\right\|$$

Since rain on Thursday is equivalent to the process being in either state 0 or state 1 on Thursday, the desired probability is given by $P_{00}^2 + P_{01}^2 = .49 + .12 = .61.$ $\diamond$

So far, all of the probabilities we have considered are conditional probabilities. For instance, P_{ij}^n is the probability that the state at time n is j *given* that the initial state at time 0 is i. If the unconditional distribution of the state at time n is desired, it is necessary to specify the probability distribution of the initial state. Let us denote this by

$$\alpha_i \equiv P\{X_0 = i\}, \quad i \geq 0 \quad \left(\sum_{i=0}^{\infty} \alpha_i = 1\right)$$

All unconditional probabilities may be computed by conditioning on the initial state. That is,

$$P\{X_n = j\} = \sum_{i=0}^{\infty} P\{X_n = j \mid X_0 = i\}P\{X_0 = i\}$$

$$= \sum_{i=0}^{\infty} P_{ij}^n \alpha_i$$

For instance, if $\alpha_0 = .4$, $\alpha_1 = .6$, in Example 2a, then the (unconditional) probability that it will rain four days after we begin keeping weather records is

$$P\{X_4 = 0\} = .4P_{00}^4 + .6P_{10}^4 = (.4)(.5749) + (.6)(.5668)$$
$$= .5700$$

3. Classification of States

State j is said to be *accessible* from state i if $P_{ij}^n > 0$ for some $n \geq 0$. Note that this implies that state j is accessible from state i if and only if, starting in i, it is possible that the process will ever enter state j. This is true since if j is not accessible from i, then

$$P\{\text{ever enter } j \,|\, \text{start in } i\} = P\left\{\bigcup_{n=0}^{\infty} \{X_n = j\} \,|\, X_0 = i\right\}$$
$$\leq \sum_{n=0}^{\infty} P\{X_n = j \,|\, X_0 = i\}$$
$$= \sum_{n=0}^{\infty} P_{ij}^n$$
$$= 0$$

Two states i and j that are accessible to each other are said to *communicate*, and we write $i \leftrightarrow j$.

Note that any state communicates with itself since, by definition,

$$P_{ii}^0 = P\{X_0 = i \,|\, X_0 = i\} = 1$$

The relation of communication satisfies the following three properties:

(i) State i communicates with state i, all $i \geq 0$.
(ii) If state i communicates with state j, then state j communicates with state i.
(iii) If state i communicates with state j, and state j communicates with state k, then state i communicates with state k.

Properties (i) and (ii) follow immediately from the definition of communication. To prove (iii) suppose that i communicates with j, and j communicates with k. Thus, there exists integers n and m such that $P_{ij}^n > 0$, $P_{jk}^m > 0$. Now by the Chapman–Kolmogorov equations, we have that

$$P_{ik}^{n+m} = \sum_{r=0}^{\infty} P_{ir}^{n} P_{rk}^{m} \geq P_{ij}^{n} P_{jk}^{m} > 0$$

Hence, state k is accessible from state i. Similarly, we can show that state i is accessible from state k. Hence, states i and k communicate.

Two states that communicate are said to be in the same *class*. It is an easy consequence of (i), (ii), and (iii) that any two classes of states are either identical or disjoint. In other words, the concept of communication divides the state space up into a number of separate classes. The Markov chain is said to be *irreducible* if there is only one class, that is, if all states communicate with each other.

Example 3a   Consider the Markov chain consisting of the three states 0, 1, 2 and having transition probability matrix

$$\mathbf{P} = \begin{Vmatrix} \frac{1}{2} & \frac{1}{2} & 0 \\ \frac{1}{2} & \frac{1}{4} & \frac{1}{4} \\ 0 & \frac{1}{3} & \frac{2}{3} \end{Vmatrix}$$

It is easy to verify that this Markov chain is irreducible. For example, it is possible to go from state 0 to state 2 since

$$0 \xrightarrow{1} 1 \xrightarrow{1} 2$$

That is, one way of getting from state 0 to state 2 is to go from state 0 to state 1 (with probability $\frac{1}{2}$) and then go from state 1 to state 2 (with probability $\frac{1}{4}$).   ◊

Example 3b   Consider a Markov chain consisting of the four states 0, 1, 2, 3 and having a transition probability matrix

$$\mathbf{P} = \begin{Vmatrix} \frac{1}{2} & \frac{1}{2} & 0 & 0 \\ \frac{1}{2} & \frac{1}{2} & 0 & 0 \\ \frac{1}{4} & \frac{1}{4} & \frac{1}{4} & \frac{1}{4} \\ 0 & 0 & 0 & 1 \end{Vmatrix}$$

The classes of this Markov chain are $\{0, 1\}$, $\{2\}$, and $\{3\}$. Note that while state 0 (or 1) is accessible from state 2, the reverse is not true. Since state 3 is an absorbing state, that is, $P_{33} = 1$, no other state is accessible from it.   ◊

For any state i we let f_i denote the probability that, starting in state i, the process will ever reenter state i. State i is said to be *recurrent* if $f_i = 1$ and *transient* if $f_i < 1$.

Suppose that the process starts in state i and i is recurrent. Hence, with probability 1, the process will eventually reenter state i. However, by the definition of a Markov chain, it follows that the process will be starting over again when it reenters state i and, therefore, state i will eventually be visited again. Continual repetition of this argument leads to the conclusion that *if state i is recurrent then, starting in state i, the process will reenter state i again and again and again—in fact, infinitely often.*

On the other hand, suppose that state i is transient. Hence, each time the process enters state i there will be a positive probability, namely $1 - f_i$, that it will never again enter that state. Therefore, starting in state i, the probability that the process will be in state i for exactly n time periods equals $f_i^{n-1}(1 - f_i)$, $n \geq 1$. In other words, *if state i is transient then, starting in state i, the number of time periods that the process will be in state i has a geometric distribution with finite mean $1/(1 - f_i)$.*

From the above two paragraphs, it follows that *state i is recurrent if and only if, starting in state i, the expected number of time periods that the process is in state i is infinite.* But, letting

$$A_n = \begin{cases} 1, & \text{if } X_n = i \\ 0, & \text{if } X_n \neq i \end{cases}$$

we have that $\sum_{n=0}^{\infty} A_n$ represents the number of periods that the process is in state i. Also

$$E\left[\sum_{n=0}^{\infty} A_n | X_0 = i \right] = \sum_{n=0}^{\infty} E[A_n | X_0 = i]$$

$$= \sum_{n=0}^{\infty} P\{X_n = i | X_0 = i\}$$

$$= \sum_{n=0}^{\infty} P_{ii}^n$$

We have thus proven the following.

Proposition 3.1. State i is

$$\text{recurrent if } \quad \sum_{n=1}^{\infty} P_{ii}^n = \infty$$

$$\text{transient if } \sum_{n=1}^{\infty} P_{ii}^n < \infty$$

The argument leading to the above proposition is doubly important for it also shows that a transient state will only be visited a finite number of times (hence the name transient). This leads to the conclusion that in a finite-state Markov chain not all states can be transient. To see this, suppose the states are 0, 1, . . . , M and suppose that they are all transient. Then after a finite amount of time (say after time T_0) state 0 will never be visited, and after a time (say T_1) state 1 will never be visited, and after a time (say T_2) state 2 will never be visited, etc. Thus, after a finite time $T = \max\{T_0, T_1, . . . , T_M\}$ no states will be visited. But as the process must be in some state after time T we arrive at a contradiction, which shows that at least one of the states must be recurrent.

Another use of Proposition 3.1 is that it enables us to show that recurrence is a class property.

Corollary 3.2. If state i is recurrent, and state i communicates with state j, then state j is recurrent.

> *Proof:*   To prove this we first note that, since state i communicates with state j, there exists integers k and m such that $P_{ij}^k > 0$, $P_{ji}^m > 0$. Now, for any integer n
>
> $$P_{jj}^{m+n+k} \geq P_{ji}^m P_{ii}^n P_{ij}^k$$
>
> This follows since the left side of the above is the probability of going from j to j in $m + n + k$ steps, while the right side is the probability of going from j to j in $m + n + k$ steps via a path that goes from j to i in m steps, then from i to i in an additional n steps, then from i to j in an additional k steps.
>
> From the above we obtain, by summing over n, that
>
> $$\sum_{n=1}^{\infty} P_{jj}^{m+n+k} \geq P_{ji}^m P_{ij}^k \sum_{n=1}^{\infty} P_{ii}^n = \infty$$
>
> since $P_{ji}^m P_{ij}^k > 0$, and $\sum_{n=1}^{\infty} P_{ii}^n$ is infinite since state i is recurrent. Thus, by Proposition 3.1 it follows that state j is also recurrent.   ◇

Remarks

(i) Corollary 3.2 also implies that transience is a class property. For if state i is transient and communicates with state j, then state j must also be transient. For if j were recurrent then, by Corollary 3.2, i would also be recurrent and hence could not be transient.

(ii) Corollary 3.2 along with our previous result that not all states in a finite Markov chain can be transient leads to the conclusion that all states of a finite irreducible Markov chain are recurrent. ◇

Example 3c Let the Markov chain consisting of the states 0, 1, 2, 3 have the transition probability matrix

$$\mathbf{P} = \begin{Vmatrix} 0 & 0 & \frac{1}{2} & \frac{1}{2} \\ 1 & 0 & 0 & 0 \\ 0 & 1 & 0 & 0 \\ 0 & 1 & 0 & 0 \end{Vmatrix}$$

Determine which states are transient and which are recurrent.

Solution: It is a simple matter to check that all states communicate and hence, since this is a finite chain, all states must be recurrent. ◇

Example 3d Consider the Markov chain having states 0, 1, 2, 3, 4 and

$$\mathbf{P} = \begin{Vmatrix} \frac{1}{2} & \frac{1}{2} & 0 & 0 & 0 \\ \frac{1}{2} & \frac{1}{2} & 0 & 0 & 0 \\ 0 & 0 & \frac{1}{2} & \frac{1}{2} & 0 \\ 0 & 0 & \frac{1}{2} & \frac{1}{2} & 0 \\ \frac{1}{4} & \frac{1}{4} & 0 & 0 & \frac{1}{2} \end{Vmatrix}$$

Determine the recurrent state.

Solution: This chain consists of the three classes $\{0, 1\}$, $\{2, 3\}$, and $\{4\}$. The first two classes are recurrent and the third transient. ◇

Example 3e (A Random Walk): Consider a Markov chain whose state space consists of the integers $i = 0, \pm 1, \pm 2, \ldots$, and have transition probabilities given by

$$P_{i, i+1} = p = 1 - P_{i, i-1}, \quad i = 0, \pm 1, \pm 2, \ldots$$

where $0 < p < 1$. In other words, on each transition the process either moves one step to the right (with probability p) or one step to the left (with probability $1 - p$). One colorful interpretation of this process is that it represents the wanderings of a drunken man as he walks along a straight line. Another is that it represents the winnings of a gambler who on each play of the game either wins or loses one dollar.

Since all states clearly communicate, it follows from Corollary 3.2 that they are either all transient or all recurrent. So let us consider state 0 and attempt to determine if $\sum_{n=1}^{\infty} P_{00}^n$ is finite or infinite.

Since it is impossible to be even (using the gambling model interpretation) after an odd number of plays we must, of course, have that

$$P_{00}^{2n+1} = 0, \quad n = 1, 2, \ldots$$

On the other hand, we would be even after $2n$ trials if and only if we won n of these and lost n of these. As each play of the game results in a win with probability p and a loss with probability $1 - p$, the desired probability is thus the binomial probability

$$P_{00}^{2n} = \binom{2n}{n} p^n (1 - p)^n = \frac{(2n)!}{n!n!} (p(1 - p))^n, \qquad n = 1, 2, 3, \ldots$$

By using an approximation, due to Stirling, which asserts that

$$n! \sim n^{n+1/2} e^{-n} \sqrt{2\pi}, \tag{3.1}$$

where we say that $a_n \sim b_n$ when $\lim_{n \to \infty} a_n / b_n = 1$, we obtain

$$P_{00}^{2n} \sim \frac{(4p(1 - p))^n}{\sqrt{\pi n}}$$

Now it is easy to verify that if $a_n \sim b_n$, then $\sum_n a_n < \infty$ if and only if $\sum_n b_n < \infty$. Hence, $\sum_{n=1}^{\infty} P_{00}^n$ will converge if and only if

$$\sum_{n=1}^{\infty} \frac{(4p(1 - p))^n}{\sqrt{\pi n}}$$

does. However, $4p(1 - p) \leq 1$ with equality holding if and only if $p = \frac{1}{2}$. Hence, $\sum_{n=1}^{\infty} P_{00}^n = \infty$ if and only if $p = \frac{1}{2}$. Thus, the chain is recurrent when $p = \frac{1}{2}$ and transient if $p \neq \frac{1}{2}$.

When $p = \frac{1}{2}$, the above process is called a *symmetric random walk*. We could also look at symmetric random walks in more than one dimension. For instance, in the two-dimensional symmetric random walk the process would, at each transition, either take one step to the left, right, up, or down, each having probability $\frac{1}{4}$. That is, the state is the pair of integers (i, j) and the transition probabilities are given by

$$P_{(i,j),(i+1,j)} = P_{(i,j),(i-1,j)} = P_{(i,j),(i,j+1)} = P_{(i,j),(i,j-1)} = \tfrac{1}{4}$$

By using the same method as in the one-dimensional case, we now show that this Markov chain is also recurrent.

Since the above chain is irreducible, it follows that all states will be recurrent if state $0 = (0,0)$ is recurrent. So consider P_{00}^{2n}. Now after $2n$ steps, the chain will be back in its original location if for some i, $0 \leq i \leq n$, the $2n$ steps consist of i steps to the left, i to the right, $n - i$ up, and $n - i$ down. Since each step will be either of these four types with probability $\tfrac{1}{4}$, it follows that the desired probability is a multinominal probability. That is,

$$
\begin{aligned}
P_{00}^{2n} &= \sum_{i=0}^{n} \frac{(2n)!}{i!i!(n-i)!(n-i)!} \left(\frac{1}{4}\right)^{2n} \\
&= \sum_{i=0}^{n} \frac{(2n)!}{n!n!} \frac{n!}{(n-i)!i!} \frac{n!}{(n-i)!i!} \left(\frac{1}{4}\right)^{2n} \\
&= \left(\frac{1}{4}\right)^{2n} \binom{2n}{n} \sum_{i=0}^{n} \binom{n}{i}\binom{n}{n-i} \\
&= \left(\frac{1}{4}\right)^{2n} \binom{2n}{n}\binom{2n}{n}
\end{aligned}
\tag{3.2}
$$

where the last equality uses the combinatorial identity

$$\binom{2n}{n} = \sum_{i=0}^{n} \binom{n}{i}\binom{n}{n-i}$$

which follows upon noting that both sides represent the number of subgroups of size n one can select from a set of n white and n black objects. Now,

$$
\begin{aligned}
\binom{2n}{n} &= \frac{(2n)!}{n!n!} \\
&\sim \frac{(2n)^{2n+1/2}e^{-2n}\sqrt{2\pi}}{n^{2n+1}\,e^{-2n}(2\pi)} \quad \text{by Stirling's approximation} \\
&= \frac{4^n}{\sqrt{\pi n}}
\end{aligned}
$$

Hence, from Equation (3.2) we see that

$$P_{00}^{2n} \sim \frac{1}{\pi n}$$

which shows that $\Sigma_n P_{00}^{2n} = \infty$, and thus all states are recurrent.

Interestingly enough, whereas the symmetric random walks in one and two dimensions are both recurrent, all higher dimensional symmetric random walks turn out to be transient. (For instance, the three-dimensional symmetric random walk is at each transition equally likely to move in any of 6 ways—either to the left, right, up, down, in, or out.) ◇

Example 3f (On the Ultimate Instability of the Aloha Protocol): Consider a communications facility in which the numbers of messages arriving during each of the time periods $n = 1, 2, \ldots$ are independent and identically distributed random variables. Let $a_i = P\{i \text{ arrivals}\}$, and suppose that $a_0 + a_1 < 1$. Each arriving message will transmit at the end of the period in which it arrives. If exactly one message is transmitted, then the transmission is successful and the message leaves the system. However, if at any time two or more messages simultaneously transmit, then a collision is deemed to occur and these messages remain in the system. Once a message is involved in a collision it will, independently of all else, transmit at the end of each additional period with probability p—the so-called Aloha protocol (because it was first instituted at the University of Hawaii). We will show below that such a system is asymptotically unstable in the sense that the number of successful transmissions will, with probability 1, be finite.

To begin let X_n denote the number of messages in the facility at the beginning of the nth period, and note that $\{X_n, n \geq 0\}$ is a Markov chain. Now for $k \geq 0$ define the indicator variables I_k by

$$I_k = \begin{cases} 1 & \text{if the first time that the chain departs state } k \text{ it} \\ & \text{directly goes to state } k - 1 \\ 0 & \text{otherwise} \end{cases}$$

and let it be 0 if the system is never in state k, $k \geq 0$. (For instance, if the successive states are $0, 1, 3, 3, 4, \ldots$, then $I_3 = 0$ since when the chain first departs state 3 it goes to state 4; whereas, if they are $0, 3, 3, 2, \ldots$, then $I_3 = 1$ since this time it goes to state 2.) Now,

$$E\left[\sum_{k=0}^{\infty} I_k\right] = \sum_{k=0}^{\infty} E[I_k]$$

$$= \sum_{k=0}^{\infty} P\{I_k = 1\}$$

$$\le \sum_{k=0}^{\infty} P\{I_k = 1 \mid k \text{ is ever visited}\} \tag{3.3}$$

Now, $P\{I_k = 1 \mid k \text{ is ever visited}\}$ is the probability that when state k is departed the next state is $k - 1$. That is, it is the conditional probability that a transition from k is to $k - 1$ given that it is not back into k, and so

$$P\{I_k = 1 \mid k \text{ is ever visited}\} = \frac{P_{k,k-1}}{1 - P_{kk}}.$$

As,

$$P_{k,k-1} = a_0 kp(1 - p)^{k-1}$$
$$P_{k,k} = a_0[1 - kp(1 - p)^{k-1}] + a_1(1 - p)^k$$

which is seen by noting that if there are k messages present on the beginning of a day, then (a) there will be $k - 1$ at the beginning of the next day if there are no new messages that day and exactly one of the k messages transmits; and (b) there will be k at the beginning of the next day if either

(i) there are no new messages and it is not the case that exactly one of the existing k messages transmits, or
(ii) there is exactly one new message (which automatically transmits) and none of the other k messages transmits.

Substitution of the above into (3.3) yields that

$$E\left[\sum_{k=0}^{\infty} I_k\right] \le \sum_{k=0}^{\infty} \frac{a_0 kp(1 - p)^{k-1}}{1 - a_0[1 - kp(1 - p)^{k-1}] - a_1(1 - p)^k}$$
$$< \infty$$

where the convergence follows by noting that when k is large the denominator of the expression in the above sum converges to $1 - a_0$ and so the convergence or divergence of the sum is determined by whether or not the sum of the terms in the numerator converge and $\Sigma_{k=0}^{\infty} k(1 - p)^{k-1} < \infty$.

Hence, $E[\Sigma_{k=0}^{\infty} I_k] < \infty$, which implies that $\Sigma_{k=0}^{\infty} I_k < \infty$ with probability 1 (for if there was a positive probability that $\Sigma_{k=0}^{\infty} I_k$ could be ∞, then its mean would be ∞). Hence, with probability 1, there will be only a finite number of states that are initially departed via a successful transmission; of equivalently, there will be some finite integer N such that whenever there are N or more messages in the system, there will never again be a successful transmission. From this (and the fact that such higher states will eventually be reached—why?) it follows that, with probability 1, there will only be a finite number of successful transmissions. $\Diamond$

Remark   For a (slightly less than rigorous) probabilistic proof of Stirling's approximation, let $X_1, X_2, \ldots$ be independent Poisson random variables each having mean 1. Let $S_n = \Sigma_{i=1}^{n} X_i$, and note that both the mean and variance of S_n are equal to n. Now,

$$
\begin{aligned}
P\{S_n = n\} &= P\{n - 1 < S_n \leq n\} \\
&= P\{-1/\sqrt{n} < (S_n - n)/\sqrt{n} \leq 0\} \\
&\approx \int_{-1/\sqrt{n}}^{0} (2\pi)^{-1/2} e^{-x^2/2} \, dx \qquad \text{when n is large by the} \\
&\qquad\qquad\qquad\qquad\qquad\qquad\qquad\quad \text{Central Limit Theorem} \\
&\approx (2\pi)^{-1/2} (1/\sqrt{n}) \\
&= (2\pi n)^{-1/2}
\end{aligned}
$$

But S_n is Poisson with mean n, and so

$$
P\{S_n = n\} = \frac{e^{-n} n^n}{n!}
$$

Hence, for n large

$$
\frac{e^{-n} n^n}{n!} \approx (2\pi n)^{-1/2}
$$

or, equivalently

$$
n! \approx n^{n+1/2} e^{-n} \sqrt{2\pi}
$$

which is Stirling's approximation. $\Diamond$

4. Limiting Probabilities

In Example 2a, we calculated $\mathbf{P}^{(4)}$ for a two-state Markov chain; it turned out to be

$$\mathbf{P}^{(4)} = \left\| \begin{array}{cc} .5749 & .4251 \\ .5668 & .4332 \end{array} \right\|$$

From this it follows that $\mathbf{P}^{(8)} = \mathbf{P}^{(4)} \cdot \mathbf{P}^{(4)}$ is given (to three significant places) by

$$\mathbf{P}^{(8)} = \left\| \begin{array}{cc} .572 & .428 \\ .570 & .430 \end{array} \right\|$$

Note that the matrix $\mathbf{P}^{(8)}$ is almost identical to the matrix $\mathbf{P}^{(4)}$, and secondly, that each of the rows of $\mathbf{P}^{(8)}$ have almost identical entries. In fact it seems that P_{ij}^n is converging to some value (as $n \to \infty$) which is the same for all i. In other words, there seems to exist a limiting probability that the process will be in state j after a large number of transitions, and this value is independent of the initial state.

To make the above heuristics more precise there are two additional properties of the states of a Markov chain that we need consider. State i is said to have *period* d if $P_{ii}^n = 0$ whenever n is not divisible by d, and d is the largest integer with this property. For instance, starting in i, it may be possible for the process to enter state i only at the times 2, 4, 6, 8, . . . , in which case state i has period 2. A state with period 1 is said to be *aperiodic*. It can be shown that periodicity is a class property. That is, if state i has period d, and states i and j communicate, then state j also has period d.

If state i is recurrent, then it is said to be *positive recurrent* if, starting in i, the expected time until the process returns to state i is finite. It can be shown that positive recurrence is a class property. While there exist recurrent states that are not positive recurrent,* it can be shown that *in a finite-state Markov chain all recurrent states are positive recurrent*. Positive recurrent, aperiodic states are called *ergodic*.

We are now ready for the following important theorem which we state without proof.

Theorem 4.1. For an irreducible ergodic Markov chain $\lim_{n \to \infty} P_{ij}^n$ exists and is independent of i. Furthermore, letting

---

*Such states are called *null recurrent*.

$$\pi_j = \lim_{n \to \infty} P_{ij}^n, \qquad j \geq 0$$

then π_j is the unique nonnegative solution of

$$\pi_j = \sum_{i=0}^{\infty} \pi_i P_{ij}, \qquad j \geq 0$$

$$\sum_{j=0}^{\infty} \pi_j = 1 \quad \Diamond \tag{4.1}$$

Remarks

(i) Given that $\pi_j = \lim_{n \to \infty} P_{ij}^n$ exists and is independent of the initial state i, it is not difficult to (heuristically) see that the π's must satisfy Equation (4.1). For let us derive an expression for $P\{X_{n+1} = j\}$ by conditioning on the state at time n. That is,

$$P\{X_{n+1} = j\} = \sum_{i=0}^{\infty} P\{X_{n+1} = j | X_n = i\} P\{X_n = i\}$$

$$= \sum_{i=0}^{\infty} P_{ij} P\{X_n = i\}$$

Letting $n \to \infty$, and assuming that we can bring the limit inside the summation, leads to

$$\pi_j = \sum_{i=0}^{\infty} P_{ij} \pi_i$$

(ii) It can be shown that π_j, the limiting probability that the process will be in state j at time n, also equals the long-run proportion of time that the process will be in state j.

(iii) In the irreducible, positive recurrent, *periodic* case we still have that the $\pi_j, j \geq 0$, are the unique nonnegative solution of

$$\pi_j = \sum_i \pi_i P_{ij},$$

$$\sum_j \pi_j = 1$$

But now π_j must be interpreted as the long-run proportion of time that the Markov chain is in state j. $\Diamond$

Example 4a   Consider Example 1a in which we assume that if it rains today, then it will rain tomorrow with probability α; and if it does not rain today, then it will rain tomorrow with probability β. If we say that the state is 0 when it rains and 1 when it does not rain, then by Equation (4.1) the limiting probabilities π_0 and π_1 are given by

$$\pi_0 = \alpha\pi_0 + \beta\pi_1$$
$$\pi_1 = (1 - \alpha)\pi_0 + (1 - \beta)\pi_1$$
$$\pi_0 + \pi_1 = 1$$

which yields that

$$\pi_0 = \frac{\beta}{1 + \beta - \alpha}, \qquad \pi_1 = \frac{1 - \alpha}{1 + \beta - \alpha}$$

For example if $\alpha = .7$ and $\beta = .4$, then the limiting probability of rain is $\pi_0 = \frac{4}{7} = .571$. $\Diamond$

Example 4b Consider Example 1c in which the mood of an individual is considered as a three-state Markov chain having a transition probability matrix

$$\mathbf{P} = \begin{Vmatrix} .5 & .4 & .1 \\ .3 & .4 & .3 \\ .2 & .3 & .5 \end{Vmatrix}$$

In the long run, what proportion of time is the process in each of the three states?

Solution: The limiting probabilities π_i, $i = 0, 1, 2$, are obtained by solving the set of equations in Equation (4.1). In this case these equations are

$$\pi_0 = .5\pi_0 + .3\pi_1 + .2\pi_2$$
$$\pi_1 = .4\pi_0 + .4\pi_1 + .3\pi_2$$
$$\pi_2 = .1\pi_0 + .3\pi_1 + .5\pi_2$$
$$\pi_0 + \pi_1 + \pi_2 = 1$$

Solving yields that

$$\pi_0 = \tfrac{21}{62}, \qquad \pi_1 = \tfrac{23}{62}, \qquad \pi_2 = \tfrac{18}{62} \quad \Diamond$$

Example 4c (A Model of Class Mobility): A problem of interest to sociologists is to determine the proportion of society that has an upper- or lower-class occupation. One possible mathematical model would be to assume that transitions between social classes of the successive generations in a family can be regarded as transitions of a Markov chain. That is, we assume that the occupation of a son will depend only on his father's occupation and not on his grandfather's occupation. Let us suppose that such a model is appropriate and that the transition probability matrix is given by

$$\mathbf{P} = \begin{Vmatrix} .45 & .48 & .07 \\ .05 & .70 & .25 \\ .01 & .50 & .49 \end{Vmatrix} \tag{4.2}$$

That is, for instance, we suppose that the son of a middle-class worker will attain an upper-, middle-, or lower-class occupation with respective probabilities .05, .70, .25.

The limiting probabilities π_i, thus satisfy

$$\pi_0 = .45\pi_0 + .05\pi_1 + .01\pi_2$$
$$\pi_1 = .48\pi_0 + .70\pi_1 + .50\pi_2$$
$$\pi_2 = .07\pi_0 + .25\pi_1 + .49\pi_2$$
$$\pi_0 + \pi_1 + \pi_2 = 1$$

Hence,

$$\pi_0 = .07, \qquad \pi_1 = .62, \qquad \pi_2 = .31$$

In other words, a society in which social mobility between classes can be described by a Markov chain with transition probability matrix given by Equation (4.2) has, in the long run, 7 percent of its people in upper-class jobs, 62 percent of its people in middle-class jobs, and 31 percent in lower-class jobs.   ◊

Example 4d (The Hardy–Weinberg Law and a Markov Chain in Genetics): Consider a large population of individuals each of whom possesses a particular pair of genes, of which each individual gene is classified as being of type *A* or type *a*. Assume that the percentages of individuals whose gene pairs are *AA*, *aa*, or *Aa* are respectively p_0, q_0, and $r_0 (p_0 + q_0 + r_0 = 1)$. When two individuals mate, each contributes one of his or her genes, chosen at random, to the resultant offspring. Assuming that the mating occurs at random, in that each individual is equally likely to mate with any other individual, we are interested in determining the percentages of individuals in the next generation whose genes are *AA*, *aa*, or *Aa*. Calling these percentages p, q, and r, they are easily obtained by focusing attention on an individual of the next generation and then conditioning on the genes of his parents. This yields

$$
\begin{aligned}
p &= P\{AA\} \\
&= P\{AA|\text{both parents are } AA\}p_0^2 \\
&\quad + P\{AA|\text{one parent is } AA \text{ and one is } Aa\}2p_0r_0 \\
&\quad + P\{AA|\text{both parents are } Aa\}r_0^2 \\
&\quad + P\{AA|\text{at least one parent is } aa\}[1 - (1 - q_0)^2] \\
&= p_0^2 + \tfrac{1}{2}2p_0r_0 + \tfrac{1}{4}r_0^2 \\
&= \left(p_0 + \frac{r_0}{2}\right)^2
\end{aligned}
$$

Similarly we obtain

$$
\begin{aligned}
q &= q_0^2 + \tfrac{1}{2}2q_0r_0 + \tfrac{1}{4}r_0^2 \\
&= \left(q_0 + \frac{r_0}{2}\right)^2 \\
r &= 2p_0q_0 + \tfrac{1}{2}2p_0r_0 + \tfrac{1}{2}2q_0r_0 + \tfrac{1}{2}r_0^2 \\
&= 2\left(p_0 + \frac{r_0}{2}\right)\left(q_0 + \frac{r_0}{2}\right)
\end{aligned}
$$

As a check of the above, note that

$$
\begin{aligned}
p + q + r &= \left(p_0 + \frac{r_0}{2}\right)\left[p_0 + \frac{r_0}{2} + q_0 + \frac{r_0}{2}\right] \\
&\quad + \left(q_0 + \frac{r_0}{2}\right)\left[q_0 + \frac{r_0}{2} + p_0 + \frac{r_0}{2}\right] \\
&= p_0 + \frac{r_0}{2} + q_0 + \frac{r_0}{2} \\
&= 1
\end{aligned}
$$

If we now investigate the genetic makeup of the following generation, we find that by the same analysis the percentages will be

$$\text{percentage } (AA) = \left(p + \frac{r}{2}\right)^2$$

$$= \left[\left(p_0 + \frac{r_0}{2}\right)^2 + \left(p_0 + \frac{r_0}{2}\right)\left(q_0 + \frac{r_0}{2}\right)\right]^2$$

$$= \left[\left(p_0 + \frac{r_0}{2}\right)\left(p_0 + \frac{r_0}{2} + q_0 + \frac{r_0}{2}\right)\right]^2$$

$$= \left(p_0 + \frac{r_0}{2}\right)^2$$

$$= p \tag{4.3}$$

Similarly,

$$\text{percentage } (aa) = q$$

and, as the sum must equal 1, we have

$$\text{percentage } (Aa) = r$$

Hence, from the first generation onward the proportion of gene pairs in the population remains stable. This is known as the *Hardy-Weinberg law.* ◊

Remarks Perhaps the Hardy-Weinberg law can be understood most easily by looking at the total gene pool. Initially the fraction of genes in the population that are type A is $\bar{p} = p_0 + r_0/2$ and the fraction that are type a is $\bar{q} = q_0 + r_0/2$. Under random mating it follows that the probability that an arbitrary offspring will receive a type A gene from both parents is $\bar{p}^2$; the probability that it will receive a type a gene from both parents is $\bar{q}^2$; and the probability that it will receive one type a and one type A gene is $2\bar{p}\bar{q}$. Hence, in the first generation the proportion of individuals having gene pairs AA, aa, Aa will be $\bar{p}^2$, $\bar{q}^2$, $2\bar{p}\bar{q}$. As the fraction of type A and type a genes will still be $\bar{p}$ and $\bar{q}$, it follows by the same argument that all subsequent generations will also have gene pairs in the proportions $\bar{p}^2$, $\bar{q}^2$, $2\bar{p}\bar{q}$. ◊

Suppose now that the gene pair population has stabilized in the percentages p, q, r, and let us follow the genetic history of a single individual and his descendants. (For simplicity, assume that each individual has exactly one offspring.) So, for a given individual, let X_n denote the genetic state of his descendent in the nth generation. The transition probability matrix of this Markov chain, namely

$$
\begin{array}{c c c c}
 & AA & aa & Aa \\
AA & p + \dfrac{r}{2} & 0 & q + \dfrac{r}{2} \\
aa & 0 & q + \dfrac{r}{2} & p + \dfrac{r}{2} \\
Aa & \dfrac{p}{2} + \dfrac{r}{4} & \dfrac{q}{2} + \dfrac{r}{4} & \dfrac{p}{2} + \dfrac{q}{2} + \dfrac{r}{2}
\end{array}
$$

is easily verified by conditioning on the state of the randomly chosen mate. It is quite intuitive (why?) that the limiting probabilities for this Markov chain (which also equal the fractions of the individual's descendants that are in each of the three genetic states) should just be p, q, and r. To verify this we must show that they satisfy Equation (4.1). As one of the equations in Equation (4.1) is redundant, it suffices to show that

$$
p = p\left(p + \frac{r}{2}\right) + r\left(\frac{p}{2} + \frac{r}{4}\right) = \left(p + \frac{r}{2}\right)^2
$$

$$
q = q\left(q + \frac{r}{2}\right) + r\left(\frac{q}{2} + \frac{r}{4}\right) = \left(q + \frac{r}{2}\right)^2
$$

$$
p + q + r = 1
$$

But this follows from Equation (4.3), and thus the result is established.

Example 4e Suppose that a production process changes states in accordance with a Markov chain having transition probabilities P_{ij}, $i, j = 1, \ldots, n$, and suppose that certain of the states are considered acceptable and the remaining unacceptable. Let A denote the acceptable states and A^c the unacceptable ones. If the production process is said to be "up" when in an acceptable state and "down" when in an unacceptable state, determine

   1. the rate at which the production process goes from up to down (that is, the rate of breakdowns);
   2. the average length of time the process remains down when it goes down; and
   3. the average length of time the process remains up when it goes up.

Solution: Let π_k, $k = 1, \ldots n$, denote the limiting probabilities. Now for $i \in A$ and $j \in A^c$ the rate at which the process enters state j from state i is

$$\text{rate enter } j \text{ from } i = \pi_i P_{ij}$$

and so the rate at which the production process enters state j from an acceptable state is

$$\text{Rate enter } j \text{ from } A = \sum_{i \in A} \pi_i P_{ij}$$

Hence, the rate at which it enters an unacceptable state from an acceptable one (which is the rate at which breakdowns occur) is

$$\text{Rate breakdowns occur} = \sum_{j \in A^c} \sum_{i \in A} \pi_i P_{ij} \qquad (4.4)$$

Now let $\overline{U}$ and $\overline{D}$ denote the average time the process remains up when it goes up and down when it goes down. As there is a single breakdown every $\overline{U} + \overline{D}$ time units on the average, it follows heuristically that

$$\text{Rate at which breakdowns occur} = \frac{1}{\overline{U} + \overline{D}}$$

and so from Equation (4.4)

$$\frac{1}{\overline{U} + \overline{D}} = \sum_{j \in A^c} \sum_{i \in A} \pi_i P_{ij} \qquad (4.5)$$

To obtain a second equation relating $\overline{U}$ and $\overline{D}$, consider the percentage of time the process is up, which, of course, is equal to $\sum_{i \in A} \pi_i$. However, since the process is up on the average $\overline{U}$ out of every $\overline{U} + \overline{D}$ time units, it follows (again somewhat heuristically) that the

$$\text{Proportion of up time} = \frac{\overline{U}}{\overline{U} + \overline{D}}$$

and so

$$\frac{\overline{U}}{\overline{U} + \overline{D}} = \sum_{i \in A} \pi_i \qquad (4.6)$$

Hence, from Equations (4.5) and (4.6) we obtain

$$\overline{U} = \frac{\displaystyle\sum_{i \in A} \pi_i}{\displaystyle\sum_{j \in A^c} \sum_{i \in A} \pi_i P_{ij}}$$

$$\overline{D} = \frac{1 - \displaystyle\sum_{i \in A} \pi_i}{\displaystyle\sum_{j \in A^c} \sum_{i \in A} \pi_i P_{ij}}$$

$$= \frac{\displaystyle\sum_{i \in A^c} \pi_i}{\displaystyle\sum_{j \in A^c} \sum_{i \in A} \pi_i P_{ij}}$$

For example, suppose the transition probability matrix is

$$P = \begin{Vmatrix} \frac{1}{4} & \frac{1}{4} & \frac{1}{2} & 0 \\ 0 & \frac{1}{4} & \frac{1}{2} & \frac{1}{4} \\ \frac{1}{4} & \frac{1}{4} & \frac{1}{4} & \frac{1}{4} \\ \frac{1}{4} & \frac{1}{4} & 0 & \frac{1}{2} \end{Vmatrix}$$

where the acceptable (up) states are 1, 2 and the unacceptable (down) ones are 3, 4. The limiting probabilities satisfy

$$\pi_1 = \pi_1 \tfrac{1}{4} + \pi_3 \tfrac{1}{4} + \pi_4 \tfrac{1}{4}$$

$$\pi_2 = \pi_1 \tfrac{1}{4} + \pi_2 \tfrac{1}{4} + \pi_3 \tfrac{1}{4} + \pi_4 \tfrac{1}{4}$$

$$\pi_3 = \pi_1 \tfrac{1}{2} + \pi_2 \tfrac{1}{2} + \pi_3 \tfrac{1}{4}$$

$$\pi_1 + \pi_2 + \pi_3 + \pi_4 = 1$$

These solve to yield

$$\pi_1 = \tfrac{3}{16}, \qquad \pi_2 = \tfrac{1}{4}, \qquad \pi_3 = \tfrac{14}{48}, \qquad \pi_4 = \tfrac{13}{48}$$

and thus

$$\text{Rate of breakdowns} = \pi_1(P_{13} + P_{14}) + \pi_2(P_{23} + P_{24})$$

$$= \tfrac{9}{32}$$

$$\overline{U} = \tfrac{14}{9} \text{ and } \overline{D} = 2$$

Hence, on the average, breakdowns occur about $\tfrac{9}{32}$ (or 28 percent) of the time. They last, on the average, 2 time units, and then there follows a stretch of (on the average) $\tfrac{14}{9}$ time units when the system is up. ◇

Remarks

   (i) The long run proportions π_j, $j \geq 0$, are often called *stationary* probabilities. The reason being that if the initial state is chosen according to the probabilities $\pi_j, j \geq 0$, then the probability of being in state j at any time n is also equal to π_j. That is, if

$$P\{X_0 = j\} = \pi_j, \qquad j \geq 0$$

then

$$P\{X_n = j\} = \pi_j \quad \text{for all } n, j \geq 0$$

The above is easily proven by induction, for if we suppose it true for $n - 1$, then writing

$$P\{X_n = j\} = \sum_i P\{X_n = j | X_{n-1} = i\} \, P\{X_{n-1} = i\}$$

$$\qquad\quad = \sum_i P_{ij}\pi_i \quad \text{by the induction hypothesis}$$

$$\qquad = \pi_j \quad \text{by Equation (4.1)}$$

   (ii) For state j, define m_{jj} to be the expected number of transitions until a Markov chain, starting in state j, returns to that state. Since, on the average, the chain will spend 1 unit of time in state j for every m_{jj} units of time, it follows that

$$\pi_j = \frac{1}{m_{jj}}$$

In words, the proportion of time in state j equals the inverse of the mean time between visits to j. (The above is a special case of a general result, sometimes called the strong law for renewal processes, which will be presented in Chapter 7.)   ◇

Example 4f   Consider independent tosses of a coin that, on each toss, lands on heads (H) with probability p and on tails (T) with probability $q = 1 - p$. What is the expected number of tosses needed for the pattern HTHT to appear?

Solution: To answer the above, let us imagine that the coin tossing does not stop when the pattern appears, but rather it goes on indefinitely. If we define the state at time n to be the most recent

4 outcomes when $n \geq 4$, and the most recent n outcomes when $n < 4$, then it is easy to see that the successive states constitute a Markov chain. For instance, if the first 5 outcomes are TTTHH, then the successive states of the Markov chain are $X_1 = $ T, $X_2 = $ TT, $X_3 = $ TTT, $X_4 = $ TTTH, and $X_5 = $ TTHH. It therefore follows from remark (ii) above that π_{HTHT}, the limiting probability of state HTHT, is equal to the inverse of the mean time to go from state HTHT to HTHT. However, for any $n \geq 4$, the probability that the state at time n is HTHT is just the probability that the toss at n is T, the one at $n - 1$ is H, the one at $n - 2$ is T, and the one at $n - 3$ is H. Since the successive tosses are independent, it follow that

$$P\{X_n = \text{HTHT}\} = p^2q^2, \qquad n \geq 4$$

and so

$$\pi_{\text{HTHT}} = \lim_{n \to \infty} P\{X_n = \text{HTHT}\} = p^2q^2$$

Hence, $1/(p^2q^2)$ is the mean time to go from HTHT to HTHT. But this means that starting with HT the expected number of additional trials to obtain HTHT is $1/(p^2q^2)$. Therefore, since in order to obtain HTHT one must first obtain HT, it follows that

$$E[\text{time to pattern HTHT}] = E[\text{time to the pattern HT}] + \frac{1}{p^2q^2}$$

To determine the expected time to the pattern HT, we can reason in the same way and let the state be the most recent 2 tosses. By the same argument as used above, it follows that the expected time between appearances of HT is equal to $1/\pi_{\text{HT}} = 1/(pq)$. As this is the same as the expected time until HT first appears, we finally obtain that

$$E[\text{time until HTHT appears}] = \frac{1}{pq} + \frac{1}{p^2q^2}$$

The same approach can be used to obtain the mean time until any given pattern appears. For instance, reasoning as above, we obtain that

$$E[\text{time until HTHHTHTHH}] = E[\text{time until HTHH}] + \frac{1}{p^6 q^3}$$

$$= E[\text{time until H}] + \frac{1}{p^3 q} + \frac{1}{p^6 q^3}$$

$$= \frac{1}{p} + \frac{1}{p^3 q} + \frac{1}{p^6 q^3}$$

Also, it is not necessary that the basic experiment has only 2 possible outcomes (which we designated above as H and T). For instance, if the successive values are independently and identically distributed with p_j denoting the probability that any given value is equal to j, $j \geq 0$, then

$$E[\text{time until } 012301] = E[\text{time until } 01] + \frac{1}{p_0^2 p_1^2 p_2 p_3}$$

$$= \frac{1}{p_0 p_1} + \frac{1}{p_0^2 p_1^2 p_2 p_3} \quad \Diamond$$

5. Some Applications

5.1. The Gambler's Ruin Problem

Consider a gambler who at each play of the game has probability p of winning one unit and probability $q = 1 - p$ of losing one unit. Assuming that successive plays of the game are independent, what is the probability that, starting with i units, the gambler's fortune will reach N before reaching 0?

If we let X_n denote the players fortune at time n, then the process $\{X_n, n = 0, 1, 2, \ldots\}$ is a Markov chain with transition probabilities

$$P_{00} = P_{NN} = 1$$
$$P_{i,i+1} = p = 1 - P_{i,i-1}, \qquad i = 1, 2, \ldots, N - 1$$

This Markov chain has three classes, namely $\{0\}$, $\{1, 2, \ldots, N - 1\}$, and $\{N\}$; the first and third class being recurrent and the second transient. Since each transient state is only visited finitely often, it follows that, after some finite amount of time, the gambler will either attain his goal of N or go broke.

Let P_i, $i = 0, 1, \ldots, N$, denote the probability that, starting with i, the gambler's fortune will eventually reach N. By conditioning on the outcome of the initial play of the game we obtain

$$P_i = pP_{i+1}, + qP_{i-1} \qquad i = 1, 2, \ldots, N - 1$$

or equivalently, since $p + q = 1$,

$$pP_i + qP_i = pP_{i+1} + qP_{i-1}$$

or

$$P_{i+1} - P_i = \frac{q}{p}(P_i - P_{i-1}), \qquad i = 1, 2, \ldots, N - 1$$

Hence, since $P_0 = 0$, we obtain from the above line that

$$P_2 - P_1 = \frac{q}{p}(P_1 - P_0) = \frac{q}{p}P_1$$

$$P_3 - P_2 = \frac{q}{p}(P_2 - P_1) = \left(\frac{q}{p}\right)^2 P_1$$

$$\vdots$$

$$P_i - P_{i-1} = \frac{q}{p}(P_{i-1} - P_{i-2}) = \left(\frac{q}{p}\right)^{i-1} P_1$$

$$\vdots$$

$$P_N - P_{N-1} = \left(\frac{q}{p}\right)(P_{N-1} - P_{N-2}) = \left(\frac{q}{p}\right)^{N-1} P_1$$

Adding the first $i - 1$ of these equations yields

$$P_i - P_1 = P_1\left[\left(\frac{q}{p}\right) + \left(\frac{q}{p}\right)^2 + \cdots + \left(\frac{q}{p}\right)^{i-1}\right]$$

or

$$P_i = \begin{cases} \dfrac{1 - (q/p)^i}{1 - (q/p)}P_1, & \text{if } \dfrac{q}{p} \neq 1 \\[2ex] iP_1, & \text{if } \dfrac{q}{p} = 1 \end{cases}$$

Now, using the fact that $P_N = 1$, we obtain that

$$P_1 = \begin{cases} \dfrac{1 - (q/p)}{1 - (q/p)^N}, & \text{if } p \neq \dfrac{1}{2} \\[2ex] \dfrac{1}{N}, & \text{if } p = \dfrac{1}{2} \end{cases}$$

and hence

$$P_i = \begin{cases} \dfrac{1 - (q/p)^i}{1 - (q/p)^N}, & \text{if } p \neq \dfrac{1}{2} \\[2ex] \dfrac{i}{N}, & \text{if } p = \dfrac{1}{2} \end{cases} \tag{5.1}$$

Note that as $N \to \infty$

$$P_i \to \begin{cases} 1 - \left(\dfrac{q}{p}\right)^i, & \text{if } p > \dfrac{1}{2} \\[2ex] 0, & \text{if } p \leq \dfrac{1}{2} \end{cases}$$

Thus, if $p > \frac{1}{2}$, there is a positive probability that the gamblers fortune will increase indefinitely; while if $p \leq \frac{1}{2}$, the gambler will, with probability 1, go broke against an infinitely rich adversary.

Example 5a Suppose Max and Patty decide to flip pennies, the one coming closest to the wall wins. Patty, being the better player, has a probability .6 of winning on each flip. If Patty starts with five pennies and Max with ten, then what is the probability that Patty will wipe Max out? What if Patty starts with ten and Max with 20?

Solution:

    (a) The desired probability is obtained from Equation (5.1) by letting $i = 5$, $N = 15$, and $p = .6$. Hence, the desired probability is

$$\frac{1 - \left(\frac{2}{3}\right)^5}{1 - \left(\frac{2}{3}\right)^{15}} \approx .87$$

    (b) The desired probability is

$$\frac{1 - \left(\frac{2}{3}\right)^{10}}{1 - \left(\frac{2}{3}\right)^{30}} \approx .98 \qquad \diamond$$

For an application of the gambler's ruin problem to drug testing, suppose that two new drugs have been developed for treating a certain disease. Drug i has a cure rate P_i, $i = 1, 2$, in the sense that each patient treated with drug i will be cured with probability P_i. These cure rates are, however, not known, and suppose we are interested in a method for deciding whether $P_1 > P_2$ or $P_2 > P_1$. To decide upon one of these alternatives, consider the following test: Pairs of patients are treated sequentially with one member of the pair receiving drug 1 and the other drug 2. The results for each pair are determined, and the testing stops when the cumulative number of cures using one of the drugs exceeds the cumulative number of cures when using the other by some fixed predetermined number. More formally let

$$X_j = \begin{cases} 1 & \text{if the patient in the } j^{\text{th}} \text{ pair to receive drug number 1 is cured} \\ 0 & \text{otherwise} \end{cases} \quad P_1$$

$$Y_j = \begin{cases} 1 & \text{if the patient in the } j^{\text{th}} \text{ pair to receive drug number 2 is cured} \\ 0 & \text{otherwise} \end{cases} \quad P_2$$

For a predetermined positive integer M the test stops after pair N where N is the first value of n such that either

$$X_1 + \cdots + X_n - (Y_1 + \cdots + Y_n) = M$$

or

$$X_1 + \cdots + X_n - (Y_1 + \cdots + Y_n) = -M$$

In the former case we then assert that $P_1 > P_2$ and in the latter that $P_2 > P_1$.

In order to help ascertain whether the above is a good test, one thing we would like to know is the probability of it leading to an incorrect decision. That is, for given P_1 and P_2 where $P_1 > P_2$, what is the probability that the test will incorrectly assert that $P_2 > P_1$? To determine this probability, note that after each pair is checked the cumulative difference of cures using drug 1 versus drug 2 will either go up by 1 with probability $P_1(1 - P_2)$—since this is the probability that drug 1 leads to a cure and drug 2 does not—or go down by 1 with probability $(1 - P_1)P_2$, or remain the same with probability $P_1 P_2 + (1 - P_1)(1 - P_2)$. Hence, if we only consider those pairs in which the cumulative difference changes, then the difference will go up 1 with probability

$$p = P\{\text{up } 1 \mid \text{up } 1 \text{ or down } 1\}$$
$$= \frac{P_1(1 - P_2)}{P_1(1 - P_2) + (1 - P_1)P_2}$$

and down 1 with probability

$$1 - p = \frac{P_2(1 - P_1)}{P_1(1 - P_2) + (1 - P_1)P_2}$$

Hence, the probability that the test will assert that $P_2 > P_1$ is equal to the probability that a gambler who wins each (one unit) bet with probability p will go down M before going up M. But Equation (5.1) with $i = M$, $N = 2M$, shows that this probability is given by

$$P\{\text{test asserts that } P_2 > P_1\} = 1 - \frac{1 - \left(\dfrac{1 - p}{p}\right)^M}{1 - \left(\dfrac{1 - p}{p}\right)^{2M}}$$

$$= \frac{1}{1 + \left(\dfrac{p}{1 - p}\right)^M}$$

Thus, for instance, if $P_1 = .6$ and $P_2 = .4$ then the probability of an incorrect decision is .017 when $M = 5$ and reduces to .0003 when $M = 10$.

5.2 A Model for Algorithmic Efficiency

The following optimization problem is called a linear program:

$$\text{minimize } \mathbf{cx}$$
$$\text{subject to } \mathbf{Ax} = \mathbf{b}$$
$$\mathbf{x} \geq 0$$

where $\mathbf{A}$ is an $m \times n$ matrix of fixed constants; $\mathbf{c} = (c_1, \ldots, c_n)$ and $\mathbf{b} = (b_1, \ldots, b_m)$ are vectors of fixed constants, and $\mathbf{x} = (x_1, \ldots, x_n)$ is the n-vector of nonnegative values that is to be chosen to minimize $\mathbf{cx} \equiv \sum_{i=1}^{n} c_i x_i$. Supposing that $n > m$, it can be shown that the optimal $\mathbf{x}$ can always be chosen to have at least $n - m$ components equal to 0—that is, it can always be taken to be one of the so-called extreme points of the feasibility region.

The simplex algorithm solves this linear program by moving from an extreme point of the feasibility region to a better (in terms of the objective function $\mathbf{cx}$) extreme point (via the pivot operation) until the optimal is reached. As there can be as many as $N \equiv \binom{n}{m}$ such extreme points, it would seem that this method might take many iterations, but, surprisingly to some, this does not appear to be the case in practice.

In order to obtain a feel for whether or not the above is surprising, let us consider a simple probabilistic (Markov chain) model as to how the algorithm moves along the extreme points. Specifically, we will suppose that if at any time the algorithm is at the jth best extreme point then after the next pivot the resulting extreme point is equally likely to be any of the $j - 1$ best. Under this assumption, we show that the time to get from the Nth best to the best extreme point has approximately, for large N, a normal distribution with mean and variance equal to the logarithm (base e) of N.

Consider a Markov chain for which $P_{11} = 1$ and

$$P_{ij} = \frac{1}{i - 1}, \qquad j = 1, \ldots, i - 1, i > 1$$

and let T_i denote the number of transitions needed to go from state i to state 1. A recursive formula for $E[T_i]$ can be obtained by conditioning on the initial transition:

$$E[T_i] = 1 + \frac{1}{i - 1} \sum_{j=1}^{i-1} E[T_j]$$

Starting with $E[T_1] = 0$, we successively see that

$$E[T_2] = 1$$
$$E[T_3] = 1 + \tfrac{1}{2}$$
$$E[T_4] = 1 + \tfrac{1}{3}(1 + 1 + \tfrac{1}{2}) = 1 + \tfrac{1}{2} + \tfrac{1}{3}$$

and it is not difficult to guess and then prove inductively that

$$E[T_i] = \sum_{j=1}^{i-1} 1/j$$

However, to obtain a more complete description of T_N, we will use the representation

$$T_N = \sum_{j=1}^{N-1} I_j$$

where

$$I_j = \begin{cases} 1, & \text{if the process ever enters } j \\ 0, & \text{otherwise} \end{cases}$$

The importance of the above representation stems from the following:

Proposition 5.1. $I_1, \ldots, I_{N-1}$ are independent and

$$P\{I_j = 1\} = 1/j, \qquad 1 \le j \le N - 1$$

Proof: Given $I_{j+1}, \ldots, I_N$, let $n = \min\{i : i > j, I_i = 1\}$ denote the lowest numbered state, greater than j, that is entered. Thus we know that the process enters state n and the next state entered is one of the states $1, 2, \ldots, j$. Hence, as the next state from state n is equally likely to be any of the lower number states $1, 2, \ldots,$ $n - 1$ we see that

$$P\{I_j = 1 | I_{j+1}, \ldots, I_N\} = \frac{1/(n-1)}{j/(n-1)} = 1/j$$

Hence, $P\{I_j = 1\} = 1/j$, and independence follows since the above conditional probability does not depend on $I_{j+1}, \ldots, I_N$.

Corollary 5.2.

(i) $E[T_N] = \sum_{j=1}^{N-1} 1/j$
(ii) $\text{Var}(T_N) = \sum_{j=1}^{N-1} 1/j \, (1 - 1/j)$
(iii)   For N large, T_N has approximately a normal distribution with mean $\log N$ and variance $\log N$.

Proof: Parts (i) and (ii) follow from Proposition 5.1 and the representation $T_N = \sum_{j=1}^{N=1} I_j$. Part (iii) follows from the central limit theorem since

$$\int_1^N \frac{dx}{x} < \sum_1^{N-1} 1/j < 1 + \int_1^{N-1} \frac{dx}{x}$$

or

$$\log N < \sum_1^{N-1} 1/j < 1 + \log(N - 1)$$

and so

$$\log N \approx \sum_{j=1}^{N-1} 1/j \qquad \diamond$$

Returning to the simplex algorithm, if we assume that n, m, and $n - m$ are all large, we have by Stirling's approximation that

$$N = \binom{n}{m} \sim \frac{n^{n+1/2}}{(n - m^{\,n-m+1/2} m^{m+1/2} \sqrt{2\pi}}$$

and so letting $c = n/m$

$$\log N \sim (mc + \tfrac{1}{2}) \log(mc) - (m(c - 1) + \tfrac{1}{2}) \log(m(c - 1))$$
$$- (m + \tfrac{1}{2}) \log m - \tfrac{1}{2} \log(2\pi)$$

or

$$\log N \sim m \left[c \log \frac{c}{c - 1} + \log(c - 1) \right]$$

Now, as $\lim\limits_{x \to \infty} x \log[x/(x - 1)] = 1$, it follows that when c is large

$$\log N \sim m[1 + \log(c - 1)]$$

Thus, for instance, if $n = 8000$, $m = 1000$, then the number of necessary transitions is approximately normally distributed with mean and variance equal to $1000(1 + \log 7) \approx 3000$. Hence, the number of necessary transitions would be roughly between

$$3000 \pm 2\sqrt{3000} \quad \text{or, roughly, } 3000 \pm 110$$

95 percent of the time.*

6. Branching Processes

In this section we consider a class of Markov chains, known as *branching processes*, which have a wide variety of applications in the biological, sociological, and engineering sciences.

Consider a population consisting of individuals able to produce offspring of the same kind. Suppose that each individual will, by the end of its lifetime, have produced j new offspring with probability P_j, $j \geq 0$, independently of the number produced by any other individual. We suppose that $P_j < 1$ for all $j \geq 0$. The number of individuals initially present, denoted by X_0, is called the size of the zeroth generation. All offspring of the zeroth generation constitute the first generation and their number is denoted by X_1. In general, let X_n denote the size of the nth generation. It follows that $\{X_n, n = 0, 1, \ldots\}$ is a Markov chain having as its state space the set of nonnegative integers.

Note that state 0 is a recurrent state, since clearly $P_{00} = 1$. Also, if $P_0 > 0$, all other states are transient. This follows since $P_{i0} = P_0^i$, which implies that starting with i individuals there is a positive probability of

---

*The material of this section is taken from S. M. Ross, "A Heuristic Approach to Simpler Efficiency," *European Jour. of Operational Research* **9**, 344–348 (1982).

at least P_0^i that no later generation will ever consist of i individuals. Moreover, since any finite set of transient states $\{1, 2, \ldots, n\}$ will be visited only finitely often, this leads to the important conclusion that, *if $P_0 > 0$, then the population with either die out or its size will converge to infinity.*

Let

$$\mu = \sum_{j=0}^{\infty} j P_j$$

denote the mean number of offspring of a single individual, and let

$$\sigma^2 = \sum_{j=0}^{\infty} (j - \mu)^2 P_j$$

be the variance of the number of offspring produced by a single individual.

Let us suppose that $X_0 = 1$, that is, initially there is a single individual present. We calculate $E[X_n]$ and $\text{Var}(X_n)$ by first noting that we may write

$$X_n = \sum_{i=1}^{X_{n-1}} Z_i$$

where Z_i represents the number of offspring of the ith individual of the $(n - 1)$st generation. By conditioning on X_{n-1}, we obtain

$$E[X_n] = E[E[X_n|X_{n-1}]]$$

$$= E\left[E\left[\sum_{i=1}^{X_{n-1}} Z_i \Big| X_{n-1} \right] \right]$$

$$= E[X_{n-1}\mu]$$

$$= \mu E[X_{n-1}] \qquad\qquad (6.1)$$

where we have used the fact that $E[Z_i] = \mu$. Since $E[X_0] = 1$, Equation (6.1) yields that

$$E[X_1] = \mu$$
$$E[X_2] = \mu E[X_1] = \mu^2$$
$$\vdots$$
$$E[X_n] = \mu E[X_{n-1}] = \mu^n$$

Similarly, $\text{Var}(X_n)$ may be obtained by using the conditional variance formula

$$\text{Var}(X_n) = E[\text{Var}(X_n|X_{n-1})] + \text{Var}(E[X_n|X_{n-1}])$$

Now, given X_{n-1}, X_n is just the sum of X_{n-1} independent random variables each having the distribution $\{P_j, j \geq 0\}$. Hence,

$$\text{Var}(X_n|X_{n-1}) = X_{n-1}\sigma^2$$

Thus, the conditional variance formula yields

$$\text{Var}(X_n) = E[X_{n-1}\sigma^2] + \text{Var}(X_{n-1}\mu)$$
$$= \sigma^2\mu^{n-1} + \mu^2 \text{Var}(X_{n-1})$$

Using the fact that $\text{Var}(X_0) = 0$ we can show by mathematical induction that the above implies that

$$\text{Var}(X_n) = \begin{cases} \sigma^2\mu^{n-1}\left(\dfrac{\mu^n - 1}{\mu - 1}\right), & \text{if } \mu \neq 1 \\[2mm] n\sigma^2, & \text{if } \mu = 1 \end{cases} \tag{6.2}$$

Let π_0 denote the probability that the population will eventually die out (under the assumption that $X_0 = 1$). More formally

$$\pi_0 = \lim_{n \to \infty} P\{X_n = 0 | X_0 = 1\}$$

The problem of determining the value of π_0 was first raised in connection with the extinction of family surnames by Galton in 1889.

We first note that $\pi_0 = 1$ if $\mu < 1$. This follows since

$$\mu^n = E[X_n] = \sum_{j=1}^{\infty} jP\{X_n = j\}$$
$$\geq \sum_{j=1}^{\infty} 1 \cdot P\{X_n = j\}$$
$$= P\{X_n \geq 1\}$$

Since $\mu^n \to 0$ when $\mu < 1$, it follows that $P\{X_n \geq 1\} \to 0$, and hence $P\{X_n = 0\} \to 1$.

In fact, it can be shown that $\pi_0 = 1$ even when $\mu = 1$. When $\mu > 1$, it turns out that $\pi_0 < 1$, and an equation determining π_0 may be derived by conditioning on the number of offspring of the initial individual, as follows:

$$\pi_0 = P\{\text{population dies out}\}$$

$$= \sum_{j=0}^{\infty} P\{\text{population dies out}|X_1 = j\}P_j$$

Now, given that $X_1 = j$, the population will eventually die out if and only if each of the j families started by the members of the first generation eventually die out. Since each family is assumed to act independently, and since the probability that any particular family dies out is just π_0, this yields that

$$P\{\text{population dies out}|X_1 = j\} = \pi_0^j$$

and thus π_0 satisfies

$$\pi_0 = \sum_{j=0}^{\infty} \pi_0^j P_j \tag{6.3}$$

In fact when $\mu > 1$, it can be shown that π_0 is the smallest positive number satisfying Equation (6.3).

Example 6a If $P_0 = \frac{1}{2}$, $P_1 = \frac{1}{4}$, $P_2 = \frac{1}{4}$, then determine π_0.

    Solution: Since $\mu = \frac{3}{4} \leq 1$, it follows that $\pi_0 = 1$. ◇

Example 6b If $P_0 = \frac{1}{4}$, $P_1 = \frac{1}{4}$, $P_2 = \frac{1}{2}$, then determine π_0.

    Solution: π_0 satisfies

$$\pi_0 = \frac{1}{4} + \frac{1}{4}\pi_0 + \frac{1}{2}\pi_0^2$$

or

$$2\pi_0^2 - 3\pi_0 + 1 = 0$$

The smallest positive solution of this quadratic equation is $\pi_0 = \frac{1}{2}$. ◇

Example 6c In Examples 6a and 6b, what is the probability that the population will die out if it initially consists of n individuals?

    Solution: Since the population will die out if and only if the families of each of the members of the initial generation die out, the desired probability is π_0^n. For Example 6a this yields $\pi_0^n = 1$, and for Example 6b, $\pi_0^n = (\frac{1}{2})^n$. ◇

7. Time Reversible Markov Chains

Consider a stationary ergodic Markov chain (that is, an ergodic Markov chain that has been in operation for a long time) having transition probabilities P_{ij} and stationary probabilities π_i, and suppose that starting at some time we trace the sequence of states going backwards in time. That is, starting at time n, consider the sequence of states X_n, X_{n-1}, X_{n-2}, ... It turns out that this sequence of states is itself a Markov chain with transition probabilities Q_{ij} defined by

$$
\begin{aligned}
Q_{ij} &= P\{X_m = j | X_{m+1} = i\} \\
&= \frac{P\{X_m = j, X_{m+1} = i\}}{P\{X_{m+1} = i\}} \\
&= \frac{P\{X_m = j\}P\{X_{m+1} = i | X_m = j\}}{P\{X_{m+1} = i\}} \\
&= \frac{\pi_j P_{ij}}{\pi_i}
\end{aligned}
$$

To prove that the reversed process is indeed a Markov chain, we must verify that $P\{X_m = j | X_{m+1} = i, X_{m+2} = i_2, X_{m+3} = i_3, \ldots, X_{m+k} = i_k\}$ also is equal to Q_{ij}.

$$
\begin{aligned}
&P\{X_m = j | X_{m+1} = i, X_{m+2} = i_2, \ldots, X_{m+k} = i_k\} \\
&= \frac{P\{X_m = j, X_{m+1} = i, X_{m+2} = i_2, \ldots, X_{m+k} = i_k\}}{P\{X_{m+1} = i, X_{m+2} = i_2, \ldots, X_{m+k} = i_k\}} \\
&= \frac{\pi_j P_{ji} P\{X_{m+2} = i_2, \ldots, X_{m+k} = i_k | X_{m+1} = i\}}{\pi_i P\{X_{m+2} = i_2, \ldots, X_{m+k} = i_k | X_{m+1} = i\}} \\
&= Q_{ij}
\end{aligned}
$$

Thus the reversed process is also a Markov chain with transition probabilities given by

$$
Q_{ij} = \frac{\pi_j P_{ji}}{\pi_i}
$$

If $Q_{ij} = P_{ij}$ for all i, j, then the Markov chain is said to be *time reversible*. The condition for time reversibility, namely $Q_{ij} = P_{ij}$, can also be expressed as

$$
\pi_i P_{ij} = \pi_j P_{ji} \quad \text{for all } i, j \tag{7.1}
$$

The condition in Equation (7.1) can be stated that, for all states i and j, the rate at which the process goes from i to j namely $\pi_i P_{ij}$) is equal to the rate at which it goes from j to i (namely $\pi_j P_{ji}$). It is worth noting that this is an obvious and necessary condition for time reversibility since a transition from i to j going backward in time is equivalent to a transition from j to i going forward in time; i.e., if $X_m = i$ and $X_{m-1} = j$, then a transition from i to j is observed if we are looking backward, and one from j to i if we are looking forward in time.

If we can find nonnegative numbers, summing to one, which satisfy Equation (7.1), then it follows that the Markov chain is time reversible and the numbers represent the limiting probabilities. This is so since if

$$x_i P_{ij} = x_j P_{ji} \quad \text{for all} \quad i,j, \quad \sum_i x_i = 1 \tag{7.2}$$

Then summing over i yields

$$\sum_i x_i P_{ij} = x_j \sum_i P_{ji} = x_j, \quad \sum_i x_i = 1$$

and, as the limiting probabilities π_i are the unique solution of the above, it follows that $x_i = \pi_i$ for all i.

Example 7a Consider a random walk with states $0, 1, \ldots, M$ and transition probabilities

$$P_{i, i+1} = \alpha_i = 1 - P_{i, i-1}, \quad i = 1, \ldots, \quad M - 1$$

$$P_{0, 1} = \alpha_0 = 1 - P_{0, 0}$$

$$P_{M, M} = \alpha_M = 1 - P_{M, M-1}$$

Without the need of any computations, it is possible to argue that this Markov chain, which can only make transitions from a state to one of its two nearest neighbors, is time reversible. This follows by noting that the number of transitions from i to $i + 1$ must at all times be within 1 of the number from $i + 1$ to i. This is so since between any two transitions from i to $i + 1$ there must be one from $i + 1$ to i (and conversely) since the only way to reenter i from a higher state is via state $i + 1$. Hence, it follows that the rate of transitions from i to $i + 1$ equals the rate from $i + 1$ to i, and so the process is time reversible.

We can easily obtain the limiting probabilities by equating for each state $i = 0, 1, \ldots, M - 1$ the rate at which the process goes from i to $i + 1$ with the rate at which it goes from $i + 1$ to i. This yields

$$\pi_0 \alpha_0 = \pi_1(1 - \alpha_1)$$

$$\pi_1 \alpha_1 = \pi_2(1 - \alpha_2)$$

$$\vdots$$

$$\pi_i \alpha_i = \pi_{i+1}(1 - \alpha_{i+1}), \quad i = 0, 1, \ldots, \; M - 1$$

Solving in terms of π_0 yields

$$\pi_1 = \frac{\alpha_0}{1 - \alpha_1} \pi_0$$

$$\pi_2 = \frac{\alpha_1}{1 - \alpha_2} \pi_1$$

$$= \frac{\alpha_1 \alpha_0}{(1 - \alpha_2)(1 - \alpha_1)} \pi_0$$

and, in general,

$$\pi_i = \frac{\alpha_{i-1} \cdot \cdot \cdot \alpha_0}{(1 - \alpha_i) \cdot \cdot \cdot (1 - \alpha_1)} \pi_0, \qquad i = 1, 2, \ldots, M$$

Since $\Sigma_0^M \pi_i = 1$, we obtain

$$\pi_0 \left[1 + \sum_{j=1}^{M} \frac{\alpha_{i-1} \cdot \cdot \cdot \alpha_0}{(1 - \alpha_j) \cdot \cdot \cdot (1 - \alpha_1)} \right] = 1$$

or

$$\pi_0 = \left[1 + \sum_{j=1}^{M} \frac{\alpha_{j-1} \cdot \cdot \cdot \alpha_0}{(1 - \alpha_j) \cdot \cdot \cdot (1 - \alpha_1)} \right]^{-1} \tag{7.3}$$

and

$$\pi_i = \frac{\alpha_{i-1} \cdot \cdot \cdot \alpha_0}{(1 - \alpha_1) \cdot \cdot \cdot (1 - \alpha_1)} \pi_0, \qquad i = 1, \ldots, M \tag{7.4}$$

For instance, if $\alpha_i \equiv \alpha$, then

$$\pi_0 = \left[1 + \sum_{j=1}^{M} \left(\frac{\alpha}{1 - \alpha} \right)^{j} \right]^{-1}$$

$$= \frac{1 - \beta}{1 - \beta^{M+1}}$$

and, in general,

$$\pi_i = \frac{\beta^i(1 - \beta)}{1 - \beta^{M+1}}, \qquad i = 0, 1, \ldots, M$$

where

$$\beta = \frac{\alpha}{1 - \alpha}$$

Another special case of Example 7a is the following urn model, proposed by the physicists P. and T. Ehrenfest to describe the movements of molecules. Suppose that M molecules are distributed among two urns; and at each time point one of the molecules is chosen at random and is then removed from its urn and placed in the other one. The number of molecules in urn I is a special case of the Markov chain of Example 7a, having

$$\alpha_i = \frac{M - i}{M}, \qquad i = 1, \ldots, M$$

Hence, using Equations (7.3) and (7.4) the limiting probabilities in this case are

$$\pi_0 = \left[1 + \sum_{j=1}^{M} \frac{(M - j + 1) \cdot\cdot\cdot (M - 1)M}{j(j - 1) \cdot\cdot\cdot 1} \right]^{-1}$$

$$= \left[\sum_{j=0}^{M} \binom{M}{j} \right]^{-1}$$

$$= (\tfrac{1}{2})^M$$

where we have used the identity

$$1 = (\tfrac{1}{2} + \tfrac{1}{2})^M$$

$$= \sum_{j=0}^{M} \binom{M}{j} (\tfrac{1}{2})^M$$

Hence, from Equation (7.4)

$$\pi_i = \binom{M}{i}(\tfrac{1}{2})^M, \qquad i = 0, 1, \ldots, M$$

As the above are just the binomial probabilities, it follows that in the long run, the positions of each of the M balls are independent and each one is equally likely to be in either urn. This, however, is quite intuitive, for if we focus on any one ball, it becomes quite clear that its position will be independent of the positions of the other balls (since no matter where the other $M - 1$ balls are, the ball under consideration at each stage will be moved with probability $1/M$) and by symmetry, it is equally likely to be in either urn. ◇

Example 7b Consider an arbitrary connected graph (see Section 6 of Chapter 3 for definitions) having a number w_{ij} associated with arc (i, j) for each arc. One instance of such a graph is given by Figure 4.1. Now consider a particle moving from node to node in this manner: If at any time the particle resides at node i, then it will next move to node j with probability P_{ij} where

$$P_{ij} = \frac{w_{ij}}{\sum_j w_{ij}}$$

and where w_{ij} is 0 if (i, j) is not an arc. For instance, for the graph of Figure 4.1, $P_{12} = 3/(3 + 1 + 2) = \frac{1}{2}$.

The time reversibility equations

$$\pi_i P_{ij} = \pi_j P_{ji}$$

reduce to

$$\pi_i \frac{w_{ij}}{\sum_j w_{ij}} = \pi_j \frac{w_{ji}}{\sum_j w_{ji}}$$

or, equivalently, since $w_{ij} = w_{ji}$

$$\frac{\pi_i}{\sum_j w_{ij}} = \frac{\pi_j}{\sum_i w_{ji}}$$

which is equivalent to

$$\frac{\pi_i}{\sum_j w_{ij}} = c$$

or

$$\pi_i = c \sum_j w_{ij}$$

or, since $1 = \sum_i \pi_i$

$$\pi_i = \frac{\sum_j w_{ij}}{\sum_i \sum_j w_{ij}}$$

As the π_i's given by this equation satisfy the time reversibility equations, it follows that the process is time reversible with these limiting probabilities.

For the graph of Figure 4.1 we have that

$$\pi_1 = \tfrac{6}{32}, \quad \pi_2 = \tfrac{3}{32}, \quad \pi_3 = \tfrac{6}{32}, \quad \pi_4 = \tfrac{5}{32}, \quad \pi_5 = \tfrac{12}{32}$$

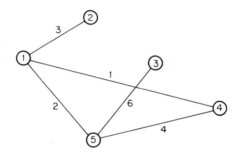

Figure 4.1   A connected graph with arc weights.   ◊

If we try to solve Equation (7.2) for an arbitrary Markov chain with states $0, 1, \ldots, M$, it will usually turn out that no solution exists. For example, from Equation (7.2)

$$x_i P_{ij} = x_j P_{ji}$$
$$x_k P_{kj} = x_j P_{jk}$$

implying (if $P_{ij} P_{jk} > 0$) that

$$\frac{x_i}{x_k} = \frac{P_{ji} P_{kj}}{P_{ij} P_{jk}}$$

which in general need not equal P_{ki}/P_{ik}. Thus, we see that a necessary condition for time reversibility is that

$$P_{ik} P_{kj} P_{ji} = P_{ij} P_{jk} P_{ki} \quad \text{for all} \quad i, j, k \tag{7.5}$$

which is equivalent to the statement that, starting in state i, the path

$i \to k \to j \to i$ has the same probability as the reversed path $i \to j \to k \to i$. To understand the necessity of this note that time reversibility implies that the rate at which a sequence of transitions from i to k to j to i occurs must equal the rate of ones from i to j to k to i (why?), and so we must have

$$\pi_i P_{ik} P_{kj} P_{ji} = \pi_i P_{ij} P_{jk} P_{ki}$$

implying Equation (7.5) when $\pi_i > 0$.

In fact we can show

Theorem 7.1. A finite ergodic Markov chain for which $P_{ij} = 0$ whenever $P_{ji} = 0$ is time reversible if and only if starting in state i, any path back to i has the same probability as the reversed path. That is, if

$$P_{i, i_1} P_{i_1, i_2} \cdot \cdot \cdot P_{i_k, i} = P_{i, i_k} P_{i_k, i_{k-1}} \cdot \cdot \cdot P_{i_1, i} \qquad (7.6)$$

for all states $i, i_1, \ldots, i_k$.

Proof: We have already proven necessity. To prove sufficiency, fix states i and j and rewrite (7.6) as

$$P_{i, i_1} P_{i_1, i_2} \cdot \cdot \cdot P_{i_k, j} P_{ji} = P_{ij} P_{j, i_k} \cdot \cdot \cdot P_{i_1, i}$$

Summing the above over all states $i_1, \ldots, i_k$ yields

$$P_{ij}^{k+1} P_{ji} = P_{ij} P_{ji}^{k+1}$$

Letting $k \to \infty$ yields that

$$\pi_j P_{ji} = P_{ij} \pi_i$$

which proves the theorem. $\Diamond$

Example 7c Suppose we are given a set of n elements, numbered 1 through n, which are to be arranged in some ordered list. At each unit of time a request is made to retrieve one of these elements, element i being requested (independently of the past) with probability P_i. After being requested, the element then is put back but not necessarily in the same position. In fact, let us suppose that the element requested is moved one closer to the front of the list; for instance, if the present list ordering is 1, 3, 4, 2, 5 and element 2 is requested, then the new ordering becomes 1, 3, 2, 4, 5. We are interested in the long-run average position of the element requested.

For any given probability vector $P = (P_1, \ldots, P_n)$, the above can be modeled as a Markov chain with $n!$ states, with the state at any time being the list order at that time. We shall show that this Markov chain is time reversible and then use this to show that the average position of the element requested when this one-closer rule is in effect is less than when the rule of always moving the requested element to the front of the line is used. The time reversibility of the resulting Markov chain when the one-closer reordering rule is in effect easily follows from Theorem 7.1. For instance, suppose $n = 3$ and consider the following path from state $(1, 2, 3)$ to itself

$$(1, 2, 3) \rightarrow (2, 1, 3) \rightarrow (2, 3, 1) \rightarrow (3, 2, 1) \rightarrow$$
$$(3, 1, 2) \rightarrow (1, 3, 2) \rightarrow (1, 2, 3)$$

The product of the transition probabilities in the forward direction is

$$P_2 P_3 P_3 P_1 P_1 P_2 = P_1^2 P_2^2 P_3^2$$

whereas in the reverse direction, it is

$$P_3 P_3 P_2 P_2 P_1 P_1 = P_1^2 P_2^2 P_3^2$$

As the general result follows in much the same manner, the Markov chain is indeed time reversible. (For a formal argument note that if f_i denotes the number of times element i moves forward in the path, then as the path goes from a fixed state back to itself, it follows that element i will also move backwards f_i times. Therefore, since the backwards moves of element i are precisely the times that it moves forward in the reverse path, it follows that the product of the transition probabilities for both the path and its reversal will equal

$$\prod_i P_i^{f_i + r_i}$$

where r_i is equal to the number of times that element i is in the first position and the path (or the reverse path) does not change states.)

For any permutation $i_1, i_2, \ldots, i_n$ of $1, 2, \ldots, n$, let $\pi(i_1, i_2, \ldots, i_n)$ denote the limiting probability under the one-closer rule. By time reversibility we have

$$P_{i_{j+1}} \pi(i_1, \ldots, i_j, i_{j+1}, \ldots, i_n)$$
$$= P_{i_j} \pi(i_1, \ldots, i_{j+1}, i_j, \ldots, i_n) \quad (7.8)$$

for all permutations.

Now the average position of the element requested can be expressed (as in Section 6.1 of Chapter 3) as

Average position $= \sum_i P_i E[\text{Position of element } i]$

$$= \sum_i P_i \left[1 + \sum_{j \neq i} P\{\text{element } j \text{ precedes element } i\} \right]$$

$$= 1 + \sum_i \sum_{j \neq i} P_i \, P\{e_j \text{ precedes } e_i\}$$

$$= 1 + \sum_{i<j} [P_i \, P\{e_j \text{ precedes } e_i\} + P_j \, P\{e_i \text{ precedes } e_j\}]$$

$$= 1 + \sum_{i<j} [P_i \, P\{e_j \text{ precedes } e_i\} + P_j(1 - P\{e_j \text{ precedes } e_i\})]$$

$$= 1 + \sum_{i<j} (P_i - P_j) \, P\{e_j \text{ precedes } e_i\} + \sum_{i<j} P_j$$

Hence, to minimize the average position of the element requested, we would want to make $P\{e_j \text{ precedes } e_i\}$ as large as possible when $P_j > P_i$ and as small as possible when $P_i > P_j$. Now under the front-of-the-line rule we showed in Section 6.1 of Chapter 3 that

$$P\{e_j \text{ precedes } e_i\} = \frac{P_j}{P_j + P_i}$$

(since under the front-of-the-line rule element j will precede element i if and only if the last request for either i or j was for j).

Therefore, to show that the one-closer rule is better than the front-of-the-line rule, it suffices to show that under the one-closer rule

$$P\{e_j \text{ precedes } e_i\} > \frac{P_j}{P_j + P_i} \quad \text{when} \quad P_j > P_i$$

Now consider any state where element i precedes element j, say $(. \ . \ . \ , i, i_1, . \ . \ . \ , i_k, j, . \ . \ .)$. By successive transpositions using Equation (7.8), we have

$$\pi(. \ . \ . \ , i, i_1, . \ . \ . \ , i_k, j, . \ . \ .)$$

$$= \left(\frac{P_i}{P_j} \right)^{k+1} \pi(. \ . \ . \ , j, i_1, . \ . \ . \ , i_k, i, . \ . \ .) \tag{7.9}$$

For instance,

$$\pi(1, 2, 3) = \frac{P_2}{P_3} \pi(1, 3, 2)$$

$$= \frac{P_2}{P_3} \frac{P_1}{P_3} \pi(3, 1, 2)$$

$$= \frac{P_2}{P_3} \frac{P_1}{P_3} \frac{P_1}{P_2} \pi(3, 2, 1)$$

$$= \left(\frac{P_1}{P_3}\right)^2 \pi(3, 2, 1)$$

Now when $P_j > P_i$, Equation (7.9) implies that

$$\pi(. \ . \ ., i, i_1, . \ . \ ., i_k, j, . \ . \ .) < \frac{P_i}{P_j} \pi(. \ . \ ., j, i_1, . \ . \ ., i_k, i, . \ . \ .)$$

Letting $\alpha(i, j) = P\{e_i \text{ precedes } e_j\}$, we see by summing over all states for which i precedes j and by using the above that

$$\alpha(i, j) < \frac{P_i}{P_j} \alpha(j, i)$$

which, since $\alpha(i, j) = 1 - \alpha(j, i)$, yields

$$\alpha(j, i) > \frac{P_j}{P_j + P_i}$$

Hence, the average position of the element requested is indeed smaller under the one-closer rule than under the front-of-the-line rule. $\diamond$

The concept of the reversed chain is useful even when the process is not time reversible. To illustrate this, we start with the following proposition whose proof is left as an exercise.

Proposition 7.1. Consider an irreducible Markov chain with transition probabilities P_{ij}. If one can find positive numbers π_i, $i \geq 0$, summing to one, and a transition probability matrix $\mathbf{Q} = [Q_{ij}]$ such that

$$\pi_i P_{ij} = \pi_j Q_{ji} \tag{7.10}$$

then the Q_{ij} are the transition probabilities of the reversed chain and the π_i are the stationary probabilities both for the original and reversed chain.

The importance of the above proposition is that, by thinking backwards, we can sometimes guess at the nature of the reversed chain and then use the set of equations (7.10) to obtain both the stationary probabilities and the Q_{ij}.

Example 7d A single bulb is necessary to light a given room. When the bulb in use fails, it is replaced by a new one at the beginning of the next day. Let X_n equal i if the bulb in use at the beginning of day n is in its ith day of use (that is, if its present age is i). For instance, if a bulb fails on day $n - 1$, then a new bulb will be put in use at the beginning of day n and so $X_n = 1$. If we suppose that each bulb, independently, fails on its ith day of use with probability p_i, $i \geq 1$, then it is easy to see that $\{X_n, n \geq 1\}$ is a Markov chain whose transition probabilities are as follows:

$$
\begin{aligned}
P_{i,1} &= P\{\text{bulb, on its } i\text{th day of use, fails}\} \\
&= P\{\text{life of bulb} = i \mid \text{life of bulb} \geq i\} \\
&= \frac{P\{L = i\}}{P\{L \geq i\}}
\end{aligned}
$$

where L, a random variable representing the lifetime of a bulb, is such that $P\{L = i\} = p_i$. Also,

$$
P_{i,i+1} = 1 - P_{i,1}
$$

Suppose now that this chain has been in operation for a long (in theory, an infinite) time and consider the sequence of states going backwards in time. Since, in the forward direction, the state is always increasing by 1 until it reaches the age at which the item fails, it is easy to see that the reverse chain will always decrease by 1 until it reaches 1 and then it will jump to a random value representing the lifetime of the (in real time) previous bulb. Thus, it seems that the reverse chain should have transition probabilities given by

$$
\begin{aligned}
Q_{i,i-1} &= 1, & i > 1 \\
Q_{1,i} &= p_i, & i \geq 1
\end{aligned}
$$

To check this, and at the same time determine the stationary probabilities, we must see if we can find, with the $Q_{i,j}$ as given above, positive numbers $\{\pi_i\}$ such that

$$
\pi_i P_{i,j} = \pi_j Q_{j,i}
$$

To begin, let $j = 1$ and consider the resulting equations

$$\pi_i P_{i,1} = \pi_1 Q_{1,i}$$

This is equivalent to

$$\pi_i \frac{P\{L = i\}}{P\{L \geq i\}} = \pi_1 \, P\{L = i\}$$

or

$$\pi_i = \pi_1 \, P\{L \geq i\}$$

Summing over all i yields

$$1 = \sum_{i=1}^{\infty} \pi_i = \pi_1 \sum_{i=1}^{\infty} P\{L \geq i\} = \pi_1 \, E[L]$$

and so, for the Q_{ij} above to represent the reverse transition probabilities, it is necessary that the stationary probabilities are

$$\pi_i = \frac{P\{L \geq i\}}{E[L]}, \qquad i \geq 1$$

To finish the proof that the reverse transition probabilities and stationary probabilities are as given all that remains is to show that they satisfy

$$\pi_i \, P_{i,i+1} = \pi_{i+1} \, Q_{i+1,i}$$

which is equivalent to

$$\frac{P\{L \geq i\}}{E[L]} \left(1 - \frac{P\{L = i\}}{P\{L \geq i\}} \right) = \frac{P\{L \geq i + 1\}}{E[L]}$$

and which is true since $P\{L \geq i\} - P\{L = i\} = P\{L \geq i + 1\}$.   ◊

8. Markov Decision Processes

Consider a process that is observed at discrete time points to be in any one of M possible states, which we number by $1, 2, \ldots, M$. After observing the state of the process, an action must be chosen, and we let A, assumed finite, denote the set of all possible actions.

If the process is in state i at time n and action a is chosen, then the next state of the system is determined according to the transition probabilities $P_{ij}(a)$.

If we let X_n denote the state of the process at time n and a_n the action chosen at time n, then the above is equivalent to stating that

$$P\{X_{n+1} = j | X_0, a_0, X_1, a_1, \ldots, X_n = i, a_n = a\} = P_{ij}(a)$$

Thus, the transition probabilities are functions only of the present state and the subsequent action.

By a policy, we mean a rule for choosing actions. We shall restrict ourselves to policies which are of the form that the action they prescribe at any time depends only on the state of the process at that time (and not on any information concerning prior states and actions). However, we shall allow the policy to be "randomized" in that its instructions may be to choose actions according to a probability distribution. In other words, a policy β is a set of numbers $\beta = \{\beta_i(a), a \in A, i = 1, \ldots, M\}$ with the interpretation that if the process is in state i, then action a is to be chosen with probability $\beta_i(a)$. Of course, we need have that

$$0 \leq \beta_i(a) \leq 1, \quad \text{for all } i, a$$

$$\sum_a \beta_i(a) = 1, \quad \text{for all } i$$

Under any given policy β, the sequence of states $\{X_n, n = 0, 1, \ldots\}$ constitutes a Markov chain with transition probabilities $P_{ij}(\beta)$ given by

$$P_{ij}(\beta) = P_\beta\{X_{n+1} = j | X_n = i\}^*$$

$$= \sum_a P_{ij}(a)\beta_i(a)$$

where the last equality follows by conditioning on the action chosen when in state i. Let us suppose that for every choice of a policy β, the resultant Markov chain $\{X_n, n = 0, 1, \ldots\}$ is ergodic.

For any policy β, let π_{ia} denote the limiting (or steady-state) probability that the process will be in state i and action a will be chosen if policy β is employed. That is,

$$\pi_{ia} = \lim_{n \to \infty} P_\beta\{X_n = i, a_n = a\}$$

---

*We use the notation P_β to signify that the probability is conditional on the fact that policy β is used.

The vector $\boldsymbol{\pi} = (\pi_{ia})$ must satisfy

    (i) $\pi_{ia} \geq 0$ for all i, a
    (ii) $\Sigma_i \Sigma_a \pi_{ia} = 1$
    (iii) $\Sigma_a \pi_{ja} = \Sigma_i \Sigma_a \, \pi_{ia} P_{ij}(a)$ for all j (8.1)

Equations (i) and (ii) are obvious, and Equation (iii) which is an analogue of Equation (4.1) follows as the left-hand side equals the steady-state probability of being in state j and the right-hand side is the same probability computed by conditioning on the state and action chosen one stage earlier.

Thus for any policy $\boldsymbol{\beta}$, there is a vector $\boldsymbol{\pi} = (\pi_{ia})$ which satisfies (i)–(iii) and with the interpretation that π_{ia} is equal to the steady-state probability of being in state i and choosing action a when policy $\boldsymbol{\beta}$ is employed. Moreover, it turns out that the reverse is also true. Namely, for any vector $\boldsymbol{\pi} = (\pi_{ia})$ which satisfies (i)–(iii), there exists a policy $\boldsymbol{\beta}$ such that if $\boldsymbol{\beta}$ is used, then the steady-state probability of being in i and choosing action a equals π_{ia}. To verify this last statement, suppose that $\boldsymbol{\pi} = (\pi_{ia})$ is a vector which satisfies (i)–(iii). Then, let the policy $\boldsymbol{\beta} = (\beta_i(a))$ be

$$\beta_i(a) = P\{\boldsymbol{\beta} \text{ chooses } a | \text{state is } i\}$$

$$= \frac{\pi_{ia}}{\displaystyle\sum_a \pi_{ia}}$$

Now let P_{ia} denote the limiting probability of being in i and choosing a when policy $\boldsymbol{\beta}$ is employed. We need to show that $P_{ia} = \pi_{ia}$. To do so, first note that $\{P_{ia}, \ i = 1, \ldots, M, a \ \varepsilon \ A\}$ are the limiting probabilities of the two-dimensional Markov chain $\{(X_n, a_n), n \geq 0\}$. Hence, by the fundamental Theorem 4.1, they are the unique solution of

    (i') $P_{ia} \geq 0$
    (ii') $\Sigma_i \Sigma_a P_{ia} = 1$
    (iii') $P_{ja} = \Sigma_i \Sigma_{a'} P_{ia'} P_{ij}(a') \beta_j(a)$

where (iii') follows since

$$P\{X_{n+1} = j, a_{n+1} = a | X_n = i, a_n = a'\} = P_{ij}(a')\beta_j(a)$$

Since

$$\beta_j(a) = \frac{\pi_{ja}}{\displaystyle\sum_a \pi_{ja}}$$

we see that (P_{ia}) is the unique solution of

$$P_{ia} \geq 0$$

$$\sum_i \sum_a P_{ia} = 1$$

$$P_{ja} = \sum_i \sum_{a'} P_{ia'} P_{ij}(a') \frac{\pi_{ja}}{\sum_a \pi_{ja}}$$

Hence, to show that $P_{ia} = \pi_{ia}$, we need show that

$$\pi_{ia} \geq 0$$

$$\sum_i \sum_a \pi_{ia} = 1$$

$$\pi_{ja} = \sum_i \sum_{a'} \pi_{ia'} P_{ij}(a') \frac{\pi_{ja}}{\sum_a \pi_{ja}}$$

The top two equations follow from (i) and (ii) of Equation (8.1), and the third which is equivalent to

$$\sum_a \pi_{ja} = \sum_i \sum_{a'} \pi_{ia'} P_{ij}(a')$$

follows from Condition (iii) of Equation (8.1).

Thus we have shown that a vector $\boldsymbol{\pi} = (\pi_{ia})$ will satisfy (i), (ii), and (iii) of Equation (8.1) if and only if there exists a policy $\boldsymbol{\beta}$ such that π_{ia} is equal to the steady-state probability of being in state i and choosing action a when $\boldsymbol{\beta}$ is used. In fact, the policy $\boldsymbol{\beta}$ is defined by $\beta_i(a) = \pi_{ia}/\sum_a \pi_{ia}$.

The preceding is quite important in the determination of "optimal" policies. For instance, suppose that a reward $R(i, a)$ is earned whenever action a is chosen in state i. Since $R(X_i, a_i)$ would then represent the reward earned at time i, the expected average reward per unit time under policy $\boldsymbol{\beta}$ can be expressed as

$$\text{Expected average reward under } \boldsymbol{\beta} = \lim_{n \to \infty} E_{\boldsymbol{\beta}} \left[\frac{\sum_{i=1}^{n} R(X_i, a_i)}{n} \right]$$

Now, if π_{ia} denotes the steady-state probability of being in state i and choosing action a, it follows that the limiting expected reward at time n equals

$$\lim_{n \to \infty} E[R(X_n, a_n)] = \sum_i \sum_a \pi_{ia} R(i, a)$$

which implies (see Problem 41) that

$$\text{Expected average reward under } \boldsymbol{\beta} = \sum_i \sum_a \pi_{ia} R(i, a)$$

Hence, the problem of determining the policy that maximizes the expected average reward is

$$\underset{\boldsymbol{\pi} = (\pi_{ia})}{\text{Maximize}} \sum_i \sum_a \pi_{ia} R(i, a)$$

$$\text{subject to} \quad \pi_{ia} \geq 0 \quad \text{for all} \quad i, a$$

$$\sum_i \sum_a \pi_{ia} = 1$$

$$\sum_a \pi_{ja} = \sum_i \sum_a \pi_{ia} P_{ij}(a) \quad \text{for all} \quad j \qquad (8.2)$$

However, the above maximization problem is a special case of what is known as a *linear program** and can thus be solved by a standard linear programming algorithm known as the *simplex algorithm*. If $\boldsymbol{\pi}^* = (\pi_{ia}^*)$ maximizes the above, then the optimal policy will be given by $\boldsymbol{\beta}^*$ where

$$\beta_i^*(a) = \frac{\pi_{ia}^*}{\displaystyle\sum_a \pi_{ia}^*}$$

Remarks

(i) It can be shown that the $\boldsymbol{\pi}^*$ maximizing Equation (8.2) has the property that π_{ia}^* is zero for all but one value of a, which implies that the optimal policy is nonrandomized, that is, the action it prescribes when in state i is a deterministic function of i.

(ii) The linear programming formulation also often works when there are restrictions placed on the class of allowable policies. For instance, suppose there is a restriction on the fraction of time the process spends in some state, say state 1. Specifically, suppose that we are only allowed to consider policies having the property that their use results in the process being in state 1 less than 100α percent of time. To determine the optimal policy subject to this re-

---

* It is called a linear program since the objective function $\Sigma_i \Sigma_q R(i, a)\pi_{ia}$ and the constraints are all linear functions of the π_{ia}.

quirement, we add to the linear programming problem the additional constraint

$$\sum_a \pi_{1a} \leq \alpha$$

since $\Sigma_a \pi_{1a}$ represents the proportion of time that the process is in state 1. ◇

Problems

1. Three white and three black balls are distributed in two urns in such a way that each contains three balls. We say that the system is in state i, $i = 0, 1, 2, 3$, if the first urn contains i white balls. At each step, we draw one ball from each urn and place the ball drawn from the first urn into the second, and conversely with the ball from the second urn. Let X_n denote the state of the system after the nth step. Explain why $\{X_n, n = 0, 1, 2, \ldots\}$ is a Markov chain and calculate its transition probability matrix.

2. Suppose that whether or not it rains today depends on previous weather conditions through the last three days. Show how this system may be analyzed by using a Markov chain. How many states are needed?

3. In Problem 2, suppose that if it has rained for the past three days, then it will rain today with probability .8; if it did not rain for any of the past three days, then it will rain today with probability .2; and in any other case the weather today will, with probability .6, be the same as the weather yesterday. Determine **P** for this Markov chain.

4. Consider a process $\{X_n, n = 0, 1, \ldots\}$ which takes on the values 0, 1, or 2. Suppose

$$P\{X_{n+1} = j | X_n = i, X_{n-1} = i_{n-1}, \ldots, X_0 = i_0\} = \begin{cases} P_{ij}^{\mathrm{I}}, & \text{when } n \text{ is even} \\ P_{ij}^{\mathrm{II}}, & \text{when } n \text{ is odd} \end{cases}$$

where $\Sigma_{j=0}^2 P_{ij}^{\mathrm{I}} = \Sigma_{j=0}^2 P_{ij}^{\mathrm{II}} = 1$, $i = 0, 1, 2$. Is $\{X_n, n \geq 0\}$ a Markov chain? If not, then show how, by enlarging the state space, we may transform it into a Markov chain.

5. Let the transition probability matrix of a two-state Markov chain be given, as in Example 1b, by

$$\mathbf{P} = \left\| \begin{matrix} p & 1 - p \\ 1 - p & p \end{matrix} \right\|$$

Show by mathematical induction that

$$\mathbf{P}^{(n)} = \left\| \begin{array}{cc} \frac{1}{2} + \frac{1}{2}(2p-1)^n & \frac{1}{2} - \frac{1}{2}(2p-1)^n \\ \frac{1}{2} - \frac{1}{2}(2p-1)^n & \frac{1}{2} + \frac{1}{2}(2p-1)^n \end{array} \right\|$$

6. In Example 1d suppose that it has rained neither yesterday nor the day before yesterday. What is the probability that it will rain tomorrow?

7. Suppose that coin 1 has probability .7 of coming up heads, and coin 2 has probability .6 of coming up heads. If the coin flipped today comes up heads, then we select coin 1 to flip tomorrow, and if it comes up tails, then we select coin 2 to flip tomorrow. If the coin initially flipped is equally likely to be coin 1 or coin 2, then what is the probability that the coin flipped on the third day after the initial flip is coin 1?

8. Specify the classes of the following Markov chains and determine whether they are transient or recurrent.

$$\mathbf{P}_1 = \left\| \begin{array}{ccc} 0 & \frac{1}{2} & \frac{1}{2} \\ \frac{1}{2} & 0 & \frac{1}{2} \\ \frac{1}{2} & \frac{1}{2} & 0 \end{array} \right\| \qquad\qquad \mathbf{P}_2 = \left\| \begin{array}{cccc} 0 & 0 & 0 & 1 \\ 0 & 0 & 0 & 1 \\ \frac{1}{2} & \frac{1}{2} & 0 & 0 \\ 0 & 0 & 1 & 0 \end{array} \right\|$$

$$\mathbf{P}_3 = \left\| \begin{array}{ccccc} \frac{1}{2} & 0 & \frac{1}{2} & 0 & 0 \\ \frac{1}{4} & \frac{1}{2} & \frac{1}{4} & 0 & 0 \\ \frac{1}{2} & 0 & \frac{1}{2} & 0 & 0 \\ 0 & 0 & 0 & \frac{1}{2} & \frac{1}{2} \\ 0 & 0 & 0 & \frac{1}{2} & \frac{1}{2} \end{array} \right\| \qquad \mathbf{P}_4 = \left\| \begin{array}{ccccc} \frac{1}{4} & \frac{3}{4} & 0 & 0 & 0 \\ \frac{1}{2} & \frac{1}{2} & 0 & 0 & 0 \\ 0 & 0 & 1 & 0 & 0 \\ 0 & 0 & \frac{1}{3} & \frac{2}{3} & 0 \\ 1 & 0 & 0 & 0 & 0 \end{array} \right\|$$

9. Prove that if the number of states in a Markov chain is M, and if state j can be reached from state i, then it can be reached in M steps or less.

10. Show that if state i is recurrent and state i does not communicate with state j, then $P_{ij} = 0$. This implies that once a process enters a recurrent class of states it can never leave that class. For this reason, a recurrent class is often referred to as a *closed* class.

11. For the random walk of Example 3e use the strong law of large numbers to give another proof that the Markov chain is transient when $p \neq \frac{1}{2}$.

Hint: Note that the state at time n can be written as $\sum_{i=1}^{n} Y_i$ where the Y_i's are independent and $P\{Y_i = 1\} = p = 1 - P\{Y_i = -1\}$.

Argue that if $p > \frac{1}{2}$, then, by the strong law of large numbers, $\Sigma_1^n Y_i \to \infty$ as $n \to \infty$ and hence the initial state 0 can be visited only finitely often, and hence must be transient. A similar argument holds when $p < \frac{1}{2}$.

12. For Example 1d, calculate the proportion of days that it rains.

13. A transition probability matrix **P** is said to be doubly stochastic if the sum over each column equals one; that is,

$$\sum_i P_{ij} = 1, \quad \text{for all } j$$

If such a chain is irreducible and aperiodic and consists of $M + 1$ states $0, 1, \ldots, M$, show that the limiting probabilities are given by

$$\pi_j = \frac{1}{M + 1}, \quad j = 0, 1, \ldots, M$$

14. A particle moves on a circle through points which have been marked 0, 1, 2, 3, 4 (in a clockwise order). At each step it has a probability p of moving to the right (clockwise) and $1 - p$ to the left (counterclockwise). Let X_n denote its location on the circle after the nth step. The process $\{X_n, n \geq 0\}$ is a Markov chain.
  (a) Find the transition probability matrix.
  (b) Calculate the limiting probabilities.

15. Let Y_n be the sum of n independent rolls of a fair die. Find

$$\lim_{n \to \infty} P\{X_n \text{ is a multiple of 13}\}$$

Hint: Define an appropriate Markov chain and apply the results of Problem 13.

16. Each morning an individual leaves his house and goes for a run. He is equally likely to leave either from his front or back door. Upon leaving the house, he chooses a pair of running shoes (or goes running barefoot if there are no shoes at the door from which he departed). On his return he is equally likely to enter, and leave his running shoes, either by the front or back door. If he owns a total of k pairs of running shoes, what proportion of the time does he run barefooted?

17. Consider the following approach to shuffling a deck of n cards. Starting with any initial ordering of the cards, one of the numbers 1,2, . . . , n is randomly chosen in such a manner that each one is equally likely to be selected. If it is number i that is chosen, then we take the

card that is in position i and put it on top of the deck—that is, we put that card in position 1. We then repeatedly perform the same operation. Show that, in the limit, the deck is perfectly shuffled in the sense that the resultant ordering is equally likely to be any of the $n!$ possible orderings.

18. Determine the limiting probabilities π_j for the model presented in Problem 1.

19. For a series of dependent trials the probability of success on any trial is $(k + 1)/(k + 2)$ where k is equal to the number of successes on the previous two trials. Compute $\lim_{n\to\infty} P\{\text{success on the } n\text{th trial}\}$.

20. An organization has N employees where N is a large number. Each employee has one of three possible job classifications and changes classifications (independently) according to a Markov chain with transition probabilities

$$\begin{bmatrix} .7 & .2 & .1 \\ .2 & .6 & .2 \\ .1 & .4 & .5 \end{bmatrix}$$

What percentage of employees are in each classification?

21. Each day, one of n possible elements is requested, the ith one with probability P_i, $i \geq 1$, $\Sigma_1^n P_i = 1$. These elements are at all times arranged in an ordered list which is revised as follows: The element selected is moved to the front of the list with the relative positions of all the other elements remaining unchanged. Define the state at any time to be the list ordering at that time and note that there are $n!$ possible states.

(a) Argue that the above is a Markov chain.
(b) For any state $i_1, \ldots, i_n$ (which is a permutation of 1, 2, $\ldots, n$), let $\pi(i_1, \ldots, i_n)$ denote the limiting probability. In order for the state to be $i_1, \ldots, i_n$, it is necessary that the last request was for i_1, the last non-i_1 request was for i_2, the last non-i_1 or i_2 request was for i_3, and so on. Hence, it appears intuitive that

$$\pi(i_1, \ldots, i_n) = P_{i_1} \frac{P_{i_2}}{1 - P_{i_1}} \frac{P_{i_3}}{1 - P_{i_1} - P_{i_2}} \cdots \frac{P_{i_{n-1}}}{1 - P_{i_1} - \cdots - P_{i_{n-2}}}$$

Verify when $n = 3$ that the above are indeed the limiting probabilities.

22. Suppose that a population consists of a fixed number, say m, of genes in any generation. Each gene is one of two possible genetic

types. If any generation has exactly i (of its m) genes being type 1, then the next generation will have j type 1 (and $m - j$ type 2) genes with probability

$$\binom{m}{j}\left(\frac{i}{m}\right)^{j}\left(\frac{m-i}{m}\right)^{m-j}, \qquad j = 0, 1, \ldots, m$$

Let X_n denote the number of type 1 genes in the nth generation, and assume that $X_0 = i$.

(a) Find $E[X_n]$.

(b) What is the probability that eventually all the genes will be type 1?

23. Consider an irreducible finite Markov chain with states 0, 1, . . . , N.

(a) Starting in state i, what is the probability the process will ever visit state j? Explain!

(b) Let $x_i = P\{$visit state N before state $0|$start in $i\}$. Compute a set of linear equations which the x_i satisfy, $i = 0, 1, \ldots, N$.

(c) If $\sum_j j\, p_{ij} = i$ for $i = 1, \ldots, N - 1$, show that $x_i = i/N$ is

a solution to the equations in part (b).

24. An individual possesses r umbrellas which he employs in going from his home to office, and vice versa. If he is at home (the office) at the beginning (end) of a day and it is raining, then he will take an umbrella with him to the office (home), provided there is one to be taken. If it is not raining, then he never takes an umbrella. Assume that, independent of the past, it rains at the beginning (end) of a day with probability p.

(i) Define a Markov chain with $r + 1$ states which will help us to determine the proportion of time that our man gets wet. (Note: He gets wet if it is raining, and all umbrellas are at his other location.)

(ii) Show that the limiting probabilities are given by

$$\pi_i = \begin{cases} \dfrac{q}{r + q}, & \text{if } i = 0 \\[2ex] \dfrac{1}{r + q}, & \text{if } i = 1, \ldots, r \end{cases} \qquad \text{where} \quad q = 1 - p$$

(iii) What fraction of time does our man get wet?

(iv) When $r = 3$, what value of p maximizes the fraction of time he gets wet?

25. Let $\{X_n, n \geq 0\}$ denote an ergodic Markov chain with limiting probabilities π. Define the process $\{Y_n, n \geq 1\}$ by $Y_n = (X_{n-1}, X_n)$. That is, Y_n keeps track of the last two states of the original chain. Is $\{Y_n, n \geq 1\}$ a Markov chain? If so, determine its transition probabilities and find

$$\lim_{n \to \infty} P\{Y_n = (i, j)\}$$

26. Verify the transition probability matrix given in Example 4d.

27. Let $P^{(1)}$ and $P^{(2)}$ denote transition probability matrices for ergodic Markov chains having the same state space. Let π^1 and π^2 denote the stationary (limiting) probability vectors for the two chains. Consider a process defined as

(i) $X_0 = 1$. A coin is then flipped and if it comes up heads, then the remaining states $X_1, \ldots$ are obtained from the transition probability matrix $P^{(1)}$ and if tails from the matrix $P^{(2)}$. Is $\{X_n, n \geq 0\}$ a Markov chain? If $p = P\{\text{coin comes up heads}\}$, what is $\lim_{n \to \infty} P(X_n = i)$?

(ii) $X_0 = 1$. At each stage the coin is flipped and if it comes up heads, then the next state is chosen according to P_1 and if tails comes up, then it is chosen according to P_2. In this case do the successive states constitute a Markov chain? If so, determine the transition probabilities. Show by a counterexample that the limiting probabilities are not the same as in Part (i).

28. A fair coin is continually flipped. Compute the expected number of flips until the following patterns appear.
    (a) HHTTHT.
    (b) HHTTHH.
    (c) HHTHHT.

29. Consider the Ehrenfest urn model in which M molecules are distributed among 2 urns, and at each time point one of the molecules is chosen at random and is then removed from its urn and placed in the other one. Let X_n denote the number of molecules in urn 1 after the nth switch and let $\mu_n = E[X_n]$. Show that
    (i) $\mu_{n+1} = 1 + (M - 2/M)\mu_n$
    (ii) Use (i) to prove that

$$\mu_n = \frac{M}{2} + \left(\frac{M-2}{M}\right)^n \left(E[X_0] - \frac{M}{2}\right)$$

30. Consider a population of individuals each of whom possesses two genes which can be either type A or type a. Suppose that in outward

appearance type A is dominant and type a is recessive. (That is, an individual will only have the outward characteristics of the recessive gene if its pair is aa.) Suppose that the population has stabilized, and the percentages of individuals having respective gene pairs AA, aa, and Aa are p, q, and r. Call an individual dominant or recessive depending on the outward characteristics it exhibits. Let S_{11} denote the probability that an offspring of two dominant parents will be recessive; and let S_{10} denote the probability that the offspring of one dominant and one recessive parent will be recessive. Compute S_{11} and S_{10} to show that $S_{11} = S_{10}^2$. (The quantities S_{10} and S_{11} are known in the genetics literature as *Snyder's ratios*.

31. In the gambler's ruin problem of Section 5, suppose the gambler's fortune is presently i, and suppose that we know that the gambler's fortune will eventually reach N (before it goes to 0). Given this information, show that the probability he wins the next gamble is

$$\frac{p[1 - (q/p)^{i+1}]}{[1 - (q/p)^i]}, \quad \text{if} \quad p \neq \tfrac{1}{2}$$

$$\frac{i + 1}{2i}, \quad \text{if} \quad p = \tfrac{1}{2}$$

Hint: The probability we want is

$$P\{X_{n+1} = i + 1 | X_n = i, \quad \lim_{m \to \infty} X_m = N\}$$

$$= \frac{P\{X_{n+1} = i + 1, \quad \lim_m X_m = N | X_n = i\}}{P\{\lim_m X_m = N | X_n = i\}}$$

32. For the gambler's ruin model of Section 5, let M_i denote the mean number of games that must be played until the gambler either goes broke or reaches a fortune of N, given that he starts with i, $i = 0, 1, \ldots, N$. Show that M_i satisfies

$$M_0 = M_N = 0; \quad M_i = 1 + pM_{i+1} + qM_{i-1}, \quad i = 1, \ldots, N - 1$$

33. Solve the equations given in Problem 32 to obtain

$$M_i = i(N - i), \quad \text{if} \quad p = \tfrac{1}{2}$$

$$= \frac{i}{q - p} - \frac{N}{q - p} \frac{1 - (q/p)^i}{1 - (q/p)^N}, \quad \text{if} \quad p \neq \tfrac{1}{2}$$

34. Consider a branching process having $\mu < 1$. Show that if $X_0 = 1$, then the expected number of individuals that ever exist in this population is given by $1/(1 - \mu)$. What if $X_0 = n$?

35. In a branching process having $X_0 = 1$ and $\mu > 1$, prove that π_0 is the *smallest* positive number satisfying Equation (6.3).

Hint: Let π be any solution of $\pi = \sum_{j=0}^{\infty} \pi^j P_j$. Show by mathematical induction that $\pi \geq P\{X_n = 0\}$ for all n, and let $n \to \infty$. In using the induction argue that

$$P\{X_n = 0\} = \sum_{j=0}^{\infty} (P\{X_{n-1} = 0\})^j P_j$$

36. For a branching process, calculate π_0 when
(a) $P_0 = \frac{1}{4}, P_2 = \frac{3}{4}$
(b) $P_0 = \frac{1}{4}, P_1 = \frac{1}{2}, P_2 = \frac{1}{4}$
(c) $P_0 = \frac{1}{6}, P_1 = \frac{1}{2}, P_3 = \frac{1}{3}$

37. At all times, an urn contains N balls—some white balls and some black balls. At each stage, a coin having probability p, $0 < p < 1$, of landing heads is flipped. If heads appears, then a ball is chosen at random from the urn and is replaced by a white ball; if tails appears, then a ball is chosen from the urn and is replaced by a black ball. Let X_n denote the number of white balls in the urn after the nth stage.

(a) Is $\{X_n, n \geq 0\}$ a Markov chain? If so, explain why.
(b) What are its classes? What are their periods? Are they transient or recurrent?
(c) Compute the transition probabilities P_{ij}.
(d) Let $N = 2$. Find the proportion of time in each state.
(e) Based on your answer in part (d) and your intuition, guess the answer for the limiting probability in the general case.
(f) Prove your guess in part (e) either by showing that Equation (4.1) is satisfied or by using the results of Example 7a.
(g) If $p = 1$, what is the expected time until there are only white balls in the urn if initially there are i white and $N - i$ black?

38. (a) Show that the limiting probabilities of the reversed Markov chain are the same as for the forward chain by showing that they satisfy the equations

$$\pi_j = \sum_i \pi_i Q_{ij}$$

(b) Give an intuitive explanation for the result of part (a).

39. M balls are initially distributed among m urns. At each stage one of the balls is selected at random, taken from whichever urn it is

in, and then placed, at random, in one of the other $M - 1$ urns. Consider the Markov chain whose state at any time is the vector $(n_1, \ldots, n_m)$ where n_i denotes the number of balls in urn i. Guess at the limiting probabilities for this Markov chain and then verify your guess and show at the same time that the Markov chain is time reversible.

40. It follows from Theorem 7.1 that for a time reversible Markov chain

$$P_{ij}P_{jk}P_{ki} = P_{ik}P_{kj}P_{ji}, \quad \text{for all} \quad i, j, k$$

It turns out that if the state space is finite and $P_{ij} > 0$ for all i, j, then the above is also a sufficient condition for time reversibility. (That is, in this case, we need only check Equation (7.6) for paths from i to i which have only two intermediate states.) Prove this.

Hint: Fix i and show that the equations

$$\pi_j P_{jk} = \pi_k P_{kj}$$

are satisfied by $\pi_j = (cP_{ij})/(P_{ji})$ where c is chosen so that $\Sigma_j \pi_j = 1$.

41. For a time reversible Markov chain, show that the rate at which transitions from i to j to k occur must equal the rate at which transitions from k to j to i occur.

42. Consider an ergodic Markov chain which need not be time reversible. Show that the limiting probabilities for the reversed Markov chain are the same as for the original chain. Give an intuitive explanation for this.

43. A Markov chain is said to be a tree process if
(i) $P_{ij} > 0$ whenever $P_{ji} > 0$.
(ii) for every pair of states i and j, $i \neq j$, there is a unique sequence of distinct states $i = i_0, i_1, \ldots, i_{n-1}, i_n = j$ such that

$$P_{i_k, i_{k+1}} > 0, \quad K = 0, 1, \ldots, n - 1$$

In other words, a Markov chain is a tree process if for every pair of distinct states i and j there is a unique way for the process to go from i to j without reentering a state (and this path is the reverse of the unique path from j to i). Argue that an ergodic tree process is time reversible.

44. On a chessboard compute the expected number of plays it takes a knight, starting in one of the four corners of the chessboard, to return to its initial position if we assume that at each play it is equally likely to choose any of its legal moves. (No other pieces are on the board.)

Hint: Make use of Example 7b.

45. In a Markov decision problem, another criterion often used, different than the expected average return per unit time, is that of the expected discounted return. In this criterion we choose a number α, $0 < \alpha < 1$, and try to choose a policy so as to maximize $E[\sum_{i=0}^{\infty} \alpha^i R(X_i, a_i)]$. (That is, rewards at time n are discounted at rate α^n.) Suppose that the initial state is chosen according to the probabilities b_i. That is,

$$P\{X_0 = i\} = b_i, \qquad i = 1, \ldots, n$$

For a given policy β let y_{ja} denote the expected discounted time that the process is in state j and action a is chosen. That is,

$$y_{ja} = E_\beta \left[\sum_{n=0}^{\infty} \alpha^n I_{\{X_n=j, a_n=a\}} \right]$$

where for any event A the indicator variable I_A is defined by

$$I_A = \begin{cases} 1, & \text{if } A \text{ occurs} \\ 0, & \text{otherwise} \end{cases}$$

(a) Show that

$$\sum_a y_{ja} = E \left[\sum_{n=0}^{\infty} \alpha^n I_{\{X_n=j\}} \right]$$

or, in other words, $\sum_a y_{ja}$ is the expected discounted time in state j under β.

(b) Show that

$$\sum_j \sum_a y_{ja} = \frac{1}{1-\alpha}$$

$$\sum_a y_{ja} = b_j + \alpha \sum_i \sum_a y_{ia} P_{ij}(a)$$

Hint: For the second equation, use the identity

$$I_{\{X_{n+1}=j\}} = \sum_i \sum_a I_{\{X_n=i, a_n=a\}} I_{\{X_{n+1}=j\}}$$

Take expectations of the above to obtain

$$E[I_{\{X_{n+1}=j\}}] = \sum_i \sum_a E[I_{\{X_n=i, a_n=a\}}] P_{ij}(a).$$

(c) If $\{y_{ja}\}$ are a set of numbers satisfying

$$\sum_j \sum_a y_{ja} = \frac{1}{1-\alpha}$$

$$\sum_a y_{ja} = b_j + \alpha \sum_i \sum_a y_{ia} P_{ij}(a) \qquad (9.1)$$

Argue that y_{ja} can be interpreted as the expected discounted time that the process is in state j and action a is chosen when the initial state is chosen according to the probabilities b_j and the policy β, given by

$$\beta_i(a) = \frac{y_{ia}}{\sum_a y_{ia}}$$

is employed.

Hint: Derive a set of equations for the expected discounted times when policy β is used and show that they are equivalent to Equation (9.1).

(d) Argue that optimal policy with respect to the expected discounted return criterion can be obtained by first solving the linear program

$$\text{maximize} \sum_j \sum_a y_{ja} R(j, a),$$

$$\text{such that} \sum_j \sum_a y_{ja} = \frac{1}{1-\alpha}$$

$$\sum_a y_{ja} = b_j + \alpha \sum_i \sum_a y_{ia} P_{ij}(a)$$

$$y_{ja} \geq 0, \quad \text{all} \quad j, a;$$

and then defining the policy β^* by

$$\beta_i^*(a) = \frac{y_{ia}^*}{\sum_a y_{ia}^*}$$

where the y_{ja}^* are the solutions of the linear program.

46. Consider an N-state Markov chain that is ergodic and let π_i, $i = 1, \ldots, N$, denote the limiting probabilities. Suppose that a reward $R(i)$ is earned whenever the process is in state i. Then $\sum_{j=0}^n R(X_j)$ is the reward earned during the first $n + 1$ time periods. (Of course, X_j is the state of the Markov chain at time j.) Show that

$$\frac{E\left[\sum_{j=0}^n R(X_j)\right]}{n + 1} \to \sum_{i=1}^n \pi_i R(i) \quad \text{as} \quad n \to \infty$$

References

1. K. L. Chung, "Markov Chains with Stationary Transition Probabilities," Springer, Berlin, 1960.

2. S. Karlin and H. Taylor, "A First Course in Stochastic Processes," Second Edition. Academic Press, New York, 1975.

3. J. G. Kemeny and J. L. Snell, "Finite Markov Chains," Van Nostrand Reinhold, Princeton, New Jersey, 1960.

4. S. M. Ross, "Stochastic Processes," John Wiley, New York, 1983.

Chapter 5

The Exponential Distribution and the Poisson Process

1. Introduction

In making a mathematical model for a real-world phenomenon it is always necessary to make certain simplifying assumptions so as to render the mathematics tractable. On the other hand, however, we cannot make too many simplifying assumptions, for then our conclusions, obtained from the mathematical model, would not be applicable to the real-world phenomenon. Thus, in short, we must make enough simplifying assumptions to enable us to handle the mathematics but not so many that the mathematical model no longer resembles the real-world phenomenon. One simplifying assumption that is often made is to assume that certain random variables are exponentially distributed. The reason for this is that the exponential distribution is both relatively easy to work with and is often a good approximation to the actual distribution.

The property of the exponential distribution which makes it easy to analyze is that it does not deteriorate with time. By this we mean that if the lifetime of an item is exponentially distributed, then an item which has been in use for ten (or any number of) hours is as good as a new item in regards to the amount of time remaining until the item fails. This will be formally defined in Section 2 where it will be shown that the exponential is the only distribution which possesses this property.

In Section 3 we shall study counting processes with an emphasis on a kind of counting process known as the Poisson process. Among other things we shall discover about this process is its intimate connection with the exponential distribution.

2. The Exponential Distribution

2.1. Definition

A continuous random variable X is said to have an *exponential distribution* with parameter λ, $\lambda > 0$, if its probability density function is given by

$$f(x) = \begin{cases} \lambda e^{-\lambda x}, & x \geq 0 \\ 0, & x < 0 \end{cases}$$

or, equivalently, if its cdf is given by

$$F(x) = \int_{-\infty}^{x} f(y)dy = \begin{cases} 1 - e^{-\lambda x}, & x \geq 0 \\ 0, & x < 0 \end{cases}$$

The mean of the exponential distribution, $E[X]$, is given by

$$E[X] = \int_{-\infty}^{\infty} xf(x)dx$$

$$= \int_{0}^{\infty} \lambda x e^{-\lambda x}dx$$

Integrating by parts ($u = x$, $dv = \lambda e^{-\lambda x}dx$) yields

$$E[X] = -xe^{-\lambda x}\big|_{0}^{\infty} + \int_{0}^{\infty} e^{-\lambda x}dx$$

$$= \frac{1}{\lambda}$$

The moment generating function $\phi(t)$ of the exponential distribution is given by

$$\phi(t) = E[e^{tX}]$$

$$= \int_{0}^{\infty} e^{tx}\lambda e^{-\lambda x}dx$$

$$= \frac{\lambda}{\lambda - t} \quad \text{for} \quad t < \lambda \tag{2.1}$$

All the moments of X can now be obtained by differentiating Equation (2.1). For example,

$$E[X^2] = \frac{d^2}{dt^2}\phi(t)\Big|_{t=0}$$

$$= \frac{2\lambda}{(\lambda - t)^3}\Big|_{t=0}$$

$$= \frac{2}{\lambda^2}$$

Also, from the above, we obtain

$$\text{Var}(X) = E[X^2] - (E[X])^2$$

$$= \frac{2}{\lambda^2} - \frac{1}{\lambda^2}$$

$$= \frac{1}{\lambda^2}$$

2.2. Properties of the Exponential Distribution

A random variable X is said to be without memory, or *memoryless*, if

$$P\{X > s + t | X > t\} = P\{X > s\} \quad \text{for all} \quad s, t \geq 0 \qquad (2.2)$$

If we think of X as being the lifetime of some instrument, then Equation (2.2) states that the probability that the instrument lives for at least $s + t$ hours given that it has survived t hours is the same as the initial probability that it lives for at least s hours. In other words, if the instrument is alive at time t, then the distribution of the remaining amount of time that it survives is the same as the original lifetime distribution, that is, the instrument does not remember that it has already been in use for a time t.

The condition in Equation (2.2) is equivalent to

$$\frac{P\{X > s + t, X > t\}}{P\{X > t\}} = P\{X > s\}$$

or

$$P\{X > s + t\} = P\{X > s\}P\{X > t\} \qquad (2.3)$$

Since Equation (2.3) is satisfied when X is exponentially distributed (for

$e^{-\lambda(s+t)} = e^{-\lambda s}e^{-\lambda t}$), it follows that exponentially distributed random variables are memoryless.

Example 2a   Suppose that the amount of time one spends in a bank is exponentially distributed with mean ten minutes, that is, $\lambda = \frac{1}{10}$. What is the probability that a customer will spend more than fifteen minutes in the bank? What is the probability that a customer will spend more than fifteen minutes in the bank given that he is still in the bank after ten minutes?

Solution: If X represents the amount of time that the customer spends in the bank, then the first probability is just

$$P\{X > 15\} = e^{-15\lambda} = e^{-3/2} \approx .220$$

The second question asks for the probability that a customer who has spent ten minutes in the bank will have to spend at least five more minutes. However, since the exponential distribution does not "remember" that the customer has already spent ten minutes in the bank, this must equal the probability that an entering customer spends at least five minutes in the bank. That is, the desired probability is just

$$P\{X > 5\} = e^{-5\lambda} = e^{-1/2} \approx .604 \quad \Diamond$$

Example 2b   Consider a post office which is manned by two clerks. Suppose that when Mr. Smith enters the system he discovers that Mr. Jones is being served by one of the clerks and Mr. Brown by the other. Suppose also that Mr. Smith is told that his service will begin as soon as either Jones or Brown leave. If the amount of time that a clerk spends with a customer is exponentially distributed with mean $1/\lambda$, what is the probability that, of the three customers, Mr. Smith is the last to leave the post office?

Solution: The answer is obtained by this reasoning: Consider the time at which Mr. Smith first finds a free clerk. At this point either Mr. Jones or Mr. Brown would have just left and the other one would still be in service. However, by the lack of memory of the exponential, it follows that the amount of time that this other man (either Jones or Brown) would still have to spend in the post office is exponentially distributed with mean $1/\lambda$. That is, it is the same as if he was just starting his service at this point. Hence, by symmetry, the probability that he finishes before Smith must equal $\frac{1}{2}$. $\Diamond$

It turns out that not only is the exponential distribution "memoryless," but it is the unique distribution possessing this property. To see this, suppose that X is memoryless and let $\bar{F}(x) = P\{X > x\}$. Then by Equation (2.3) it follows that

$$\bar{F}(s + t) = \bar{F}(s)\bar{F}(t)$$

That is, $\bar{F}(x)$ satisfies the functional equation

$$g(s + t) = g(s)g(t)$$

However, it turns out that the only right continuous solution of this functional equation is

$$g(x) = e^{-\lambda x}*$$

and since a distribution function is always right continuous we must have

$$\bar{F}(x) = e^{-\lambda x}$$

or

$$F(x) = P\{X \le x\} = 1 - e^{-\lambda x}$$

which shows that X is exponentially distributed.

Example 2c Suppose that the amount of time that a lightbulb works before burning itself out is exponentially distributed with mean ten hours. Suppose that a person enters a room in which a lightbulb is burning. If this person desires to work for five hours, then what is the probability that he will be able to complete his work without the bulb burning out? What can be said about this probability when the distribution is not exponential?

---

*This is proven as follows: If $g(s + t) = g(s)g(t)$, then

$$g\left(\frac{2}{n}\right) = g\left(\frac{1}{n} + \frac{1}{n}\right) = g^2\left(\frac{1}{n}\right)$$

and repeating this yields $g(m/n) = g^m(1/n)$. Also

$$g(1) = g\left(\frac{1}{n} + \frac{1}{n} + \cdots + \frac{1}{n}\right) = g^n\left(\frac{1}{n}\right) \quad \text{or} \quad g\left(\frac{1}{n}\right) = (g(1))^{1/n}$$

Hence $g(m/n) = (g(1))^{m/n}$ which implies, since g is right continuous, that $g(x) = (g(1))^x$. Since $g(1) = (g(\frac{1}{2}))^2 \ge 0$ we obtain $g(x) = e^{-\lambda x}$, where $\lambda = -\log(g(1))$.

Solution: Since the bulb is burning when the person enters the room it follows, by the memoryless property of the exponential, that its remaining lifetime is exponential with mean ten. Hence the desired probability is

$$P\{\text{remaining lifetime} > 5\} = 1 - F(5) = e^{-5\lambda} = e^{-1/2}$$

However, if the lifetime distribution F is not exponential, then the relevant probability is

$$P\{\text{lifetime} > t + 5 | \text{lifetime} > t\} = \frac{1 - F(t + 5)}{1 - F(t)}$$

where t is the amount of time that the bulb had been in use prior to the person entering the room. That is, if the distribution is not exponential then additional information is needed (namely t) before the desired probability can be calculated. In fact, it is for this reason, namely that the distribution of the remaining lifetime is independent of the amount of time that the object has already survived, that the assumption of an exponential distribution is so often made.   ◊

The memoryless property is further illustrated by the failure rate function (also called the hazard rate function) of the exponential distribution.

Consider a continuous random variable X having distribution function F and density f. The *failure* (or *hazard*) *rate* function $r(t)$ is defined by

$$r(t) = \frac{f(t)}{1 - F(t)} \tag{2.4}$$

To interpret $r(t)$, suppose that X has survived for t hours, and we desire the probability that X will not survive for an additional time dt. That is, consider $P\{X \in (t, t + dt) | X > t\}$. Now

$$P\{X \in (t, t + dt) | X > t\} = \frac{P\{X \in (t, t + dt), X > t\}}{P\{X > t\}}$$

$$= \frac{P\{X \in (t, t + dt)\}}{P\{X > t\}}$$

$$\approx \frac{f(t)dt}{1 - F(t)}$$

$$= r(t)dt$$

That is, $r(t)$ represents the conditional probability density that a t-year-old item will fail.

Suppose now that the lifetime distribution is exponential. Then, by the memoryless property, it follows that the distribution of remaining life for a t-year-old item is the same as for a new item. Hence $r(t)$ should be constant. This checks out since

$$r(t) = \frac{f(t)}{1 - F(t)}$$

$$= \frac{\lambda e^{-\lambda t}}{e^{-\lambda t}}$$

$$= \lambda$$

Thus, the failure rate function for the exponential distribution is constant. The parameter λ is often referred to as the *rate* of the disribution. (Note that the rate is the reciprocal of the mean, and vice versa.)

It turns out that the failure rate function $r(t)$ uniquely determines the distribution F. To prove this, we note by Equation (2.4) that

$$r(t) = \frac{d/dt\, F(t)}{1 - F(t)}$$

Integrating both sides yields

$$\log(1 - F(t)) = -\int_0^t r(t)dt + k$$

or

$$1 - F(t) = e^k \exp\left\{-\int_0^t r(t)dt\right\}$$

Letting $t = 0$ shows that $k = 0$ and thus

$$F(t) = 1 - \exp\left\{-\int_0^t r(t)dt\right\}$$

2.3. Further Properties of the Exponential Distribution

Let $X_1, \ldots, X_n$ be independent and identically distributed exponential random variables having mean $1/\lambda$. It follows from the results of *Example 5i* of Chapter 2 that $X_1 + \cdots + X_n$ has a gamma distribution with parameters n and λ. Let us now give a second verification of this result by using mathematical induction. As there is nothing to prove when $n = 1$, let us start by assuming that $X_1 + \cdots + X_{n-1}$ has density given by

$$f_{X_1 + \cdots + X_{n-1}}(t) = \lambda e^{-\lambda t} \frac{(\lambda t)^{n-2}}{(n-2)!}$$

Hence,

$$f_{X_1 + \cdots + X_{n-1} + X_n}(t) = \int_0^\infty f_{X_n}(t-s) f_{X_1 + \cdots + X_{n-1}}(s) ds$$

$$= \int_0^t \lambda e^{-\lambda(t-s)} \lambda e^{-\lambda s} \frac{(\lambda s)^{n-2}}{(n-2)!} ds$$

$$= \lambda e^{-\lambda t} \frac{(\lambda t)^{n-1}}{(n-1)!}$$

which proves the result.

Another useful calculation is to determine the probability that one exponential random variable is smaller than another. That is, suppose that X_1 and X_2 are independent exponential random variables with respective means $1/\lambda_1$ and $1/\lambda_2$; then what is $P\{X_1 < X_2\}$? This probability is calculated by conditioning on X_2:

$$P\{X_1 < X_2\} = \int_0^\infty P\{X_1 < X_2 | X_2 = x\} \lambda_2 e^{-\lambda_2 x} dx$$

$$= \int_0^\infty P\{X_1 < x\} \lambda_2 e^{-\lambda_2 x} dx$$

$$= \int_0^\infty (1 - e^{-\lambda_1 x}) \lambda_2 e^{-\lambda_2 x} dx$$

$$= \int_0^\infty \lambda_2 e^{-\lambda_2 x} dx - \lambda_2 \int_0^\infty e^{-(\lambda_1 + \lambda_2)x} dx$$

$$= 1 - \frac{\lambda_2}{\lambda_1 + \lambda_2}$$

$$= \frac{\lambda_1}{\lambda_1 + \lambda_2} \tag{2.5}$$

Example 2d   Suppose one has a stereo system consisting of two main parts, a radio and a speaker. If the lifetime of the radio is exponential with mean 1000 hours and the lifetime of the speaker is exponential with mean 500 hours independent of the radio's lifetime, then what is the probability that the systems failure (when it occurs) will be caused by the radio failing?

Solution: From Equation (2.5) (with $\lambda_1 = 1/1000$, $\lambda_2 = 1/500$) we see that the answer is

$$\frac{1/1000}{1/1000 + 1/500} = \frac{1}{3} \quad \diamond$$

3. The Poisson Process

3.1. Counting Processes

A stochastic process $\{N(t), t \geq 0\}$ is said to be a *counting process* if $N(t)$ represents the total number of "events" that have occurred up to time t. Some examples of counting processes are the following:

(a) If we let $N(t)$ equal the number of persons who have entered a particular store at or prior to time t, then $\{N(t), t \geq 0\}$ is a counting process in which an event corresponds to a person entering the store. Note that if we had let $N(t)$ equal the number of persons in the store at time t, then $\{N(t), t \geq 0\}$ would *not* be a counting process (why not?).

(b) If we say that an event occurs whenever a child is born, then $\{N(t), t \geq 0\}$ is a counting process when $N(t)$ equals the total number of people who were born by time t. (Does $N(t)$ include persons who have died by time t? Explain why it must.)

(c) If $N(t)$ equals the number of goals that a given soccer player has scored by time t, then $\{N(t), t \geq 0\}$ is a counting process. An event of this process will occur whenever the soccer player scores a goal.

From its definition we see that for a counting process $N(t)$ must satisfy.

(i) $N(t) \geq 0$.
(ii) $N(t)$ is integer valued.
(iii) If $s < t$, then $N(s) \leq N(t)$.
(iv) For $s < t$, $N(t) - N(s)$ equals the number of events that have occurred in the interval (s, t).

A counting process is said to possess *independent increments* if the numbers of events which occur in disjoint time intervals are independent. For example, this means that the number of events which have occurred by time 10 (that is, $N(10)$) must be independent of the number of events occurring between times 10 and 15 (that is, $N(15) - N(10)$).

The assumption of independent increments might be reasonable for example (a), but it probably would be unreasonable for example (b). The reason for this is that if in example (b) $N(t)$ is very large, then it is probable that there are many people alive at time t; this would lead us to believe that the number of new births between time t and time $t + s$ would also tend to be large (that is, it does not seem reasonable that $N(t)$ is independent of $N(t + s) - N(t)$, and so $\{N(t), t \geq 0\}$ would not have independent increments in example (b)). The assumption of independent increments in example (c) would be justified if we believed that the soccer player's chances of scoring a goal today does not depend on "how he's been going." It would not be justified if we believed in "hot streaks" or "slumps."

A counting process is said to possess *stationary increments* if the distribution of the number of events which occur in any interval of time depends only on the length of the time interval. In other words, the process has stationary increments if the number of events in the interval $(t_1 + s, t_2 + s)$ (that is, $N(t_2 + s) - N(t_1 + s)$) has the same distribution as the number of events in the interval (t_1, t_2) (that is, $N(t_2) - N(t_1)$) for all $t_1 < t_2$, and $s > 0$.

The assumption of stationary increments would only be reasonable in example (a) if there were no times of day at which people were more likely to enter the store. Thus, for instance, if there was a rush hour (say between 12 A.M. and 1 P.M.) each day, then the stationarity assumption would not be justified. If we believed that the earth's population is basically constant (a belief not held at present by most scientists), then the assumption of stationary increments might be reasonable in example (b). Stationary increments do not seem to be a reasonable assumption in example (c) since, for one thing, most people would agree that the soccer player would probably score more goals while in the age bracket 25–30 than he would while in the age bracket 35–40.

3.2. Definition of the Poisson Process

One of the most important counting processes is the Poisson process which is defined as follows:

Definition 3.1.   The counting process $\{N(t), t \geq 0\}$ is said to be a *Poisson process having rate* λ, $\lambda > 0$, if

(i) $N(0) = 0$.
(ii) The process has independent increments.

(iii) The number of events in any interval of length t is Poisson distributed with mean λt. That is, for all $s, t \geq 0$

$$P\{N(t + s) - N(s) = n\} = e^{-\lambda t} \frac{(\lambda t)^n}{n!}, \qquad n = 0, 1, \ldots$$

Note that it follows from condition (iii) that a Poisson process has stationary increments and also that

$$E[N(t)] = \lambda t$$

which explains why λ is called the rate of the process.

In order to determine if an arbitrary counting process is actually a Poisson process we must show that conditions (i), (ii), and (iii) are satisfied. Condition (i), which simply states that the counting of events begins at time $t = 0$, and condition (ii) can usually be directly verified from our knowledge of the process. However, it is not at all clear how we would determine that condition (iii) is satisfied, and for this reason an equivalent definition of a Poisson process would be useful.

As a prelude to giving a second definition of a Poisson process we shall define the concept of a function $f(\cdot)$ being $o(h)$.

Definition The function $f(\cdot)$ is said to be $o(h)$ if

$$\lim_{h \to 0} \frac{f(h)}{h} = 0$$

Example 3a

(i) The function $f(x) = x^2$ is $o(h)$ since

$$\lim_{h \to 0} \frac{f(h)}{h} = \lim_{h \to 0} \frac{h^2}{h} = \lim_{h \to 0} h = 0$$

(ii) The function $f(x) = x$ is not $o(h)$ since

$$\lim_{h \to 0} \frac{f(h)}{h} = \lim_{h \to 0} \frac{h}{h} = \lim_{h \to 0} 1 = 1 \neq 0$$

(iii) If $f(\cdot)$ is $o(h)$ and $g(\cdot)$ is $o(h)$, then so is $f(\cdot) + g(\cdot)$. This follows since

$$\lim_{h \to 0} \frac{f(h) + g(h)}{h} = \lim_{h \to 0} \frac{f(h)}{h} + \lim_{h \to 0} \frac{g(h)}{h} = 0 + 0 = 0$$

(iv) If $f(\cdot)$ is $o(h)$, then so is $g(\cdot) = cf(\cdot)$. This follows since

$$\lim_{h \to 0} \frac{cf(h)}{h} = c \lim \frac{f(h)}{h} = c \cdot 0 = 0$$

(v) From (iii) and (iv) it follows that any finite linear combination of functions, each of which is $o(h)$, is $o(h)$.   ◇

In order for the function $f(\cdot)$ to be $o(h)$ it is necessary that $f(h)/h$ go to zero as h goes to zero. But if h goes to zero, the only way for $f(h)/h$ to go to zero is for $f(h)$ to go to zero faster than h does. That is, for h small, $f(h)$ must be small compared to h.

We are now in a position to give an alternative definition of a Poisson process.

Definition 3.2.   The counting process $\{N(t), t \geq 0\}$ is said to be a Poisson process having rate λ, $\lambda > 0$, if

(i) $N(0) = 0$.
(ii) The process has stationary and independent increments.
(iii) $P\{N(h) = 1\} = \lambda h + o(h)$.
(iv) $P\{N(h) \geq 2\} = o(h)$.

Theorem 3.1.   Definitions 3.1 and 3.2 are equivalent.

Proof: We first show that Definition 3.2 implies Definition 3.1. To do this let

$$P_n(t) = P\{N(t) = n\}$$

We derive a differential equation for $P_0(t)$ in the following manner:

$$
\begin{aligned}
P_0(t + h) &= P\{N(t + h) = 0\} \\
&= P\{N(t) = 0, N(t + h) - N(t) = 0\} \\
&= P\{N(t) = 0\}P\{N(t + h) - N(t) = 0\} \\
&= P_0(t)[1 - \lambda h + o(h)]
\end{aligned}
$$

where the final two equations follow from Assumption (ii) plus the fact that Assumptions (iii) and (iv) imply that $P\{N(h) = 0\} = 1 - \lambda h + o(h)$. Hence,

$$\frac{P_0(t + h) - P_0(t)}{h} = -\lambda P_0(t) + \frac{o(h)}{h}$$

Now, letting $h \to 0$ we obtain

$$P_0'(t) = -\lambda P_0(t)$$

or equivalently

$$\frac{P_0'(t)}{P_0(t)} = -\lambda$$

which implies, by integration, that

$$\log P_0(t) = -\lambda t + c$$

or

$$P_0(t) = Ke^{-\lambda t}$$

Since $P_0(0) = P\{N(0) = 0\} = 1$, we arrive at

$$P_0(t) = e^{-\lambda t} \tag{3.1}$$

Similarly, for $n > 0$,

$$
\begin{aligned}
P_n(t + h) &= P\{N(t + h) = n\} \\
&= P\{N(t) = n, N(t + h) - N(t) = 0\} \\
&\quad + P\{N(t) = n - 1, N(t + h) - N(t) = 1\} \\
&\quad + \sum_{k=2}^{n} P\{N(t) = n - k, N(t + h) - N(t) = k\}
\end{aligned}
$$

However, by Assumption (iv), the last term in the above is $o(h)$; hence, by using Assumption (ii), we obtain

$$
\begin{aligned}
P_n(t + h) &= P_n(t)P_0(h) + P_{n-1}(t)P_1(h) + o(h) \\
&= (1 - \lambda h)P_n(t) + \lambda h P_{n-1}(t) + o(h)
\end{aligned}
$$

Thus,

$$\frac{P_n(t + h) - P_n(t)}{h} = -\lambda P_n(t) + \lambda P_{n-1}(t) + \frac{o(h)}{h}$$

and letting $h \to 0$ yields

$$P_n'(t) = -\lambda P_n(t) + \lambda P_{n-1}(t)$$

or equivalently,

$$e^{\lambda t}[P_n'(t) + \lambda P_n(t)] = \lambda e^{\lambda t}P_{n-1}(t)$$

Hence,

$$\frac{d}{dt}(e^{\lambda t}P_n(t)) = \lambda e^{\lambda t}P_{n-1}(t) \tag{3.2}$$

Now by Equation (3.1), we have

$$\frac{d}{dt}(e^{t\lambda}P_1(t)) = \lambda$$

or

$$P_1(t) = (\lambda t + c)e^{-\lambda t}$$

which, since $P_1(0) = 0$, yields

$$P_1(t) = \lambda t e^{-\lambda t}$$

To show that $P_n(t) = e^{-\lambda t}(\lambda t)^n/n!$, we use mathematical induction and hence first assume it for $n = 1$. Then by Equation (3.2),

$$\frac{d}{dt}(e^{\lambda t}P_n(t)) = \frac{\lambda^n t^{n-1}}{(n-1)!}$$

or

$$e^{\lambda t}P_n(t) = \frac{(\lambda t)^n}{n!} + c$$

which implies the result (since $P_n(0) = 0$). This proves that Definition 3.2 implies Definition 3.1.

We shall leave it to the reader to prove the reverse.   ◇

Remarks The result that $N(t)$ has a Poisson distribution is a consequence of the Poisson approximation to the binomial distribution (see Section 2.4 of Chapter 2). To see this subdivide the interval $[0, t]$ into k equal parts where k is very large (Figure 5.1). Now it can be shown using Axiom (iv) of Definition 3.2 that as k increases to ∞ the probability of having two or more events in any of the k subintervals goes to 0. Hence, $N(t)$ will (with a probability going to 1) just equal the number of subintervals in which an event occurs. However, by stationary and independent increments this number will have a binomial distribution with parameters k and $p = \lambda t/k + o(t/k)$. Hence, by the Poisson approximation to the binomial we see by letting k approach ∞ that $N(t)$ will have a

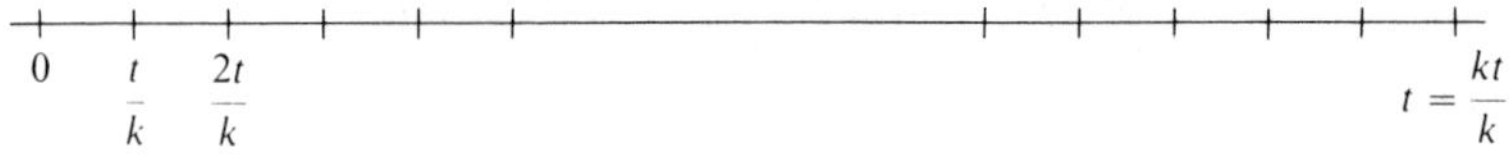

Figure 5.1.

Poisson distribution with mean equal to

$$\lim_{k \to 0} k \left[\lambda \frac{t}{k} + o\left(\frac{t}{k}\right) \right] = \lambda t + \lim_{k \to \infty} \frac{[to(t/k)]}{t/k}$$

$$= \lambda t$$

by using the definition of $o(h)$ and the fact that $t/k \to 0$ as $t \to \infty$.

3.3. Interarrival and Waiting Time Distributions

Consider a Poisson process, and let us denote the time of the first event by T_1. Further, for $n > 1$, let T_n denote the elapsed time between the $(n - 1)$st and the nth event. The sequence $\{T_n, n = 1, 2, . . .\}$ is called the *sequence of interarrival times*. For instance, if $T_1 = 5$ and $T_2 = 10$, then the first event of the Poisson process would have occurred at time 5 and the second at time 15.

We shall now determine the distribution of the T_n. To do so, we first note that the event $\{T_1 > t\}$ takes place if and only if no events of the Poisson process occur in the interval $[0, t]$ and thus,

$$P\{T_1 > t\} = P\{N(t) = 0\} = e^{-\lambda t}$$

Hence, T_1 has an exponential distribution with mean $1/\lambda$. Now

$$P\{T_2 > t\} = E[P\{T_2 > t | T_1\}] \qquad (3.3)$$

However,

$$P\{T_2 > t | T_1 = s\} = P\{0 \text{ events in } (s, s + t] | T_1 = s\}$$

$$= P\{0 \text{ events in } (s, s + t]\}$$

$$= e^{-\lambda t}$$

where the last two equations followed from independent and stationary increments. Therefore, from Equation (3.3) we conclude that T_2 is also an exponential random variable with mean $1/\lambda$, and furthermore, that T_2 is independent of T_1. Repeating the same argument yields

Proposition 3.1. $T_n, n = 1, 2, . . .$, are independent identically distributed exponential random variables having mean $1/\lambda$. ◊

Remarks The proposition should not surprise us. The assumption of stationary and independent increments is basically equivalent to asserting that, at any point in time, the process *probabilistically*

restarts itself. That is, the process from any point on is independent of all that has previously occurred (by independent increments), and also has the same distribution as the original process (by stationary increments). In other words, the process has no *memory*, and hence exponential interarrival times are to be expected. ◇

Another quantity of interest is S_n, the arrival time of the nth event, also called the *waiting time* until the nth event. It is easily seen that

$$S_n = \sum_{i=1}^{n} T_i, \qquad n \geq 1$$

and hence from Proposition 3.1 and the results of Section 2.2 it follows that S_n has a gamma distribution with parameters n and λ. That is, the probability density of S_n is given by

$$f_{S_n}(t) = \lambda e^{-\lambda t} \frac{(\lambda t)^{n-1}}{(n-1)!}, \qquad t \geq 0 \tag{3.4}$$

Equation (3.4) may also have been derived by noting that the nth event will occur prior or at time t if and only if the number of events occurring by time t is at least n. That is,

$$N(t) \geq n \leftrightarrow S_n \leq t$$

and hence,

$$F_{S_n}(t) = P\{S_n \leq t\} = P\{N(t) \geq n\} = \sum_{j=n}^{\infty} e^{-\lambda t} \frac{(\lambda t)^j}{j!}$$

which, upon differentiation, yields

$$f_{S_n}(t) = -\sum_{j=n}^{\infty} \lambda e^{-\lambda t} \frac{(\lambda t)^j}{j!} + \sum_{j=n}^{\infty} \lambda e^{-\lambda t} \frac{(\lambda t)^{j-1}}{j-1!}$$

$$= \lambda e^{-\lambda t} \frac{(\lambda t)^{n-1}}{(n-1)!} + \sum_{j=n+1}^{\infty} \lambda e^{-\lambda t} \frac{(\lambda t)^{j-1}}{(j-1)!} - \sum_{j=n}^{\infty} \lambda e^{-\lambda t} \frac{(\lambda t)^j}{j!}$$

$$= \lambda e^{-\lambda t} \frac{(\lambda t)^{n-1}}{(n-1)!}$$

Example 3b Suppose that people immigrate into a territory at a Poisson rate $\lambda = 1$ per day.

(a) What is the expected time until the tenth immigrant arrives?

(b) What is the probability that the elapsed time between the tenth and the eleventh arrival exceeds two days?

Solution:

(a) $E[S_{10}] = 10/\lambda = 10$ days.
(b) $P\{T_{11} > 2\} = e^{-2\lambda} = e^{-2} \approx .133$. ◇

Proposition 3.1 also gives us another way of defining a Poisson process. For suppose that we start out with a sequence $\{T_n, n \geq 1\}$ of independent identically distributed exponential random variables each having mean $1/\lambda$. Now let us define a counting process by saying that the nth event of this process occurs at time

$$S_n \equiv T_1 + T_2 + \cdots + T_n$$

The resultant counting process $\{N(t), t \geq 0\}^*$ will be Poisson with rate λ.

3.4. Further Properties of Poisson Processes

Consider a Poisson process $\{N(t), t \geq 0\}$ having rate λ, and suppose that each time an event occurs it is classified as either a type I or a type II event. Suppose further that each event is classified as a type I event with probability p and a type II event with probability $1 - p$ independently of all other events. For example, suppose that customers arrive at a store in accordance with a Poisson process having rate λ; and suppose that each arrival is male with probability $\frac{1}{2}$ and female with probability $\frac{1}{2}$. Then a type I event would correspond to a male arrival and a type II event to a female arrival.

Let $N_1(t)$ and $N_2(t)$ denote respectively the number of type I and type II events occurring in $[0, t]$. Note that $N(t) = N_1(t) + N_2(t)$.

Proposition 3.2. $\{N_1(t), t \geq 0\}$ and $\{N_2(t), t \geq 0\}$ are both Poisson processes having respective rates λp and $\lambda(1 - p)$. Furthermore, the two processes are independent.

---

*A formal definition of $N(t)$ is given by

$$N(t) \equiv \max\{n : S_n \leq t\}$$

(where $S_0 \equiv 0$).

Proof: Let us calculate the joint probability $P\{N_1(t) = n, N_2(t) = m\}$. To do this we first condition on $N(t)$ to obtain

$$P\{N_1(t) = n, N_2(t) = m\}$$
$$= \sum_{k=0}^{\infty} P\{N_1(t) = n, N_2(t) = m | N(t) = k\} P\{N(t) = k\}$$

Now in order for there to have been n type I events and m type II events there must have been a total of $n + m$ events occurring in $[0, t]$. That is,

$$P\{N_1(t) = n, N_2(t) = m | N(t) = k\} = 0 \quad \text{when} \quad k \neq n + m$$

Hence,

$$P\{N_1(t) = n, N_2(t) = m\}$$
$$= P\{N_1(t) = n, N_2(t) = m | N(t) = n + m\} P\{N(t) = n + m\}$$
$$= P\{N_1(t) = n, N_2(t) = m | N(t) = n + m\} e^{-\lambda t} \frac{(\lambda t)^{n+m}}{(n+m)!}$$

However, given that $n + m$ events occurred, since each event has probability p of being a type I event and probability $1 - p$ of being a type II event, it follows that the probability that n of them will be type I and m of them type II events is just the binomial probability $\binom{n+m}{n} p^n (1 - p)^m$. Thus,

$$P\{N_1(t) = n, \quad N_2(t) = m\} = \binom{n+m}{n} p^n (1 - p)^m e^{-\lambda t} \frac{(\lambda t)^{n+m}}{(n+m)!}$$
$$= e^{-\lambda t p} \frac{(\lambda t p)^n}{n!} e^{-\lambda t(1-p)} \frac{(\lambda t(1 - p))^m}{m!} \quad (3.5)$$

Hence,

$$P\{N_1(t) = n\} = \sum_{m=0}^{\infty} P\{N_1(t) = n, N_2(t) = m\}$$
$$= e^{-\lambda t p} \frac{(\lambda t p)^n}{n!} \sum_{m=0}^{\infty} e^{-\lambda t(1-p)} \frac{(\lambda t(1 - p))^m}{m!}$$
$$= e^{-\lambda t p} \frac{(\lambda t p)^n}{n!}$$

That is, $\{N_1(t), t \geq 0\}$ is a Poisson process having rate λp. (How do we know that the other conditions of Definition 3.1 are satisfied? Argue it out!)

Similarly,

$$P\{N_2(t) = m\} = e^{-\lambda t(1-p)} \frac{(\lambda t(1-p))^m}{m!}$$

and so $\{N_2(t), t \geq 0\}$ is a Poisson process having rate $\lambda(1-p)$. Finally, it follows from Equation (3.5) that the two processes are independent (since the joint distribution factors). $\diamond$

Remarks   It is not surprising that $\{N_1(t), t \geq 0\}$ and $\{N_2(t), t \geq 0\}$ are Poisson processes. What is somewhat surprising is the fact that they are independent. For assume that customers arrive at a bank at a Poisson rate of $\lambda = 1$ per hour and suppose that each customer is a man with probability $\frac{1}{2}$ and a woman with probability $\frac{1}{2}$. Now suppose that 100 men arrived in the first ten hours. Then how many women would we expect to have arrived in that time? One might argue that as the number of male arrivals is 100 and as each arrival is male with probability $\frac{1}{2}$, then the expected number of total arrivals should be 200 and hence the expected number of female arrivals should also be 100. But, as shown by the previous proposition, such reasoning is spurious and the expected number of female arrivals in the first ten hours is five, independent of the number of male arrivals in that period.

To obtain an intuition as to why Proposition 3.2 is true reason as follows: If we divide the interval $(0, t)$ into n subintervals of equal length t/n, where n is very large, then (neglecting events having probability "little o") each subinterval will have a small probability $\lambda t/n$ of containing a single event. As each event has probability p of being of type I, it follows that each of the n subintervals will have either no events, a type I event, a type II event with respective probabilities

$$1 - \frac{\lambda t}{n}, \qquad \frac{\lambda t}{n} p, \qquad \frac{\lambda t}{n}(1 - p)$$

Hence from the result which gives the Poisson as the limit of binomials, we can immediately conclude that $N_1(t)$ and $N_2(t)$ are Poisson distributed with respective means $\lambda t p$ and $\lambda t(1 - p)$. To see that they are independent, suppose, for instance, that $N_1(t) = k$. Then of the n subintervals, k will contain a type I event, and thus the other $n - k$ will each contain a type II event with probability

$$P\{\text{type II}|\text{no type I}\} = \frac{\dfrac{\lambda t}{n}(1 - p)}{1 - \dfrac{\lambda t}{n}p}$$

$$= \frac{\lambda t}{n}(1 - p) + o\left(\frac{t}{n}\right)$$

Hence, as $n - k$ will still be a very large number, we see again from the Poisson limit theorem that, given $N_1(t) = k$, $N_2(t)$ will be Poisson with mean $\lim_{n \to \infty} [(n - k)\lambda t(1 - p)/n] = \lambda t(1 - p)$, and so independence is established. ◇

Example 3c   If immigrants to area A arrive at a Poisson rate of ten per week, and if each immigrant is of English descent with probability $\frac{1}{12}$, then what is the probability that no people of English descent will emigrate to area A during the month of February?

Solution: By the previous proposition it follows that the number of Englishmen emigrating to area A during the month of February is Poisson distributed with mean $4 \cdot 10 \cdot \frac{1}{12} = \frac{10}{3}$. Hence the desired probability is $e^{-10/3}$.   ◇

It follows from Proposition 3.2 that if each of a Poisson number of individuals is independently classified into one of two possible groups with respective probabilities p and $1 - p$, then the number of individuals in each of the two groups will be independent Poisson random variables. As this result easily generalizes to the case where the classification is into any one of r possible groups, we have the following application to a model of employees moving about in an organization.

Example 3d   Consider a system in which individuals at any time are classified as being in one of r possible states, and assume that an individual changes states in accordance with a Markov chain having transition probabilities P_{ij}, $i, j = 1, \ldots, r$. That is, if an individual is in state i during a time period then, independently of its previous states, it will be in state j during the next time period with probability P_{ij}. The individuals are assumed to move through the system independently of each other. Suppose that the numbers of people initially in states $1, 2, \ldots, r$ are independent Poisson random variables with respective means $\lambda_1, \lambda_2, \ldots, \lambda_r$. We are interested in determining the joint distribution of the numbers of individuals in states 1, 2, . . . , r at some time n.

Solution: For fixed i, let $N_j(i), j = 1, \ldots, r$, denote the number of those individuals, initially in state i, that are in state j at time n. Now each of the (Poisson distributed) number of people initially in state i will, independently of each other, be in state j at time n with probability P_{ij}^n, where P_{ij}^n is the n-stage transition probability for the Markov chain having transition probabilities P_{ij}. Hence, the $N_j(i), j = 1, \ldots, r$, will be independent Poisson random variables with respective means $\lambda_i P_{ij}^n, j = 1, \ldots, r$. As the sum of independent Poisson random variables is itself a Poisson random variable, it follows that the number of individuals in state j at time n—namely $\Sigma_{i=1}^r N_j(i)$—will be independent Poisson random variables with respective means $\Sigma_i \lambda_i P_{ij}^n$, for $j = 1, \ldots, r$. $\Diamond$

Example 3e (A Multinomial Limit): Consider n independent trials each of which results in one of the outcomes $1, \ldots, r, r + 1$ with respective probabilities $P_1, \ldots, P_r, P_{r+1} = 1 - \Sigma_0^r P_i$. Let X_i denote the number of trials that result in outcome $i, i = 1, \ldots, r$. Assuming that n is very large and $\Sigma_{i=1}^r P_i$ is very small, what is the approximate joint distribution of $(X_1, \ldots, X_r)$?

To answer the above, let $\lambda_i = nP_i, i = 1, \ldots, r$ and note first that the number of trials that result in any of the first r outcomes is, by the Poisson approximation to the binomial, (approximately) Poisson distributed with mean $\Sigma_{i=1}^r \lambda_i$. As each of these trials will independently result in outcome i with probability $P_i/\Sigma_{i=1}^r P_i = \lambda_i/\Sigma_1^r \lambda_i$ it thus follows from Proposition 3.2 that $X_1, \ldots, X_r$ will be (approximately) independent Poisson random variables with respective means $\lambda_i, i = 1, \ldots, r$. $\Diamond$

The next probability calculation related to Poisson processes that we shall determine is the probability that n events occur in one Poisson process before m events have occurred in a second and independent Poisson process. More formally let $\{N_1(t), t \geq 0\}$ and $\{N_2(t), t \geq 0\}$ be two independent Poisson processes having respective rates λ_1 and λ_2. Also, let S_n^1 denote the time of the nth event of the first process, and S_m^2 the time of the mth event of the second process. We seek

$$P\{S_n^1 < S_m^2\} \tag{3.6}$$

Before attempting to calculate Equation (3.6) for general n and m, let us consider the special case $n = m = 1$. Since S_1^1, the time of the first event of the $N_1(t)$ process, and S_1^2, the time of the first event of the $N_2(t)$ process, are both exponentially distributed random variables (by Proposition 3.1) with respective means $1/\lambda_1$ and $1/\lambda_2$, it follows from Section 2.3 that

$$P\{S_1^1 < S_1^2\} = \frac{\lambda_1}{\lambda_1 + \lambda_2} \tag{3.7}$$

Let us now consider the probability that two events occur in the $N_1(t)$ process before a single event has occurred in the $N_2(t)$ process. That is, $P\{S_2^1 < S_1^2\}$. To calculate this we reason as follows: In order for the $N_1(t)$ process to have two events before a single event occurs in the $N_2(t)$ process it is first necessary that the initial event that occurs must be an event of the $N_1(t)$ process (and this occurs, by Equation (3.7), with probability $\lambda_1/(\lambda_1 + \lambda_2)$). Now given that the initial event is from the $N_1(t)$ process, the next thing that must occur for S_2^1 to be less than S_1^2 is for the second event to also be an event of the $N_1(t)$ process. However, when the first event occurs both processes start all over again (by the memoryless property of Poisson processes) and hence this conditional probability is also $\lambda_1/(\lambda_1 + \lambda_2)$, and hence the desired probability is given by

$$P\{S_2^1 < S_1^2\} = \left(\frac{\lambda_1}{\lambda_1 + \lambda_2}\right)^2$$

In fact this reasoning shows that *each event that occurs is going to be an event of the $N_1(t)$ process with probability $\lambda_1/(\lambda_1 + \lambda_2)$ and an event of the $N_2(t)$ process with probability $\lambda_2/(\lambda_1 + \lambda_2)$, independent of all that has previously occurred.* In other words, the probability that the $N_1(t)$ process reaches n before the $N_2(t)$ process reaches m is just the probability that n heads will appear before m tails if one flips a coin having probability $p = \lambda_1/(\lambda_1 + \lambda_2)$ of a head appearing. But by noting that this event will occur if and only if the first $n + m - 1$ tosses result in n or more heads then we see that our desired probability is given by

$$P\{S_n^1 < S_m^2\} = \sum_{k=n}^{n+m-1} \binom{n+m-1}{k}\left(\frac{\lambda_1}{\lambda_1 + \lambda_2}\right)^k \left(\frac{\lambda_2}{\lambda_1 + \lambda_2}\right)^{n+m-1-k}$$

3.5. Conditional Distribution of the Arrival Times

Suppose we are told that exactly one event of a Poisson process has taken place by time t, and we are asked to determine the distribution of the time at which the event occurred. Now, since a Poisson process possesses stationary and independent increments it seems reasonable that each interval in $[0, t]$ of equal length should have the same probability of containing the event. In other words, the time of the event should be uniformly distributed over $[0, t]$. This is easily checked since, for $s \leq t$

$$
\begin{aligned}
P[T_1 < s | N(t) = 1] &= \frac{P\{T_1 < s, N(t) = 1\}}{P\{N(t) = 1\}} \\
&= \frac{P\{1 \text{ event in } [0, s), 0 \text{ events in } [s, t)\}}{P\{N(t) = 1\}} \\
&= \frac{P\{1 \text{ event in } [0, s)\}P\{0 \text{ events in } [s, t)\}}{P\{N(t) = 1\}} \\
&= \frac{\lambda s e^{-\lambda s} e^{-\lambda(t-s)}}{\lambda t e^{-\lambda t}} \\
&= \frac{s}{t}
\end{aligned}
$$

This result may be generalized, but before doing so we need to introduce the concept of order statistics.

Let $Y_1, Y_2, \ldots, Y_n$ be n random variables. We say that $Y_{(1)}$, $Y_{(2)}, \ldots, Y_{(n)}$ are the order statistics corresponding to $Y_1, Y_2, \ldots, Y_n$ if $Y_{(k)}$ is the kth smallest value among $Y_1, \ldots, Y_n$, $k = 1, 2, \ldots, n$. For instance if $n = 3$ and $Y_1 = 4$, $Y_2 = 5$, $Y_3 = 1$ then $Y_{(1)} = 1$, $Y_{(2)} = 4$, $Y_{(3)} = 5$. If the $Y_i, i = 1, \ldots, n$, are independent identically distributed continuous random variables with probability density f, then the joint density of the order statistics $Y_{(1)}, Y_{(2)}, \ldots, Y_{(n)}$ is given by

$$
f(y_1, y_2, \ldots, y_n) = n! \prod_{i=1}^{n} f(y_i), \qquad y_1 < y_2 < \cdots < y_n
$$

The above follows since

(i) $(Y_{(1)}, Y_{(2)}, \ldots, Y_{(n)})$ will equal $(y_1, y_2, \ldots, y_n)$ if $(Y_1, Y_2, \ldots, Y_n)$ is equal to any of the $n!$ permutations of $(y_1, y_2, \ldots, y_n)$.

and

(ii) the probability density that $(Y_1, Y_2, \ldots, Y_n)$ is equal to $y_{i_1}, \ldots, y_{i_n}$ is $\Pi_{j=1}^{n} f(y_{i_j}) = \Pi_{j=1}^{n} f(y_j)$ when $i_1, \ldots, i_n$ is a permutation of $1, 2, \ldots, n$.

If the Y_i, $i = 1, \ldots, n$, are uniformly distributed over $(0, t)$, then we obtain from the above that the joint density function of the order statistics $Y_{(1)}, Y_{(2)}, \ldots, Y_{(n)}$ is

$$f(y_1, y_2, \ldots, y_n) = \frac{n!}{t^n}, \qquad 0 < y_1 < y_2 < \cdots < y_n < t$$

We are now ready for the following useful theorem.

Theorem 3.2. Given that $N(t) = n$, the n arrival times $S_1, \ldots, S_n$ have the same distribution as the order statistics corresponding to n independent random variables uniformly distributed on the interval $(0, t)$.

> *Proof:* To obtain the conditional density of $S_1, \ldots, S_n$ given that $N(t) = n$ note that for $0 < s_1 < \cdots < s_n < t$ the event that $S_1 = s_1, S_2 = s_2, \ldots, S_n = s_n, N(t) = n$ is equivalent to the event that the first $n + 1$ interarrival times satisfy $T_1 = s_1, T_2 = s_2 - s_1, \ldots, T_n = s_n - s_{n-1}, T_{n+1} > t - s_n$. Hence, using Proposition 3.1, we have that the conditional joint density of $S_1, \ldots, S_n$ given that $N(t) = n$ is as follows:
>
> $$\begin{aligned} f(s_1, \ldots, s_n | n) &= \frac{f(s_1, \ldots, s_n, n)}{P\{N(t) = n\}} \\ &= \frac{\lambda e^{-\lambda s_1} \lambda e^{-\lambda(s_2 - s_1)} \ldots \lambda e^{-\lambda(s_n - s_{n-1})} e^{-\lambda(t - s_n)}}{e^{-\lambda t}(\lambda t)^n / n!} \\ &= \frac{n!}{t^n}, \qquad 0 < s_1 < \cdots < s_n < t \end{aligned}$$
>
> which proves the result.   ◊

Remarks   The above result is usually paraphrased as stating that, under the condition that n events have occurred in $(0, t)$, the times $S_1, \ldots, S_n$ at which events occur, considered as unordered random variables, are distributed independently and uniformly in the interval $(0, t)$.

Application of Theorem 3.2. (Sampling a Poisson Process)

In Proposition 3.2 we showed that if each event of a Poisson process is independently classified as a type I event with probability p and as a type II event with probability $1 - p$ then the counting processes of type I and type II events are independent Poisson processes with respective rates λp and $\lambda(1 - p)$. Suppose now, however, that there are k possible types of events and that the probability that an event is classified as a type i event, $i = 1, \ldots, k$, depends on the time the event occurs. Specifically, suppose that if an event occurs at time y then it will be classified as a type i event, independently of anything that has previously occurred, with probability $P_i(y)$, $i = 1, \ldots, k$ where $\Sigma_{i=1}^k P_i(y) = 1$. Upon using Theorem 3.2 we can prove the following useful proposition.

Proposition 3.3. If $N_i(t)$, $i = 1, \ldots, k$, represents the number of type i events occurring by time t then $N_i(t)$, $i = 1, \ldots, k$, are independent Poisson random variables having means

$$E[N_i(t)] = \lambda \int_0^t P_i(s)ds$$

Before proving this proposition let us first illustrate its use.

Example 3f (An Infinite Server Queue): Suppose that customers arrive at a service station in accordance with a Poisson process with rate λ. Upon arrival the customer is immediately served by one of an infinite number of possible servers, and the service times are assumed to be independent with a common distribution G. What is the distribution of $X(t)$, the number of customers that have completed service by time t? What is the distribution of $Y(t)$, the number of customers that are being served at time t?

To answer the above questions let us agree to call an entering customer a type I customer if he completes his service by time t and a type II customer if he does not complete his service by time t. Now, if the customer enters at time s, $s \leq t$, then he will be a type I customer if his service time is less than $t - s$. Since the service time distribution is G, the probability of this will be $G(t - s)$. Similarly, a customer entering at time s, $s \leq t$, will be a type II customer with probability $1 - G(t - s)$. Hence,

$$P(s) = G(t - s), \qquad s \leq t$$

and thus from Proposition 3.3 we obtain that the distribution of $X(t)$, the number of customers that have completed service by time t, is Poisson distributed with mean

$$E[X(t)] = \lambda \int_0^t G(t - s)ds = \lambda \int_0^t G(y)dy \qquad (3.8)$$

Similarly, the distribution of $Y(t)$, the number of customers being served at time t is Poisson with mean

$$E[Y(t)] = \lambda \int_0^t (1 - G(t - s))ds = \lambda \int_0^t (1 - G(y))dy \qquad (3.9)$$

Furthermore, $X(t)$ and $Y(t)$ are independent.

Suppose now that we are interested in computing the joint distribution of $Y(t)$ and $Y(t + s)$—that is the joint distribution of the number in the system at time t and at time $t + s$. To accomplish this, say that an arrival is

type 1: if he arrives before time t and completes service between t and $t + s$,
type 2: if he arrives before t and completes service after $t + s$,
type 3: if he arrives between t and $t + s$ and completes service after $t + s$,
type 4: otherwise.

Hence an arrival at time y will be type i with probability $P_i(y)$ given by

$$P_1(y) = \begin{cases} G(t + s - y) - G(t - y) & \text{if } y < t \\ 0 & \text{otherwise} \end{cases}$$

$$P_2(y) = \begin{cases} 1 - G(t + s - y) & \text{if } y < t \\ 0 & \text{otherwise} \end{cases}$$

$$P_3(y) = \begin{cases} 1 - G(t + s - y) & \text{if } t < y < t + s \\ 0 & \text{otherwise} \end{cases}$$

$$P_4(y) = 1 - P_1(y) - P_2(y) - P_3(y)$$

Hence, if $N_i = N_i(s + t)$, $i = 1, 2, 3$ denotes the number of type i events that occur, then from Proposition 3.3, N_i, $i = 1, 2, 3$ are independent Poisson random variables with respective means

$$E[N_i] = \lambda \int_0^{t+s} P_i(y)dy, \qquad i = 1, 2, 3$$

As

$$Y(t) = N_1 + N_2$$
$$Y(t + s) = N_2 + N_3$$

it is now an easy matter to compute the joint distribution of $Y(t)$ and $Y(t + s)$. For instance,

$\text{Cov}[Y(t), Y(t + s)]$

$\qquad = \text{Cov}\,(N_1 + N_2, N_2 + N_3)$

$\qquad = \text{Cov}\,(N_2, N_2)$ by independence of N_1, N_2, N_3

$\qquad = \text{Var}\,(N_2)$

$\qquad = \lambda \int_0^t [1 - G(t + s - y)]dy = \lambda \int_0^t [1 - G(u + s)]du$

where the last equality follows since the variance of a Poisson random variable equals its mean, and from the substitution $u = t - y$. Also, the joint distribution of $Y(t)$ and $Y(t + s)$ is as follows:

$P\{Y(t) = i, \quad Y(t + s) = j\} = P\{N_1 + N_2 = i, \quad N_2 + N_3 = j\}$

$\qquad = \sum_{\ell = 0}^{\min(i,j)} P\{N_2 = \ell, \quad N_1 = i - \ell, \quad N_3 = j - \ell\}$

$\qquad = \sum_{\ell = 0}^{\min(i,j)} P\{N_2 = \ell\}\, P\{N_1 = i - \ell\}\, P\{N_3 = j - \ell\} \quad \Diamond$

Example 3g (An Immigration Model): Suppose immigrants arrive at an area at a Poisson rate λ, and suppose that the amount of time that they spend in the area before leaving has distribution G. We shall leave it to the reader to check that $N_1(t)$, the number of immigrants that have moved out of the area by time t, and $N_2(t)$, the number of immigrants living in the area at time t have distributions given respectively by Equations (3.8) and (3.9).

Let us now prove Proposition 3.3.

Proof of Proposition 3.3.
Let us compute the joint probability $P\{N_i(t) = n_i, i = 1, \ldots, k\}$. To do so note first that in order for there to have been n_i type i events for $i = 1, \ldots, k$ there must have been a total of $\sum_{i=1}^k n_i$ events. Hence, conditioning on $N(t)$ yields

$P\{N_1(t) = n_1, \ldots, N_k(t) = n_k\}$

$\qquad = P\left\{N_1(t) = n_1, \ldots, N_k(t) = n_k | N(t) = \sum_{i=1}^k n_i\right\} \times$

$\qquad \quad P\left\{N(t) = \sum_{i=1}^k n_i\right\}$

Now consider an arbitrary event that occurred in the interval $[0, t]$. If it had occurred at time s, then the probability that it would be a type i event would be $P_i(s)$. Hence, since by Theorem 3.2 this event will have occurred at some time uniformly distributed on $(0, t)$, it follows that the probability that this event will be a type i event is

$$P_i = \frac{1}{t} \int_0^t P_i(s)\,ds$$

independently of the other events. Hence, $P\{N_i(t) = n_i, i = 1, \ldots, k | N(t) = \Sigma_{i=1}^k n_i\}$ will just equal the multinomial probability of n_i type i outcomes for $i = 1, \ldots, k$ when each of $\Sigma_{i=1}^k n_i$ independent trials result in outcome i with probability $P_i, i = 1, \ldots, k$. That is,

$$P\left\{ N_1(t) = n_1, \ldots, N_k(t) = n_k | N(t) = \sum_{i=1}^k n_i \right\}$$

$$= \frac{\left(\sum_{i=1}^k n_i \right)!}{n_1! \ldots n_k!} P_1^{n_1} \ldots P_k^{n_k}$$

Consequently,

$$P\{N_1(t) = n_1, \ldots, N_k(t) = n_k\}$$

$$= \frac{\left(\sum_i n_i \right)!}{n_1! \cdots n_k!} P_1^{n_1} \cdots P_k^{n_k} e^{-\lambda t} \frac{(\lambda t)^{\Sigma_i n_i}}{\left(\sum_i n_i \right)!}$$

$$= \prod_{i=1}^k e^{-\lambda t P_i} (\lambda t P_i)^{n_i} / n_i!$$

and the proof is complete.

We now present some additional examples of the usefulness of Theorem 3.2.

Example 3h (An Electronic Counter): Suppose that electrical pulses having random amplitudes arrive at a counter in accordance with a Poisson process with rate λ. The amplitude of a pulse is assumed to decrease with time at an exponential rate. That is, we suppose that if a pulse has an amplitude of A units upon arrival, then its amplitude at a time t units later will be $Ae^{-\alpha t}$. We further suppose that the initial amplitudes of the pulses are independent and have a common distribution F.

Let $S_1, S_2, \ldots$ be the arrival times of the pulses and let $A_1, A_2, \ldots$ be their respective amplitudes. Then

$$A(t) = \sum_{i=1}^{N(t)} A_i e^{-\alpha(t - S_i)}$$

represents the total amplitude at time t. We can determine the expected value of $A(t)$ by conditioning on $N(t)$, the number of pulses to arrive by time t. This yields

$$E[A(t)] = \sum_{n=0}^{\infty} E[A(t)|N(t) = n]e^{-\lambda t} \frac{(\lambda t)^n}{n!}$$

Now, conditioned on $N(t) = n$, the unordered arrival times $(S_1, \ldots, S_n)$ are distributed as independent uniform $(0, t)$ random variables. Hence, given $N(t) = n$, $A(t)$ has the same distribution as $\sum_{j=1}^{n} A_j e^{-\alpha(t - Y_j)}$, where $Y_j, j = 1, \ldots, n$, are independent and uniformly distributed on $(0, t)$. Thus,

$$E[A(t)|N(t) = n] = E\left[\sum_{j=1}^{n} A_j e^{-\alpha(t - Y_j)} \right]$$

$$= nE[A]E[e^{-\alpha(t - Y)}]$$

where $E[A]$ is the mean initial amplitude of a pulse, and Y is a uniform $(0, t)$ random variable. Hence,

$$E[e^{-\alpha(t - Y)}] = \int_0^t e^{-\alpha(t - y)} \frac{dy}{t}$$

$$= \frac{e^{-\alpha t}}{\alpha t} e^{\alpha y} \Big|_{y=0}^{y=t}$$

$$= \frac{1 - e^{-\alpha t}}{\alpha t}$$

and thus,

$$E[A(t)|N(t) = n] = nE[A] \frac{(1 - e^{-\alpha t})}{\alpha t}$$

or

$$E[A(t)|N(t)] = N(t)E[A] \frac{(1 - e^{-\alpha t})}{\alpha t}$$

Taking expectations and using the fact that $E[N(t)] = \lambda t$, we have

$$E[A(t)] = \frac{\lambda E[A]}{\alpha}(1 - e^{-\alpha t}) \quad \Diamond$$

Example 3i   (An Optimization Example): Suppose that items arrive at a processing plant in accordance with a Poisson process with rate λ. At a fixed time T, all items are dispatched from the system. The problem is to choose an intermediate time, $t \in (0, T)$, at which all items in the system are dispatched, so as to minimize the total expected wait of all items.

If we dispatch at time t, $0 < t < T$, then the expected total wait of all items will be

$$\frac{\lambda t^2}{2} + \frac{\lambda(T - t)^2}{2}$$

To see why the above is true, we reason as follows: The expected number of arrivals in $(0, t)$ is λt, and each arrival is uniformly distributed on $(0, t)$, and hence has expected wait $t/2$. Thus, the expected total wait of items arriving in $(0, t)$ is $\lambda t^2/2$. Similar reasoning holds for arrivals in (t, T), and the above follows. To minimize this quantity, we differentiate with respect to t to obtain

$$\frac{d}{dt}\left[\lambda \frac{t^2}{2} + \lambda \frac{(T - t)^2}{2}\right] = \lambda t - \lambda(T - t)$$

and equating to 0 shows that the dispatch time that minimizes the expected total wait is $t = T/2$.

We end this section with a result, quite similar in spirit to Theorem 3.2, which states that given S_n, the time of the nth event, then the first $n - 1$ event times are distributed as the ordered values of a set of $n - 1$ random variables uniformly distributed on $(0, S_n)$.

Proposition 3.4.   Given that $S_n = t$, the set $S_1, \ldots, S_{n-1}$ has the distribution of a set of $n - 1$ independent uniform $(0, t)$ random variables.

Proof: We can prove the above in the same manner as we did Theorem 3.2, or we can argue more loosely as follows:

$$S_1, \ldots, S_{n-1}|S_n = t \sim S_1, \ldots, S_{n-1}|S_n = t, \quad N(t^-) = n - 1$$
$$\sim S_1, \ldots, S_{n-1}|N(t^-) = n - 1$$

where $\sim$ means "has the same distribution as" and t^- is infinitesimally smaller than t. The result now follows from Theorem 3.2. ◊

3.6 Estimating Software Reliability

When a new computer software package is developed, a testing procedure is often put into effect to eliminate the faults, or bugs, in the package. One common procedure is to try the package on a set of well-known problems to see if any errors result. This goes on for some fixed time, with all resulting errors being noted. Then the testing stops and the package is carefully checked to determine the specific bugs that were responsible for the observed errors. The package is then altered to remove these bugs. As we cannot be certain that all the bugs in the package have been eliminated, however, a problem of great importance is the estimation of the error rate of the revised software package.

To model the above, let us suppose that initially the package contains an unknown number, m, of bugs, which we will refer to as bug 1, bug 2, . . . , bug m. Suppose also that bug i will cause errors to occur in accordance with a Poisson process having an unknown rate λ_i, $i = 1, \ldots, m$. Then, for instance, the number of errors due to bug i that occur in any s units of operating time is Poisson distributed with mean $\lambda_i s$. Also suppose that these Poisson processes caused by bugs i, $i = 1, \ldots, m$ are independent. In addition, suppose that the package is to be run for t time units with all resulting errors being noted. At the end of this time a careful check of the package is made to determine the specific bugs that caused the errors (that is, a *debugging,* takes place). These bugs are removed, and the problem is then to determine the error rate for the revised package.

If we let

$$\psi_i(t) = \begin{cases} 1 & \text{if bug } i \text{ has not caused an error by } t \\ 0 & \text{otherwise} \end{cases}$$

then the quantity we wish to estimate is

$$\Lambda(t) = \sum_i \lambda_i \psi_i(t) \tag{3.10}$$

the error rate of the final package. To start, note that

$$E[\Lambda(t)] = \sum_i \lambda_i E[\psi_i(t)]$$

$$= \sum_i \lambda_i e^{-\lambda_i t}$$

Now each of the bugs that are discovered would have been responsible for a certain number of errors. Let us denote by $M_j(t)$ the number of bugs that were responsible for j errors, $j \geq 1$. That is, $M_1(t)$ is the number of bugs that caused exactly 1 error, $M_2(t)$ is the number that caused 2 errors, and so on, with $\Sigma_j j M_j(t)$ equalling the total number of errors that resulted. To compute $E[M_1(t)]$, let us define the indicator variables, $I_i(t)$, $i \geq 1$, by

$$I_i(t) = \begin{cases} 1 & \text{bug } i \text{ causes exactly 1 error} \\ 0 & \text{otherwise} \end{cases}$$

Then,

$$M_1(t) = \sum_i I_i(t)$$

and so

$$E[M_1(t)] = \sum_i E[I_i(t)] = \sum_i \lambda_i t e^{-\lambda_i t} \qquad (3.11)$$

Thus, from (3.10) and (3.11) we obtain the intriguing result that

$$E\left[\Lambda(t) - \frac{M_1(t)}{t}\right] = 0 \qquad (3.12)$$

This suggests the possible use of $M_1(t)/t$ as an estimate of $\Lambda(t)$. To determine whether or not $M_1(t)/t$ constitutes a "good" estimate of $\Lambda(t)$ we shall look at how far apart these two quantities tend to be. That is, we will compute

$$E\left[\left(\Lambda(t) - \frac{M_1(t)}{t}\right)^2\right]$$

$$= \text{Var}\left(\Lambda(t) - \frac{M_1(t)}{t}\right) \quad \text{from (3.12)}$$

$$= \text{Var}(\Lambda(t)) - \frac{2}{t}\text{Cov}(\Lambda(t), M_1(t)) + \frac{1}{t^2}\text{Var}(M_1(t))$$

Now

$$\text{Var}(\Lambda(t)) = \sum_i \lambda_i^2 \text{Var}(\psi_i(t)) = \sum_i \lambda_i^2 e^{-\lambda_i t}(1 - e^{-\lambda_i t})$$

$$\text{Var}(M_1(t)) = \sum_i \text{Var}(I_i(t)) = \sum_i \lambda_i t e^{-\lambda_i t}(1 - \lambda_i t e^{-\lambda_i t})$$

$$\text{Cov}\left(\Lambda(t), M_1(t)\right) = \text{Cov}\left(\sum_i \lambda_i \psi_i(t), \sum_j I_j(t)\right)$$

$$= \sum_i \sum_j \text{Cov}(\lambda_i \psi_i(t), I_j(t))$$

$$= \sum_i \lambda_i \text{Cov}(\psi_i(t), I_i(t))$$

$$= -\sum_i \lambda_i e^{-\lambda_i t} \lambda_i t e^{-\lambda_i t}$$

where the last two equalities follow since $\psi_i(t)$ and $I_j(t)$ are independent when $i \neq j$ as they refer to different Poisson processes and $\psi_i(t)I_i(t) = 0$. Hence we obtain that

$$E\left[\left(\Lambda(t) - \frac{M_1(t)}{t}\right)^2\right] = \sum_i \lambda_i^2 e^{-\lambda_i t} + \frac{1}{t}\sum_i \lambda_i e^{-\lambda_i t}$$

$$= \frac{E[M_1(t) + 2M_2(t)]}{t^2}$$

where the last equality follows from (3.11) and the identity (which we leave as an exercise)

$$E[M_2(t)] = \frac{1}{2}\sum_i (\lambda_i t)^2 e^{-\lambda_i t} \tag{3.13}$$

Thus we can estimate the average square of the difference between $\Lambda(t)$ and $M_1(t)/t$ by the observed value of $M_1(t) + 2M_2(t)$ divided by t^2.

Example 3j Suppose that in 100 units of operating time 20 bugs are discovered of which two resulted in exactly one, and three resulted in exactly two, errors. Then we would estimate that $\Lambda(100)$ is something akin to the value of a random variable whose mean is equal to $1/50$ and whose variance is equal to $8/10000$. ◇

4. Generalizations of the Poisson Process

4.1. Nonhomogeneous Poisson Process

In this section we consider two generalizations of the Poisson process. The first of these is the nonhomogeneous, also called the nonstationary, Poisson process, which is obtained by allowing the arrival rate at time t to be a function of t.

Definition 4.1.   The counting process $\{N(t), t \geq 0\}$ is said to be a *nonhomogeneous Poisson process with intensity function* $\lambda(t), t \geq 0$, if

(i) $N(0) = 0$
(ii) $\{N(t), t \geq 0\}$ has independent increments
(iii) $P\{N(t + h) - N(t) \geq 2\} = o(h)$
(iv) $P\{N(t + h) - N(t) = 1\} = \lambda(t)h + o(h)$.

If we let $m(t) = \int_0^t \lambda(s)ds$, then it can be shown that

$$P\{N(t + s) - N(t) = n\}$$

$$= e^{-[m(t+s)-m(t)]} \cdot \frac{[m(t + s) - m(t)]^n}{n!}, \quad n \geq 0 \quad (4.1)$$

Or, in other words, $N(t + s) - N(t)$ is Poisson distributed with mean $m(t + s) - m(t)$. Thus, for instance, $N(t)$ is Poisson distributed with mean $m(t)$, and for this reason $m(t)$ is called the *mean value function* of the process. Note that if $\lambda(t) = \lambda$ (that is, if we have a Poisson process), then $m(t) = \lambda t$ and so Equation (4.1) reduces to the fact that for a Poisson process $N(t + s) - N(t)$ is Poisson distributed with mean λs.

The proof of Equation (4.1) follows along the lines of the proof of Theorem 3.1 with a slight modification. That is, we fix t and define

$$P_n(s) = P\{N(t + s) - N(t) = n\}$$

Now,

$$P_0(s + h) = P\{N(t + s + h) - N(t) = 0\}$$
$$= P\{0 \text{ events in } (t, t + s), 0 \text{ events in } [t + s, t + s + h]\}$$
$$= P\{0 \text{ events in } (t, t + s)\}P\{0 \text{ events in } [t + s, t + s + h]\}$$
$$= P_0(s)[1 - \lambda(t + s)h + o(h)]$$

where the last two equations follow from independent increments plus the fact that (iii) and (iv) imply that $P\{N(t + s + h) - N(t + s) = 0\} = 1 - \lambda(t + s)h + o(h)$. Hence,

$$\frac{P_0(s + h) - P_0(s)}{h} = -\lambda(t + s)P_0(s) + \frac{o(h)}{h}$$

letting $h \to 0$ yields

$$P_0'(s) = -\lambda(t + s)P_0(s)$$

or

$$\log P_0(s) = - \int_0^s \lambda(t + u)\,du$$

$$= - \int_t^{t+s} \lambda(y)\,dy$$

or

$$P_0(s) = e^{-[m(t+s)-m(t)]}$$

The remainder of the verification of Equation (4.1) follows similarly and is left as an exercise.

The importance of the nonhomogeneous Poisson process resides in the fact that we no longer require the condition of stationary increments. Thus we now allow for the possibility that events may be more likely to occur during certain times during the day than during other times.

Example 4a Norbert runs a hot dog stand which opens at 8 A.M. From 8 until 11 A.M. customers seem to arrive, on the average, at a steadily increasing rate that starts with an initial rate of 5 customers per hour at 8 A.M. and reaches a maximum of 20 customers per hour at 11 A.M. From 11 A.M. until 1 P.M. the (average) rate seems to remain constant at 20 customers per hour. However, the (average) arrival rate then drops steadily from 1 P.M. until closing time at 5 P.M. at which time it has the value of 12 customers per hour. If we assume that the number of customers arriving at Norbert's stand during disjoint time periods is independent, then what is a good probability model for the above? What is the probability that no customers arrive between 8:30 A.M. and 9:30 A.M. on Monday morning? What is the expected number of arrivals in this period?

Solution: A good model for the above would be to assume that arrivals constitute a nonhomogeneous Poisson process with intensity function $\lambda(t)$ given by

$$\lambda(t) = \begin{cases} 5 + 5t, & 0 \le t \le 3 \\ 20, & 3 \le t \le 5 \\ 20 - 2(t - 5), & 5 \le t \le 9 \end{cases}$$

and

$$\lambda(t) = \lambda(t - 9) \quad \text{for} \quad t > 9$$

Note that $N(t)$ represents the number of arrivals during the first t hours that the store is open. That is, we do not count the hours between 5 P.M. and 8 A.M. If for some reason we wanted $N(t)$ to represent the number of arrivals during the first t hours regardless of whether the store was open or not, then, assuming that the process begins at midnight we would let

$$\lambda(t) = \begin{cases} 0, & 0 \leq t < 8 \\ 5 + 5(t - 8), & 8 \leq t \leq 11 \\ 20, & 11 \leq t \leq 13 \\ 20 - 2(t - 13), & 13 \leq t \leq 17 \\ 0, & 17 < t \leq 24 \end{cases}$$

and

$$\lambda(t) = \lambda(t - 24) \quad \text{for} \quad t > 24$$

As the number of arrivals between 8:30 A.M. and 9:30 A.M. will be Poisson with mean $m(\frac{3}{2}) - m(\frac{1}{2})$ in the first representation (and $m(\frac{19}{2}) - m(\frac{17}{2})$ in the second representation), we have that the probability that this number is zero is

$$\exp\left\{ - \int_{1/2}^{3/2} (5 + 5t)dt \right\} = e^{-10}$$

and the mean number of arrivals is

$$\int_{1/2}^{3/2} (5 + 5t)dt = 10 \quad \diamond$$

When the intensity function $\lambda(t)$ is bounded, we can think of the nonhomogeneous process as being a random sample from a homogeneous Poisson process. Specifically, let λ be such that

$$\lambda(t) \leq \lambda \quad \text{for all} \quad t \geq 0$$

and consider a Poisson process with rate λ. Now if we suppose that an event of the Poisson process that occurs at time t is counted with probability $\lambda(t)/\lambda$, then the process of counted events is a nonhomogeneous Poisson process with intensity function $\lambda(t)$. This last statement easily follows from Definition 4.1. For instance (i), (ii), and (iii) follow since they are also true for the homogeneous Poisson process. Axiom (iv) follows since

$$P\{\text{one counted event in } (t, t + h)\} = P\{\text{one event in } (t, t + h)\} \frac{\lambda(t)}{\lambda} + o(h)$$

$$= \lambda h \frac{\lambda(t)}{\lambda} + o(h)$$

$$= \lambda(t)h + o(h)$$

Example 4b (The Output Process of an Infinite Server Poisson Queue $(M/G/\infty)$): It turns out that the output process of the $M/G/\infty$ queue—that is, of the infinite server queue having Poisson arrivals and general service distribution G—is a nonhomogeneous Poisson process having intensity function $\lambda(t) = \lambda G(t)$. To prove this claim, note first that the (joint) probability (density) that a customer arrives at time s and departs at time t is equal to λ, the probability (intensity) of an arrival at time s, multiplied by $g(t - s)$, the probability (density) that its service time is $t - s$. Hence,

$$\bar{\lambda}(t) = \int_0^t \lambda g(t - s)ds = \lambda G(t)$$

is the probability intensity of a departure at time t. Now suppose we are told that a departure occurs at time t—how does this affect the probabilities of other departure times? Well, even if we knew the arrival time of the customer who departed at time t, this would not affect the arrival times of other customers (because of the independent increment assumption of the Poisson arrival process). Hence, as there are always servers available for arrivals this information cannot affect the probabilities of other departure times. Thus, the departure process has independent increments, and departures occur at t with intensity $\lambda(t)$, which verifies the claim. $\diamond$

4.2. Compound Poisson Process

A stochastic process $\{X(t), t \geq 0\}$ is said to be a *compound Poisson process* if it can be represented as

$$X(t) = \sum_{i=1}^{N(t)} Y_i, \qquad t \geq 0 \tag{4.2}$$

where $\{N(t), t \geq 0\}$ is a Poisson process, and $\{Y_n, n \geq 0\}$ is a family of independent and identically distributed random variables which are also independent of $\{N(t), t \geq 0\}$.

Examples of Compound Poisson Processes

(i) If $Y_i \equiv 1$, then $X(t) = N(t)$, and so we have the usual Poisson process.

(ii) Suppose that buses arrive at a sporting event in accordance with a Poisson process, and suppose that the numbers of customers in each bus are assumed to be independent and identically distributed. Then $\{X(t), t \geq 0\}$ is a compound Poisson process where $X(t)$ denotes the number of customers who have arrived by t. In Equation (4.2) Y_i represents the number of customers in the ith bus.

(iii) Suppose customers leave a supermarket in accordance with a Poisson process. If Y_i, the amount spent by the ith customer, $i = 1, 2, \ldots$, are independent and identically distributed, then $\{X(t), t \geq 0\}$ is a compound Poisson process when $X(t)$ denotes the total amount of money spent by time t. $\Diamond$

Let us calculate the mean and variance of $X(t)$. To calculate $E[X(t)]$, we first condition on $N(t)$ to obtain

$$E[X(t)] = E(E[X(t)|N(t)])$$

Now

$$E[X(t)|N(t) = n] = E\left[\sum_{i=1}^{N(t)} Y_i \Big| N(t) = n\right]$$

$$= E\left[\sum_{i=1}^{n} Y_i \Big| N(t) = n\right]$$

$$= E\left[\sum_{i=1}^{n} Y_i\right]$$

$$= nE[Y_1]$$

where we have used the assumed independence of the Y_i's and $N(t)$. Hence,

$$E[X(t)|N(t)] = N(t)E[Y_1] \tag{4.3}$$

and therefore

$$E[X(t)] = \lambda t E[Y_1] \tag{4.4}$$

To calculate $\text{Var}[X(t)]$ we use the conditional variance formula (see Problem 22 of Chapter 3)

$$\text{Var}[X(t)] = E[\text{Var}(X(t)|N(t))] + \text{Var}(E[X(t)|N(t)]) \tag{4.5}$$

Now,

$$\text{Var}[X(t)|N(t) = n] = \text{Var}\left[\sum_{i=1}^{N(t)} Y_i \Big| N(t) = n\right]$$

$$= \text{Var}\left[\sum_{i=1}^{n} Y_i\right]$$

$$= n\,\text{Var}(Y_1)$$

and thus

$$\text{Var}[X(t)|N(t)] = N(t)\,\text{Var}(Y_1). \qquad (4.6)$$

Therefore using Equations (4.3) and (4.6), we obtain by Equation (4.5) that

$$\text{Var}[X(t)] = E[N(t)\,\text{Var}(Y_1)] + \text{Var}[N(t)EY_1]$$

$$= \lambda t\,\text{Var}(Y_1) + (EY_1)^2\lambda t$$

$$= \lambda t[\text{Var}(Y_1) + (EY_1)^2]$$

$$= \lambda t E[Y_1^2] \qquad (4.7)$$

where we have used the fact that $\text{Var}[N(t)] = \lambda t$ since $N(t)$ is Poisson distributed with mean λt.

Example 4c Suppose that families migrate to an area at a Poisson rate $\lambda = 2$ per week. If the number of people in each family is independent and takes on the values 1, 2, 3, 4 with respective probabilities $\frac{1}{6}, \frac{1}{3}, \frac{1}{3}, \frac{1}{6}$, then what is the expected value and variance of the number of individuals migrating to this area during a fixed five-week period?

Solution: Letting Y_i denote the number of people in the ith family, we have that

$$E[Y_i] = 1 \cdot \tfrac{1}{6} + 2 \cdot \tfrac{1}{3} + 3 \cdot \tfrac{1}{3} + 4 \cdot \tfrac{1}{6} = \tfrac{5}{2}$$

$$E[Y_i^2] = 1^2 \cdot \tfrac{1}{6} + 2^2 \cdot \tfrac{1}{3} + 3^2 \cdot \tfrac{1}{3} + 4^2 \cdot \tfrac{1}{6} = \tfrac{43}{6}$$

Hence, letting $X(5)$ denote the number of immigrants during a five-week period, we obtain from Equations (4.4) and (4.7) that

$$E[X(5)] = 2 \cdot 5 \cdot \tfrac{5}{2} = 25$$

and

$$\text{Var}[X(5)] = 2 \cdot 5 \cdot \tfrac{43}{6} = \tfrac{215}{3} \quad \Diamond$$

Problems

1. The time T required to repair a machine is an exponentially distributed random variable with mean $\frac{1}{2}$ (hours).
   (a) What is the probability that a repair time exceeds $\frac{1}{2}$ hour?
   (b) What is the probability that a repair takes at least $12\frac{1}{2}$ hours given that its duration exceeds 12 hours?

2. Consider a post office with two clerks. Three people, A, B, and C, enter simultaneously. A and B go directly to the clerks, and C waits until either A or B leaves before he begins service. What is the probability that A is still in the post office after the other two have left when
   (a) the service time for each clerk is exactly (nonrandom) ten minutes?
   (b) the service times are i with probability $\frac{1}{3}$, $i = 1, 2, 3$?
   (c) the service times are exponential with mean $1/\mu$?

3. The lifetime of a radio is exponentially distributed with a mean of ten years. If Jones buys a ten-year-old radio, what is the probability that it will be working after an additional ten years?

4. In Example 2b if server i serves at an exponential rate λ_i, $i = 1, 2$, show that

$$P\{\text{Smith is not last}\} = \left(\frac{\lambda_1}{\lambda_1 + \lambda_2}\right)^2 + \left(\frac{\lambda_2}{\lambda_1 + \lambda_2}\right)^2$$

5. If X_1 and X_2 are independent nonnegative continuous random variables, show that

$$P\{X_1 < X_2 | \min(X_1, X_2) = t\} = \frac{r_1(t)}{r_1(t) + r_2(t)}$$

where $r_i(t)$ is the failure rate function of X_i.

6. Show that the failure rate function of a gamma distribution with parameters n and λ is increasing when $n \geq 1$.

7. Norb and Nat enter a barbershop simultaneously—Norb to get a shave and Nat a haircut. If the amount of time it takes to receive a haircut (shave) is exponentially distributed with mean 20 (15) minutes, and if Norb and Nat are immediately served, what is the probability that Nat finishes before Norb?

8. If X and Y are independent exponential random variables with respective mean $1/\lambda_1$ and $1/\lambda_2$, then compute the distribution of $Z =$

min(X, Y). What is the conditional distribution of Z given that Z = X?

9. (Problem 8 continued). Give a heuristic argument that the conditional distribution of $Y - Z$, given that $Z = X$, is exponential with mean $1/\lambda_2$.

10. A flashlight needs two batteries to be operational. Consider such a flashlight along with a set of n functional batteries—battery 1, battery 2, . . . , battery n. Initially, battery 1 and 2 are installed. Whenever a battery fails, it is immediately replaced by the lowest numbered functional battery that has not yet been put in use. Suppose that the lifetimes of the different batteries are independent exponential random variables each having rate μ. At a random time, call it T, a battery will fail and our stockpile will be empty. At that moment exactly one of the batteries—which we call battery X—will not yet have failed.
   (a) What is $P\{X = n\}$?
   (b) What is $P\{X = 1\}$?
   (c) What is $P\{X = i\}$?
   (d) Find $E[T]$.
   (e) What is the distribution of T?

11. Let X and Y be independent exponential random variables having respective rates λ and μ. Let I, independent of X, Y, be such that

$$I = \begin{cases} 1 & \text{with probability } \dfrac{\mu}{\lambda + \mu} \\ 0 & \text{with probability } \dfrac{\lambda}{\lambda + \mu} \end{cases}$$

and define Z by

$$Z = \begin{cases} X & \text{if } I = 1 \\ -Y & \text{if } I = 0 \end{cases}$$

(a) Show, by computing their moment generating functions, that $X - Y$ and Z have the same distribution.
(b) Using the lack of memory property of exponential random variables, give a simple explanation of the result of part (a).

12. Two individuals, A and B, both require kidney transplants. If she does not receive a new kidney, then A will die after an exponential time with rate μ_A, and B after an exponential time with rate μ_B. New kidneys arrive in accordance with a Poisson process having rate λ. It

has been decided that the first kidney will go to A (or to B if B is alive and A is not at that time) and the next one to B (if still living).

(a) What is the probability A obtains a new kidney?

(b) What is the probability B obtains a new kidney?

13. Show that Definition 1 of a Poisson process implies Definition 2.

14. Show that Assumption (iv) of Definition 2 follows from Assumptions (ii) and (iii).

Hint: Derive a functional equation for $g(t) = P\{N(t) = 0\}$.

15. Cars cross a certain point in the highway in accordance with a Poisson process with rate $\lambda = 3$ per minute. If Al blindly runs across the highway, then what is the probability that he will be uninjured if the amount of time that it takes him to cross the road is s seconds? (Assume that if he is on the highway when a car passes by, then he will be injured.) Do it for $s = 2, 5, 10, 20$.

16. Suppose in Problem 15, that Al is agile enough to escape from a single car, but if he encounters two or more cars while attempting to cross the road, then he will be injured. What is the probability that he will be unhurt if it takes him s seconds to cross. Do it for $s = 5, 10, 20, 30$.

17. Show that if $\{N_i(t), t \geq 0\}$ are independent Poisson processes with rate λ_i, $i = 1, 2$, then $\{N(t), t \geq 0\}$ is a Poisson process with rate $\lambda_1 + \lambda_2$ where $N(t) = N_1(t) + N_2(t)$.

18. In Problem 17 what is the probability that the first event of the combined process is from the N_1 process?

19. For a Poisson process show, for $s < t$, that

$$P\{N(s) = k | N(t) = n\} = \binom{n}{k}\left(\frac{s}{t}\right)^k \left(1 - \frac{s}{t}\right)^{n-k}, \qquad k = 0, 1, \ldots, n$$

20. Men and women enter a supermarket according to independent Poisson processes having respective rates two and four per minute. Starting at an arbitrary time, compute the probability that at least two men arrive before three women arrive.

21. Events occur according to a Poisson process with rate $\lambda = 2$ per hour.

(a) What is the probability that no event occurs between 8 P.M. and 9 P.M.?

(b) Starting at noon, what is the expected time at which the fourth event occurs?

(c) What is the probability that two or more events occur between 6 P.M. and 8 P.M.?

22. Pulses arrive at a Geiger counter in accordance with a Poisson process at a rate of three arrivals per minute. Each particle arriving at the counter has a probability $\frac{2}{3}$ of being recorded. Let $X(t)$ denote the number of pulses recorded by time t minutes.

(a) $P\{X(t) = 0\} = ?$

(b) $E[X(t)] = ?$

23. Cars pass a point on the highway at a Poisson rate of one per minute. If five percent of the cars on the road are Dodges, then

(a) what is the probability that at least one Dodge passes by during an hour?

(b) given that ten Dodges have passed by in an hour, what is the expected number of cars to have passed by in that time?

(c) if 50 cars have passed by in an hour, what is the probability that five of them were Dodges?

24. Customers arrive at a bank at a Poisson rate λ. Suppose two customers arrived during the first hour. What is the probability that

(a) both arrived during the first 20 minutes?

(b) at least one arrived during the first 20 minutes?

25. A system has a random number of flaws that we will suppose is Poisson distributed with mean c. Each of these flaws will, independently, cause the system to fail at a random time having distribution G. When a system failure occurs, suppose that the flaw causing the failure is immediately located and fixed.

(a) What is the distribution of the number of failures by time t?

(b) What is the distribution of the number of flaws that remain in the system at time t?

(c) Are the random variables in (a) and (b) dependent or independent?

26. Suppose that the times between successive arrivals of customers at a single-server station are independent random variables having a common distribution F. Suppose that when a customer arrives, it either immediately enters service if the server is free or else it joins the end of the waiting line if the server is busy with another customer. When the server completes work on a customer that customer leaves the system and the next waiting customer, if there are any, enters ser-

vice. Let X_n denote the number of customers in the system immediately before the nth arrival, and let Y_n denote the number of customers that remain in the system when the nth customer departs. The successive service times of customers are independent random variables (which are also independent of the interarrival times) having a common distribution G.

(a) If F is the exponential distribution with rate λ, which, if any, of the processes $\{X_n\}$, $\{Y_n\}$ is a Markov chain?

(b) If G is the exponential distribution with rate μ, which, if any, of the processes $\{X_n\}$, $\{Y_n\}$ is a Markov chain?

(c) Give the transition probabilities of any Markov chains in (a) and (b).

27. Verify Equation (4.1).

28. Events occur according to a nonhomogeneous Poisson process whose mean value function is given by

$$m(t) = t^2 + 2t, \qquad t \geq 0$$

What is the probability that n events occur between times $t = 4$ and $t = 5$?

29. A store opens at 8 A.M. From 8 until 10 customers arrive at a Poisson rate of four an hour. Between 10 and 12 they arrive at a Poisson rate of eight an hour. From 12 to 2 the arrival rate increases steadily from eight per hour at 12 to ten per hour at 2; and from 2 to 5 the arrival rate drops steadily from ten per hour at 2 to four per hour at 5. Determine the probability distribution of the number of customers that enter the store on a given day.

30. Consider a nonhomogeneous Poisson process whose intensity function $\lambda(t)$ is bounded and continuous. Show that such a process is equivalent to a process of counted events from a (homogeneous) Poisson process having rate λ, where an event at time t is counted (independent of the past) with probability $\lambda(t)/\lambda$; and where λ is chosen so that $\lambda(s) < \lambda$ for all s.

31. Let T_1, T_2, . . . denote the interarrival times of events of a nonhomogeneous Poisson process having intensity function $\lambda(t)$.

(i) Are the T_i independent?

(ii) Are the T_i identically distributed?

(iii) Find the distribution of T_1.

32.

(i) Let $\{N(t), t \geq 0\}$ be a nonhomogeneous Poisson process with mean value function $m(t)$. Given $N(t) = n$, show that the unordered set of arrival times has the same distribution as n independent and identically distributed random variables having distribution function

$$F(x) = \begin{cases} \dfrac{m(x)}{m(t)}, & x \le t \\ 1, & x > t \end{cases}$$

(ii) Suppose that workmen incur accidents in accordance with a nonhomogeneous Poisson process with mean value function $m(t)$. Suppose further that each injured man is out of work for a random amount of time having distribution F. Let $X(t)$ be the number of workers who are out of work at time t. By using part (i), find $E[X(t)]$.

33. An insurance company pays out claims on its life insurance policies in accordance with a Poisson process having rate $\lambda = 5$ per week. If the amount of money paid on each policy is exponentially distributed with mean \$2000, what is the mean and variance of the amount of money paid by the insurance company in a four-week span?

34. Some components of a two-component system fail after receiving a shock. Shocks of three types arrive independently and in accordance with Poisson processes. Shocks of the first type arrive at a Poisson rate λ_1 and cause the first component to fail. Those of the second type arrive at a Poisson rate λ_2 and cause the second component to fail. The third type of shock arrives at a Poisson rate λ_3 and causes both components to fail. Let X_1 and X_2 denote the survival times for the two components. Show that the joint distribution of X_1 and X_2 is given by

$$P\{X_1 > s, X_2 > t\} = \exp\{-\lambda_1 s - \lambda_2 t - \lambda_3 \max(s, t)\}$$

This distribution is known as the *bivariate exponential distribution*.

35. In Problem 34 show that X_1 and X_2 both have exponential distributions.

36. Let $X_1, X_2, \ldots, X_n$ be independent and identically distributed exponential random variables. Show that the probability that the largest of them is greater than the sum of the others is $n/2^{n-1}$. That is if

$$M = \max_j X_j$$

then show

$$P\left\{M > \sum_{i=1}^{n} X_i - M\right\} = \frac{n}{2^{n-1}}$$

Hint: What is $P\{X_1 > \Sigma_{i=2}^n X_i\}$?

37. Prove Equation (3.13).

38. Prove that

(i) $\max(X_1, X_2) = X_1 + X_2 - \min(X_1, X_2)$ and, in general,

(ii) $\max(X_1, \ldots, X_n) = \sum_{1}^{n} X_i - \sum\sum_{i<j} \min(X_i, X_j)$

$$+ \sum\sum\sum_{i<j<k} \min(X_i, X_j, X_k) + \cdots$$

$$+ (-1)^{n-1} \min(X_1, X_2, \ldots, X_n)$$

Show by defining appropriate random variables X_i, $i = 1, \ldots, n$, and by taking expectations in (ii) how to obtain the well-known formula

$$P\left(\bigcup_{1}^{n} A_i\right) = \sum_{i} P(A_i) - \sum\sum_{i<j} P(A_i A_j)$$

$$+ \cdots + (-1)^{n-1} P(A_1 \cdots A_n)$$

(iii) Consider n independent Poisson processes—the ith having rate λ_i. Derive an expression for the expected time until an event has occurred in all n processes.

39. A two-dimensional Poisson process is a process of randomly occurring events in the plane such that

(a) for any region of area A the number of events in that region has a Poisson distribution with mean λA and

(b) the number of events in nonoverlapping regions are independent.

For such a process consider an arbitrary point in the plane and let X denote its distance from its nearest event (where distance is measured in the usual Euclidean manner). Show that

(i) $$P\{X > t\} = e^{-\lambda \pi t^2}$$

(ii) $$E[X] = \frac{1}{2\sqrt{\lambda}}$$

References

1. H. Cramér and M. Leadbetter, "Stationary and Related Stochastic Processes," John Wiley, New York, 1966.

2. S. Ross, "Stochastic Processes," John Wiley, New York, 1983.

Chapter 6

Continuous-Time
Markov Chains

1. Introduction

In this chapter we consider a class of probability models that have a wide variety of applications in the real world. They are the continuous-time analogue of the Markov chains of Chapter 4 and as such are characterized by the Markovian property that given the present state, the future is independent of the past.

One example of a continuous-time Markov chain has already been met. This is, of course, the Poisson process of Chapter 5. For if we let the total number of arrivals by time t[that is, $N(t)$] be the state of the process at time t, then the Poisson process is a continuous-time Markov chain having states $0, 1, 2, \ldots$ and which always proceeds from state n to state $n + 1$, where $n \geq 0$. Such a process is known as a *pure birth process* since the state of the system is always increased by one. More generally, an exponential model which can go (in one transition) only from state n to either state $n - 1$ or state $n + 1$ is called a *birth and death model*. For such a model, transitions from state n to state $n + 1$ are designated as births, and those from n to $n - 1$ as deaths. Birth and death models have wide applicability in the study of biological systems and in the study of waiting line systems in which the state represents the number of customers in the system. These models will be studied extensively in this chapter.

In Section 2 we define continuous-time Markov chains and then relate them to the discrete-time Markov chains of Chapter 4. In Section 3 we consider birth and death processes and in Section 4 we derive two sets of differential equations—the forward and backward equations—which describe the probability laws for the system. The material in Section 5 is concerned with determining the limiting (or long-run) probabilities connected with a continuous-time Markov chain. In Section 6 we consider the topic of time reversibility. We show that all birth and death processes are time reversible, and then illustrate the importance of this observation to queueing systems. In the final section we show how to "uniformize" Markov chains, a technique useful for numerical computations.

2. Continuous-Time Markov Chains

Suppose we have a continuous-time stochastic process $\{X(t), t \geq 0\}$ taking on values in the set of nonnegative integers. In analogy with the definition of a discrete-time Markov chain, given in Chapter 4, we say that the process $\{X(t), t \geq 0\}$ is a *continuous-time Markov chain* if for all $s, t \geq 0$ and nonnegative integers $i, j, x(u), 0 \leq u < s$

$$P\{X(t + s) = j | X(s) = i, X(u) = x(u), 0 \leq u < s\}$$
$$= P\{X(t + s) = j | X(s) = i\}$$

In other words, a continuous-time Markov chain is a stochastic process having the Markovian property that the conditional distribution of the future $X(t + s)$ given the present $X(s)$ and the past $X(u)$, $0 \leq u < s$, depends only on the present and is independent of the past. If, in addition,

$$P\{X(t + s) = j | X(s) = i\}$$

is independent of s, then the continuous-time Markov chain is said to have stationary or homogeneous transition probabilities.

All Markov chains considered in this text will be assumed to have stationary transition probabilities.

Suppose that a continuous-time Markov chain enters state i at some time, say time 0, and suppose that the process does not leave state i (that is, a transition does not occur) during the next ten minutes. What is the probability that the process will not leave state i during the following five minutes? Now since the process is in state i at time 10 it follows, by the Markovian property, that the probability that it remains in that state during the interval $[10, 15]$ is just the (unconditional) probability that it stays in state i for at least five minutes. That is, if we let

T_i denote the amount of time that the process stays in state i before making a transition into a different state, then

$$P\{T_i > 15 | T_i > 10\} = P\{T_i > 5\}$$

or, in general, by the same reasoning.

$$P\{T_i > s + t | T_i > s\} = P\{T_i > t\}$$

for all s, $t \geq 0$. Hence, the random variable T_i is *memoryless* and must thus (see Section 2.2 of Chapter 5) be *exponentially* distributed.

In fact, the above gives us another way of defining a continuous-time Markov chain. Namely, it is a stochastic process having the properties that each time it enters state i

(i) the amount of time it spends in that state before making a transition into a different state is exponentially distributed with mean, say $1/v_i$, and

(ii) when the process leaves state i, it next enters state j with some probability, say P_{ij}. Of course, the P_{ij} must satisfy

$$P_{ii} = 0, \quad \text{all} \quad i$$

$$\sum_j P_{ij} = 1, \quad \text{all} \quad i$$

In other words, a continuous-time Markov chain is a stochastic process that moves from state to state in accordance with a (discrete-time) Markov chain, but is such that the amount of time it spends in each state, before proceeding to the next state, is exponentially distributed. In addition, the amount of time the process spends in state i, and the next state visited, must be independent random variables. For if the next state visited were dependent on T_i, then information as to how long the process has already been in state i would be relevant to the prediction of the next state—and this contradicts the Markovian assumption.

Example 2a (A Shoeshine Shop): Consider a shoeshine establishment consisting of two chairs—chair 1 and chair 2. A customer upon arrival goes initially to chair 1 where his shoes are cleaned and polish is applied. After this is done the customer moves on to chair 2 where the polish is buffed. The service times at the two chairs are assumed to be independent random variables which are exponentially distributed with respective rates μ_1 and μ_2. Suppose that potential customers arrive in accordance with a Poisson process having rate λ, and that a potential customer will only enter the system if both chairs are empty.

The above model can be analyzed as a continuous-time Markov chain, but first we must decide upon an appropriate state space. Since a potential customer will enter the system only if there are no other customers present, it follows that there will always either be 0 or 1 customers in the system. However, if there is 1 customer in the system, then we would also need to know which chair he was presently in. Hence, an appropriate state space might consist of the three states 0, 1, and 2 where the states have the following interpretation:

State	Interpretation
0	system is empty
1	a customer is in chair 1
2	a customer is in chair 2

We leave it as an exercise for the reader to verify that

$$v_0 = \lambda, \qquad v_1 = \mu_1, \qquad v_2 = \mu_2$$
$$P_{01} = P_{12} = P_{20} = 1 \quad \Diamond$$

3. Birth and Death Processes

Consider a system whose state at any time is represented by the number of people in the system at that time. Suppose that whenever there are n people in the system, then (i) new arrivals enter the system at an exponential rate λ_n, and (ii) people leave the system at an exponential rate μ_n. That is, whenever there are n persons in the system, then the time until the next arrival is exponentially distributed with mean $1/\lambda_n$ and is independent of the time until the next departure which is itself exponentially distributed with mean $1/\mu_n$. Such a system is called a birth and death process. The parameters $\{\lambda_n\}_{n=0}^{\infty}$ and $\{\mu_n\}_{n=1}^{\infty}$ are called respectively the arrival (or birth) and departure (or death) rates.

Thus, a birth and death process is a continuous-time Markov chain with states $\{0, 1, \ldots\}$ for which transitions from state n may go only to either state $n - 1$ or state $n + 1$. The relation between the birth and death rates and the state transition rates and probabilities are

$$v_0 = \lambda_0,$$
$$v_i = \lambda_i + \mu_i, \qquad i > 0$$
$$P_{01} = 1,$$

$$P_{i,\,i+1} = \frac{\lambda_i}{\lambda_i + \mu_i}, \qquad i > 0$$

$$P_{i,\,i-1} = \frac{\mu_i}{\lambda_i + \mu_i}, \qquad i > 0$$

The preceding follows, since when there are i in the system, then the next state will be $i + 1$ if a birth occurs before a death; and the probability that an exponential random variable with rate λ_i will occur earlier than an (independent) exponential with rate μ_i is $\lambda_i/(\lambda_i + \mu_i)$ (and so, $P_{i,\,i+1} = \lambda_i/(\lambda_i + \mu_i)$), and the time until either occurs is exponentially distributed with rate $\lambda_i + \mu_i$ (and so, $v_i = \lambda_i + \mu_i$).

Example 3a (The Poisson Process): Consider a birth and death process for which

$$\mu_n = 0, \quad \text{for all} \quad n \geq 0$$

$$\lambda_n = \lambda, \quad \text{for all} \quad n \geq 0$$

This is a process in which departures never occur, and the time between successive arrivals is exponential with mean $1/\lambda$. Hence, this is just the Poisson process.

A birth and death process for which $\mu_n = 0$ for all n is called a pure birth process. Another pure birth process is given by the next example.

Example 3b (A Birth Process with Linear Birthrate): Consider a population whose members can give birth to new members but cannot die. If each member acts independently of the others and takes an exponentially distributed amount of time, with mean $1/\lambda$, to give birth, then if $X(t)$ is the population size at time t, then $\{X(t), t \geq 0\}$ is a pure birth process with $\lambda_n = n\lambda$, $n \geq 0$. This follows since if the population consists of n persons and each gives birth at an exponential rate λ, then the total rate at which births occur is $n\lambda$. This pure birth process is known as a Yule process after G. Yule who used it in his mathematical theory of evolution. ◇

Example 3c (A Linear Growth Model): Consider a population in which each individual gives birth at an exponential rate λ and dies at an exponential rate μ. This is a birth and death process having

$$\mu_n = n\mu, \qquad n \geq 1$$

$$\lambda_n = n\lambda, \qquad n \geq 0$$

Let $M_i(t)$ denote the expected size at time t of a population that consists of i individuals at time 0. That is,

$$M_i(t) = E[X(t) \mid X(0) = i]$$

Before determining $M_i(t)$, let us note that since each individual gives birth at an exponential rate λ and dies at an exponential rate μ no matter what is happening with other individuals in the population, it follows that starting with i individuals in the population is equivalent to having i independent processes each starting with a single individual. Hence, we see that

$$M_i(t) = i\, M_1(t)$$

Let $M(t) = M_1(t)$. To compute $M(t)$, we condition on T, the time of the first event. Since, when there is a single individual in the population, the birth rate is λ and the death rate μ, it follows that T is exponential with rate $\lambda + \mu$. Hence,

$$M(t) = E[X(t) \mid X(0) = 1]$$
$$= \int_0^\infty E[X(t) \mid T = s](\lambda + \mu)e^{-(\lambda+\mu)s}\, ds \qquad (3.1)$$

Now, if the first event occurs at a time s that is greater than t, then clearly the population size at t is equal to 1, its initial population size. Suppose now that $s < t$ and that the event at time s is a birth. Therefore, there are now, at time s, 2 individuals in the population and so, by the Markovian property, it follows that the population size at t has the same distribution as would the population size at time $t - s$ if there were initially 2 members of the population at time 0. That is,

$$E[X(t) \mid T = s, \text{ event at } s \text{ is a birth}] = E[X(t - s) \mid X(0) = 2]$$
$$= M_2(t - s)$$
$$= 2M(t - s)$$

On the other hand, if the event occurring at time s, $s < t$, is a death, then the population becomes extinct at time s and so its size at t is 0. Hence, upon conditioning upon whether the event at time T is a birth or death, we obtain from (3.1) that

$$M(t) = \int_0^t \frac{\lambda}{\lambda + \mu}\, 2M(t-s)\,(\lambda+\mu)e^{-(\lambda+\mu)s}\, ds + \int_t^\infty (\lambda+\mu)e^{-(\lambda+\mu)s}\, ds$$

$$= 2\lambda \int_0^t M(t - s) \, e^{-(\lambda + \mu)s} \, ds + e^{-(\lambda + \mu)t}$$
$$= 2\lambda e^{-(\lambda + \mu)t} \int_0^t M(y) \, e^{(\lambda + \mu)y} \, dy + e^{-(\lambda + \mu)t} \qquad (3.2)$$

where the last equality follows upon making the substitution $y = t - s$. Letting

$$A(t) = 2\lambda \int_0^t M(y) \, e^{(\lambda + \mu)y} \, dy + 1$$

we see from (3.2) that

$$M(t) = e^{-(\lambda + \mu)t} A(t)$$

Differentiation yields

$$\begin{aligned} M'(t) &= -(\lambda + \mu) \, e^{-(\lambda + \mu)t} A(t) + e^{-(\lambda + \mu)t} A'(t) \\ &= -(\lambda + \mu) M(t) + e^{-(\lambda + \mu)t} 2\lambda M(t) e^{(\lambda + \mu)t} \\ &= (\lambda - \mu) M(t) \end{aligned}$$

implying that

$$\log M(t) = (\lambda - \mu)t + C$$

or

$$M(t) = K e^{(\lambda - \mu)t}$$

Since, $M(0) = E[X(0) \mid X(0) = 1] = 1$, we see that $K = 1$ and so

$$M(t) = e^{(\lambda - \mu)t}$$

and, in general

$$M_i(t) = E[X(t) \mid X(0) = i] = i \, e^{(\lambda - \mu)t}$$

It is interesting to note that the mean population size at t either increases to infinity at an exponential rate when an individual's birth rate exceeds his death rate or decreases to zero when the reverse inequality holds.   ◇

Example 3d   (A Linear Growth Model with Immigration): A model in which

$$\mu_n = n\mu, \qquad n \geq 1$$
$$\lambda_n = n\lambda + \theta, \qquad n \geq 0$$

is called a linear growth process with immigration. Such processes occur naturally in the study of biological reproduction and population growth. Each individual in the population is assumed to give birth at an exponential rate λ; in addition, there is an exponential rate of increase θ of the population due to an external source such as immigration. Hence, the total birth rate where there are n persons in the system is $n\lambda + \theta$. Deaths are assumed to occur at an exponential rate μ for each member of the population, and hence $\mu_n = n\mu$.

To compute $E[X(t) \mid X(0) = i]$, the expected population size at time t given that it is of size i at time 0, think of the i individuals present at time 0 as being the initial ancestors of i separate families. Since each family member will (independently) give birth at an exponential rate λ and die at an exponential rate μ, it follows from the results of Example 3c that the expected size at time t of each of the i original families is $e^{(\lambda-\mu)t}$. Also, think of each immigrant as the initial ancestor of a new family. Since the expected size at time t of a family that originates at time s is $e^{(\lambda-\mu)(t-s)}$ (why?), it therefore follows, upon conditioning on the number of immigrants by time t, and using the result (Theorem 3.2 of Chapter 5) that conditional on this number the set of arrival times are independent and uniform on $(0,t)$ that

$$E\begin{bmatrix}\text{number of members of} \\ \text{immigrant families at } t\end{bmatrix} = \theta t \int_0^t \frac{e^{(\lambda-\mu)(t-s)}}{t}\,ds$$

$$= \theta \int_0^t e^{(\lambda-\mu)y}\,dy$$

$$= \frac{\theta}{\lambda-\mu}[e^{(\lambda-\mu)t} - 1]$$

where the last equality supposes that $\lambda \neq \mu$. (When $\lambda = \mu$, the correct answer is θt.)

Hence, combining this with the non-immigrant family members at time t, we see that

$$E[X(t) \mid X(0) = i] = ie^{(\lambda-\mu)t} + \frac{\theta}{\lambda-\mu}[e^{(\lambda-\mu)t} - 1]$$

with the second term being replaced by θt when $\lambda = \mu$. $\diamond$

Example 3e (The Queueing System $M/M/1$): Suppose that customers arrive at a single-server service station in accordance with a Poisson process having rate λ. That is, the times between successive arrivals are independent exponential random variables having mean $1/\lambda$. Upon arrival, each customer goes directly into service if the server is free, and if not, then the customer joins the queue (that is, he waits in line). When the server finishes serving a customer, the customer leaves the system and the next customer in line, if there are any waiting, enters the service. The successive service times are assumed to be independent exponential random variables having mean $1/\mu$.

The above is known as the $M/M/1$ queueing system. The first M refers to the fact that the interarrival process is Markovian (since it is a Poisson process) and the second to the fact that the service distribution is exponential (and, hence, Markovian). The 1 refers to the fact that there is a single server.

If we let $X(t)$ denote the number in the system at time t then $\{X(t), t \geq 0\}$ is a birth and death process with

$$\mu_n = \mu, \qquad n \geq 1$$
$$\lambda_n = \lambda, \qquad n \geq 0$$

Example 3f (A Multiserver Exponential Queueing System): Consider an exponential queueing system in which there are s servers available. An entering customer first waits in line and then goes to the first free server. This is a birth and death process with parameters

$$\mu_n = \begin{cases} n\mu, & 1 \leq n \leq s \\ s\mu, & n > s \end{cases}$$
$$\lambda_n = \lambda, \qquad n \geq 0$$

To see why this is true, reason as follows: If there are n customers in the system, where $n \leq s$, then n servers will be busy. Since each of these servers works at a rate μ, the total departure rate will be $n\mu$. On the other hand, if there are n customers in the system, where $n > s$, then all s of the servers will be busy, and thus the total departure rate will be $s\mu$. This is known as an $M/M/s$ queueing model (why?). $\diamond$

Consider now a general birth and death process with birth rates $\{\lambda_n\}$ and death rates $\{\mu_n\}$, where $\mu_0 = 0$, and let T_i denote the time, starting from state i, it takes for the process to enter state $i + 1$, $i \geq 0$.

We will recursively compute $E[T_i]$, $i \geq 0$, by starting with $i = 0$. Since T_0 is exponential with rate λ_0, we have that

$$E[T_0] = \frac{1}{\lambda_0}$$

For $i > 0$, we condition whether the first transition takes the process into state $i - 1$ or $i + 1$. That is, let

$$I_i = \begin{cases} 1 & \text{if the first transition from } i \text{ is to } i + 1 \\ 0 & \text{if the first transition from } i \text{ is to } i - 1 \end{cases}$$

and note that

$$E[T_i \mid I_i = 1] = \frac{1}{\lambda_i + \mu_i}$$

$$E[T_i \mid I_i = 0] = \frac{1}{\lambda_i + \mu_i} + E[T_{i-1}] + E[T_i] \qquad (3.3)$$

The above follows since, independent of whether the first transition is from a birth or death, the time until it occurs is exponential with rate $\lambda_i + \mu_i$; now if this first transition is a birth, then the population size is at $i + 1$ and so no additional time is needed, whereas if it is a death, then the population size becomes $i - 1$ and the additional time needed to reach $i + 1$ is equal to the time it takes to return to state i (and this has mean $E[T_{i-1}]$) plus the additional time it then takes to reach $i + 1$ (and this has mean $E[T_i]$). Hence, since the probability that the first transition is a birth is $\lambda_i/(\lambda_i + \mu_i)$, we see that

$$E[T_i] = \frac{1}{\lambda_i + \mu_i} + \frac{\mu_i}{\lambda_i + \mu_i} (E[T_{i-1}] + E[T_i])$$

or, equivalently,

$$E[T_i] = \frac{1}{\lambda_i} + \frac{\mu_i}{\lambda_i} E[T_{i-1}], \qquad i \geq 1$$

Starting with $E[T_0] = 1/\lambda_0$, the above yields an efficient method to successively compute $E[T_1]$, $E[T_2]$, and so on.

Suppose now that we wanted to determine the expected time to go from state i to state j where $i < j$. This can be accomplished using the above, by noting that this quantity will equal $E[T_i] + E[T_{i+1}] + \cdots + E[T_{j-1}]$.

Example 3g   For the birth and death process having parameters $\lambda_1 \equiv \lambda$, $\mu_1 \equiv \mu$,

$$E[T_i] = \frac{1}{\lambda} + \frac{\mu}{\lambda} E[T_{i-1}]$$

$$= \frac{1}{\lambda}(1 + \mu E[T_{i-1}])$$

Starting with $E[T_0] = 1/\lambda$, we see that

$$E[T_1] = \frac{1}{\lambda}\left(1 + \frac{\mu}{\lambda}\right)$$

$$E[T_2] = \frac{1}{\lambda}\left[1 + \frac{\mu}{\lambda} + \left(\frac{\mu}{\lambda}\right)^2\right]$$

and, in general,

$$E[T_i] = \frac{1}{\lambda}\left[1 + \frac{\mu}{\lambda} + \left(\frac{\mu}{\lambda}\right)^2 + \cdots + \left(\frac{\mu}{\lambda}\right)^i\right]$$

$$= \frac{1 - (\mu/\lambda)^{i+1}}{\lambda - \mu}, \qquad i \geq 0$$

The expected time to reach state j, starting at state k, $k < j$, is

$$E[\text{time to go from } k \text{ to } j] = \sum_{i=k}^{j-1} E[T_i]$$

$$= \frac{j - k}{\lambda - \mu} - \frac{(\mu/\lambda)^{k+1}}{\lambda - \mu}\frac{[1 - (\mu/\lambda)^{j-k}]}{1 - \mu/\lambda}$$

The above assumes that $\lambda \neq \mu$. If $\lambda = \mu$, then

$$E[T_i] = \frac{i + 1}{\lambda}$$

$$E[\text{time to go from } k \text{ to } j] = \frac{j(j + 1) - k(k + 1)}{2\lambda} \qquad \diamond$$

We can also compute the variance of the time to go from 0 to $i + 1$ by utilizing the conditional variance formula (see Problem 22 of Chapter 3). First note that Equation (3.3) can be written as

$$E[T_i \mid I_i] = \frac{1}{\lambda_i + \mu_i} + (1 - I_i)(E[T_{i-1}] + E[T_i])$$

and so

$$\begin{aligned} \text{Var}(E[T_i \mid I_i]) &= (E[T_{i-1}] + E[T_i])^2 \, \text{Var}(I_i) \\ &= (E[T_{i-1}] + E[T_i])^2 \, \frac{\mu_i \lambda_i}{(\mu_i + \lambda_i)^2} \end{aligned} \tag{3.4}$$

where $\text{Var}(I_i)$ is as shown since I_i is a Bernoulli random variable with parameter $p = \lambda_i/(\lambda_i + \mu_i)$. Also, note that if we let X_i denote the time until the transition from i occurs, then

$$\begin{aligned} \text{Var}(T_i \mid I_i = 1) &= \text{Var}(X_i \mid I_i = 1) \\ &= \text{Var}(X_i) \\ &= \frac{1}{(\lambda_i + \mu_i)^2} \end{aligned} \tag{3.5}$$

where the above uses the fact that the time until transition is independent of the next state visited. Also,

$$\begin{aligned} \text{Var}(T_i \mid I_i = 0) &= \text{Var}(X_i + \text{time to get back to } i \\ &\quad + \text{time to then reach } i + 1) \\ &= \text{Var}(X_i) + \text{Var}(T_{i-1}) + \text{Var}(T_i) \end{aligned} \tag{3.6}$$

where the above uses the fact that the three random variables above are independent. We can rewrite (3.5) and (3.6) as

$$\text{Var}(T_i \mid I_i) = \text{Var}(X_i) + (1 - I_i)[\text{Var}(T_{i-1}) + \text{Var}(T_i)]$$

and so,

$$E[\text{Var}(T_i \mid I_i)] = \frac{1}{(\mu_i + \lambda_i)^2} + \frac{\mu_i}{\mu_i + \lambda_i}[\text{Var}(T_{i-1}) + \text{Var}(T_i)] \tag{3.7}$$

Hence, using the conditional variance formula, which states that $\text{Var}(T_i)$ is the sum of (3.7) and (3.4), we obtain

$$\text{Var}(T_i) = \frac{1}{(\mu_i + \lambda_i)^2} + \frac{\mu_i}{\mu_i + \lambda_i} [\text{Var}(T_{i-1}) + \text{Var}(T_i)]$$

$$+ \frac{\mu_i \lambda_i}{(\mu_i + \lambda_i)^2} (E[T_{i-1}] + E[T_i])^2$$

or, equivalently,

$$\text{Var}(T_i) = \frac{1}{\lambda_i(\lambda_i + \mu_i)} + \frac{\mu_i}{\lambda_i} \text{Var}(T_{i-1}) + \frac{\mu_i}{\mu_i + \lambda_i} (E[T_{i-1}] + E[T_i])^2$$

Starting with $\text{Var}(T_0) = 1/\lambda_0^2$ and using the former recursion to obtain the expectations, we can recursively compute $\text{Var}(T_i)$. In addition, if we want the variance of the time to reach state j, starting from state k, $k < j$, then as this can be expressed as the time to go from k to $k + 1$ plus the additional time to go from $k + 1$ to $k + 2$, and so on. Since, by the Markovian property, these successive random variables are independent, it follows that

$$\text{Var(time to go from } k \text{ to } j) = \sum_{i=k}^{j-1} \text{Var}(T_i)$$

4. The Kolmogorov Differential Equations

For any pair of states i and j let

$$q_{ij} = v_i P_{ij}$$

Since v_i is the rate at which the process makes a transition when in state i and P_{ij} is the probability that this transition is into state j, it follows that q_{ij} is the rate, when in i, that the process makes a transition into j.

The quantities q_{ij} are called the instantaneous transition rates. Since

$$v_i = \sum_j v_i P_{ij} = \sum_j q_{ij}$$

and

$$P_{ij} = \frac{q_{ij}}{v_i} = \frac{q_{ij}}{\sum_j q_{ij}}$$

it follows that specifying the instantaneous rates determines the parameters of the process.

Let

$$P_{ij}(t) = P\{X(t + s) = j | X(s) = i\}$$

represent the probability that a process presently in state i will be in state j a time t later. We shall attempt to derive a set of differential equations for these transition probabilities $P_{ij}(t)$. However, first we will need the following two lemmas.

Lemma 4.1.

(a) $\displaystyle\lim_{h \to 0} \frac{1 - P_{ii}(h)}{h} = v_i$

(b) $\displaystyle\lim_{h \to 0} \frac{P_{ij}(h)}{h} = q_{ij}$ when $i \neq j$

Proof: We first note that since the amount of time until a transition occurs is exponentially distributed it follows that the probability of two or more transitions in a time h is $o(h)$. Thus, $1 - P_{ii}(h)$, the probability that a process in state i at time 0 will not be in state i at time h, equals the probability that a transition occurs within time h plus something small compared to h. Therefore,

$$1 - P_{ii}(h) = v_i h + o(h)$$

and part (a) is proven. To prove part (b), we note that $P_{ij}(h)$, the probability that the process goes from state i to state j in a time h, equals the probability that a transition occurs in this time multiplied by the probability that the transition is into state j, plus something small compared to h. That is,

$$P_{ij}(h) = h v_i P_{ij} + o(h)$$

and part (b) is proven. $\diamond$

Lemma 4.2. For all $s \geq 0$, $t \geq 0$,

$$P_{ij}(t + s) = \sum_{k=0}^{\infty} P_{ik}(t) P_{kj}(s) \tag{4.1}$$

Proof: In order for the process to go from state i to state j in time $t + s$, it must be somewhere at time t and thus

$$P_{ij}(t + s) \equiv P\{X(t + s) = j | X(0) = i\}$$

$$= \sum_{k=0}^{\infty} P\{X(t + s) = j, X(t) = k | X(0) = i\}$$

$$= \sum_{k=0}^{\infty} P\{X(t + s) = j | X(t) = k, X(0) = i\} \cdot P\{X(t) = k | X(0) = i\}$$

$$= \sum_{k=0}^{\infty} P\{X(t + s) = j | X(t) = k\} \cdot P\{X(t) = k | X(0) = i\}$$

$$= \sum_{k=0}^{\infty} P_{kj}(s) P_{ik}(t)$$

and the proof is completed.   ◇

The set of equations (4.1) is known as the *Chapman–Kolmogorov* equations. From Lemma 4.2, we obtain

$$P_{ij}(h + t) - P_{ij}(t) = \sum_{k=0}^{\infty} P_{ik}(h) P_{kj}(t) - P_{ij}(t)$$

$$= \sum_{k \neq i} P_{ik}(h) P_{kj}(t) - [1 - P_{ii}(h)] P_{ij}(t)$$

and thus

$$\lim_{h \to 0} \frac{P_{ij}(t + h) - P_{ij}(t)}{h} = \lim_{h \to 0} \left\{ \sum_{k \neq i} \frac{P_{ik}(h)}{h} P_{kj}(t) - \left[\frac{1 - P_{ii}(h)}{h} \right] P_{ij}(t) \right\}$$

Now assuming that we can interchange the limit and the summation in the above and applying Lemma 4.1, we obtain that

$$P'_{ij}(t) = \sum_{k \neq i} q_{ik} P_{kj}(t) - v_i P_{ij}(t)$$

It turns out that the above interchange can indeed be justified and, hence, we have the following theorem.

Theorem 4.1.   (Kolmogorov's Backward Equations) For all states i, j, and times $t \geq 0$,

$$P'_{ij}(t) = \sum_{k \neq i} q_{ik} P_{kj}(t) - v_i P_{ij}(t)$$

Example 4a   The backward equations for the pure birth process become

$$P'_{ij}(t) = \lambda_i P_{i+1, j}(t) - \lambda_i P_{ij}(t) \quad \diamond$$

Example 4b The backward equations for the birth and death process become

$$P'_{0j}(t) = \lambda_0 P_{1j}(t) - \lambda_0 P_{0j}(t)$$

$$P'_{ij}(t) = (\lambda_i + \mu_i)\left[\frac{\lambda_i}{\lambda_i + \mu_i} P_{i+1, j}(t) + \frac{\mu_i}{\lambda_i + \mu_i} P_{i-1, j}(t) \right] - (\lambda_i + \mu_i)P_{ij}(t)$$

or equivalently

$$P'_{0j}(t) = \lambda_0[P_{1j}(t) - P_{0j}(t)]$$
$$P'_{ij}(t) = \lambda_i P_{i+1, j}(t) + \mu_i P_{i-1, j}(t) - (\lambda_i + \mu_i)P_{ij}(t), \qquad i > 0 \quad \diamond \quad (4.2)$$

Example 4c (A Continuous-Time Markov Chain Consisting of Two States): Consider a machine that works for an exponential amount of time having mean $1/\lambda$ before breaking down; and suppose that it takes an exponential amount of time having mean $1/\mu$ to repair the machine. If the machine is in working condition at time 0, then what is the probability that it will be working at time $t = 10$?

To answer the above question, we note that the process is a birth and death process (with state 0 meaning that the machine is working and state 1 that it is being repaired) having parameters

$$\lambda_0 = \lambda \qquad\qquad \mu_1 = \mu$$
$$\lambda_i = 0, \quad i \neq 0, \qquad \mu_i = 0, i \neq 1$$

We shall derive the desired probability, namely $P_{00}(10)$ by solving the set of differential equations given in Example 4b. From Equation (4.2), we obtain

$$P'_{00}(t) = \lambda[P_{10}(t) - P_{00}(t)] \tag{4.3}$$
$$P'_{10}(t) = \mu P_{00}(t) - \mu P_{10}(t) \tag{4.4}$$

Multiplying Equation (4.3) by μ and Equation (4.4) by λ and then adding the two equations yields

$$\mu P'_{00}(t) + \lambda P'_{10}(t) = 0$$

By integrating, we obtain that

$$\mu P_{00}(t) + \lambda P_{10}(t) = c$$

However, since $P_{00}(0) = 1$ and $P_{10}(0) = 0$, we obtain that $c = \mu$ and hence

$$\mu P_{00}(t) + \lambda P_{10}(t) = \mu \qquad (4.5)$$

or equivalently

$$\lambda P_{10}(t) = \mu[1 - P_{00}(t)]$$

By substituting this result in Equation (4.3), we obtain

$$P'_{00}(t) = \mu[1 - P_{00}(t)] - \lambda P_{00}(t)$$
$$= \mu - (\mu + \lambda)P_{00}(t)$$

Letting

$$h(t) = P_{00}(t) - \frac{\mu}{\mu + \lambda}$$

we have

$$h'(t) = \mu - (\mu + \lambda)\left[h(t) + \frac{\mu}{\mu + \lambda}\right]$$
$$= -(\mu + \lambda)h(t)$$

or

$$\frac{h'(t)}{h(t)} = -(\mu + \lambda)$$

By integrating both sides, we obtain

$$\log h(t) = -(\mu + \lambda)t + c$$

or

$$h(t) = Ke^{-(\mu+\lambda)t}$$

and thus

$$P_{00}(t) = Ke^{-(\mu+\lambda)t} + \frac{\mu}{\mu + \lambda}$$

which finally yields, by setting $t = 0$ and using the fact that $P_{00}(0) = 1$, that

$$P_{00}(t) = \frac{\lambda}{\mu + \lambda}e^{-(\mu+\lambda)t} + \frac{\mu}{\mu + \lambda}$$

From Equation (4.5), this also implies that

$$P_{10}(t) = \frac{\mu}{\mu + \lambda} - \frac{\mu}{\mu + \lambda}e^{-(\mu+\lambda)t}$$

Hence, our desired probability $P_{00}(10)$ equals

$$P_{00}(10) = \frac{\lambda}{\mu + \lambda} e^{-10(\mu + \lambda)} + \frac{\mu}{\mu + \lambda} \quad \diamond$$

Another set of differential equations, different from the backward equations, may also be derived. This set of equations, known as *Kolmogorov's forward equations* is derived as follows. From the Chapman–Kolmogorov equations (Lemma 4.2), we have

$$P_{ij}(t + h) - P_{ij}(t) = \sum_{k=0}^{\infty} P_{ik}(t)P_{kj}(h) - P_{ij}(t)$$

$$= \sum_{k \neq j} P_{ik}(t)P_{kj}(h) - [1 - P_{jj}(h)]P_{ij}(t)$$

and thus

$$\lim_{h \to 0} \frac{P_{ij}(t + h) - P_{ij}(t)}{h} = \lim_{h \to 0} \left\{ \sum_{k \neq j} P_{ik}(t) \frac{P_{kj}(h)}{h} - \left[\frac{1 - P_{jj}(h)}{h} \right] P_{ij}(t) \right\}$$

and, assuming that we can interchange limit with summation, we obtain by Lemma 4.1 that

$$P_{ij}'(t) = \sum_{k \neq j} q_{kj}P_{ik}(t) - v_jP_{ij}(t)$$

Unfortunately, we cannot always justify the interchange of limit and summation and thus the above is not always valid. However, they do hold in most models, including all birth and death processes and all finite state models. We thus have

Theorem 4.2. (Kolmogorov's Forward Equations): Under suitable regularity conditions,

$$P_{ij}'(t) = \sum_{k \neq j} q_{kj}P_{ik}(t) - v_jP_{ij}(t) \tag{4.6}$$

We shall now attempt to solve the forward equations for the pure birth process. For this process, Equation (4.6) reduces to

$$P_{ij}'(t) = \lambda_{j-1}P_{i,\,j-1}(t) - \lambda_jP_{ij}(t)$$

However, by noting that $P_{ij}(t) = 0$ whenever $j < i$ (since no deaths can occur), we can rewrite the above equation to obtain

$$P_{ii}'(t) = -\lambda_iP_{ii}(t)$$

$$P_{ij}'(t) = \lambda_{j-1}P_{i,\,j-1}(t) - \lambda_jP_{ij}(t), \qquad j \geq i + 1 \tag{4.7}$$

Proposition 4.1. For a pure birth process,

$$P_{ii}(t) = e^{-\lambda_i t}, \qquad\qquad i \geq 0$$

$$P_{ij}(t) = \lambda_{j-1} e^{-\lambda_j t} \int_0^t e^{\lambda_j s} P_{i,\,j-1}(s)\,ds, \qquad j \geq i + 1$$

Proof: The fact that $P_{ii}(t) = e^{-\lambda_i t}$ follows from Equation (4.7) by integrating and using the fact that $P_{ii}(0) = 1$. To prove the corresponding result for $P_{ij}(t)$, we note by Equation (4.7) that

$$e^{\lambda_j t}[P'_{ij}(t) + \lambda_j P_{ij}(t)] = e^{\lambda_j t}\lambda_{j-1}P_{i,\,j-1}(t)$$

or

$$\frac{d}{dt}[e^{\lambda_j t}P_{ij}(t)] = \lambda_{j-1}e^{\lambda_j t}P_{i,\,j-1}(t)$$

Hence, since $P_{ij}(0) = 0$, we obtain the desired results. $\diamond$

Example 4d (The Yule Process): The pure birth process having

$$\lambda_j = j\lambda, \qquad j \geq 0$$

which was first encountered in Section 3 is known as a Yule process. We shall use Proposition 4.1 to obtain $P_{ij}(t)$ for the Yule process.

Corollary 4.1. For a Yule process,

$$P_{ij}(t) = \binom{j-1}{j-i} e^{-\lambda i t}(1 - e^{-\lambda t})^{j-i}, \qquad j \geq i \geq 1 \qquad (4.8)$$

Proof: Fix i. We show, by mathematical induction, that Equation (4.8) holds for all $j \geq i$. It holds for $j = i$ by Proposition 4.1. So suppose that Equation (4.8) is true when $j = n$ where $n \geq i$. Now, by Proposition 4.1,

$$P_{i,\,n+1}(t) = \lambda_n e^{-\lambda_{n+1} t} \int_0^t e^{\lambda_{n+1} s} P_{in}(s)\,ds$$

$$= n\lambda e^{-(n+1)\lambda t} \int_0^t e^{(n+1)\lambda s} \binom{n-1}{n-i} e^{-i\lambda s}(1 - e^{-\lambda s})^{n-i}\,ds$$

$$= n \binom{n-1}{n-i} e^{-(n+1)\lambda t} \int_0^t \lambda e^{(n+1-i)\lambda s} (1 - e^{-\lambda s})^{n-i} ds$$

Now, it can be shown that

$$\int_0^t \lambda e^{(n+1-i)\lambda s} (1 - e^{-\lambda s})^{n-i} ds = e^{(n+1-i)\lambda t} \frac{(1 - e^{-\lambda t})^{n+1-i}}{n+1-i}$$

and hence the above can be written

$$\frac{n}{n+1-i} \binom{n-1}{n-i} e^{-i\lambda t} (1 - e^{-\lambda t})^{n+1-i}$$

$$= \binom{n+1-1}{n+1-i} e^{-i\lambda t} (1 - e^{-\lambda t})^{n+1-i}$$

which shows that Equation (4.8) holds for $j = n + 1$ and thus completes the induction.* $\Diamond$

Example 4e (Forward Equations for Birth and Death Process): The forward equation (Equation 4.6) for the general birth and death process becomes

$$P'_{i0}(t) = \sum_{k \neq 0} (\lambda_k + \mu_k) P_{k0} P_{ik}(t) - \lambda_0 P_{i0}(t)$$

$$= (\lambda_1 + \mu_1) \frac{\mu_1}{\lambda_1 + \mu_1} P_{i1}(t) - \lambda_0 P_{i0}(t)$$

$$= \mu_1 P_{i1}(t) - \lambda_0 P_{i0}(t) \tag{4.9}$$

$$P'_{ij}(t) = \sum_{k \neq j} (\lambda_k + \mu_k) P_{kj} P_{ik}(t) - (\lambda_j + \mu_j) P_{ij}(t)$$

$$= (\lambda_{j-1} + \mu_{j-1}) \frac{\lambda_{j-1}}{\lambda_{j-1} + \mu_{j-1}} P_{i,\,j-1}(t)$$

$$+ (\lambda_{j+1} + \mu_{j+1}) \frac{\mu_{j+1}}{\lambda_{j+1} + \mu_{j+1}} P_{i,\,j+1}(t) - (\lambda_j + \mu_j) P_{ij}(t)$$

$$= \lambda_{j-1} P_{i,\,j-1}(t) + \mu_{j+1} P_{i,\,j+1}(t) - (\lambda_j + \mu_j) P_{ij}(t) \quad \Diamond \tag{4.10}$$

---

*For a probabilistic proof of this result see Problem 8.

5. Limiting Probabilities

In analogy with a basic result in discrete-time Markov chains, the probability that a continuous-time Markov chain will be in state j at time t often converges to a limiting value which is independent of the initial state. That is, if we call this value P_j, then

$$P_j \equiv \lim_{t \to \infty} P_{ij}(t)$$

where we are assuming that the limit exists and is independent of the initial state i.

To derive a set of equations for the P_j, consider first the set of forward equations

$$P'_{ij}(t) = \sum_{k \neq j} q_{kj} P_{ik}(t) - v_j P_{ij}(t)$$

Now, if we let t approach ∞, then assuming that we can interchange limit and summation, we obtain

$$\lim_{t \to \infty} P'_{ij}(t) = \lim_{t \to \infty} \left[\sum_{k \neq j} q_{kj} P_{ik}(t) - v_j P_{ij}(t) \right]$$

$$= \sum_{k \neq j} q_{kj} P_k - v_j P_j$$

However, as $P_{ij}(t)$ is a bounded function (being a probability it is always between 0 and 1), it follows that if $P'_{ij}(t)$ converges, then it must converge to 0 (why is this?). Hence, we must have that

$$0 = \sum_{k \neq j} q_{kj} P_k - v_j P_j$$

or

$$v_j P_j = \sum_{k \neq j} q_{kj} P_k, \quad \text{all states} \quad j \tag{5.1}$$

The preceding set of equations, along with this equation

$$\sum_j P_j = 1 \tag{5.2}$$

can be used to solve for the limiting probabilities.

Remarks

(i) We have assumed that the limiting probabilities P_j exist. A sufficient condition for this is that

(a) all states of the Markov chain communicate in the sense that starting in state i there is a positive probability of ever being in state j, for all i, j and *(b)* the Markov chain is positive recurrent in the sense that, starting in any state, the mean time to return to that state is finite.

If conditions *(a)* and *(b)* hold, then the limiting probabilities will exist and satisfy Equations (5.1) and (5.2). In addition, P_j also will have the interpretation of being the long-run proportion of time that the process is in state j.

(ii) Equations (5.1) and (5.2) have a nice interpretation: In any interval $(0, t)$ the number of transitions into state j must equal to within 1 the number of transitions out of state j (why?). Hence, in the long-run, the rate at which transitions into state j occur must equal the rate at which transitions out of state j occur. Now when the process is in state j, it leaves at rate v_j, and, as P_j is the proportion of time it is in state j, it thus follows that

$$v_j P_j = \text{rate at which the process leaves state } j$$

Similarly, when the process is in state k, it enters j at a rate q_{kj}. Hence, as P_k is the proportion of time in state k, we see that the rate at which transitions from k to j occur in just $q_{kj}P_k$; and thus

$$\sum_{k \neq j} q_{kj} P_k = \text{rate at which the process enters state } j$$

So, Equation (5.1) is just a statement of the equality of the rates at which the process enters and leaves state j. Because it balances (that is, equates) these rates, the equations (5.1) are sometimes referred to as "balance equations."

(iii) When the limiting probabilities P_j exist, we say that the chain is ergodic. The P_j are sometimes called stationary probabilities since it can be shown that (as in the discrete-time case) if the initial state is chosen according to the distribution $\{P_j\}$, then the probability of being in state j at time t is P_j, for all t. ◇

Let us now determine the limiting probabilities for a birth and death process. From Equation (5.1) or equivalently, by equating the rate at which the process leaves a state with the rate at which it enters that state, we obtain

State	Rate at which leave = rate at which enter
0	$\lambda_0 P_0 = \mu_1 P_1$
1	$(\lambda_1 + \mu_1) P_1 = \mu_2 P_2 + \lambda_0 P_0$
2	$(\lambda_2 + \mu_2) P_2 = \mu_3 P_3 + \lambda_1 P_1$
$n, n \geq 1$	$(\lambda_n + \mu_n) P_n = \mu_{n+1} P_{n+1} + \lambda_{n-1} P_{n-1}$

By adding to each equation the equation above it, we obtain

$$\lambda_0 P_0 = \mu_1 P_1$$
$$\lambda_1 P_1 = \mu_2 P_2$$
$$\lambda_2 P_2 = \mu_3 P_3$$
$$\vdots$$
$$\lambda_n P_n = \mu_{n+1} P_{n+1}, \qquad n \geq 0$$

Solving in terms of P_0 yields

$$P_1 = \frac{\lambda_0}{\mu_1} P_0$$

$$P_2 = \frac{\lambda_1}{\mu_2} P_1 = \frac{\lambda_1 \lambda_0}{\mu_2 \mu_1} P_0$$

$$P_3 = \frac{\lambda_2}{\mu_3} P_2 = \frac{\lambda_2 \lambda_1 \lambda_0}{\mu_3 \mu_2 \mu_1} P_0$$

$$\vdots$$

$$P_n = \frac{\lambda_{n-1}}{\mu_n} P_{n-1} = \frac{\lambda_{n-1} \lambda_{n-2} \cdots \lambda_1 \lambda_0}{\mu_n \mu_{n-1} \cdots \mu_2 \mu_1} P_0$$

And by using the fact that $\sum_{n=0}^{\infty} P_n = 1$, we obtain

$$1 = P_0 + P_0 \sum_{n=1}^{\infty} \frac{\lambda_{n-1} \cdots \lambda_1 \lambda_0}{\mu_n \cdots \mu_2 \mu_1}$$

or

$$P_0 = \frac{1}{1 + \displaystyle\sum_{n=1}^{\infty} \frac{\lambda_0 \lambda_1 \cdots \lambda_{n-1}}{\mu_1 \mu_2 \cdots \mu_n}}$$

and so

$$P_n = \frac{\lambda_0 \lambda_1 \cdots \lambda_{n-1}}{\mu_1 \mu_2 \cdots \mu_n \left(1 + \displaystyle\sum_{n=1}^{\infty} \frac{\lambda_0 \lambda_1 \cdots \lambda_{n-1}}{\mu_1 \mu_2 \cdots \mu_n}\right)}, \qquad n \geq 1 \qquad (5.3)$$

The foregoing equations also show us what condition is necessary for these limiting probabilities to exist. Namely, it is necessary that

$$\sum_{n=1}^{\infty} \frac{\lambda_0\lambda_1 \cdots \lambda_{n-1}}{\mu_1\mu_2 \cdots \mu_n} < \infty \tag{5.4}$$

This condition also may be shown to be sufficient.

In the multiserver exponential queueing system (Example 7 of a birth and death process), Condition (5.4) reduces to

$$\sum_{n=s+1}^{\infty} \frac{\lambda^n}{(s\mu)^n} < \infty$$

which is equivalent to $\lambda/s\mu < 1$.

For the linear growth model with immigration (Example 4 of a birth and death process), Condition (5.4) reduces to

$$\sum_{n=1}^{\infty} \frac{\theta(\theta + \lambda) \cdots (\theta + (n - 1)\lambda)}{n!\mu^n} < \infty$$

Using the ratio test, the above will converge when

$$\lim_{n\to\infty} \frac{\theta(\theta + \lambda) \cdots (\theta + n\lambda)}{(n + 1)!\mu^{n+1}} \frac{n!\mu^n}{\theta(\theta + \lambda) \cdots (\theta + (n - 1)\lambda)} \equiv \lim_{n\to\infty} \frac{\theta + n\lambda}{(n + 1)\mu}$$

$$\equiv \frac{\lambda}{\mu} < 1$$

That is, the condition is satisfied when $\lambda < \mu$. When $\lambda \geq \mu$ it is easy to show that Condition (5.4) is not satisfied.

Example 5a (A Machine Repair Model): Consider a job shop which consists of M machines and one serviceman. Suppose that the amount of time each machine runs before breaking down is exponentially distributed with mean $1/\lambda$, and suppose that the amount of time that it takes for the serviceman to fix a machine is exponentially distributed with mean $1/\mu$. We shall attempt to answer these questions: (a) What is the average number of machines not in use? (b) What proportion of time is each machine in use?

Solution: If we say that the system is in state n whenever n machines are not in use, then the above is a birth and death process having parameters

$$\mu_n = \mu, \qquad n \geq 1$$

$$\lambda_n = \begin{cases} (M - n)\lambda, & n \leq M \\ 0, & n > M \end{cases}$$

This is so in the sense that a failing machine is regarded as an arrival and a fixed machine as a departure. If any machines are broken down, then since the serviceman's rate is μ, $\mu_n = \mu$. On the other hand, if n machines are not in use, then since the $M - n$ machines in use each fail at a rate λ, it follows that $\lambda_n = (M - n)\lambda$. From Equation (5.3) we have that P_n, the probability that n machines will not be in use, is given by

$$P_0 = \cfrac{1}{1 + \sum_{n=1}^{M} [M\lambda(M-1)\lambda \cdot \cdot \cdot (M-n+1)\lambda/\mu^n]}$$

$$= \cfrac{1}{1 + \sum_{n=1}^{M} (\lambda/\mu)^n M!/(M-n)!}$$

$$P_n = \cfrac{\dfrac{M!}{(M-n)!}\left(\dfrac{\lambda}{\mu}\right)^n}{1 + \sum_{n=1}^{M} \left(\dfrac{\lambda}{\mu}\right)^n \dfrac{M!}{(M-n)!}}, \qquad n = 0, 1, \ldots, M$$

Hence, the average number of machines not in use is given by

$$\sum_{n=0}^{M} nP_n = \sum_{n=0}^{M} n \frac{M!}{(M-n)!}\left(\frac{\lambda}{\mu}\right)^n \bigg/ \left(1 + \sum_{n=1}^{M} \left(\frac{\lambda}{\mu}\right)^n \frac{M!}{(M-n)!}\right) \quad (5.5)$$

To obtain the long-run proportion of time that a given machine is working we will compute the equivalent limiting probability of its working. To do so, we condition the number of machines that are not working to obtain

$$P\{\text{Machine is working}\} = \sum_{n=0}^{M} P\{\text{Machine is working}|n \text{ not working}\}P_n$$

$$= \sum_{n=0}^{M} \frac{M-n}{M} P_n \qquad \begin{array}{l}\text{(since if } n \text{ are not working,}\\ \text{then } M-n \text{ are working!)}\end{array}$$

$$= 1 - \sum_{0}^{M} \frac{nP_n}{M}$$

where $\sum_{0}^{M} nP_n$ is given by Equation (5.5).   ◇

Example 5b (The M/M/1 Queue): In the M/M/1 queue $\lambda_n = \lambda$, $\mu_n = \mu$ and thus from Equation (5.3)

$$P_n = \frac{\left(\dfrac{\lambda}{\mu}\right)^n}{1 + \displaystyle\sum_{n=1}^{\infty} \left(\dfrac{\lambda}{\mu}\right)^n}$$

$$= \left(\frac{\lambda}{\mu}\right)^n (1 - \lambda/\mu), \qquad n \geq 0$$

provided that $\lambda/\mu < 1$. It is intuitive that λ must be less than μ for limiting probabilities to exist. For customers arrive at rate λ and are served at rate μ, and thus if $\lambda > \mu$, then they arrive at a faster rate than they can be served and the queue size will go to infinity. The case $\lambda = \mu$ behaves much like the symmetric random walk of Section 3 of Chapter 4 which is null recurrent and thus has no limiting probabilities.   ◇

Example 5c   Let us reconsider the shoeshine shop of Example 2a, and determine the proportion of time the process is in each of the states 0, 1, 2. As this is not a birth and death process (since the process can go directly from state 2 to state 0), we start with the balance equations for the limiting probabilities.

State	Rate that the process leaves = rate that the process enters
0	$\lambda P_0 = \mu_2 P_2$
1	$\mu_1 P_1 = \lambda P_0$
2	$\mu_2 P_2 = \mu_1 P_1$

Solving in terms of P_0 yields

$$P_2 = \frac{\lambda}{\mu_2} P_0, \qquad P_1 = \frac{\lambda}{\mu_1} P_0$$

which implies, since $P_0 + P_1 + P_2 = 1$, that

$$P_0 \left[1 + \frac{\lambda}{\mu_2} + \frac{\lambda}{\mu_1} \right] = 1$$

or

$$P_0 = \frac{\mu_1\mu_2}{\mu_1\mu_2 + \lambda(\mu_1 + \mu_2)}$$

and

$$P_1 = \frac{\lambda\mu_2}{\mu_1\mu_2 + \lambda(\mu_1 + \mu_2)}$$

$$P_2 = \frac{\lambda\mu_1}{\mu_1\mu_2 + \lambda(\mu_1 + \mu_2)} \qquad \Diamond$$

Example 5d   Consider a set of n components along with a single repairman. Suppose that component i functions for an exponentially distributed time with rate λ_i and then fails: The time it then takes to repair component i is exponential with rate μ_i, $i = 1, \ldots, n$. Suppose that when there is more than one failed component the repairman always works on the most recent failure. For instance, if there are at present two failed components—say components one and two of which one has failed most recently—then the repairman will be working on component one. However, if component three should fail before one's repair is completed, then the repairman would stop working on component one and switch to component three, (that is, a newly failed component preempts service).

To analyze the above as a continuous-time Markov chain, the state must represent the set of failed components in the order of failure. That is, the state will be $i_1, i_2, \ldots, i_k$ if $i_1, i_2, \ldots, i_k$ are the k failed components (all the other $n - k$ being functional) with i_1 having been the most recent failure (and is thus presently being repaired), i_2 the second most recent, and so on. As there are $k!$ possible orderings for a fixed set of k failed components and $\binom{n}{k}$ choices of that set, it follows that there are $\Sigma_{k=0}^{n} \binom{n}{k}k! = \Sigma_{k=0}^{n} n!/(n - k)! = n! \Sigma_{i=0}^{n} 1/i!$ possible states.

The balance equations for the limiting probabilities are as follows:

$$\left(\mu_{i_1} + \sum_{\substack{i \neq i_j \\ j=1,\ldots,k}} \lambda_i\right) P(i_1, \ldots, i_k)$$

$$= \sum_{\substack{i \neq i_j \\ j=1,\ldots,k}} P(i, i_1, \ldots, i_k)\mu_i + P(i_2, \ldots, i_k)\lambda_{i_1}$$

$$\sum_{i=1}^{n} \lambda_i P(\phi) = \sum_{i=1}^{n} P(i)\mu_i \tag{5.6}$$

where ϕ is the state when all components are working. The above equations follow because state $i_1, \ldots, i_k$ can be left either by a failure of any of the additional components or by a repair completion of component i_1. Also that state can be entered either by a repair completion of component i when the state is $i, i_1, \ldots, i_k$ or by a failure of component i_1 when the state is $i_2, \ldots, i_k$.

However if we take

$$P(i_1, \ldots, i_k) = \frac{\lambda_{i_1}\lambda_{i_2} \cdots \lambda_{i_k}}{\mu_{i_1}\mu_{i_2} \cdots \mu_{i_k}} P(\phi) \tag{5.7}$$

then it is easily seen that the Equations (5.6) are satisfied. Hence by uniqueness these must be the limiting probabilities with $P(\phi)$ determined to make their sum equal 1. That is,

$$P(\phi) = \left[1 + \sum_{i_1, \ldots, i_k} \frac{\lambda_{i_1} \cdots \lambda_{i_k}}{\mu_{i_1} \cdots \mu_{i_k}} \right]^{-1}$$

As an illustration, suppose $n = 2$ and so there are 5 states ϕ, 1, 2, 12, 21. Then from the above we would have

$$P(\phi) = \left[1 + \frac{\lambda_1}{\mu_1} + \frac{\lambda_2}{\mu_2} + \frac{2\lambda_1\lambda_2}{\mu_1\mu_2} \right]^{-1}$$

$$P(1) = \frac{\lambda_1}{\mu_1} P(\phi)$$

$$P(2) = \frac{\lambda_2}{\mu_2} P(\phi)$$

$$P(1, 2) = P(2, 1) = \frac{\lambda_1\lambda_2}{\mu_1\mu_2} P(\phi)$$

It is interesting to note, using (5.7), that given the set of failed components, each of the possible orderings of these components is equally likely. ◇

6. Time Reversibility

Consider a continuous-time Markov chain that is ergodic and let us consider the limiting probabilities P_i from a different point of view than previously. If we consider the sequence of states visited, ignoring the amount of time spent in each state during a visit, then this sequence constitutes a discrete-time Markov chain with transition probabilities

P_{ij}. Let us assume that this discrete-time Markov chain, called the embedded chain, is ergodic and denote by π_i its limiting probabilities. That is, the π_i are the unique solution of

$$\pi_i = \sum_j \pi_j P_{ji}, \quad \text{all} \quad i$$

$$\sum_i \pi_i = 1$$

Now since π_i represents the proportion of transitions that take the process into state i, and as $1/v_i$ is the mean time spent in state i during a visit, it seems intuitive that P_i, the proportion of time in state i, should be a weighted average of the π_i where π_i is weighted proportionately to $1/v_i$. That is, it is intuitive that

$$P_i = \frac{\dfrac{\pi_i}{v_i}}{\displaystyle\sum_j \frac{\pi_j}{v_j}} \tag{6.1}$$

To check the above, recall that the limiting probabilities P_i must satisfy

$$v_i P_i = \sum_{j \neq i} P_j q_{ji}, \quad \text{all} \quad i$$

or equivalently, since $P_{ii} = 0$

$$v_i P_i = \sum_j P_j q_{ji}, \quad \text{all} \quad i$$

Hence, for the P_i's to be given by Equation (6.1), it would be necessary that

$$\pi_i = \sum_j \pi_j P_{ji}, \quad \text{all} \quad i$$

But this, of course, follows since it is in fact the defining equation for the π_i's.

Suppose now that the continuous-time Markov chain has been in operation for a long time, and suppose that starting at some (large) time T we trace the process going backward in time. To determine the probability structure of this reversed process, we first note that given we are in state i at some time—say t—the probability that we have been in this state for an amount of time greater than s is just $e^{-v_i s}$. This is so, since

$P\{\text{Process is in state } i \text{ throughout } [t - s, t] | X(t) = i\}$

$$= \frac{P\{\text{Process is in state } i \text{ throughout } [t - s, t]\}}{P\{X(t) = i\}}$$

$$= \frac{P\{X(t - s) = i\}e^{-v_i s}}{P\{X(t) = i\}}$$

$$= e^{-v_i s}$$

since for t large $P\{X(t - s) = i\} = P\{X(t) = i\} = P_i$.

In other words, going backward in time, the amount of time the process spends in state i is also exponentially distributed with rate v_i. In addition, as was shown in Section 7 of Chapter 4, the sequence of states visited by the reversed process constitutes a discrete-time Markov chain with transition probabilities Q_{ij} given by

$$Q_{ij} = \frac{\pi_j P_{ji}}{\pi_i}$$

Hence, we see from the above that the reversed process is a continuous-time Markov chain with the same transition rates as the forward-time process and with one-stage transition probabilities Q_{ij}. Therefore, the continuous-time Markov chain will be *time reversible,* in the sense that the process reversed in time has the same probabilistic structure as the original process, if the embedded chain is time reversible. That is, if

$$\pi_i P_{ij} = \pi_j P_{ji}, \quad \text{for all} \quad i, j$$

Now using the fact that $P_i = (\pi_i/v_i)/(\Sigma_j \pi_j/v_j)$, we see that the above condition is equivalent to

$$P_i q_{ij} = P_j q_{ji}, \quad \text{for all} \quad i, j \tag{6.2}$$

Since P_i is the proportion of time in state i and q_{ij} is the rate when in state i that the process goes to j, the condition of time reversibility is that *the rate at which the process goes directly from state i to state j is equal to the rate at which it goes directly from state i to state j is equal to the rate at which it goes directly from j to i.* It should be noted that this is exactly the same condition needed for an ergodic discrete-time Markov chain to be time reversible (see Section 7 of Chapter 4).

An application of the above condition for time reversibility yields the following proposition concerning birth and death processes.

Proposition 6.1. An ergodic birth and death process is time reversible.

Proof: We must show that the rate at which a birth and death process goes from state i to state $i + 1$ is equal to the rate at which it goes from $i + 1$ to i. Now in any length of time t the number of transitions from i to $i + 1$ must equal to within 1 the number from $i + 1$ to i (since between each transition from i to $i + 1$ the process must return to i, and this can only occur through $i + 1$, and vice versa). Hence, as the number of such transitions goes to infinity as $t \to \infty$, it follows that the rate of transitions from i to $i + 1$ equals the rate from $i + 1$ to i. $\diamond$

Proposition 6.1 can be used to prove the important result that the output process of an $M/M/s$ queue is a Poisson process. We state this as a corollary.

Corollary 6.1.   Consider an $M/M/s$ queue in which customers arrive in accordance with a Poisson process having rate λ and are served by any one of s servers—each having an exponentially distributed service time with rate μ. If $\lambda < s\mu$, then the output process of customers departing is, after the process has been in operation for a long time, a Poisson process with rate λ.

Proof: Let $X(t)$ denote the number of customers in the system at time t. Since the $M/M/s$ process is a birth and death process, it follows from Proposition 6.1 that $\{X(t), t \geq 0\}$ is time reversible. Now going forward in time, the time points at which $X(t)$ increases by 1 constitute a Poisson process since these are just the arrival times of customers. Hence, by time reversibility the time points at which the $X(t)$ increases by 1 when we go backward in time also constitute a Poisson process. But these latter points are

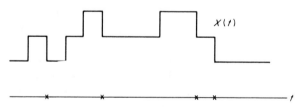

x = times at which going backward in time, $X(t)$ increases
= times at which going forward in time, $X(t)$ decreases

Figure 6.1.

exactly the points of time when customers depart. (See Figure 6.1). Hence, the departure times constitute a Poisson process with rate λ. $\diamond$

Let

$$q_{ij} = v_i P_{ij}, \qquad j \neq i$$

denote the rate, when in state i, that the process makes a transition into state j. The quantity q_{ij} is commonly called the instantaneous transition rate from i to j. Note that a process is time reversible if and only if

$$P_i q_{ij} = P_j q_{ji} \quad \text{for all} \quad i \neq j$$

Analogous to the result for discrete time Markov chains if one can find a probability vector $\mathbf{P}$ that satisfies the above then the Markov chain is time reversible and the P_i's are the long run probabilities. That is, we have the following proposition.

Proposition 6.2. If for some set $\{P_i\}$

$$\sum_i P_i = 1, \qquad P_i \geq 0$$

and

$$P_i q_{ij} = P_j q_{ji} \qquad \text{for all} \quad i \neq j \tag{6.3}$$

then the continuous-time Markov chain is time reversible and P_i represents the limiting probability of being in state i.

Proof: For fixed i we obtain upon summing (6.3) over all $j: j \neq i$

$$\sum_{j \neq i} P_i q_{ij} = \sum_{j \neq i} P_j q_{ji}$$

or, since $\sum_{j \neq i} q_{ij} = v_i$,

$$v_i P_i = \sum_{j \neq i} P_j q_{ji}$$

Hence, the P_i's satisfy the balance equations and thus represent the limiting probabilities. As (6.3) holds, the chain is time reversible.

Example 6a Consider a set of n machines and a single repair facility to service them. Suppose that when machine i, $i = 1, \ldots, n$, goes down it requires an exponentially distributed amount of work with rate μ_i to get it back up. The repair facility divides its efforts equally among all down components in the sense that whenever there are k down machines $1 \leq k \leq n$ each receives work at a rate of $1/k$ per unit time. Finally, suppose that each time machine i goes back up it remains up for an exponentially distributed time with rate λ_i.

The above can be analyzed as a continuous-time Markov chain

having 2^n states where the state at any time corresponds to the set of machines that are down at that time. Thus, for instance, the state will be $(i_1, i_2, \ldots, i_k)$ when machines $i_1, \ldots, i_k$ are down and all the others are up. The instantaneous transition rates are as follows:

$$q_{(i_1, \ldots, i_{k-1}), (i_1, \ldots, i_k)} = \lambda_{i_k}$$

$$q_{(i_1, \ldots, i_k), (i_1, \ldots, i_{k-1})} = \mu_{i_k}/k$$

where $i_1, \ldots, i_k$ are all distinct. The above follows since the failure rate of machine i_k is always λ_{i_k} and the repair rate of machine i_k when there are k failed machines is μ_{i_k}/k.

Hence the time reversible equations (6.3) are

$$P(i_1, \ldots, i_k)\, \mu_{i_k}/k = P(i_1, \ldots, i_{k-1})\lambda_{i_k}$$

or

$$
\begin{aligned}
P(i_1, \ldots, i_k) &= \frac{k\lambda_{i_k}}{\mu_{i_k}} P(i_1, \ldots, i_{k-1})\\
&= \frac{k\lambda_{i_k}}{\mu_{i_k}} \frac{(k-1)\lambda_{i_{k-1}}}{\mu_{i_{k-1}}} P(i_1, \ldots, i_{k-2}) \quad \text{upon iterating}\\
&=\\
&\;\;\vdots\\
&= k! \prod_{j=1}^{k} (\lambda_{i_j}/\mu_{i_j})P(\phi)
\end{aligned}
$$

where ϕ is the state in which all components are working. As

$$P(\phi) + \Sigma P(i_1, \ldots, i_k) = 1$$

we see that

$$P(\phi) = \left[1 + \sum_{i_1, \ldots, i_k} k! \prod_{j=1}^{k} (\lambda_{i_j}/\mu_{i_j}) \right]^{-1} \tag{6.4}$$

where the above sum is over all the $2^n - 1$ nonempty subsets $\{i_1, \ldots, i_k\}$ of $\{1, 2, \ldots, n\}$. Hence as the time reversible equations are satisfied for this choice of probability vector it follows from Proposition 6.2 that the chain is time reversible and

$$P(i_1, \ldots, i_k) = k! \prod_{j=1}^{k} (\lambda_{i_j}/\mu_{i_j})P(\phi)$$

with $P(\phi)$ being given by (6.4).

For instance, suppose there are two machines. Then from the above we would have

$$P(\phi) = \cfrac{1}{1 + \cfrac{\lambda_1}{\mu_1} + \cfrac{\lambda_2}{\mu_2} + \cfrac{2\lambda_1\lambda_2}{\mu_1\mu_2}}$$

$$P(1) = \cfrac{\lambda_1/\mu_1}{1 + \cfrac{\lambda_1}{\mu_1} + \cfrac{\lambda_2}{\mu_2} + \cfrac{2\lambda_1\lambda_2}{\mu_1\mu_2}}$$

$$P(2) = \cfrac{\lambda_2/\mu_2}{1 + \cfrac{\lambda_1}{\mu_1} + \cfrac{\lambda_2}{\mu_2} + \cfrac{2\lambda_1\lambda_2}{\mu_1\mu_2}}$$

$$P(1, 2) = \cfrac{2\lambda_1\lambda_2}{\mu_1\mu_2\left[1 + \cfrac{\lambda_1}{\mu_1} + \cfrac{\lambda_2}{\mu_2} + \cfrac{2\lambda_1\lambda_2}{\mu_1\mu_2}\right]} \qquad \diamond$$

7. Uniformization

Consider a continuous-time Markov chain in which the mean time spent in a state is the same for all states. That is, suppose that $v_i = v$, for all states i. In this case since the amount of time spent in each state during a visit is exponentially distributed with rate v, it follows that if we let $N(t)$ denote the number of state transitions by time t, then $\{N(t), t \geq 0\}$ will be a Poisson process with rate v.

To compute the transition probabilities $P_{ij}(t)$, we can condition on $N(t)$:

$$P_{ij}(t) = P\{X(t) = j | X(0) = i\}$$

$$= \sum_{n=0}^{\infty} P\{X(t) = j | X(0) = i, N(t) = n\} P\{N(t) = n | X(0) = i\}$$

$$= \sum_{n=0}^{\infty} P\{X(t) = j | X(0) = i, N(t) = n\} e^{-vt} \frac{(vt)^n}{n!}$$

Now the fact that there have been n transitions by time t tells us something about the amounts of time spent in each of the first n states visited, but since the distribution of time spent in each state is the same for all states, it follows that knowing that $N(t) = n$ gives us no information about which states were visited. Hence,

$$P\{X(t) = j | X(0) = i, N(t) = n\} = P_{ij}^n$$

where P_{ij}^n is just the n-stage transition probability associated with the discrete-time Markov chain with transition probabilities P_{ij}; and so when $v_i \equiv v$

$$P_{ij}(t) = \sum_{n=0}^{\infty} P_{ij}^n e^{-vt} \frac{(vt)^n}{n!} \tag{7.1}$$

Equation (7.1) is quite useful from a computational point of view since it enables us to approximate $P_{ij}(t)$ by taking a partial sum and then computing (by matrix multiplication of the transition probability matrix) the relevant n stage probabilities P_{ij}^n.

Whereas the applicability of Equation (7.1) would appear to be quite limited since it supposes that $v_i \equiv v$, it turns out that most Markov chains can be put in that form by the trick of allowing fictitious transitions from a state to itself. To see how this works, consider any Markov chain for which the v_i are bounded, and let v be any number such that

$$v_i \leq v, \quad \text{for all} \quad i \tag{7.2}$$

Now when in state i, the process actually leaves at rate v_i; but this is equivalent to supposing that transitions occur at rate v, but only the fraction v_i/v of transitions are real ones (and thus real transitions occur at rate v_i) and the remaining fraction $1 - v_i/v$ are fictitious transitions which leave the process in state i. In other words, any Markov chain satisfying Condition (7.2) can be thought of as being a process which spends an exponential amount of time with rate v in state i and then makes a transition to j with probability P_{ij}^*, where

$$P_{ij}^* = \begin{cases} 1 - \dfrac{v_i}{v}, & j = i \\ \dfrac{v_i}{v} P_{ij}, & j \neq i \end{cases} \tag{7.3}$$

Hence, from Equation (7.1) we have that the transition probabilities can be computed by

$$P_{ij}(t) = \sum_{n=0}^{\infty} P_{ij}^{*\,n} e^{-vt} \frac{(vt)^n}{n!}$$

where P_{ij}^* are the n-stage transition probabilities corresponding to Equation (7.3). This technique of uniformizing the rate in which a transition occurs from each state by introducing transitions from a state to itself is known as *uniformization*.

Example 7a   Let us reconsider Example 4c which models the workings of a machine—either on or off—as a two-state continuous-time Markov chain with

$$P_{01} = P_{10} = 1$$
$$v_0 = \lambda, \qquad v_1 = \mu$$

Letting $v = \lambda + \mu$, the uniformized version of the above is to consider it a continuous-time Markov chain with

$$P_{00} = \frac{\mu}{\lambda + \mu} = 1 - P_{01}$$

$$P_{10} = \frac{\mu}{\lambda + \mu} = 1 - P_{11}$$
$$v_i = \lambda + \mu, \qquad i = 1, 2$$

As $P_{00} = P_{10}$ it follows that the probability of a transition into state 0 is equal to $\mu/(\lambda + \mu)$ no matter what the present state. As a similar result is true for state 1, it follows that the n-stage transition probabilities are given by

$$P_{i0}^n = \frac{\mu}{\lambda + \mu}, \qquad n \geq 1, \quad i = 0, 1$$

$$P_{i1}^n = \frac{\lambda}{\lambda + \mu}, \qquad n \geq 1, \quad i = 0, 1$$

Hence,

$$P_{00}(t) = \sum_{n=0}^{\infty} P_{00}^n e^{-(\lambda+\mu)t} \frac{[(\lambda + \mu)t]^n}{n!}$$

$$= e^{-(\lambda+\mu)t} + \sum_{n=1}^{\infty} \left(\frac{\mu}{\lambda + \mu}\right) e^{-(\lambda+\mu)t} \frac{[(\lambda + \mu)t]^n}{n!}$$

$$= e^{-(\lambda+\mu)t} + [1 - e^{-(\lambda+\mu)t}] \frac{\mu}{\lambda + \mu}$$

$$= \frac{\mu}{\lambda + \mu} + \frac{\lambda}{\lambda + \mu} e^{-(\lambda+\mu)t}$$

Similarly,

$$P_{11}(t) = \sum_{n=0}^{\infty} P_{11}^n e^{-(\lambda + \mu)t} \frac{[(\lambda + \mu)t]^n}{n!}$$

$$= e^{-(\lambda + \mu)t} + [1 - e^{-(\lambda + \mu)t}] \frac{\lambda}{\lambda + \mu}$$

$$= \frac{\lambda}{\lambda + \mu} + \frac{\mu}{\lambda + \mu} e^{-(\lambda + \mu)t}$$

The remaining probabilities are

$$P_{01}(t) = 1 - P_{00}(t)$$

$$= \frac{\lambda}{\lambda + \mu}[1 - e^{-(\lambda + \mu)t}]$$

$$P_{10}(t) = 1 - P_{11}(t)$$

$$= \frac{\mu}{\lambda + \mu}[1 - e^{-(\lambda + \mu)t}] \qquad \diamond$$

8. Computing the Transition Probabilities

For any pair of states i and j let

$$r_{ij} = \begin{cases} q_{ij} & \text{if } i \neq j \\ -v_i & \text{if } i = j \end{cases}$$

Using this notation, we can rewrite the Kolmogorov backward equations

$$P'_{ij}(t) = \sum_{k \neq i} q_{ik} P_{kj}(t) - v_i P_{ij}(t)$$

and the forward equations

$$P'_{ij}(t) = \sum_{k \neq j} q_{kj} P_{ik}(t) - v_j P_{ij}(t)$$

as follows:

$$P'_{ij}(t) = \sum_k r_{ik} P_{kj}(t) \qquad \text{(backward)}$$

$$P'_{ij}(t) = \sum_k r_{kj} P_{ik}(t) \qquad \text{(forward)}$$

This representation is especially revealing when we introduce matrix notation. Define the matrices $\mathbf{R}$, $\mathbf{P}(t)$, and $\mathbf{P}'(t)$ by letting the element in row i, column j of these matrices be, respectively, r_{ij}, $P_{ij}(t)$, and $P'_{ij}(t)$. Since the backward equations say that the element in row i, column j of the matrix $\mathbf{P}'(t)$ can be obtained by multiplying the ith row of the matrix $\mathbf{R}$ by the jth column of the matrix $\mathbf{P}(t)$, it is equivalent to the matrix equation

$$\mathbf{P}'(t) = \mathbf{R}\,\mathbf{P}(t) \tag{8.1}$$

Similarly, the forward equations can be written as

$$\mathbf{P}'(t) = \mathbf{P}(t)\,\mathbf{R} \tag{8.2}$$

Now, just as the solution of the scalar differential equation

$$f'(t) = c\,f(t)$$

(or, equivalently, $f'(t) = f(t)c$) is

$$f(t) = f(0)e^{ct}$$

it can be shown that the solution of the matrix differential equations (8.1) and (8.2) is given by

$$\mathbf{P}(t) = \mathbf{P}(0)\,e^{\mathbf{R}t}$$

Since $\mathbf{P}(0) = \mathbf{I}$ (the identity matrix), this yields that

$$\mathbf{P}(t) = e^{\mathbf{R}t} \tag{8.3}$$

where the matrix $e^{\mathbf{R}t}$ is defined by

$$e^{\mathbf{R}t} = \sum_{n=0}^{\infty} \mathbf{R}^n \frac{t^n}{n!} \tag{8.4}$$

with $\mathbf{R}^n$ being the (matrix) multiplication of $\mathbf{R}$ by itself n times.

The direct use of Equation (8.4) to compute $\mathbf{P}(t)$ turns out to be very inefficient for two reasons. First, since the matrix $\mathbf{R}$ contains both positive and negative elements (remember the off-diagonal elements are the q_{ij} while the ith diagonal element is $-v_i$), there is the problem of computer round-off error when we compute the powers of $\mathbf{R}$. Second, we often have to compute many of the terms in the infinite sum (8.4) to arrive at a good approximation. However, there are certain indirect ways that we can utilize the relation (8.3) to efficiently approximate the matrix $\mathbf{P}(t)$. We now present two of these methods.

Approximation Method 1

Rather than using (8.3) to compute $e^{\mathbf{R}t}$, we can use the matrix equivalent of the identity

$$e^x = \lim_{n \to \infty} \left(1 + \frac{x}{n} \right)^n$$

which states that

$$e^{\mathbf{R}t} = \lim_{n \to \infty} \left(\mathbf{I} + \mathbf{R}\frac{t}{n} \right)^n$$

Thus, if we let n be a power of 2, say $n = 2^k$, then we can approximate $\mathbf{P}(t)$ by raising the matrix $\mathbf{M} = \mathbf{I} + \mathbf{R}t/n$ to the nth power, which can be accomplished by k matrix multiplications (by first multiplying $\mathbf{M}$ by itself to obtain $\mathbf{M}^2$ and then multiplying that by itself to obtain $\mathbf{M}^4$ and so on). In addition, since only the diagonal elements of $\mathbf{R}$ are negative (and the diagonal elements of the identity matrix $\mathbf{I}$ are equal to 1) by choosing n large enough, we can guarantee that the matrix $\mathbf{I} + \mathbf{R}t/n$ has all non-negative elements.

Approximation Method 2

A second approach to approximating $e^{\mathbf{R}t}$ uses the identity

$$e^{-\mathbf{R}t} = \lim_{n \to \infty} \left(\mathbf{I} - \mathbf{R}\frac{t}{n} \right)^n$$

$$\approx \left(\mathbf{I} - \mathbf{R}\frac{t}{n} \right)^n \qquad \text{for } n \text{ large}$$

and thus

$$\mathbf{P}(t) = e^{\mathbf{R}t} \approx \left(\mathbf{I} - \mathbf{R}\frac{t}{n}\right)^{-n}$$

$$= \left[\left(\mathbf{I} - \mathbf{R}\frac{t}{n}\right)^{-1}\right]^{n}$$

Hence, if we again choose n to be a large power of 2, say $n = 2^k$, we can approximate $\mathbf{P}(t)$ by first computing the inverse of the matrix $\mathbf{I} - \mathbf{R}t/n$ and then raising that matrix to the nth power (by utilizing k matrix multiplications). It can be shown that the matrix $\mathbf{I} - \mathbf{R}t/n$ will have only non-negative elements.

Remark Both of the above computational approaches for approximating $\mathbf{P}(t)$ have probabilistic interpretations (see Problems 25 and 26). ◇

Problems

1. A population of organisms consists of both male and female members. In a small colony any particular male is likely to mate with any particular female in any time interval of length h, with probability $\lambda h + o(h)$. Each mating immediately produces one offspring, equally likely to be male or female. Let $N_1(t)$ and $N_2(t)$ denote the number of males and females in the population at t. Derive the parameters of the continuous-time Markov chain $\{N_1(t), N_2(t)\}$, i.e., the v_i, P_{ij} of Section 2.

2. Suppose that a one-celled organism can be in one of two states—either A or B. An individual in state A will change to state B at an exponential rate α; an individual in state B divides into two new individuals of type A at an exponential rate β. Define an appropriate continuous-time Markov chain for a population of such organisms and determine the appropriate parameters for this model.

3. Consider two machines that are maintained by a single repairman. Machine i functions for an exponential time with rate μ_i before breaking down, $i = 1, 2$. The repair times (for either machine) are exponential with rate μ. Can we analyze this as a birth and death process? If so, what are the parameters? If not, how can we analyze it?

4. Potential customers arrive at a single-server station in accordance with a Poisson process with rate λ. However, if the arrival finds n customers already in the station, then he will enter the system with

probability α_n. Assuming an exponential service rate μ, set this up as a birth and death process and determine the birth and death rates.

5. Consider a birth and death process with birth rates $\lambda_i = (i + 1)\lambda$, $i \geq 0$, and death rates $\mu_i = i\mu$, $i \geq 0$.
   (a) Determine the expected time to go from state 0 to state 4.
   (b) Determine the expected time to go from state 2 to state 5.
   (c) Determine the variances in parts (a) and (b).

6. Consider two machines, both of which have an exponential lifetime with mean $1/\lambda$. There is a single repairman that can service machines at an exponential rate μ. Set up the Kolmogorov backward equations; you need not solve them.

7. The birth and death process with parameters $\lambda_n = 0$ and $\mu_n = \mu$, $n > 0$ is called a pure death process. Find $P_{ij}(t)$.

8. Consider two machines. Machine i operates for an exponential time with rate λ_i and then fails; its repair time is exponential with rate μ_i, $i = 1, 2$. The machines act independently of each other. Define a four-state continuous-time Markov chain which jointly describes the condition of the two machines. Use the assumed independence to compute the transition probabilities for this chain and then verify that these transition probabilities satisfy the forward and backward equations.

9. Consider a Yule process starting with a single individual—that is, suppose $X(0) = 1$. Let T_i denote the time it takes the process to go from a population of size i to one of size $i + 1$.
   (a) Argue that T_i, $i = 1, \ldots, j$, are independent exponentials with respective rates $i\lambda$.
   (b) Let $X_1, \ldots, X_j$ denote independent exponential random variables each having rate λ, and interpret X_i as the lifetime of component i. Argue that $\max(X_1, \ldots, X_j)$ can be expressed as

$$\max(X_1, \ldots, X_j) = \varepsilon_1 + \varepsilon_2 + \cdots + \varepsilon_j$$

where $\varepsilon_1, \varepsilon_2, \ldots, \varepsilon_j$ are independent exponentials with respective rates $j\lambda$, $(j - 1)\lambda, \ldots, \lambda$.
Hint: Interpret ε_i as the time between the $i - 1$ and the ith failure.
   (c) Using (a) and (b) argue that

$$P\{T_1 + \cdots + T_j \leq t\} = (1 - e^{-\lambda t})^j$$

   (d) Use (c) to obtain that

$$P_{1j}(t) = (1 - e^{-\lambda t})^{j-1} - (1 - e^{-\lambda t})^j = e^{-\lambda t}(1 - e^{-\lambda t})^{j-1}$$

and hence, given $X(0) = 1$, $X(t)$ has a geometric distribution with parameter $p = e^{-\lambda t}$.

(e) Use (d) to obtain that

$$P_{ij}(t) = \binom{j-1}{i-1} e^{-\lambda t i}(1 - e^{-\lambda t})^{j-i}$$

Hint: What is the distribution of the sum of i independent geometrics each having parameter p?

10. Each individual in a biological population is assumed to give birth at an exponential rate λ, and to die at an exponential rate μ. In addition, there is an exponential rate of increase θ due to immigration. However, immigration is not allowed when the population size is N or larger.

(a) Set this up as a birth and death model.

(b) If $N = 3$, $1 = \theta = \lambda$, $\mu = 2$, determine the proportion of time that immigration is restricted.

11. A small barbershop, operated by a single barber, has room for at most two customers. Potential customers arrive at a Poisson rate of three per hour, and the successive service times are independent exponential random variables with mean $\frac{1}{4}$ hour. What is

(a) the average number of customers in the shop?

(b) the proportion of potential customers that enter the shop?

(c) If the barber could work twice as fast, how much more business would he do?

12. Potential customers arrive at a full-service, one-pump gas station at a Poisson rate of 20 cars per hour. However, customers will only enter the station for gas if there are no more than two cars (including the one currently being attended to) at the pump. Suppose the amount of time required to service a car is exponentially distributed with a mean of five minutes.

(a) What fraction of the attendant's time will be spent servicing cars?

(b) What fraction of potential customers are lost?

13. A service center consists of two servers, each working at an exponential rate of two services per hour. If customers arrive at a Poisson rate of three per hour, then, assuming a system capacity of at most three customers,

(a) what fraction of potential customers enter the system?

(b) what would the value of (a) be if there was only a single server, and his rate was twice as fast (that is, $\mu = 4$)?

14. The following problem arises in molecular biology. The surface of a bacterium is supposed to consist of several sites at which foreign molecules—some acceptable and some not—become attached. We consider a particular site and assume that molecules arrive at the site according to a Poisson process with parameter λ. Among these molecules a proportion α is acceptable. Unacceptable molecules stay at the site for a length of time which is exponentially distributed with parameter μ_1, whereas an acceptable molecule remains at the site for an exponential time with rate μ_2. An arriving molecule will become attached only if the site is free of other molecules. What percentage of time is the site occupied with an acceptable (unacceptable) molecule?

15. A job shop consists of three machines and two repairmen. The amount of time a machine works before breaking down is exponentially distributed with mean 10. If the amount of time it takes a single repairman to fix a machine is exponentially distributed with mean 8, then
(a) what is the average number of machines not in use?
(b) what proportion of time are both repairmen busy?

16. Consider a taxi station where taxis and customers arrive in accordance with Poisson processes with respective rates of one and two per minute. A taxi will wait no matter how many other taxis are present. However, if an arriving customer does not find a taxi waiting, he leaves. Find
(a) the average number of taxis waiting, and
(b) the proportion of arriving customers that get taxis.

17. Customers arrive at a service station, manned by a single server who serves at an exponential rate μ_1, at a Poisson rate λ. After completion of service the customer then joins a second system where the server serves at an exponential rate μ_2. Such a system is called a *tandem* or *sequential* queueing system. Assuming that $\lambda < \mu_i$, $i = 1, 2$, determine the limiting probabilities.
Hint: Try a solution of the form $P_{n,m} = C\alpha^n\beta^m$, and determine C, α, β.

18. Consider an ergodic $M/M/s$ queue in steady state (that is, after a long time) and argue that the number presently in the system is independent of the sequence of past departure times. That is, for instance, knowing that there have been departures 2, 3, 5, and 10 time units ago does not affect the distribution of the number presently in the system.

19. In the *M/M/s* queue if you allow the service rate to depend on the number in the system (but in such a way so that it is ergodic), what can you say about the output process? What can you say when the service rate μ remains unchanged but $\lambda > s\mu$?

20. If $\{X(t)\}$ and $\{Y(t)\}$ are independent continuous-time Markov chains, both of which are time reversibles, how that the process $\{X(t), Y(t)\}$ is also a time reversible Markov chain.

21. Consider a set of n machines and a single repair facility to service these machines. Suppose that when machine i, $i = 1, \ldots, n$, fails it requires an exponentially distributed amount of work with rate μ_i to repair it. The repair facility divides its efforts equally among all failed machines in the sense that whenever there are k failed machines each one receives work at a rate of $1/k$ per unit time. Finally suppose that each time machine i is repaired, it stays up for an exponentially distributed time with rate λ_i.

(a) Define an appropriate state space so as to be able to analyze the above system as a continuous time Markov chain.
(b) Give the instantaneous transition rates (that is, give the q_{ij}).
(c) Write the time reversibility equations.
(d) Find the limiting probabilities and show that the process is time reversible.

22. Consider a graph with nodes $1, 2, \ldots, n$ and the $\binom{n}{2}$ arcs (i, j), $i \neq j$, $i, j, = 1, \ldots, n$. (See Section 6.2 of Chapter 3 for appropriate definitions.) Suppose that a particle moves along this graph as follows: Events occur along the arcs (i, j) according to independent Poisson processes with rates λ_{ij}. An event along arc (i, j) causes that arc to become excited. If the particle is at node i at the moment that (i, j) becomes excited, it instantaneously moves to node j; $i, j = 1, \ldots, n$. Let P_j denote the proportion of time that the particle is at node j. Show that

$$P_j = \frac{1}{n}$$

Hint: Use time reversibility.

23. For the continuous-time Markov chain of Problem 3 present a uniformized version.

24. Consider a system of n components such that the working times of component i, $i = 1, \ldots, n$, are exponentially distributed with rate λ_i. When failed, however, the repair rate of component i depends on how many other components are down. Specifically, suppose that the

instantaneous repair rate of component i, $i = 1, \ldots, n$, when there are a total of k failed components, is $\alpha^k \mu_i$.

(a) Explain how we can analyze the above as a continuous-time Markov chain. Define the states and give the parameters of the chain.

(b) Show that, in steady state, the chain is time reversible and compute the limiting probabilities.

25. Let Y denote an exponential random variable with rate λ that is independent of the continuous time Markov chain $\{X(t)\}$ and let

$$\bar{P}_{ij} = P\{X(Y) = j \mid X(0) = i\}$$

(a) Show that

$$\bar{P}_{ij} = \frac{1}{v_i + \lambda} \sum_k q_{ik} \bar{P}_{kj} + \frac{\lambda}{v_i + \lambda} \delta_{ij}$$

where δ_{ij} is 1 when $i = j$ and 0 when $i \neq j$.

(b) Show that the solution of the above set of equations is given by

$$\bar{P} = (I - R/\lambda)^{-1}$$

where $\bar{P}$ is the matrix of elements $\bar{P}_{ij}$, I is the identity matrix, and R the matrix specified in Section 8.

(c) Suppose now that $Y_1, \ldots, Y_n$ are independent exponentials with rate λ that are independent of $\{X(t)\}$. Show that $P\{X(Y_1 + \cdots + Y_n) = j \mid X(0) = i\}$ is equal to the element in row i, column j of the matrix $\bar{P}^n$.

(d) Explain the relationship of the above to Approximation 2 of Section 8.

26. (a) Show that Approximation 1 of Section 8 is equivalent to uniformizing the continuous time Markov chain with a value v such that $vt = n$ and then approximating $P_{ij}(t)$ by P_{ij}^{*n}.

(b) Explain why the above should make a good approximation. *Hint:* What is the standard deviation of a Poisson random variable with mean n?

References

1. D. R. Cox and H. D. Miller, "The Theory of Stochastic Processes," Methuen, London, 1965.

2. A. W. Drake, "Fundamentals of Applied Probability Theory," McGraw-Hill, New York, 1967.

3. S. Karlin and H. Taylor, "A First Course in Stochastic Processes," Second ed., Academic Press, New York, 1975.

4. E. Parzen, "Stochastic Processes," Holden-Day, San Francisco, California, 1962.

5. S. Ross, "Stochastic Processes," John Wiley, New York, 1983.

Chapter 7

Renewal Theory and Its Applications

1. Introduction

We have seen that a Poisson process is a counting process for which the times between successive events are independent and identically distributed exponential random variables. One possible generalization is to consider a counting process for which the times between successive events are independent and identically distributed with an arbitrary distribution. Such a counting process is called a *renewal process*.

Let $\{N(t), t > 0\}$ be a counting process and let X_n denote the time between the $(n - 1)$st and the nth event of this process, $n \geq 1$.

Definition 1.1. If the sequence of nonnegative random variables $\{X_1, X_2, \ldots\}$ is independent and identically distributed, then the counting process $\{N(t), t \geq 0\}$ is said to be a *renewal process*.

Thus, a renewal process is a counting process such that the time until the first event occurs has some distribution F, the time between the first and second event has, independently of the time of the first event, the same distribution F, and so on. When an event occurs, we say that a renewal has taken place.

For an example of a renewal process, suppose that we have an infinite supply of lightbulbs whose lifetimes are independent and identically distributed. Suppose also that we use a single lightbulb at a time,

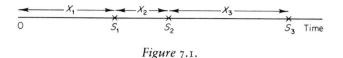

Figure 7.1.

and when it fails we immediately replace it with a new one. Under these conditions, $\{N(t), t \geq 0\}$ is a renewal process when $N(t)$ represents the number of lightbulbs that have failed by time t.

For a renewal process having interarrival times $X_1, X_2, \ldots$, let

$$S_0 = 0, \quad S_n = \sum_{i=1}^{n} X_i, \quad n \geq 1$$

That is, $S_1 = X_1$ is the time of the first renewal; $S_2 = X_1 + X_2$ is the time until the first renewal plus the time between the first and second renewal, that is, S_2 is the time of the second renewal. In general, S_n denotes the time of the nth renewal (see Figure 7.1).

We shall let F denote the interarrival distribution and to avoid trivialities, we assume that $F(0) = P\{X_n = 0\} < 1$. Furthermore, we let

$$\mu = E[X_n], \quad n \geq 1$$

be the mean time between successive renewals. It follows from the non-negativity of X_n and the fact that X_n is not identically 0 that $\mu > 0$.

The first question we shall attempt to answer is whether an infinite number of renewals can occur in a finite amount of time. That is, can $N(t)$ be infinite for some (finite) value of t? To show that this cannot occur, we first note that, as S_n is the time of the nth renewal, $N(t)$ may be written as

$$N(t) = \max\{n : S_n \leq t\} \tag{1.1}$$

To understand why Equation (1.1) is valid, suppose, for instance, that $S_4 \leq t$ but $S_5 > t$. Hence, the fourth renewal had occurred by time t but the fifth renewal occurred after time t; or in other words, $N(t)$, the number of renewals that occurred by time t, must equal 4. Now by the strong law of large numbers it follows that, with probability 1,

$$\frac{S_n}{n} \to \mu \quad \text{as} \quad n \to \infty$$

But since $\mu > 0$ this means that S_n must be going to infinity as n goes to infinity. Thus, S_n can be less than or equal to t for at most a finite number of values of n, and hence by Equation (1.1), $N(t)$ must be finite.

However, though $N(t) < \infty$ for each t, it is true that, with probability 1,

$$N(\infty) \equiv \lim_{t \to \infty} N(t) = \infty$$

This follows since the only way in which $N(\infty)$, the total number of renewals that occur, can be finite is for one of the interarrival times to be infinite. Therefore,

$$P\{N(\infty) < \infty\} = P\{X_n = \infty \text{ for some } n\}$$

$$= P\left\{ \bigcup_{n=1}^{\infty} \{X_n = \infty\} \right\}$$

$$\leq \sum_{n=1}^{\infty} P\{X_n = \infty\}$$

$$= 0$$

2. Distribution of $N(t)$

The distribution of $N(t)$ can be obtained, at least in theory, by first noting the important relationship that *the number of renewals by time t is greater than or equal to n if and only if the nth renewal occurs before or at time t.* That is,

$$N(t) \geq n \Leftrightarrow S_n \leq t \tag{2.1}$$

From Equation (2.1) we obtain

$$P\{N(t) = n\} = P\{N(t) \geq n\} - P\{N(t) \geq n + 1\}$$
$$= P\{S_n \leq t\} - P\{S_{n+1} \leq t\} \tag{2.2}$$

Now since the random variables X_i, $i \geq 1$, are independent and have a common distribution F, it follows that $S_n = \sum_{i=1}^{n} X_i$ is distributed as F_n, the n-fold convolution of F with itself (Chapter 2, Section 5). Therefore, from Equation (2.2) we obtain

$$P\{N(t) = n\} = F_n(t) - F_{n+1}(t)$$

Example 2a Suppose that $P\{X_n = i\} = p(1 - p)^{i-1}$, $i \geq 1$. That is, suppose that the interarrival distribution is geometric. Now $S_1 = X_1$ may be interpreted as the number of trials necessary to get a single success when each trial is independent and has a probability p of being a success. Similarly, S_n may be interpreted as the number of trials necessary to attain n successes, and hence has the negative binomial distribution

$$P\{S_n = k\} = \begin{cases} \binom{k-1}{n-1} p^n (1-p)^{k-n}, & k \geq n \\ 0, & k < n \end{cases}$$

Thus, from Equation (2.2) we have that

$$P\{N(t) = n\} = \sum_{k=n}^{[t]} \binom{k-1}{n-1} p^n (1-p)^{k-n}$$

$$- \sum_{k=n+1}^{[t]} \binom{k-1}{n} p^{n+1} (1-p)^{k-n-1}$$

where $[t]$ denotes the largest integer that is not larger than t, and $\sum_{k=r}^{[t]}$ is interpreted to be 0 when r is larger than $[t]$. ◊

By using Equation (2.1) we can calculate $m(t)$, the mean value of $N(t)$, as

$$m(t) = E[N(t)]$$

$$= \sum_{n=1}^{\infty} P\{N(t) \geq n\}$$

$$= \sum_{n=1}^{\infty} P\{S_n \leq t\}$$

$$= \sum_{n=1}^{\infty} F_n(t)$$

where we have used the fact that if X is nonnegative and integer valued, then

$$E[X] = \sum_{k=1}^{\infty} kP\{X = k\} = \sum_{k=1}^{\infty} \sum_{n=1}^{k} P\{X = k\}$$

$$= \sum_{n=1}^{\infty} \sum_{k=n}^{\infty} P\{X = k\} = \sum_{n=1}^{\infty} P\{X \geq n\}$$

The function $m(t)$ is known as the *mean-value* or the *renewal function*.

It can be shown that the mean-value function $m(t)$ uniquely determines the renewal process. Specifically, there is a one-to-one correspondence between the interarrival distributions F and the mean-value functions $m(t)$.

Example 2b Suppose we have a renewal process whose mean-value function is given by

$$m(t) = 2t, \qquad t \ge 0$$

What is the distribution of the number of renewals occurring by time 10?

Solution: Since $m(t) = 2t$ is the mean-value function of a Poisson process with rate 2, it follows, by the one-to-one correspondence of interarrival distributions F and mean-value functions $m(t)$, that F must be exponential with mean $\frac{1}{2}$. Thus, the renewal process is a Poisson process with rate 2 and hence

$$P\{N(10) = n\} = e^{-20} \frac{(20)^n}{n!}, \qquad n \ge 0 \quad \Diamond$$

Another interesting result that we state without proof is that

$$m(t) < \infty \quad \text{for all} \quad t < \infty$$

Remarks

(i) Since $m(t)$ uniquely determines the interarrival distribution, it follows that the Poisson process is the only renewal process having a linear mean-value function.

(ii) Some readers might think that the finiteness of $m(t)$ should follow directly from the fact that, with probability 1, $N(t)$ is finite. However, such reasoning is not valid, for consider the following: Let Y be a random variable having the following probability distribution

$$Y = 2^n \quad \text{with probability} \quad (\tfrac{1}{2})^n, \quad n \ge 1$$

Now

$$P\{Y < \infty\} = \sum_{n=1}^{\infty} P\{Y = 2^n\} = \sum_{n=1}^{\infty} (\tfrac{1}{2})^n = 1$$

But

$$E[Y] = \sum_{n=1}^{\infty} 2^n P\{Y = 2^n\} = \sum_{n=1}^{\infty} 2^n (\tfrac{1}{2})^n = \infty$$

Hence, even when Y is finite, it can still be true that $E[Y] = \infty$. $\quad \Diamond$

An integral equation satisfied by the renewal function can be obtained by conditioning on the time of the first renewal. Assuming that the interarrival distribution F is continuous with density function f this yields

$$m(t) = E[N(t)] = \int_0^\infty E[N(t) \mid X_1 = x] f(x)\, dx \qquad (2.3)$$

Now suppose that the first renewal occurs at a time x that is less than t. Then, using the fact that a renewal process probabilistically starts over when a renewal occurs, it follows that the number of renewals by time t would have the same distribution as 1 plus the number of renewals in the first $t - x$ time units. Therefore,

$$E[N(t) \mid X_1 = x] = 1 + E[N(t - x)] \qquad \text{if } x < t$$

Since, clearly

$$E[N(t) \mid X_1 = x] = 0 \qquad \text{when } x > t$$

we obtain from (2.3) that

$$
\begin{aligned}
m(t) &= \int_0^t [1 + m(t - x)] f(x)\, dx \\
&= F(t) + \int_0^t m(t - x) f(x)\, dx \qquad (2.4)
\end{aligned}
$$

The equation (2.4) is called the *renewal equation* and can sometimes be solved to obtain the renewal function.

Example 2c   One instance in which the renewal equation can be solved is when the interarrival distribution is uniform—say uniform on $(0,1)$. We will now present a solution in this case when $t < 1$. For such values of t, the renewal function becomes

$$
\begin{aligned}
m(t) &= t + \int_0^t m(t - x)\, dx \\
&= t + \int_0^t m(y)\, dy \qquad \text{by the substitution } y = t - x
\end{aligned}
$$

Differentiating the above equation yields

$$m'(t) = 1 + m(t)$$

Letting $h(t) = 1 + m(t)$ we obtain

$$h'(t) = h(t)$$

or

$$log\ b(t)\ =\ t\ +\ C$$

or

$$b(t)\ =\ Ke^t$$

or

$$m(t)\ =\ Ke^t\ -\ 1$$

Since $m(0) = 0$, we see that $K = 1$, and so we obtain

$$m(t)\ =\ e^t\ -\ 1,\qquad 0 \le t \le 1\quad \Diamond$$

3. Limit Theorems and Their Applications

We have shown previously that, with probability 1, $N(t)$ goes to infinity as t goes to infinity. However, it would be nice to know the rate at which $N(t)$ goes to infinity. That is, we would like to be able to say something about $\lim_{t\to\infty} N(t)/t$.

As a prelude to determining the rate at which $N(t)$ grows, let us first consider the random variable $S_{N(t)}$. In words, just what does this random variable represent? Proceeding inductively suppose, for instance, that $N(t) = 3$. Then $S_{N(t)} = S_3$ represents the time of the third event. Since there are only three events that have occurred by time t, S_3 also represents the time of the last event prior to (or at) time t. This is, in fact, what $S_{N(t)}$ represents—namely the time of the last renewal *prior to or at* time t. Similar reasoning leads to the conclusion that $S_{N(t)+1}$ represents the time of the first renewal *after* time t (see Figure 7.2). We now are ready to prove

Proposition 3.1.   With probability 1,

$$\frac{N(t)}{t} \to \frac{1}{\mu}\quad \text{as}\quad t \to \infty$$

Proof: Since $S_{N(t)}$ is the time of the last renewal prior to or at time t, and $S_{N(t)+1}$ is the time of the first renewal after time t, we have that

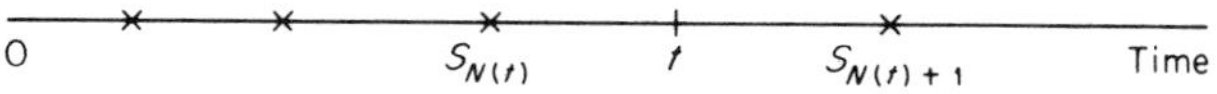

Figure 7.2.

$$S_{N(t)} \leq t < S_{N(t)+1}$$

or

$$\frac{S_{N(t)}}{N(t)} \leq \frac{t}{N(t)} < \frac{S_{N(t)+1}}{N(t)} \qquad (3.1)$$

However, since $S_{N(t)}/N(t) = \Sigma_{i=1}^{N(t)} X_i/N(t)$ is the average of $N(t)$ independent and identically distributed random variables, it follows by the strong law of large numbers that $S_{N(t)}/N(t) \to \mu$ as $N(t) \to \infty$. But since $N(t) \to \infty$ when $t \to \infty$ we obtain

$$\frac{S_{N(t)}}{N(t)} \to \mu \quad \text{as} \quad t \to \infty$$

Furthermore, writing

$$\frac{S_{N(t)+1}}{N(t)} = \left(\frac{S_{N(t)+1}}{N(t) + 1} \right) \left(\frac{N(t) + 1}{N(t)} \right)$$

we have that $S_{N(t)+1}/(N(t) + 1) \to \mu$ by the same reasoning as above and

$$\frac{N(t) + 1}{N(t)} \to 1 \quad \text{as} \quad t \to \infty$$

Hence,

$$\frac{S_{N(t)+1}}{N(t)} \to \mu \quad \text{as} \quad t \to \infty$$

The result now follows by Equation (3.1) since $t/N(t)$ is between two numbers, each of which converges to μ as $t \to \infty$. ◊

Remarks

(i) The above propositions are true even when μ, the mean time between renewals, is infinite. In this case, we interpret $1/\mu$ to be 0.

(ii) The number $1/\mu$ is called the *rate* of the renewal process. ◊

Proposition 3.1 says that the average renewal rate up to time t will, with probability 1, converge to $1/\mu$ as $t \to \infty$. What about the expected average renewal rate? Is it true that $m(t)/t$ also converges to $1/\mu$? This result, known as the *elementary renewal theorem*, will be stated without proof.

Elementary Renewal Theorem

$$\frac{m(t)}{t} \to \frac{1}{\mu} \quad \text{as} \quad t \to \infty$$

As before, $1/\mu$ is interpreted as 0 when $\mu = \infty$.

Remarks   At first glance it might seem that the elementary renewal theorem should be a simple consequence of Proposition 3.1. That is, since the average renewal rate will, with probability 1, converge to $1/\mu$, should this not imply that the expected average renewal rate also converges to $1/\mu$? We must, however, be careful; consider the next example.   ◇

Example 3a   Let U be a random variable which is uniformly distributed on $(0, 1)$; and define the random variables Y_n, $n \geq 1$, by

$$Y_n = \begin{cases} 0, & \text{if } U > \dfrac{1}{n} \\[2ex] n, & \text{if } U \leq \dfrac{1}{n} \end{cases}$$

Now, since, with probability 1, U will be greater than 0, it follows that Y_n will equal 0 for all sufficiently large n. That is, Y_n will equal 0 for all n large enough so that $1/n < U$. Hence, with probability 1,

$$Y_n \to 0 \quad \text{as} \quad n \to \infty$$

However,

$$E[Y_n] = nP\left\{ U \leq \frac{1}{n} \right\} = n\frac{1}{n} = 1$$

Therefore, even though the sequence of random variables Y_n converges to 0, the expected values of the Y_n are all identically 1.   ◇

Example 3b   Beverly has a radio which works on a single battery. As soon as the battery in use fails, Beverly immediately replaces it with a new battery. If the lifetime of a battery (in hours) is distributed uniformly over the interval (30, 60), then at what rate does Beverly have to change batteries?

Solution: If we let $N(t)$ denote the number of batteries that have failed by time t, we have by Proposition 3.1 that the rate at which Beverly replaces batteries is given by

$$\lim_{t \to \infty} \frac{N(t)}{t} = \frac{1}{\mu} = \frac{1}{45}$$

That is, in the long run, Beverly will have to replace one battery in a 45-hour period.   ◊

Example 3c   Suppose in Example 3b that Beverly does not keep any surplus batteries on hand, and so each time a failure occurs she must go and buy a new battery. If the amount of time it takes for her to get a new battery is uniformly distributed over (0, 1), then what is the average rate that Beverly changes batteries?

Solution: In this case the mean time between renewals is given by

$$\mu = EU_1 + EU_2$$

where U_1 is uniform over (30, 60) and U_2 is uniform over (0, 1). Hence,

$$\mu = 45 + \tfrac{1}{2} = 45\tfrac{1}{2}$$

and so in the long run, Beverly will be putting in a new battery at the rate of $\frac{2}{91}$. That is, she will put in two new batteries every 91 hours.   ◊

Example 3d   Suppose that potential customers arrive at a single-server bank in accordance with a Poisson process having rate λ. However, suppose that the potential customer only will enter the bank if the server is free when he arrives. That is, if there is already a customer in the bank, then our arrivee, rather than entering the bank, will go home. If we assume that the amount of time spent in the bank by an entering customer is a random variable having a distribution G, then

(a) what is the rate at which customers enter the bank? and

(b) what proportion of potential customers actually enter the bank?

Solution: In answering these questions, let us suppose that at time 0 a customer has just entered the bank. (That is, we define the process to start when the first customer enters the bank.) If we let μ_G denote the mean service time, then, by the memoryless property of the Poisson process, it follows that the mean time between entering customers is

$$\mu = \mu_G + \frac{1}{\lambda}$$

Hence, the rate at which customers enter the bank will be given by

$$\frac{1}{\mu} = \frac{\lambda}{1 + \lambda \mu_G}$$

On the other hand, since potential customers will be arriving at a rate λ, it follows that the proportion of them entering the bank will be given by

$$\frac{\lambda}{1 + \lambda \mu_G} \bigg/ \lambda = \frac{1}{1 + \lambda \mu_G}$$

In particular if $\lambda = 2$ (in hours) and $\mu_G = 2$, then only one customer out of five will actually enter the system.   ◊

A somewhat unusual application of Proposition 3.1 is provided by our next example.

Example 3e   A sequence of independent trials, each of which results in outcome number i with probability P_i, $i = 1, \ldots, n$, $\Sigma_i^n P_i = 1$, is observed until the same outcome occurs k times in a row; this outcome then is declared to be the winner of the game. For instance, if $k = 2$ and the sequence of outcomes is 1, 2, 4, 3, 5, 2, 1, 3, 3, then we stop after 9 trials and declare outcome number 3 the winner. What is the probability that i wins, $i = 1, \ldots, n$, and what is the expected number of trials?

Solution: We begin by computing the expected number of coin tosses, call in $E[T]$, until a run of k successive heads occur when the tosses are independent and each lands on heads with probability p. By conditioning on the time of the first nonhead we obtain

$$E[T] = \sum_{j=1}^{k} (1 - p)p^{j-1}(j + E[T]) + kp^k$$

Solving the above for $E[T]$ yields

$$E[T] = k + \frac{(1 - p)}{p^k} \sum_{j=1}^{k} jp^{j-1}$$

Upon simplifying, we obtain

$$E[T] = \frac{1 + p + \cdots + p^{k-1}}{p^k}$$

$$= \frac{(1 - p^k)}{p^k(1 - p)} \tag{3.2}$$

Now let us return to our example, and let us suppose that as soon as the winner of a game has been determined we immediately begin playing another game. For each i let us determine the rate at which outcome i wins. Now every time i wins, everything starts over again and thus wins by i constitute renewals. Hence, from Proposition 3.1, the rate at which i wins is equal to

$$\text{Rate at which } i \text{ wins} = \frac{1}{E[N_i]}$$

where N_i denotes the number of trials played between successive wins of outcome i. Hence from Equation (3.2) we see that

$$\text{Rate at which } i \text{ wins} = \frac{P_i^k(1 - P_i)}{(1 - P_i^k)} \tag{3.3}$$

Hence, the long-run proportion of games which are won by number i is given by

$$\text{Proportion of games } i \text{ wins} = \frac{\text{rate at which } i \text{ wins}}{\sum_{j=1}^{n} \text{rate at which } j \text{ wins}}$$

$$= \frac{\dfrac{P_i^k(1 - P_i)}{(1 - P_i^k)}}{\sum_{j=1}^{n} \dfrac{P_j^k(1 - P_j)}{(1 - P_j^k)}}$$

However, it follows from the strong law of large numbers that the long-run proportion of games that i wins will, with probability 1, be equal to the probability that i wins any given game. Hence,

$$P\{i \text{ wins}\} = \frac{\dfrac{P_i^k(1-P_i)}{(1-P_i^k)}}{\displaystyle\sum_{j=1}^{n} \frac{P_j^k(1-P_j)}{(1-P_j^k)}}$$

To compute the expected time of a game, we first note that the

$$\text{Rate at which games end} = \sum_{i=1}^{n} \text{Rate at which } i \text{ wins}$$

$$= \sum_{i=1}^{n} \frac{P_i^k(1-P_i)}{(1-P_i^k)} \quad \text{(from Equation (3.3))}$$

Now, as everything starts over when a game ends, it follows by Proposition 3.1 that the rate at which games end is equal to the reciprocal of the mean time of a game. Hence,

$$E[\text{Time of a game}] = \frac{1}{\text{Rate at which games end}}$$

$$= \frac{1}{\displaystyle\sum_{i=1}^{n} \frac{P_i^k(1-P_i)}{(1-P_i^k)}} \quad \diamond$$

A key element in the proof of the elementary renewal theorem, which is also of independent interest, is the establishment of a relationship between $m(t)$, the mean number of renewals by time t, and $E[S_{N(t)+1}]$, the expected time of the first renewal after t. Letting

$$g(t) = E[S_{N(t)+1}]$$

we will derive an integral equation, similar to the renewal equation, for $g(t)$ by conditioning on the time of the first renewal. This yields

$$g(t) = \int_0^\infty E[S_{N(t)+1} \mid X_1 = x] f(x) \, dx$$

where we have supposed that the interarrival times are continuous with density f. Now if the first renewal occurs at time x and $x > t$, then clearly the time of the first renewal after t is x. On the other hand, if the first renewal occurs at a time $x < t$, then by regarding x as the new origin, it follows that the expected time, from this origin, of the first renewal occurring after a time $t - x$ from this origin is $g(t - x)$. That is, we see that

$$E[S_{N(t)+1} \mid X_1 = x] = \begin{cases} g(t - x) + x & \text{if } x < t \\ x & \text{if } x > t \end{cases}$$

Substituting this into the above equation gives

$$\begin{aligned} g(t) &= \int_0^t (g(t - x) + x)f(x)\, dx + \int_t^\infty xf(x)\, dx \\ &= \int_0^t g(t - x)f(x)\, dx + \int_0^\infty xf(x)\, dx \end{aligned}$$

or

$$g(t) = \mu + \int_0^t g(t - x)f(x)\, dx$$

which is quite similar to the renewal equation

$$m(t) = F(t) + \int_0^t m(t - x)f(x)\, dx$$

Indeed, if we let

$$g_1(t) = \frac{g(t)}{\mu} - 1$$

we see that

$$g_1(t) + 1 = 1 + \int_0^t [g_1(t - x) + 1]f(x)\, dx$$

or,

$$g_1(t) = F(t) + \int_0^\infty g_1(t - x)f(x)\, dx$$

That is, $g_1(t) = E[S_{N(t)+1}]/\mu - 1$ satisfies the renewal equation and thus, by uniqueness, must be equal to $m(t)$. We have thus proven

Proposition 3.2.

$$E[S_{N(t)+1}] = \mu[m(t) + 1]$$

A second derivation of Proposition 3.2 is given in Problems 8 and 9. To see how Proposition 3.2 can be used to establish the elementary renewal theorem, let $Y(t)$ denote the time from t until the next renewal. $Y(t)$ is called the excess, or residual life, at t. As the first renewal after t will occur at time $t + Y(t)$, we see that

$$S_{N(t)+1} = t + Y(t)$$

Taking expectations, and utilizing Proposition 3.2 yields

$$\mu[m(t) + 1] = t + E[Y(t)] \qquad (3.4)$$

which implies that

$$\frac{m(t)}{t} = \frac{1}{\mu} - \frac{1}{t} + \frac{E[Y(t)]}{t\mu}$$

The elementary renewal theorem can now be proven by showing that

$$\lim_{t \to \infty} \frac{E[Y(t)]}{t} = 0$$

(see Problem 9).

The relation (3.4) shows that if one can determine $E[Y(t)]$, the mean excess at t, then one can compute $m(t)$ and vice-versa.

Example 3f Consider the renewal process whose interarrival distribution is the convolution of two exponentials; that is,

$$F = F_1 * F_2 \qquad \text{where } F_i(t) = 1 - e^{-\mu_i t}, \quad i = 1, 2$$

We will determine the renewal function by first determining $E[Y(t)]$. To obtain the mean excess at t, imagine that each renewal corresponds to a new machine being put in use, and suppose that each machine has two components—initially component 1 is employed and this lasts an exponential time with rate μ_1, and then component 2, which functions for an exponential time with rate μ_2, is employed. When component 2 fails, a new machine is put in use (that is, a renewal occurs). Now consider the process $\{X(t), t \geq 0\}$ where $X(t)$ is i if a type i component is in use at time t. It is easy to see that $\{X(t), t \geq 0\}$ is a two-state continuous-time Markov chain, and so, using the results of Example 4c of Chapter 6, its transition probabilities are

$$P_{11}(t) = \frac{\mu_1}{\mu_1 + \mu_2} e^{-(\mu_1 + \mu_2)t} + \frac{\mu_2}{\mu_1 + \mu_2}$$

To compute the remaining life of the machine in use at time t, we condition on whether it is using its first or second component: for if it is still using its first component, then its remaining life is $1/\mu_1 + 1/\mu_2$, whereas if it is already using its second component, then its remaining life is $1/\mu_2$. Hence, letting $p(t)$ denote the probability that the machine in use at time t is using its first component, we have that

$$E[Y(t)] = \left(\frac{1}{\mu_1} + \frac{1}{\mu_2}\right)p(t) + \frac{1 - p(t)}{\mu_2}$$

$$= \frac{1}{\mu_2} + \frac{p(t)}{\mu_1}$$

But, since at time 0 the first machine is utilizing its first component, it follows that $p(t) = P_{11}(t)$, and so, upon using the above expression of $P_{11}(t)$, we obtain that

$$E[Y(t)] = \frac{1}{\mu_2} + \frac{1}{\mu_1 + \mu_2} e^{-(\mu_1 + \mu_2)t} + \frac{\mu_2}{\mu_1(\mu_1 + \mu_2)} \tag{3.5}$$

Now it follows from (3.4) that

$$m(t) + 1 = \frac{t}{\mu} + \frac{E[Y(t)]}{\mu} \tag{3.6}$$

where μ, the mean interarrival time, is given in this case by

$$\mu = \frac{1}{\mu_1} + \frac{1}{\mu_2} = \frac{\mu_1 + \mu_2}{\mu_1\mu_2} \tag{3.7}$$

Substituting (3.5) and (3.7) into (3.6) yields, after simplifying, that

$$m(t) = \frac{\mu_1\mu_2}{\mu_1 + \mu_2} t - \frac{\mu_1\mu_2}{(\mu_1 + \mu_2)^2} [1 - e^{-(\mu_1 + \mu_2)t}] \quad \diamond$$

Remark   Using the relationship (3.6) and results from the two-state continuous-time Markov chain, the renewal function can also be obtained in the same manner as in Example 3f for the interarrival distributions

$$F(t) = pF_1(t) + (1 - p)F_2(t)$$

and

$$F(t) = pF_1(t) + (1 - p)(F_1 * F_2)(t)$$

when $F_i(t) = 1 - e^{-\mu_i t}$, $t > 0$, $i = 1,2$.   ◇

An important limit theorem is the central limit theorem for renewal processes. This states that, for large t, $N(t)$ is approximately normally distributed with mean t/μ and variance $t\sigma^2/\mu^3$, where μ and σ are, respectively, the mean and variance of the interarrival distribution. That is, we have the following theorem which we state without proof.

Central Limit Theorem for Renewal Processes

$$\lim_{t \to \infty} P\left\{ \frac{N(t) - t/\mu}{\sqrt{t\sigma^2/\mu^3}} < x \right\} = \frac{1}{\sqrt{2\pi}} \int_{-\infty}^{x} e^{-x^2/2} \, dx$$

4. Renewal Reward Processes

A large number of probability models are special cases of the following model. Consider a renewal process $\{N(t), t \geq 0\}$ having interarrival times X_n, $n \geq 1$, and suppose that each time a renewal occurs we receive a reward. We denote by R_n, the reward earned at the time of the nth renewal. We shall assume that the R_n, $n \geq 1$, are independent and identically distributed. However, we do allow for the possibility that R_n may (and usually will) depend on X_n, the length of the nth renewal interval. If we let

$$R(t) = \sum_{n=1}^{N(t)} R_n$$

then $R(t)$ represents the total reward earned by time t. Let

$$E[R] = E[R_n], \qquad E[X] = E[X_n]$$

Proposition 4.1. If $E[R] < \infty$ and $E[X] < \infty$, then with probability 1

$$\frac{R(t)}{t} \to \frac{E[R]}{E[X]} \quad \text{as} \quad t \to \infty$$

Proof: Write

$$\frac{R(t)}{t} = \frac{\displaystyle\sum_{n=1}^{N(t)} R_n}{t}$$

$$= \left(\frac{\displaystyle\sum_{n=1}^{N(t)} R_n}{N(t)}\right)\left(\frac{N(t)}{t}\right)$$

By the strong law of large numbers we obtain that

$$\frac{\displaystyle\sum_{n=1}^{N(t)} R_n}{N(t)} \to E[R] \quad \text{as} \quad t \to \infty$$

and by Proposition 3.1 that

$$\frac{N(t)}{t} \to \frac{1}{E[X]} \quad \text{as} \quad t \to \infty$$

The result thus follows. ◊

If we say that a *cycle* is completed every time a renewal occurs then Proposition 4.1 states that the long-run average reward is just the expected reward earned during a cycle divided by the expected length of a cycle.

Example 4a In Example 3d if we suppose that the amounts that the successive customers deposit in the bank are independent random variables having a common distribution H, then the rate at which deposits accumulate—that is, $\lim_{t\to\infty}$ (Total deposits by time t)/t —is given by

$$\frac{E[\text{Deposits during a cycle}]}{E[\text{Time of cycle}]} = \frac{\mu_H}{\mu_G + 1/\lambda}$$

where $\mu_G + 1/\lambda$ is the mean time of a cycle, and μ_H is the mean of the distribution H. ◊

Example 4b (A Car Buying Model): The lifetime of a car is a continuous random variable having a distribution H and probability density h. Mr. Brown has a policy that he buys a new car as soon as his old one either breaks down or reaches the age of T years. Suppose that a new car costs C_1 dollars and also that an additional cost of C_2 dollars is incurred whenever Mr. Brown's car breaks down. Under the assumption that a used car has no resale value, what is Mr. Brown's long-run average cost?

If we say that a cycle is complete every time Mr. Brown gets a new car, then it follows from Proposition 4.1 (with costs replacing rewards) that his long-run average cost equals

$$\frac{E[\text{cost incurred during a cycle}]}{E[\text{length of a cycle}]}$$

Now letting X be the lifetime of Mr. Brown's car during an arbitrary cycle, then the cost incurred during that cycle will be given by

$$C_1, \quad\quad \text{if} \quad X > T$$
$$C_1 + C_2, \quad \text{if} \quad X \le T$$

so the expected cost incurred over a cycle is

$$C_1 P\{X > T\} + (C_1 + C_2)P\{X \le T\} = C_1 + C_2 H(T)$$

Also, the length of the cycle is

$$X, \quad \text{if} \quad X \le T$$
$$T, \quad \text{if} \quad X > T$$

and so the expected length of a cycle is

$$\int_0^T xh(x)dx + \int_T^\infty Th(x)dx = \int_0^T xh(x)dx + T[1 - H(T)]$$

Therefore, Mr. Brown's long-run average cost will be

$$\frac{C_1 + C_2 H(T)}{\int_0^T xh(x)dx + T[1 - H(T)]} \tag{4.1}$$

Now, suppose that the lifetime of a car (in years) is uniformly distributed over $(0, 10)$, and suppose that C_1 is 3 (thousand) dollars and C_2 is $\frac{1}{2}$ (thousand) dollars. What value of T minimizes Mr. Brown's long-run average cost?

If Mr. Brown uses the value T, $T \leq 10$, then from (4.1) his long-run average cost equals

$$\frac{3 + \dfrac{1}{2}\dfrac{T}{10}}{\displaystyle\int_0^T \frac{x}{10}\,dx + T\left(1 - \frac{T}{10}\right)} = \frac{3 + \dfrac{T}{20}}{\dfrac{T^2}{20} + \dfrac{(10T - T^2)}{10}}$$

$$= \frac{60 + T}{20T - T^2}$$

We can now minimize this by using the calculus. Toward this end, let

$$g(T) = \frac{60 + T}{20T - T^2}$$

then

$$g'(T) = \frac{(20T - T^2) - (60 + T)(20 - 2T)}{(20T - T^2)^2}$$

Equating to 0 yields

$$20T - T^2 = (60 + T)(20 - 2T)$$

or, equivalently,

$$T^2 + 120T - 1200 = 0$$

which yields the solutions

$$T \approx 9.25 \quad \text{and} \quad T \approx -129.25$$

Since $T \leq 10$, it follows that the optimal policy for Mr. Brown would be to purchase a new car whenever his old car reaches the age of 9.25 years. ◊

Example 4c (Dispatching a Train): Suppose that customers arrive at a train depot in accordance with a renewal process having a mean interarrival time μ. Whenever there are N customers waiting in the depot, a train leaves. If the depot incurs a cost at the rate of nc dollars per unit time whenever there are n customers waiting, what is the average cost incurred by the depot?

If we say that a cycle is completed whenever a train leaves, then the above is a renewal reward process. The expected length of a cycle is the expected time required for N customers to arrive and, since the mean interarrival time is μ, this equals

$$E[\text{length of cycle}] = N\mu$$

If we let T_n denote the time between the nth and $(n + 1)$st arrival in a cycle, then the expected cost of a cycle may be expressed as

$$E[\text{cost of a cycle}] = E[cT_1 + 2cT_2 + \cdots + (N - 1)cT_{N-1}]$$

which since, $E[T_n] = \mu$, equals

$$c\mu \frac{N}{2}(N - 1)$$

Hence, the average cost incurred by the depot is

$$\frac{c\mu N(N - 1)}{2N\mu} = \frac{c(N - 1)}{2}$$

Suppose now that each time a train leaves the depot incurs a cost of six units. What value of N minimizes the depot's long-run average cost when $c = 2$, $\mu = 1$?

In this case, we have that the average cost per unit time when the depot uses N is

$$\frac{6 + c\mu \frac{1}{2} N(N - 1)}{N\mu} = N - 1 + \frac{6}{N}$$

By treating this as a continuous function of N and using the calculus, we obtain that the minimal value of N is

$$N = \sqrt{6} \approx 2.45$$

Hence, the optimal integral value of N is either 2 which yields a value 4, or 3 which also yields the value 4. Hence, either $N = 2$ or $N = 3$ minimizes the depot's average cost. ◊

Example 4d Consider a manufacturing process that sequentially produces items, each of which is either defective or acceptable. The following type of sampling scheme is often employed in an attempt to detect and eliminate most of the defective items. Initially, each item is inspected and this continues until there are k consecutive items that are acceptable. At this point 100% inspection ends and each successive item is independently inspected with probability α. This partial inspection continues until a defective item is encountered, at which time 100% inspection is reinstituted, and the process begins anew. If each item is, independently, defective with probability q

(a) what proportion of items are inspected?

(b) if defective items are removed when detected, what proportion of the remaining items are defective?

Remark Before starting our analysis, it should be noted that the above inspection scheme was devised for situations in which the probability of producing a defective item changed over time. It was hoped that 100% inspection would correlate with times at which the defect probability was large and partial inspection when it was small. However, it is still important to see how such a scheme would work in the extreme case where the defect probability remains constant throughout.

Solution: We begin our analysis by noting that we can treat the above as a renewal reward process with a new cycle starting each time 100% inspection is instituted. We then have

$$\text{proportion of items inspected} = \frac{E[\text{number inspected in a cycle}]}{E[\text{number produced in a cycle}]}$$

Let N_k denote the number of items inspected until there are k consecutive acceptable items. Once partial inspection begins—that is after N_k items have been produced—since each inspected item will be defective with probability q, it follows that the expected number that will have to be inspected to find a defective item is $1/q$. Hence,

$$E[\text{number inspected in a cycle}] = E[N_k] + \frac{1}{q}$$

In addition, since at partial inspection each item produced will, independently, be inspected and found to be defective with probability αq, it follows that the number of items produced until one is inspected and found to be defective is $1/\alpha q$, and so

$$E[\text{number produced in a cycle}] = E[N_k] + \frac{1}{\alpha q}$$

Also, as $E[N_k]$ is the expected number of trials needed to obtain k acceptable items in a row when each item is acceptable with probability $p = 1 - q$, it follows from Example 4e of Chapter 3 that

$$E[N_k] = \frac{1}{p} + \frac{1}{p^2} + \cdots + \frac{1}{p^k} = \frac{(1/p)^k - 1}{q}$$

Hence, we obtain

$$P_1 \equiv \text{proportion of items that are inspected} = \frac{(1/p)^k}{(1/p)^k - 1 + 1/\alpha}$$

To answer (b), note first that since each item produced is defective with probability q it follows that the proportion of items that are both inspected and found to be defective is qP_1. Hence, for N large, out of the first N items produced there will be (approximately) NqP_1 that are discovered to be defective and thus removed. As the first N items will contain (approximately) Nq defective items, it follows that there will be $Nq - NqP_1$ defective items not discovered. Hence,

$$\text{proportion of the non-removed items that are defective} \approx \frac{Nq(1 - P_1)}{N(1 - qP_1)}$$

As the approximation becomes exact as $N \to \infty$, we see that

$$\text{proportion of the non-removed items that are defective} = \frac{q(1 - P_1)}{(1 - qP_1)} \diamond$$

Example 4e (The Average Age of a Renewal Process): Consider a renewal process having interarrival distribution F and define $A(t)$ to be the time at t since the last renewal. If renewals represent old items failing and being replaced by new ones, then $A(t)$ represents the age of the item in use at time t. Since $S_{N(t)}$ represents the time of the first event prior to or at time t, we have that

$$A(t) = t - S_{N(t)}$$

We are interested in the average value of the age—that is, in

$$\lim_{s \to \infty} \frac{\int_0^s A(t)\,dt}{s}$$

To determine the above quantity, we use renewal reward theory in the following way: Let us assume that any time we are being paid money at a rate equal to the age of the renewal process at that time. That is, at time t, we are being paid at rate $A(t)$, and so $\int_0^s A(t)dt$ represents our total earnings by time s. As everything starts over again when a renewal occurs, it follows that

$$\frac{\int_0^s A(t)dt}{s} \to \frac{E[\text{Reward during a renewal cycle}]}{E[\text{Time of a renewal cycle}]}$$

Now since the age of the renewal process a time t into a renewal cycle is just t, we have

$$\text{Reward during a renewal cycle} = \int_0^X t\,dt$$

$$= \frac{X^2}{2}$$

where X is the time of the renewal cycle. Hence, we have that

$$\text{Average value of age} \equiv \lim_{s\to\infty} \frac{\int_0^s A(t)dt}{s}$$

$$= \frac{E[X^2]}{2E[X]} \qquad (4.2)$$

where X is an interarrival time having distribution function F. $\Diamond$

Example 4f (The Average Excess of a Renewal Process): Another quantity associated with a renewal process is $E(t)$, the excess of residual time at time t. $E(t)$ is defined to equal the time from t until the next renewal and, as such, represents the remaining (or residual) life of the item in use at time t. The average value of the excess, namely

$$\lim_{s\to\infty} \frac{\int_0^s E(t)dt}{s}$$

also can be easily obtained by renewal reward theory. To do so, suppose that we are paid at time t at a rate equal to $E(t)$. Then our average reward per unit time will, by renewal reward theory, be given by

$$\text{Average value of excess} \equiv \lim_{s \to \infty} \frac{\int_0^s E(t)dt}{s}$$

$$= \frac{E[\text{Reward during a cycle}]}{E[\text{Length of a cycle}]}$$

Now, letting X denote the length of a renewal cycle, we have that

$$\text{Reward during a cycle} = \int_0^X (X - t)dt$$

$$= \frac{X^2}{2}$$

and thus the average value of the excess is

$$\text{Average value of excess} = \frac{E[X^2]}{2E[X]}$$

which was the same result obtained for the average value of the age of renewal process.   ◊

5. Regenerative Processes

Consider a stochastic process $\{X(t), t \geq 0\}$ with state space 0, 1, 2, . . . , having the property that there exist time points at which the process (probabilistically) restarts itself. That is, suppose that with probability one, there exists a time T_1, such that the continuation of the process beyond T_1 is a probabilistic replica of the whole process starting at 0. Note that this property implies the existence of further times $T_2, T_3, . . .$, having the same property as T_1. Such a stochastic process is known as a *regenerative process*.

From the above, it follows that $T_1, T_2, . . .$ constitute the arrival times of a renewal process, and we shall say that a cycle is completed every time a renewal occurs.

Examples

(1) A renewal process is regenerative, and T_1 represents the time of the first renewal.

(2) A recurrent Markov chain is regenerative, and T_1 represents the time of the first transition into the initial state.

We are interested in determining the long-run proportion of time that a regenerative process spends in state j. To obtain this quantity, let us imagine that we earned a reward at a rate 1 per unit time when the process is in state j and at rate 0 otherwise. That is, if $I(s)$ represents the rate at which we earn at time s, then

$$I(s) = \begin{cases} 1, & \text{if } X(s) = j \\ 0, & \text{if } X(s) \neq j \end{cases}$$

and

$$\text{Total reward earned by } t = \int_0^t I(s)ds$$

As the above is clearly a renewal reward process which starts over again at the cycle time T_1, we see from Proposition 4.1 that

$$\text{Average reward per unit time} = \frac{E[\text{Reward by time } T_1]}{E[T_1]}$$

However, the average reward per unit time is just equal to the proportion of time that the process is in state j. That is, we have

Proposition 5.1. For a regenerative process, the long-run

$$\text{Proportion of time in state } j = \frac{E[\text{Amount of time in } j \text{ during a cycle}]}{E[\text{Time of a cycle}]}$$

Remarks If the cycle time T_1 is a continuous random variable, then it can be shown by using an advanced theorem called the "key renewal theorem" that the above is equal also to the limiting probability that the system is in state j at time t. That is, if T_1 is continuous, then

$$\lim_{t \to \infty} P\{X(t) = j\} = \frac{E[\text{Amount of time in } j \text{ during a cycle}]}{E[\text{Time of a cycle}]}$$

Example 5a (Markov Chains): Consider a positive recurrent Markov chain which is initially in state i. By the Markovian property each time the process reenters state i, it starts over again. Thus returns to state i are renewals and constitute the beginnings of a new cycle. By Proposition 5.1, it follows that the long-run

Proportion of time in state j

$$= \frac{E[\text{Amount of time in } j \text{ during an } i - i \text{ cycle}]}{\mu_{ii}}$$

where μ_{ii} represents the mean time to return to state i. If we take j to equal i, then we obtain

$$\text{Proportion of time in state } i = \frac{1}{\mu_{ii}} \quad \diamond$$

Example 5b (A Queueing System with Renewal Arrivals): Consider a waiting time system in which customers arrive in accordance with an arbitrary renewal process and are served one at a time by a single server having an arbitrary service distribution. If we suppose that at time 0 the initial customer has just arrived, then $\{X(t), t \geq 0\}$ is a regenerative process, where $X(t)$ denotes the number of customers in the system at time t. The process regenerates each time a customer arrives and finds the server free. $\diamond$

5.1. Alternating Renewal Processes

Another example of a regenerative process is provided by what is known as an alternating renewal process, which considers that a system can be in one of two states: on or off. Initially it is on, and it remains on for a time Z_1; it then goes off and remains off for a time Y_1. It then goes on for a time Z_2; then off for a time Y_2; then on, and so on.

We suppose that the random vectors (Z_n, Y_n), $n \geq 1$ are independent and identically distributed. That is, both the sequence of random variables $\{Z_n\}$ and the sequence $\{Y_n\}$ are independent and identically distributed; but we allow Z_n and Y_n to be dependent. In other words, each time the process goes on, everything starts over again, but when it then goes off, we allow the length of the off time to depend on the previous on time.

Let $E[Z] = E[Z_n]$ and $E[Y] = E[Y_n]$ denote respectively the mean lengths of an on and off period.

We are concerned with P_{on}, the long-run proportion of time that the system is on. If we let

$$X_n = Y_n + Z_n, \qquad n \geq 1$$

then at time X_1 the process starts over again. That is, the process starts over again after a complete cycle consisting of an on and an off interval. In other words, a renewal occurs whenever a cycle is completed. Therefore, we obtain from Proposition 5.1 that

$$\begin{aligned} P_{on} &= \frac{E[Z]}{E[Y] + E[Z]} \\ &= \frac{E[\text{on}]}{E[\text{on}] + E[\text{off}]} \end{aligned} \tag{5.1.1}$$

Also, if we let P_{off} denote the long-run proportion of time that the system is off, then

$$P_{off} = 1 - P_{on}$$
$$= \frac{E[off]}{E[on] + E[off]} \qquad (5.1.2)$$

Example 5.1a (A Production Process): One example of an alternating renewal process is a production process (or a machine) which works for a time Z_1, then breaks down and has to be repaired (which takes a time Y_1), then works for a time Z_2, then is down for a time Y_2, and so on. If we suppose that the process is as good as new after each repair, then this constitutes an alternating renewal process. It is worthwhile to note that in this example it makes sense to suppose that the repair time will depend on the amount of time the process had been working before breaking down. ◊

Example 5.1b (The Age of a Renewal Process): Suppose we are interested in determining the proportion of time that the age of a renewal process is less than some constant c. To do so, let a cycle correspond to a renewal, and say that the system is "on" at time t if the age at t is less than or equal to c, and say it is "off" if the age at t is greater than c. In other words, the system is "on" the first c time units of a renewal interval, and "off" the remaining time. Hence, letting X denote a renewal interval, we have from Equation (5.1.1)

$$\text{Proportion of time age is less than } c = \frac{E[\min(X, c)]}{E[X]}$$

$$= \frac{\int_0^\infty P\{\min(X, c) > x\}dx}{E[X]}$$

$$= \frac{\int_0^c P\{X > x\}dx}{E[X]}$$

$$= \frac{\int_0^c (1 - F(x))dx}{E[X]} \qquad (5.1.3)$$

where F is the distribution function of X and where we have used the identity that for a nonnegative random variable Y

$$E[Y] = \int_0^\infty P\{Y > x\}dx \quad \diamond$$

Example 5.1c (The Excess of a Renewal Process): We leave it as an exercise for the reader to show by an argument similar to the one used in the previous example that

Proportion of time the excess is less than c

$$= \int_0^c \frac{(1 - F(x))dx}{E[X]} \quad (5.1.4)$$

which is the same result obtained for the age. $\quad \diamond$

Example 5.1d (The Single-Server Poisson Arrival Queue): Consider a single-server service station in which customers arrive according to a Poisson process having rate λ. An arriving customer is immediately served if the server is free, and if not, then he waits in line (that is, he joins the queue). The successive service times will be assumed to be independent and identically distributed random variables having mean $1/\mu$, where $\mu > \lambda$. What proportion of time is the server busy?

The above process will alternate between busy periods when the server is working and idle periods when he is free. Hence, a cycle will consist of a busy period followed by an idle period. Thus, letting P_B denote the proportion of time that the server is busy, we obtain that

$$P_B = \frac{E[\text{Length of a busy period}]}{E[\text{Length of a busy period}] + E[\text{Length of an idle period}]} \quad (5.1.5)$$

Now an idle period begins when a customer completes his service and there are no additional customers waiting for service, and it ends when the next customer arrives. Hence, by the lack of memory of the Poisson process it follows that an idle period will be exponentially distributed with mean $1/\lambda$; that is,

$$E[\text{Length of an idle period}] = 1/\lambda \quad (5.1.6)$$

Now a busy period begins when a customer enters and finds the server free. To determine $E[L_B]$, the mean length of a busy period, we shall condition on both the number of customers entering the system during the service time of the customer initiating the busy period and the length of this service time. Letting N denote the number of customers entering the system during the initial customer's service time, and S the length of this service, we have

$$E[L_B] = E[E[L_B|N, S]]$$

Now if $N = 0$, then the busy period would end when the initial customer completes his service. Therefore,

$$E[L_B|N = 0, S] = S$$

On the other hand, suppose that $N = 1$. Then at time S there will be a single customer in the system. Furthermore, since the arrival stream of customers is a Poisson process it will be starting over again at time S. Hence, at time S the expected additional time until the server becomes free will just be the expected length of a busy period. That is,

$$E[L_B|N = 1, S] = S + E[L_B]$$

Finally, suppose that n customers arrive during the service time of the initial customer. To determine the conditional expected length of a busy period in this situation, we first note that the length of the busy period will not depend on the order in which we serve waiting customers. That is, it will not make a difference if we serve waiting customers on a first-come, first-served order or on any other order. Hence, let us suppose that our n arrivals, C_1, C_2, . . . , C_n, during the initial service time are served as follows. Customer 1 (that is, C_1) is served first, but C_2 is not served until the only customers in the system are C_2, C_3, . . . , C_n. That is, C_2 is not served until the system is free of all customers but C_2, . . . , C_n. For instance, customers arriving during C_1's service time will be served before C_2, C_3, . . . , C_n. Similarly, C_3 will not be served until the system is free of all customers but C_3, . . . , C_n, and so on. A little thought will reveal that the expected length of time between C_i and C_{i+1}'s service, $i = 1, . . . , n - 1$, will just be $E[L_B]$. Therefore,

$$E[L_B|N = n, S] = S + nE[L_B]$$

or, equivalently,

$$E[L_B|N, S] = S + NE[L_B]$$

Hence, taking expectations of the above yields

$$E[L_B] = E[S] + E[NE[L_B]]$$
$$= E[S] + E[N]E[L_B]$$

or

$$E[L_B] = \frac{E[S]}{1 - E[N]}$$

$$= \frac{1/\mu}{1 - E[N]} \tag{5.1.7}$$

To determine $E[N]$, the expected number of arrivals during the initial service time of a busy period, we condition on S, the length of the service time, to obtain

$$E[N] = E[E[N|S]]$$

However, since the arrival stream is a Poisson process with rate λ it follows that the expected number of arrivals during an interval of length S is just λS. That is, $E[N|S] = \lambda S$, and hence

$$E[N] = E[\lambda S] = \lambda E[S] = \lambda/\mu$$

Substituting in Equation (5.1.7) yields

$$E[L_B] = \frac{1/\mu}{1 - \lambda/\mu}$$

$$= \frac{1}{\mu - \lambda} \tag{5.1.8}$$

Therefore, from Equations (5.1.5), (5.1.6), and (5.1.8), we have that P_B, the proportion of time that the server is busy, is given by

$$P_B = \frac{1/(\mu - \lambda)}{1/\lambda + 1/(\mu - \lambda)}$$

$$= \frac{\lambda}{\mu}$$

From this it follows that P_I, the proportion of time that the server is idle, is given by

$$P_I = 1 - P_B$$

$$= 1 - \lambda/\mu$$

For instance, if $\lambda = 1$ and $\mu = 3$, then the server is busy one-third of the time and is idle two-thirds of the time. On the other hand, if we double λ to obtain $\lambda = 2$, then the proportion of time the server is busy doubles to two-thirds of the time.

The above has been calculated under the assumption that $\lambda < \mu$. When $\lambda \geq \mu$ it can be shown that $P_B = 1$ and $P_I = 0$. ◊

6. Semi-Markov Processes

Consider a process that can be in either state 1 or state 2 or state 3. It is initially in state 1 where it remains for a random amount of time having mean μ_1, then it goes to state 2 where it remains for a random amount of time having mean μ_2, then it goes to state 3 where it remains for a mean time μ_3, then back to state 1, and so on. What proportion of time is the process in state i, $i = 1, 2, 3$?

If we say that a cycle is completed each time the process returns to state i, and if we let the reward be the amount of time we spend in state i during that cycle, then the above is a renewal reward process. Hence, from Proposition 4.1 we obtain that P_i, the proportion of time that the process is in state i, is given by

$$P_i = \frac{\mu_i}{\mu_1 + \mu_2 + \mu_3}, \qquad i = 1, 2, 3$$

Similarly, if we had a process which could be in any of N states 1, 2, . . . , N and which moved from state $1 \to 2 \to 3 \to \cdots \to N - 1 \to N \to 1$, then the long-run proportion of time that the process spends in state i is

$$P_i = \frac{\mu_i}{\mu_1 + \mu_2 + \cdots + \mu_N}, \qquad i = 1, 2, . . . , N$$

where μ_i is the expected amount of time the process spends in state i during each visit.

Let us now generalize the above to the following situation. Suppose that a process can be in any one of N states 1, 2, . . . , N, and that each time it enters state i it remains there for a random amount of time having mean μ_i and then makes a transition into state j with probability P_{ij}. Such a process is called a *semi-Markov process*. Note that if the amount of time that the process spends in each state before making a transition is identically 1, then the semi-Markov process is just a Markov chain.

Let us calculate P_i for a semi-Markov process. To do so, we first consider π_i the proportion of transitions that take the process into state *i.* Now if we let X_n denote the state of the process after the nth transition, then $\{X_n, n \geq 0\}$ is a Markov chain with transition probabilities $P_{ij}, i, j = 1, 2, \ldots, N$. Hence π_i will just be the limiting (or stationary) probabilities for this Markov chain (Chapter 4, Section 4). That is, π_i will be the unique nonnegative solution of

$$\sum_{i=1}^{N} \pi_i = 1$$

$$\pi_i = \sum_{j=1}^{N} \pi_j P_{ji}, \qquad i = 1, 2, \ldots, N^* \tag{6.1}$$

Now since the process spends an expected time μ_i in state i whenever it visits that state, it seems intuitive that P_i should be a weighted average of the π_i where π_i is weighted proportionately to μ_i. That is,

$$P_i = \frac{\pi_i \mu_i}{\displaystyle\sum_{j=1}^{N} \pi_j \mu_j}, \qquad i = 1, 2, \ldots, N \tag{6.2}$$

where the π_i are given as the solution to Equation (6.1).

Example 6a Consider a machine that can be in one of three states: *good condition, fair condition,* or *broken down.* Suppose that a machine in good condition will remain this way for a mean time μ_1 and then will go to either the fair condition or the broken condition with respective probabilities $\frac{3}{4}$ and $\frac{1}{4}$. A machine in fair condition will remain that way for a mean time μ_2 and then will break down. A broken machine will be repaired, which takes a mean time μ_3, and when repaired will be in good condition with probability $\frac{2}{3}$ and fair condition with probability $\frac{1}{3}$. What proportion of time is the machine in each state?

Solution: Letting the states be 1, 2, 3 we have by Equation (6.1) that the π_i satisfy

$$\pi_1 + \pi_2 + \pi_3 = 1$$
$$\pi_1 = \tfrac{2}{3} \pi_3$$
$$\pi_2 = \tfrac{3}{4} \pi_1 + \tfrac{1}{3} \pi_3$$
$$\pi_3 = \tfrac{1}{4} \pi_1 + \pi_2$$

---

*We shall assume that there exists a solution of Equation (6.1). That is, we assume that all of the states in the Markov chain communicate.

The solution being

$$\pi_1 = \tfrac{4}{15}, \qquad \pi_2 = \tfrac{1}{3}, \qquad \pi_3 = \tfrac{2}{5}$$

Hence, from Equation (6.2) we obtain that P_i, the proportion of time the machine is in state i, is given by

$$P_1 = \frac{4\mu_1}{4\mu_1 + 5\mu_2 + 6\mu_3}$$

$$P_2 = \frac{5\mu_2}{4\mu_1 + 5\mu_2 + 6\mu_3}$$

$$P_3 = \frac{6\mu_3}{4\mu_1 + 5\mu_2 + 6\mu_3}$$

For instance, if $\mu_1 = 5$, $\mu_2 = 2$, $\mu_3 = 1$, then the machine will be in good condition five-ninths of the time, in fair condition five-eighteenths of the time, in broken condition one-sixth of the time. ◇

Remarks When the distributions of the amount of time spent in each state during a visit are continuous, then P_i also represents the limiting (as $t \to \infty$) probability that the process will be in state i at time t. ◇

Example 6b Consider a renewal process in which the interarrival distribution is discrete and is such that

$$P\{X = i\} = p_i, \qquad i \geq 1$$

where X represents an interarrival random variable. Let $L(t)$ denote the length of the renewal interval that contains the point t (that is, if $N(t)$ is the number of renewals by time t and X_n the nth interarrival time, then $L(t) = X_{N(t)+1}$). If we think of each renewal as corresponding to the failure of a lightbulb (which is then replaced at the beginning of the next period by a new bulb), then $L(t)$ will equal i if the bulb in use at time t is in its ith period of use.

It is easy to see that $L(t)$ is a semi-Markov process. To determine the proportion of time that $L(t) = j$, note that each time a transition occurs—that is, each time a renewal occurs—the next state will be j with probability p_j. That is, the transition probabilities of the embedded Markov chain are $P_{ij} = p_j$. Hence, the limiting probabilities of this embedded chain are given by

$$\pi_j = p_j$$

and, since the mean time the semi-Markov process spends in state j before a transition occurs is j, it follows that the long-run proportion of time the state is j is

$$P_j = \frac{jp_j}{\sum_i ip_i} \quad \Diamond$$

7. The Inspection Paradox

Suppose that a piece of equipment, say a battery, is installed and serves until it breaks down. Upon failure it is instantly replaced by a like battery, and this process continues without interruption. Letting $N(t)$ denote the number of batteries that have failed by time t, we have that $\{N(t), t \geq 0\}$ is a renewal process.

Suppose further that the distribution F of the lifetime of a battery is not known and is to be estimated by the following sampling inspection scheme. We fix some time t and observe the total lifetime of the battery that is in use at time t. Since F is the distribution of the lifetime for all batteries, it seems reasonable that it should be the distribution for this battery. However, this is the *inspection paradox* for it turns out that the *battery in use at time t tends to have a larger lifetime than an ordinary battery*.

To understand the above so-called paradox, we reason as follows. In renewal theoretic terms what we are interested in is the length of the renewal interval containing the point t. That is, we are interested in $X_{N(t)+1} = S_{N(t)+1} - S_{N(t)}$ (see Figure 7.2). To calculate the distribution of $X_{N(t)+1}$ we condition on the time of the last renewal prior to (or at) time t. That is,

$$P\{X_{N(t)+1} > x\} = E[P\{X_{N(t)+1} > x | S_{N(t)} = t - s\}]$$

where we recall (Figure 7.2) that $S_{N(t)}$ is the time of the last renewal prior to (or at) t. Since there are no renewals between $t - s$ and t, it follows that $X_{N(t)+1}$ must be larger than x if $s > x$. That is,

$$P\{X_{N(t)+1} > x | S_{N(t)} = t - s\} = 1 \quad \text{if} \quad s > x \tag{7.1}$$

On the other hand, suppose that $s \leq x$. As before, we know that a renewal occurred at time $t - s$ and no additional renewals occurred between $t - s$ and t, and we ask for the probability that no renewals occur for an additional time $x - s$. That is, we are asking for the probability that an interarrival time will be greater than x given that it is greater than s. Therefore, for $s \leq x$,

$$
\begin{aligned}
P\{X_{N(t)+1} > x | S_{N(t)} &= t - s\} \\
&= P\{\text{interarrival time} > x | \text{interarrival time} > s\} \\
&= P\{\text{interarrival time} > x\} / P\{\text{interarrival time} > s\} \\
&= \frac{1 - F(x)}{1 - F(s)} \\
&\geq 1 - F(x)
\end{aligned} \tag{7.2}
$$

Hence, from Equations (7.1) and (7.2) we see that, for all s,

$$
P\{X_{N(t)+1} > x | S_{N(t)} = t - s\} \geq 1 - F(x)
$$

Taking expectations on both sides of the above yields that

$$
P\{X_{N(t)+1} > x\} \geq 1 - F(x) \tag{7.3}
$$

However, $1 - F(x)$ is the probability that an ordinary renewal interval is larger than x, that is, $1 - F(x) = P\{X_n > x\}$, and thus Equation (7.3) is a statement of the inspection paradox that the length of the renewal interval containing the point t tends to be larger than an ordinary renewal interval.

Remarks To obtain an intuitive feel for the so-called inspection paradox, reason as follows. We think of the whole line being covered by renewal intervals, one of which covers the point t. Is it not more likely that a larger interval, as opposed to a shorter interval, covers the point t? ◊

We can actually calculate the distribution of $X_{N(t)+1}$ when the renewal process is a Poisson process. (Note that, in the general case, we did not need to calculate explicitly $P\{X_{N(t)+1} > x\}$ to show that it was at least as large as $1 - F(x)$.) To do so we write

$$
X_{N(t)+1} = A(t) + E(t)
$$

where $A(t)$ denotes the time from t since the last renewal, and $E(t)$ denotes the time from t until the next renewal (see Figure 7.3). $A(t)$ is known as the *age* of the process at time t (in our example it would be the age at time t of the battery in use at time t), and $E(t)$ is known as the *excess life* of the process at time t (it is the additional time from t until the battery fails). Of course, it is true that $A(t) = t - S_{N(t)}$, and $E(t) = S_{N(t)+1} - t$.

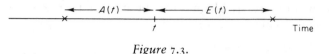

Figure 7.3.

   To calculate the distribution of $X_{N(t)+1}$ we first note the important fact that, for a Poisson process, $A(t)$ and $E(t)$ are independent. This follows since by the memoryless property of the Poisson process, the time from t until the next renewal will be exponentially distributed and will be independent of all that has previously occurred (including, in particular, $A(t)$). In fact, this shows that if $\{N(t),\ t \geq 0\}$ is a Poisson process with rate λ, then

$$P\{E(t) \leq x\} = 1 - e^{-\lambda x} \tag{7.4}$$

The distribution of $A(t)$ may be obtained as follows

$$P\{A(t) > x\} = \begin{cases} P\{0 \text{ renewals in } [t-x, t]\}, & \text{if } x \leq t \\ 0, & \text{if } x > t \end{cases}$$

$$= \begin{cases} e^{-\lambda x}, & \text{if } x \leq t \\ 0, & \text{if } x > t \end{cases}$$

or equivalently

$$P\{A(t) \leq x\} = \begin{cases} 1 - e^{-\lambda x}, & x \leq t \\ 1, & x > t \end{cases} \tag{7.5}$$

Hence, by the independence of $E(t)$ and $A(t)$ the distribution of $X_{N(t)+1}$ is just the convolution of the exponential distribution (Equation (7.4)) and the distribution Equation (7.5). It is interesting to note that for t large, $A(t)$ approximately has an exponential distribution. Thus, for t large, $X_{N(t)+1}$ has the distribution of the convolution of two identically distributed exponential random variables, which by Section 2.3 of Chapter 5, is the gamma distribution with parameters $(2, \lambda)$. In particular, for t large, the expected length of the renewal interval containing the point t is approximately *twice* the expected length of an ordinary renewal interval.

8. Computing the Renewal Function

The difficulty with attempting to use the identity

$$m(t) = \sum_{n=1}^{\infty} F_n(t)$$

to compute the renewal function is that the determination of $F_n(t) = P\{X_1 + \cdots + X_n \le t\}$ requires the computation of an n-dimensional integral. We present below an effective algorithm which requires as inputs only one-dimensional integrals.

Let Y be an exponential random variable having rate λ, and suppose that Y is independent of the renewal process $\{N(t),\ t \ge 0\}$. We start by determining $E[N(Y)]$, the expected number of renewals by the random time Y. To do so, we first condition on X_1, the time of the first renewal. This yields

$$E[N(Y)] = \int_0^{\infty} E[N(Y) \mid X_1 = x] f(x)\, dx \tag{8.1}$$

where f is the interarrival density. To determine $E[N(Y) \mid X_1 = x]$, we now condition on whether or not Y exceeds x. Now, if $Y < x$, then as the first renewal occurs at time x, it follows that the number of renewals by time Y is equal to 0. On the other hand, if we are given that $x < Y$, then the number of renewals by time Y will equal 1 (the one at x) plus the number of additional renewals between x and Y. But, by the memoryless property of exponential random variables, it follows that, given that $Y > x$, the amount by which it exceeds x is also exponential with rate λ, and so given that $Y > x$ the number of renewals between x and Y will have the same distribution as $N(Y)$. Hence,

$$E[N(Y) \mid X_1 = x,\ Y < x] = 0$$
$$E[N(Y) \mid X_1 = x,\ Y > x] = 1 + E[N(Y)]$$

and so,

$$
\begin{aligned}
E[N(Y) \mid X_1 = x] &= E[N(Y) \mid X_1 = x,\ Y < x]\, P\{Y < x \mid X_1 = x\} \\
&\quad + E[N(Y) \mid X_1 = x,\ Y > x]\, P\{Y > x \mid X_1 = x\} \\
&= E[N(Y) \mid X_1 = x,\ Y > x]\, P\{Y > x\} \\
&\quad \text{since } Y \text{ and } X_1 \text{ are independent} \\
&= (1 + E[N(Y)])\, e^{-\lambda x}
\end{aligned}
$$

Substituting this into (8.1) gives

$$E[N(Y)] = (1 + E[N(Y)]) \int_0^{\infty} e^{-\lambda x} f(x)\, dx$$

or

$$E[N(Y)] = \frac{E[e^{-\lambda X}]}{1 - E[e^{-\lambda X}]} \tag{8.2}$$

where X has the renewal interarrival distribution.

If we let $\lambda = 1/t$, then (8.2) presents an expression for the expected number of renewals (not by time t, but) by a random exponentially distributed time with mean t. However, as such a random variable need not be close to its mean (its variance is t^2) equation (8.2) need not be particularly close to $m(t)$. To obtain an accurate approximation suppose that $Y_1, Y_2, \ldots, Y_n$ are independent exponentials with rate λ and suppose they are also independent of the renewal process. Let, for $r = 1, \ldots, n$,

$$m_r = E[N(Y_1 + \cdots + Y_r)]$$

To compute an expression for m_r, we again start by conditioning on X_1, the time of the first renewal.

$$m_r = \int_0^\infty E[N(Y_1 + \cdots + Y_r) \mid X_1 = x] f(x) \, dx \tag{8.3}$$

To determine the above conditional expectation, we now condition on the number of partial sums $\sum_{i=1}^j Y_i, j = 1, \ldots, r$, that are less than x. Now, if all r partial sums are less than x—that is, if $\sum_{i=1}^r Y_i < x$—then clearly the number of renewals by time $\sum_{i=1}^r Y_i$ is 0. On the other hand, given that k, $k < r$, of these partial sums are less than x, it follows from the lack of memory property of the exponential that the number of renewals by time $\sum_{i=1}^r Y_i$ will have the same distribution as 1 plus $N(Y_{k+1} + \cdots + Y_r)]$. Hence,

$$E[N(Y_1 + \cdots + Y_r) \mid X_1 = x, k \text{ of the sums} \sum_{i=1}^j Y_i \text{ are less than } x]$$

$$= \begin{cases} 0 & \text{if } k = r \\ 1 + m_{r-k} & \text{if } k < r \end{cases} \tag{8.4}$$

To determine the distribution of the number of the partial sums that are less than x, note that the successive values of these partial sums $\sum_{i=1}^j Y_i, j = 1, \ldots, r$, have the same distribution as the first r event times of a Poisson process with rate λ (since each successive partial sum is the previous sum plus an independent exponential with rate λ). Hence, it follows that, for $k < r$,

$$P\{k \text{ of the partial sums } \sum_{i=1}^{j} Y_i \text{ are less than } x \mid X_1 = x\}$$
$$= \frac{e^{-\lambda x}(\lambda x)^k}{k!} \tag{8.5}$$

Upon substitution of (8.4) and (8.5) into Equation (8.3), we obtain that

$$m_r = \int_0^\infty \sum_{k=0}^{r-1} (1 + m_{r-k}) \frac{e^{-\lambda x}(\lambda x)^k}{k!} f(x) \, dx$$

or, equivalently,

$$m_r = \frac{\sum_{k=1}^{r-1} (1 + m_{r-k}) E[X^k e^{-\lambda X}] \frac{\lambda^k}{k!} + E[e^{-\lambda X}]}{1 - E[e^{-\lambda X}]} \tag{8.6}$$

If we set $\lambda = n/t$, then starting with m_1 given by Equation (8.2), we can use Equation (8.6) to recursively compute $m_2, \ldots, m_n$. The approximation of $m(t) = E[N(t)]$ is given by $m_n = E[N(Y_1 + \cdots + Y_n)]$. Since $Y_1 + \cdots + Y_n$ is the sum of n independent exponential random variables each with mean t/n, it follows that it is (gamma) distributed with mean t and variance $nt^2/n^2 = t^2/n$. Hence, by choosing n large, $\sum_{i=1}^n Y_i$ will be a random variable having most of its probability concentrated about t, and so $E[N(\sum_{i=1}^n Y_i)]$ should be quite close to $E[N(t)]$. (Indeed, if $m(t)$ is continuous at t, it can be shown that these approximations converge to $m(t)$ as n goes to infinity.)

Example 8a Table 7.1 compares the approximation with the exact value for the distributions F_i with densities f_i, $i = 1,2,3$, which are given by

$$f_1(x) = xe^{-x}$$
$$1 - F_2(x) = .3e^{-x} + .7e^{-2x}$$
$$1 - F_3(x) = .5e^{-x} + .5e^{-5x} \quad \diamond$$

Table 7.1

		Exact		Approximation			
F_i	t	$m(t)$	$n = 1$	$n = 3$	$n = 10$	$n = 25$	$n = 50$
1	1	0.2838	0.3333	0.3040	0.2903	0.2865	0.2852
1	2	0.7546	0.8000	0.7697	0.7586	0.7561	0.7553
1	5	2.250	2.273	2.253	2.250	2.250	2.250
1	10	4.75	4.762	4.751	4.750	4.750	4.750
2	0.1	0.1733	0.1681	0.1687	0.1689	0.1690	–
2	0.3	0.5111	0.4964	0.4997	0.5010	0.5014	–
2	0.5	0.8404	0.8182	0.8245	0.8273	0.8281	0.8283
2	1	1.6400	1.6087	1.6205	1.6261	1.6277	1.6283
2	3	4.7389	4.7143	4.7294	4.7350	4.7363	4.7367
2	10	15.5089	15.5000	15.5081	15.5089	15.5089	15.5089
3	0.1	0.2819	0.2692	0.2772	0.2804	0.2813	–
3	0.3	0.7638	0.7105	0.7421	0.7567	0.7609	–
3	1	2.0890	2.0000	2.0556	2.0789	2.0850	2.0870
3	3	5.4444	5.4000	5.4375	5.4437	5.4442	5.4443

Remark   The material of this section is taken from S. M. Ross, "Approximations in Renewal Theory," *Probability in the Engineering and Informational Sciences,* 1(2), 163–175 (1987).   ◇

Problems

1. Is it true that
   (a) $N(t) < n$ if and only if $S_n > t$?
   (b) $N(t) \leq n$ if and only if $S_n \geq t$?
   (c) $N(t) > n$ if and only if $S_n < t$?

2. Suppose that the interarrival distribution for a renewal process is Poisson distributed with mean μ. That is, suppose

$$P\{X_n = k\} = e^{-\mu} \frac{\mu^k}{k!}, \qquad k = 0, 1, \ldots$$

   (a) Find the distribution of S_n.
   (b) Calculate $P\{N(t) = n\}$.

3. If the mean-value function of the renewal process $\{N(t), t \geq 0)$ is given by $m(t) = t/2, t \geq 0$, then what is $P\{N(5) = 0\}$?

4. Consider a renewal process $\{N(t), t \geq 0\}$ having a gamma (r, λ) interarrival distribution. That is, the interarrival density is

$$f(x) = \frac{\lambda e^{-\lambda x}(\lambda x)^{r-1}}{(r-1)!}, \qquad x > 0$$

(a) Show that

$$P\{N(t) \geq n\} = \sum_{i=nr}^{\infty} \frac{e^{-\lambda t}(\lambda t)^i}{i!}$$

(b) Use (a) to show that

$$m(t) = \sum_{i=r}^{\infty} \left[\frac{i}{r} \right] \frac{e^{-\lambda t}(\lambda t)^i}{i!}$$

where $[i/r]$ is the largest integer less than or equal to i/r.
Hint: Use the relationship between the gamma (r, λ) distribution and the sum of r independent exponentials with rate λ, to define $N(t)$ in terms of a Poisson process with rate λ.

5. Mr. Smith works on a temporary basis. The mean length of each job he gets is three months. If the amount of time he spends between jobs is exponentially distributed with mean 2, then at what rate does Mr. Smith get new jobs?

6. A machine in use is replaced by a new machine either when it fails or when it reaches the age of T years. If the lifetimes of successive machines are independent with a common distribution F having density f, show that
    (a) the long-run rate at which machines are replaced equals

$$\left[\int_0^T xf(x)dx + T(1 - F(T)) \right]^{-1}$$

(b) the long-run rate at which machines in use fail equals

$$\frac{F(T)}{\int_0^T xf(x)dx + T[1 - F(T)]}$$

7. A renewal process for which the time until the initial renewal has a different distribution than the remaining interarrival times is called a delayed (or a general) renewal process. Prove that Proposition 3.1 remains valid for a delayed renewal process. (In general, it can be shown

that all of the limit theorems for a renewal process remain valid for a delayed renewal process provided that the time until the first renewal has a finite mean.)

8. Let X_1, X_2, . . . be a sequence of independent random variables. The nonnegative integer valued random variable N is said to be a *stopping time* for the sequence if the event $\{N = n\}$ is independent of X_{n+1}, X_{n+2}, . . ., the idea being that the X_i are observed one at a time—first X_1, then X_2, and so on—and N represents the number observed when we stop. Hence, the event $\{N = n\}$ corresponds to stopping after having observed X_1, . . . , X_n and thus must be independent of the values of random variables yet to come, namely, X_{n+1}, X_{n+2}, ...

(a) Let X_1, X_2, . . . be independent with

$$P\{X_i = 1\} = p = 1 - P\{X_i = 0\}, \qquad i \geq 1$$

Define

$$N_1 = \min\{n : X_1 + \cdot \cdot \cdot + X_n = 5\}$$

$$N_2 = \begin{cases} 3, & \text{if} \quad X_1 = 0 \\ 5, & \text{if} \quad X_1 = 1 \end{cases}$$

$$N_3 = \begin{cases} 3, & \text{if} \quad X_4 = 0 \\ 2, & \text{if} \quad X_4 = 1 \end{cases}$$

Which of the N_i are stopping times for the sequence X_1, . . . ? An important result, known as *Wald's equation* states that if X_1, X_2, . . . are independent and identically distributed and have a finite mean $E(X)$, and if N is a stopping time for this sequence having a finite mean, then

$$E\left[\sum_{i=1}^{N} X_i\right] = E[N]E[X]$$

To prove Wald's equation, let us define the indicator variables I_i, $i \geq 1$ by

$$I_i = \begin{cases} 1, & \text{if} \quad i \leq N \\ 0, & \text{if} \quad i > N \end{cases}$$

(b) Show that

$$\sum_{i=1}^{N} X_i = \sum_{i=1}^{\infty} X_i I_i$$

From Part (b) we see that

$$E\left[\sum_{i=1}^{N} X_i\right] = E\left[\sum_{i=1}^{\infty} X_i I_i\right]$$

$$= \sum_{i=1}^{\infty} E[X_i I_i]$$

where the last equality assumes that the expectation can be brought inside the summation (as indeed can be rigorously proven in this case).

(c) Argue that X_i and I_i are independent.

Hint: I_i equals 0 or 1 depending upon whether or not we have yet stopped after observing which random variables?

(d) From Part (c) we have

$$E\left[\sum_{i=1}^{N} X_i\right] = \sum_{i=1}^{\infty} E[X] E[I_i]$$

Complete the proof of Wald's equation.

(e) What does Wald's equation tell us about the stopping times in Part (a)?

9. Wald's equation can be used as the basis of a proof of the elementary renewal theorem. Let $X_1, X_2, \ldots$ denote the interarrival times of a renewal process and let $N(t)$ be the number of renewals by time t.

(a) Show that whereas $N(t)$ is *not* a stopping time, $N(t) + 1$ is.

Hint: Note that

$$N(t) = n \Leftrightarrow X_1 + \cdots + X_n \leq t \quad \text{and} \quad X_1 + \cdots + X_{n+1} > t$$

(b) Argue that

$$E\left[\sum_{i=1}^{N(t)+1} X_i\right] = \mu[m(t) + 1]$$

(c) Suppose that the X_i are bounded random variables. That is, suppose there is a constant M such that $P\{X_i < M\} = 1$. Argue that

$$t < \sum_{i=1}^{N(t)+1} X_i < t + M$$

(d) Use the previous parts to prove the elementary renewal theorem when the interarrival times are bounded.

10. Consider a miner trapped in a room which contains three doors. Door 1 leads him to freedom after two-day's travel; door 2 returns him to his room after four-day's journey; and door 3 returns him to his room after six-day's journey. Suppose at all times he is equally likely to choose any of the three doors, and let T denote the time it takes the miner to become free.

(a) Define a sequence of independent and identically distributed random variables $X_1, X_2, \ldots$ and a stopping time N such that

$$T = \sum_{i=1}^{N} X_i$$

Note: You may have to imagine that the miner continues to randomly choose doors even after he reaches safety.

(b) Use Wald's equation to find $E[T]$.

(c) Compute $E[\sum_{i=1}^{N} X_i | N = n]$ and note that it is not equal to $E[\sum_{i=1}^{n} X_i]$.

(d) Use Part (c) for a second derivation of $E[T]$.

11. In Example 3d suppose that potential customers arrive in accordance with a renewal process having interarrival distribution F. Would the number of events by time t constitute a (possible delayed) renewal process if an event corresponds to a customer

(i) entering the bank?

(ii) leaving the bank?

What if F were exponential?

12. Compute the renewal function when the interarrival distribution F is such that

$$1 - F(t) = pe^{-\mu_1 t} + (1 - p)e^{-\mu_2 t}$$

13. For the renewal process whose interarrival times are uniformly distributed over $(0,1)$, determine the expected time from $t = 1$ until the next renewal.

14. For a renewal reward process consider

$$W_n = \frac{R_1 + R_2 + \cdots + R_n}{X_1 + X_2 + \cdots + X_n}$$

W_n represents the average reward earned during the first n cycles. Show that $W_n \to E[R]/E[X]$ as $n \to \infty$.

15. Consider a single-server bank for which customers arrive in accordance with a Poisson process with rate λ. If a customer only will

enter the bank if the server is free when he arrives, and if the service time of a customer has the distribution G, then what proportion of time is the server busy?

16. The lifetime of a car has a distribution H and probability density h. Ms. Jones buys a new car as soon as her old car either breaks down or reaches the age of T years. A new car costs C_1 dollars and an additional cost of C_2 dollars is incurred whenever a car breaks down. Assuming that a T-year-old car in working order has an expected resale value $R(T)$, what is Ms. Jones' long-run average cost?

17. If H is the uniform distribution over $(2, 8)$ and if $C_1 = 4$, $C_2 = 1$, and $R(T) = 4 - (T/2)$, then what value of T minimizes Ms. Jones' long-run average cost in Problem 14?

18. In Problem 16 suppose that H is exponentially distributed with mean 5, $C_1 = 3$, $C_2 = \frac{1}{2}$, $R(T) = 0$. What value of T minimizes Ms. Jones' long-run average cost?

19. Consider a train station to which customers arrive in accordance with a Poisson process having rate λ. A train is summoned whenever there are N customers waiting in the station, but it takes K units of time for the train to arrive at the station. When it arrives, it picks up all waiting customers. Assuming that the train station incurs a cost at a rate of nc per unit time whenever there are n customers present, find the long-run average cost.

20. In Example 4d, what proportion of the defective items that are produced are discovered?

21. Satellites are launched according to a Poisson process with rate λ. Each satellite will, independently, orbit the earth for a random time having distribution F. Let $X(t)$ denote the number of satellites orbiting at time t.

(i) Determine $P\{X(t) = k\}$. *Hint:* Relate this to the $M/G/\infty$ queue.
(ii) If at least one satellite is orbiting, then messages can be transmitted and we say that the system is functional. If the first satellite is orbited at time $t = 0$, determine the expected time that the system remains functional.
Hint: Make use of part (i) when $k = 0$.

22. A group of n skiers continually, and independently, climb up and then ski down a particular slope. The time it takes skier i to climb up has distribution F_i, and it is independent of her time to ski down, which has distribution H_i, $i = 1, \ldots, n$. Let $N(t)$ denote the total number of times members of this group have skied down the slope by

time t. Also, let $U(t)$ denote the number of skiers climbing up the hill at time t.

(i) What is $\lim_{t \to \infty} N(t)/t$?

(ii) Find $\lim_{t \to \infty} P\{U(t) = k\}$.

(iii) If all F_i are exponential with rate λ and all G_i are exponential with rate μ, what is $P\{U(t) = k\}$?

23. Three marksmen take turns shooting at a target. Marksman 1 shoots until he misses, then Marksman 2 begins shooting until he misses, then Marksman 3 until he misses, and then back to Marksman 1, and so on. Each time Marksman 1 fires he hits the target, independently of the past, with probability P_i, $i = 1, 2, 3$. Determine the proportion of time, in the long run, that each Marksman shoots.

24. Each time a certain machine breaks down it is replaced by a new one of the same type. In the long run, what percentage of time is the machine in use less than one year old if the life distribution of a machine is

   (i) uniformly distributed over $(0, 2)$?

   (ii) exponentially distributed with mean 1?

25. For any interarrival distribution function F, we define the equilibrium distribution function of F, called F_e, by

$$F_e(x) = \int_0^x \frac{[1 - F(y)]}{\mu} \, dy$$

where μ is the mean of the distribution F. Show that if F is an exponential distribution, then $F_e = F$.

26. Consider a system which can be in either state 1 or 2 or 3. Each time the system enters state i it remains there for a random amount of time having mean μ_i and then makes a transition into state j with probability P_{ij}. Suppose

$$P_{12} = 1, \quad P_{21} = P_{23} = \tfrac{1}{2}, \quad P_{31} = 1$$

   (a) What proportion of transitions take the system into state 1?

   (b) If $\mu_1 = 1$, $\mu_2 = 2$, $\mu_3 = 3$, then what proportion of time does the system spend in each state?

27. Consider a semi-Markov process in which the amount of time that the process spends in each state before making a transition into a different state is exponentially distributed. What kind of a process is this?

28. A taxi alternates between three different locations. Whenever it reaches location i, it stops and spends a random time having mean t_i before obtaining another passenger, $i = 1,2,3$. A passenger entering the cab at location i will want to go to location j with probability P_{ij}. The time to travel from i to j is a random variable with mean m_{ij}. Suppose that $t_1 = 1$, $t_2 = 2$, $t_3 = 4$, $P_{12} = 1$, $P_{23} = 1$, $P_{31} = \frac{2}{3} = 1 - P_{32}$, $m_{12} = 10$, $m_{23} = 20$, $m_{31} = 15$, $m_{32} = 25$. Define an appropriate semi-Markov process and determine

    (a) the proportion of time the taxi is waiting at location i, and

    (b) the proportion of time the taxi is on the road from i to j, i, $j = 1,2,3$.

29. Consider a renewal process having the gamma (n,λ) interarrival distribution, and let $Y(t)$ denote the time from t until the next renewal. Use the theory of semi-Markov processes to show that

$$\lim_{t \to \infty} P\{Y(t) < x\} = \frac{1}{n} \sum_{i=1}^{n} G_{i,\lambda}(x)$$

where $G_{i,\lambda}(x)$ is the gamma (i,λ) distribution function.

30. For a renewal process, let $A(t)$ be the age at time t. Prove that if $\mu < \infty$, then with probability 1

$$\frac{A(t)}{t} \to 0 \quad \text{as} \quad t \to \infty$$

31. If $A(t)$ and $E(t)$ are respectively the age and the excess at time t of a renewal process having an interarrival distribution F, calculate

$$P\{E(t) > x | A(t) = s\}$$

32. Verify Equation (5.1.4).

33. To prove Equation (6.2), define the following notation:

    $X_i^j \equiv$ time spent in state i on the jth visit to this state;

    $N_i(m) \equiv$ number of visits to state i in the first m transitions.

In terms of the above notation, write expressions for

    (a) the amount of time during the first m transitions that the process is in state i;

    (b) the proportion of time during the first m transitions that the process is in state i.

Argue that, with probability 1,

(c) $\sum\limits_{j=1}^{N_i(m)} \dfrac{X_i^j}{N_i(m)}$ $/N_i(m)$ $\rightarrow \mu_i$ as $m \rightarrow \infty$.

(d) $N_i(m)/m \rightarrow \pi_i$ as $m \rightarrow \infty$.

Combine parts (a), (b), (c), and (d) to prove Equation (6.2).

34. Let X_i, $i = 1, 2, \ldots$, be the interarrival times of the renewal process $\{N(t)\}$, and let Y, independent of the X_i, be exponential with rate λ.

(a) Use the lack of memory property of the exponential to argue that
$$P\{X_1 + \cdots + X_n < Y\} = (P\{X < Y\})^n$$
(b) Use (a) to show that
$$E[N(Y)] = \frac{E[e^{-\lambda X}]}{1 - E[e^{-\lambda X}]}$$

where X has the interarrival distribution.

35. Write a program to approximate $m(t)$ for the interarrival distribution $F*G$, where F is exponential with mean 1 and G is exponential with mean 3.

References

For additional results and for a more rigorous treatment of renewal theory, the reader should see:

1. S. Ross, "Stochastic Processes," John Wiley, New York, 1983.
Other useful references are:

2. D. R. Cox, "Renewal Theory," Methuen, London, 1962.

3. W. Feller, "An Introduction to Probability Theory and Its Applications," Vol. II., John Wiley, New York, 1966.

4. E. Parzen, "Stochastic Processes," Holden-Day, San Francisco, California, 1962.

5. N. U. Prabhu, "Stochastic Processes," Macmillan, New York, 1963.

Chapter 8

Queueing Theory

1. Introduction

In this chapter we will study a class of models in which customers arrive in some random manner at a service facility. Upon arrival they are made to wait in queue until it is their turn to be served. Once served they are generally assumed to leave the system. For such models we will be interested in determining, among other things, such quantities as the average number of customers in the system (or in the queue) and the average time a customer spends in the system (or spends waiting in the queue).

In Section 2 we derive a series of basic queueing identities which are of great use in analyzing queueing models. We also introduce three different sets of limiting probabilities which correspond to what an arrival sees, what a departure sees, and what an outside observer would see.

In Section 3 we deal with queueing systems in which all of the defining probability distributions are assumed to be exponential. For instance, the simplest such model is to assume that customers arrive in accordance with a Poisson process (and thus the interarrival times are exponentially distributed) and are served one at a time by a single server who takes an exponentially distributed length of time for each service. These exponential queueing models are special examples of continuous-time Markov chains and so can be analyzed as in Chapter 6. However, at the cost of a (very) slight amount of repetition we shall not assume the reader to be familiar with the material of Chapter 6, but rather we shall redevelop any needed material. Specifically we shall derive anew (by a heuristic argument) the formula for the limiting probabilities.

In Section 4 we consider models in which customers move randomly among a network of servers. The model of Section 4.1 is an open system in which customers are allowed to enter and depart the system, whereas the one studied in Section 4.2 is closed in the sense that the set of customers in the system is constant over time.

In Section 5 we study the model $M/G/1$, which while assuming Poisson arrivals, allows the service distribution to be arbitrary. To analyze this model we first introduce in Section 5.1 the concept of work, and then use this concept in Section 5.2 to help analyze this system. In Section 5.3 we derive the average amount of time that a server remains busy between idle periods.

In Section 6 we consider some variations of the model $M/G/1$. In particular in Section 6.1 we suppose that bus loads of customers arrive according to a Poisson process and that each bus contains a random number of customers. In Section 6.2 we suppose that there are two different classes of customers—with type 1 customers receiving service priority over type 2.

In Section 7 we consider a model with exponential service times but where the interarrival times between customers is allowed to have an arbitrary distribution. We analyze this model by use of an appropriately defined Markov chain. We also derive the mean length of a busy period and of an idle period for this model.

In the final section of the chapter we talk about multiserver systems. We start with loss systems, in which arrivals, finding all servers busy, are assumed to depart and as such are lost to the system. This leads to the famous result known as Erlang's loss formula, which presents a simple formula for the number of busy servers in such a model when the arrival process is Poisson and the service distribution is general. We then discuss multiserver systems in which queues are allowed. However, except in the case where exponential service times are assumed, there are very few explicit formulas for these models. We end by presenting an approximation for the average time a customer waits in queue in a k-server model which assumes Poisson arrivals but allows for a general service distribution.

2. Preliminaries

In this section we will derive certain identities which are valid in the great majority of queueing models.

2.1. Cost Equations

Some fundamental quantities of interest for queueing models are

L, the average number of customers in the system;
L_Q, the average number of customers waiting in queue;
W, the average amount of time that a customer spends in the system;
W_Q, the average amount of time that a customer spends waiting in queue.

A large number of interesting and useful relationships between the above and other quantities of interest can be obtained by making use of the following idea: Imagine that entering customers are forced to pay money (according to some rule) to the system. We would then have the following basic cost identity

Average rate at which the system earns

$$= \lambda_a \times \text{average amount an entering customer pays,} \quad (2.1)$$

where λ_a is defined to be average arrival rate of entering customers. That is, if $N(t)$ denotes the number of customer arrivals by time t, then

$$\lambda_a = \lim_{t \to \infty} \frac{N(t)}{t}$$

We now present an heuristic proof of Equation (2.1).

Heuristic Proof of Equation (2.1)
Let T be a fixed large number. In two different ways, we will compute the average amount of money the system has earned by time T. On one hand, this quantity approximately can be obtained by multiplying the average rate at which the system earns by the length of time T. On the other hand, we can approximately compute it by multiplying the average amount paid by an entering customer by the average number of customers entering by time T (and this latter factor is approximately $\lambda_a T$). Hence, both sides of Equation (2.1) when multiplied by T are approximately equal to the average amount earned by T. The result then follows by letting $T \to \infty$.*

By choosing appropriate cost rules, many useful formulas can be obtained as special cases of Equation (2.1). For instance, by supposing

---

*This can be made into a rigorous proof provided we assume that the queueing process is regenerative in the sense of Section 5 of Chapter 7. Most models, including all the ones in this chapter, satisfy this condition.

that each customer pays \$1 unit time while in the system, Equation (2.1) yields the so-called Little's formula,

$$L = \lambda_a W \tag{2.2}$$

This follows since, under this cost rule, the rate at which the system earns is just the number in the system, and the amount a customer pays is just equal to its time in the system.

Similarly if we suppose that each customer pays \$1 per unit time while in queue, then Equation (2.1) yields

$$L_Q = \lambda_a W_Q \tag{2.3}$$

By supposing the cost rule that each customer pays \$1 per unit time while in service we obtain from Equation (2.1) that the

Average number of customers in service $= \lambda_a E[S]$ (2.4)

where $E[S]$ is defined as the average amount of time a customer spends in service.

It should be emphasized that Equations (2.1)–(2.4) are valid for almost all queueing models regardless of the arrival process, the number of servers, or queue discipline.

2.2. Steady-State Probabilities

Let $X(t)$ denote the number of customers in the system at time t and define P_n, $n \geq 0$, by

$$P_n = \lim_{t \to \infty} P\{X(t) = n\}$$

where we assume the above limit exists. In other words, P_n is the limiting or long-run probability that there will be exactly n customers in the system. It is sometimes referred to as the *study-state probability* of exactly n customers in the system. It also usually turns out that P_n equals the (long-run) proportion of time that the system contains exactly n customers. For example, if $P_0 = 0.3$, then in the long-run, the system will be empty of customers for 30 percent of the time. Similarly, $P_1 = 0.2$ would imply that for 20 percent of the time the system would contain exactly one customer.*

Two other sets of limiting probabilities are $\{a_n, \ n \geq 0\}$ and $\{d_n, \ n \geq 0\}$, where

---

*A sufficient condition for the validity of the dual interpretation of P_n is that the queueing process be regenerative.

a_n = proportion of customers that find n
in the system when they arrive, and
d_n = proportion of customers leaving behind n
in the system when they depart

That is, P_n is the proportion of time during which there are n in the system; a_n is the proportion of arrivals that find n; and d_n is the proportion of departures that leave behind n. That these quantities need not always be equal is illustrated by the following example.

Example 2a Consider a queueing model in which all customers have service times equal to 1, and where the times between successive customers are always greater than 1 (for instance, the interarrival times could be uniformly distributed over $(1, 2)$). Hence, as every arrival finds the system empty and every departure leaves it empty, we have

$$a_0 = d_0 = 1$$

However,

$$P_0 \neq 1$$

as the system is not always empty of customers.

It was, however, no accident that a_n equaled d_n in the previous example. That arrivals and departures always see the same number of customers is always true as is shown in the next proposition. ◊

Proposition 2.1. In any system in which customers arrive one at a time and are served one at a time

$$a_n = d_n, \qquad n \geq 0$$

Proof: An arrival will see n in the system whenever the number in the system goes from n to $n + 1$; similarly, a departure will leave behind n whenever the number in the system goes from $n + 1$ to n. Now in any interval of time T the number of transitions from n to $n + 1$ must equal to within 1 the number from $n + 1$ to n. (For instance, if transitions from 2 to 3 occur 10 times, then 10 times there must have been a transition back to 2 from a higher state (namely 3).) Hence, the rate of transitions from n to $n + 1$ equals the rate from $n + 1$ to n; or, equivalently, the rate at which arrivals find n equals the rate at which departures leave n. The result now follows since the overall arrival rate must equal the overall departure rate (what goes in eventually goes out).

Hence, on the average, arrivals and departures always see the same number of customers. However, as Example 2a illustrates, they do not, in general, see the time averages. One important exception where they do is in the case of Poisson arrivals.

Proposition 2.2.   Poisson arrivals always see time averages. In particular, for Poisson arrivals,

$$P_n = a_n$$

To understand why Poisson arrivals always see time averages, consider an arbitrary Poisson arrival. If we knew that it arrived at time t, then the conditional distribution of what it sees upon arrival is the same as the unconditional distribution of the system state at time t. For knowing that an arrival occurs at time t gives us no information about what occurred prior to t. (Since the Poisson process has independent increments, knowing that an event occurred at time t does not affect the distribution of what occurred prior to t.) Hence, an arrival would just see the system according to the limiting probabilities.

Contrast the foregoing with the situation of Example 2a where knowing that an arrival occurred at time t tells us a great deal about the past; in particular it tells us that there have been no arrivals in $(t - 1, t)$. Thus, in this case, we cannot conclude that the distribution of what an arrival at time t observes is the same as the distribution of the system state at time t.

For a second argument as to why Poisson arrivals see time averages, note that the total time the system is in state n by time T is (roughly) $P_n T$. Hence, as Poisson arrivals always arrive at rate λ no matter what the system state, it follows that the number of arrivals in $[0, T]$ that find the system in state n is (roughly) $\lambda P_n T$. In the long run, therefore, the rate at which arrivals find the system in state n is λP_n and, as λ is the overall arrival rate, it follows that $\lambda P_n / \lambda = P_n$ is the proportion of arrivals that find the system in state n.

3. Exponential Models

3.1. A Single-Server Exponential Queueing System

Suppose that customers arrive at a single-server service station in accordance with a Poisson process having rate λ. That is, the times between successive arrivals are independent exponential random variables having mean $1/\lambda$. Each customer, upon arrival, goes directly into service if the server is free and, if not, the customer joins the queue. When the server finishes serving a customer, the customer leaves the system, and the next customer in line, if there is any, enters service. The successive service times are assumed to be independent exponential random variables having mean $1/\mu$.

The above is called the $M/M/1$ queue. The two M's refer to the fact that both the interarrival and service distributions are exponential (and thus memoryless, or Markovian), and the 1 to the fact that there is a single server. To analyze it, we shall begin by determining the limiting probabilities P_n, for $n = 0, 1, \ldots$. In order to do so, think along the following lines. Suppose that we have an infinite number of rooms numbered $0, 1, 2, \ldots$, and suppose that we instruct an individual to enter room n whenever there are n customers in the system. That is, he would be in room 2 whenever there are two customers in the system; and if another were to arrive, then he would leave room 2 and enter room 3. Similarly, if a service would take place he would leave room 2 and enter room 1 (as there would now be only one customer in the system).

Now suppose that in the long-run our individual is seen to have entered room 1 at the rate of ten times an hour. Then at what rate must he have left room 1? Clearly, at this same rate of ten times an hour. For the total number of times that he enters room 1 must be equal to (or one greater than) the total number of times he leaves room 1. This sort of argument thus yields the general principle which will enable us to determine the state probabilities. Namely, for each $n \geq 0$, *the rate at which the process enters state n equals the rate at which it leaves state n.* Let us now determine these rates. Consider first state 0. When in state 0 the process can leave only by an arrival as clearly there cannot be a departure when the system is empty. Since the arrival rate is λ and the proportion of time that the process is in state 0 is P_0, it follows that the rate at which the process leaves state 0 is λP_0. On the other hand, state 0 can only be reached from state 1 via a departure. That is, if there is a single customer in the system and he completes service, then the system becomes empty. Since the service rate is μ and the proportion of time that the system has exactly one customer is P_1, it follows that the rate at which the process enters state 0 is μP_1.

Hence, from our rate-equality principle we get our first equation,

$$\lambda P_0 = \mu P_1$$

Now consider state 1. The process can leave this state either by an arrival (which occurs at rate λ) or a departure (which occurs at rate μ). Hence, when in state 1, the process will leave this state at a rate of

$\lambda + \mu$.* Since the proportion of time the process is in state 1 is P_1, the rate at which the process leaves state 1 is $(\lambda + \mu)P_1$. On the other hand, state 1 can be entered either from state 0 via an arrival from state 2 via a departure. Hence, the rate at which the process enters state 1 is $\lambda P_0 + \mu P_2$. As the reasoning for the other states is similar, we obtain the following set of equations:

State	Rate at which the process leaves = rate at which it enters
0	$\lambda P_0 = \mu P_1$
$n, n \geq 1$	$(\lambda + \mu)P_n = \lambda P_{n-1} + \mu P_{n+1}$ (3.1)

The set of equations (3.1) which balances the rate at which the process enters each state with the rate at which it leaves that state is known as *balance equations*.

In order to solve Equations (3.1), we rewrite them to obtain

$$P_1 = \frac{\lambda}{\mu} P_0$$

$$P_{n+1} = \frac{\lambda}{\mu} P_n + \left(P_n - \frac{\lambda}{\mu} P_{n-1} \right), \qquad n \geq 1$$

Solving in terms of P_0 yields

$$P_0 = P_0$$

$$P_1 = \frac{\lambda}{\mu} P_0$$

$$P_2 = \frac{\lambda}{\mu} P_1 + \left(P_1 - \frac{\lambda}{\mu} P_0 \right) = \frac{\lambda}{\mu} P_1 = \left(\frac{\lambda}{\mu} \right)^2 P_0$$

$$P_3 = \frac{\lambda}{\mu} P_2 + \left(P_2 - \frac{\lambda}{\mu} P_1 \right) = \frac{\lambda}{\mu} P_2 = \left(\frac{\lambda}{\mu} \right)^3 P_0$$

$$P_4 = \frac{\lambda}{\mu} P_3 + \left(P_3 - \frac{\lambda}{\mu} P_2 \right) = \frac{\lambda}{\mu} P_3 = \left(\frac{\lambda}{\mu} \right)^4 P_0$$

$$P_{n+1} = \frac{\lambda}{\mu} P_n + \left(P_n - \frac{\lambda}{\mu} P_{n-1} \right) = \frac{\lambda}{\mu} P_n = \left(\frac{\lambda}{\mu} \right)^{n+1} P_0$$

---

*If one event occurs at rate λ and another occurs at rate μ, then the total rate at which either event occurs is $\lambda + \mu$. For suppose one man earns \$2.00 per hour and another earns \$3.00 per hour, then together they clearly earn \$5.00 per hour.

In order to determine P_0 we use the fact that the P_n must sum to 1, and thus

$$1 = \sum_{n=0}^{\infty} P_n = \sum_{n=0}^{\infty} \left(\frac{\lambda}{\mu}\right)^n P_0 = \frac{P_0}{1 - \dfrac{\lambda}{\mu}}$$

or

$$P_0 = 1 - \frac{\lambda}{\mu}$$

$$P_n = \left(\frac{\lambda}{\mu}\right)^n \left(1 - \frac{\lambda}{\mu}\right), \qquad n \geq 1 \tag{3.2}$$

Notice that for the above equations to make sense, it is necessary for λ/μ to be less than 1. For otherwise $\sum_{n=0}^{\infty} (\lambda/\mu)^n$ would be infinite and all the P_n would be 0. Hence, we shall assume that $\lambda/\mu < 1$. Note that it is quite intuitive that there would be no limiting probabilities if $\lambda > \mu$. For suppose that $\lambda > \mu$. Since customers arrive at a Poisson rate λ, it follows that the expected total number of arrivals by time t is λt. On the other hand, what is the expected number of customers served by time t? If there were always customers present, then the number of customers served would be a Poisson process having rate μ since the time between successive services would be independent exponentials having mean $1/\mu$. Hence, the expected number of customers served by time t is no greater than μt; and, therefore, the expected number in the system at time t is at least

$$\lambda t - \mu t = (\lambda - \mu)t$$

Now if $\lambda > \mu$, then the above number goes to infinity as t becomes large. That is, $\lambda/\mu > 1$, the queue size increases without limit and there will be no limiting probabilities. It also should be noted that the condition $\lambda/\mu < 1$ is equivalent to the condition that the mean service time be less than the mean time between successive arrivals. This is the general condition that must be satisfied for limiting probabilities to exist in most single-server queueing systems.

Now let us attempt to express the quantities L, L_Q, W, and W_Q in terms of the limiting probabilities P_n. Since P_n is the long-run probability that the system contains exactly n customers, the average number of customers in the system clearly is given by

$$L = \sum_{n=0}^{\infty} nP_n$$

$$= \sum_{n=0}^{\infty} n \left(\frac{\lambda}{\mu}\right)^n \left(1 - \frac{\lambda}{\mu}\right)$$

$$= \frac{\lambda}{\mu - \lambda} \qquad (3.3)$$

where the last equation followed upon application of the algebraic identity

$$\sum_{n=0}^{\infty} nx^n = \frac{x}{(1 - x)^2}$$

The quantities W, W_Q, and L_Q now can be obtained with the help of Equations (2.2) and (2.3). That is, since $\lambda_a = \lambda$, we have from Equation (3.3) that

$$W = \frac{L}{\lambda}$$

$$= \frac{1}{\mu - \lambda}$$

$$W_Q = W - E[S]$$

$$= W - \frac{1}{\mu}$$

$$= \frac{\lambda}{\mu(\mu - \lambda)}$$

$$L_Q = \lambda W_Q$$

$$= \frac{\lambda^2}{\mu(\mu - \lambda)} \qquad (3.4)$$

Example 3a Suppose that customers arrive at a Poisson rate of one per every 12 minutes, and that the service time is exponential at a rate of one service per 8 minutes. What are L and W?

Solution: Since $\lambda = \frac{1}{12}$, $\mu = \frac{1}{8}$, we have

$$L = 2, \qquad W = 24$$

Hence, the average number of customers in the system is two, and the average time a customer spends in the system is 24 minutes.

Now suppose that the arrival rate increases 20 percent to $\lambda = \frac{1}{10}$. What is the corresponding change in L and W? Again using Equations (3.3), we get

$$L = 4, \qquad W = 40$$

Hence, an increase of 20 percent in the arrival rate *doubled* the average number of customers in the system.

To understand this better, write Equations (3.3) as

$$L = \frac{\dfrac{\lambda}{\mu}}{1 - \dfrac{\lambda}{\mu}}, \qquad W = \frac{\dfrac{1}{\mu}}{1 - \dfrac{\lambda}{\mu}}$$

From these equations we can see that when λ/μ is near 1, a slight increase in λ/μ will lead to a large increase in L and W. ◇

A Technical Remark

We have used the fact that if one event occurs at an exponential rate λ, and another independent event at an exponential rate μ, then together they occur at an exponential rate $\lambda + \mu$. To check this formally, let T_1 be the time at which the first event occurs, and T_2 the time at which the second event occurs. Then

$$P\{T_1 \le t\} = 1 - e^{-\lambda t}$$
$$P\{T_2 \le t\} = 1 - e^{-\mu t}$$

Now if we are interested in the time until either T_1 or T_2 occurs, then we are interested in $T = \min(T_1, T_2)$. Now

$$P\{T \le t\} = 1 - P\{T > t\}$$
$$= 1 - P\{\min(T_1, T_2) > t\}$$

However, $\min(T_1, T_2) > t$ if and only if both T_1 and T_2 are greater than t; hence,

$$P\{T \le t\} = 1 - P\{T_1 > t, T_2 > t\}$$
$$= 1 - P\{T_1 > t\}P\{T_2 > t\}$$
$$= 1 - e^{-\lambda t}e^{-\mu t}$$
$$= 1 - e^{-(\lambda + \mu)t}$$

Thus, T has an exponential distribution with rate $\lambda + \mu$, and we are justified in adding the rates. ◇

Let W^* denote the amount of time an arbitrary customer spends in the system. To obtain the distribution of W^*, we condition on the number in the system when the customer arrives. This yields

$$P\{W^* \le a\} = \sum_{i=0}^{\infty} P\{W^* \le a | n \text{ in the system when he arrives}\}$$

$$\times P\{n \text{ in the system when he arrives}\} \qquad (3.5)$$

Now consider the amount of time that our customer must spend in the system if there are already n customers present when he arrives. If $n = 0$, then his time in the system will just be his service time. When $n \ge 1$, there will be one customer in service and $n - 1$ waiting in line ahead of our arrival. The customer in service might have been in service for some time, but due to the lack of memory of the exponential distribution (see Section 2 of Chapter 5), it follows that our arrival would have to wait an exponential amount of time with rate μ for this customer to complete service. As he also would have to wait an exponential amount of time for each of the other $n - 1$ customers in line, it follows, upon adding his own service time, that the amount of time that a customer must spend in the system if there are already n customers present when he arrives is the sum of $n + 1$ independent and identically distributed exponential random variables with rate μ. But it is known (see Section 2.3 of Chapter 5) that such a random variable has a gamma distribution with parameters $(n + 1, \mu)$. That is,

$$P\{W^* \le a \mid n \text{ in the system when he arrives}\}$$

$$= \int_0^a \mu e^{-\mu t} \frac{(\mu t)^n}{n!} \, dt$$

As

$$P\{n \text{ in the system when he arrives}\} = P_n \qquad \text{(since Poisson arrivals)}$$

$$= \left(\frac{\lambda}{\mu}\right)^n \left(1 - \frac{\lambda}{\mu}\right)$$

we have from Equation (3.5) and the above that

$$P\{W^* \le a\} = \sum_{n=0}^{\infty} \int_0^a \mu e^{-\mu t} \frac{(\mu t)^n}{n!} \, dt \left(\frac{\lambda}{\mu}\right)^n \left(1 - \frac{\lambda}{\mu}\right)$$

$$= \int_0^a (\mu - \lambda) e^{-\mu t} \sum_{n=0}^{\infty} \frac{(\lambda t)^n}{n!} \, dt \qquad \text{(by interchanging)}$$

$$= \int_0^a (\mu - \lambda)e^{-\mu t}e^{\lambda t}dt$$

$$= \int_0^a (\mu - \lambda)e^{-(\mu - \lambda)t}dt$$

$$= 1 - e^{-(\mu - \lambda)a}$$

In other words, W^*, the amount of time a customer spends in the system, is an exponential random variable with rate $\mu - \lambda$. (As a check, we note that $E[W^*] = 1/(\mu - \lambda)$ which checks with Equation (3.4) since $W = E[W^*]$.)

3.2. A Single-Server Exponential Queueing System Having Finite Capacity

In the previous model, we assumed that there was no limit on the number of customers that could be in the system at the same time. However, in reality there is always a finite system capacity N, in the sense that there can be no more than N customers in the system at any time. By this, we mean that if an arriving customer finds that there are already N customers present, then it does not enter the system.

As before, we let P_n, $0 \le n \le N$, denote the limiting probability that there are n customers in the system. The rate equality principle yields the following set of balance equations:

State	*Rate at which* *the process leaves = rate at which it enters*
0	$\lambda P_0 = \mu P_1$
$1 \le n \le N - 1$	$(\lambda + \mu)P_n = \lambda P_{n-1} + \mu P_{n+1}$
N	$\mu P_N = \lambda P_{N-1}$

The argument for state 0 is exactly as before. Namely, when in state 0, the process will leave only via an arrival (which occurs at rate λ) and hence the rate at which the process leaves state 0 is λP_0. On the other hand, the process can enter state 0 only from state 1 via a departure; hence, the rate at which the process enters state 0 is μP_1. The equation for states n, where $1 \le n < N$, is the same as before. The equation for state N is different because now state N can only be left via a departure since an arriving customer will not enter the system when it is in state N; also, state N can now only be entered from state $N - 1$ (as there is no longer a state $N + 1$) via an arrival.

To solve, we again rewrite the above system of equations

$$P_1 = \left(\frac{\lambda}{\mu}\right) P_0$$

$$P_{n+1} = \frac{\lambda}{\mu} P_n + \left(P_n - \frac{\lambda}{\mu} P_{n-1}\right), \qquad 1 \le n \le N - 1$$

$$P_N = \left(\frac{\lambda}{\mu}\right) P_{N-1}$$

which, solving in terms of P_0, yields

$$P_1 = \frac{\lambda}{\mu} P_0$$

$$P_2 = \frac{\lambda}{\mu} P_1 + \left(P_1 - \frac{\lambda}{\mu} P_0\right) = \frac{\lambda}{\mu} P_1 = \left(\frac{\lambda}{\mu}\right)^2 P_0$$

$$P_3 = \frac{\lambda}{\mu} P_2 + \left(P_2 - \frac{\lambda}{\mu} P_1\right) = \frac{\lambda}{\mu} P_2 = \left(\frac{\lambda}{\mu}\right)^3 P_0$$

$$\vdots$$

$$P_{N-1} = \frac{\lambda}{\mu} P_{N-2} + \left(P_{N-2} - \frac{\lambda}{\mu} P_{N-3}\right) = \left(\frac{\lambda}{\mu}\right)^{N-1} P_0$$

$$P_N = \left(\frac{\lambda}{\mu}\right) P_{N-1} = \left(\frac{\lambda}{\mu}\right)^N P_0 \tag{3.6}$$

By using the fact that $\Sigma_{n=0}^N P_n = 1$, we obtain

$$1 = P_0 \sum_{n=0}^{N} \left(\frac{\lambda}{\mu}\right)^n$$

$$= P_0 \left[\frac{1 - (\lambda/\mu)^{N+1}}{1 - \lambda/\mu}\right]$$

or

$$P_0 = \frac{(1 - \lambda/\mu)}{1 - (\lambda/\mu)^{N+1}}$$

and hence from Equation (3.6) we obtain

$$P_n = \frac{(\lambda/\mu)^n (1 - \lambda/\mu)}{1 - (\lambda/\mu)^{N+1}}, \qquad n = 0, 1, \ldots, N \tag{3.7}$$

Note that in this case, there is no need to impose the condition that

$\lambda/\mu < 1$. The queue size is, by definition, bounded so there is no possibility of its increasing indefinitely.

As before, L may be expressed in terms of P_n to yield

$$L = \sum_{n=0}^{N} nP_n$$

$$= \frac{(1 - \lambda/\mu)}{1 - (\lambda/\mu)^{N+1}} \sum_{n=0}^{N} n\left(\frac{\lambda}{\mu}\right)^n$$

which after some algebra yields

$$L = \frac{\lambda[1 + N(\lambda/\mu)^{N+1} - (N+1)(\lambda/\mu)^N]}{(\mu - \lambda)(1 - (\lambda/\mu)^{N+1})} \tag{3.8}$$

In deriving W, the expected amount of time a customer spends in the system, we must be a little careful about what we mean by a customer. Specifically, are we including those "customers" who arrive to find the system full and thus do not spend any time in the system? Or, do we just want the expected time spent in the system by a customer that actually entered the system? The two questions lead, of course, to different answers. In the first case, we have $\lambda_a = \lambda$; whereas in the second case, since the fraction of arrivals that actually enter the system is $1 - P_N$, it follows that $\lambda_a = \lambda(1 - P_N)$. Once it is clear what we mean by a customer, W can be obtained from

$$W = \frac{L}{\lambda_a}$$

Example 3b Suppose that it costs $c\mu$ dollars per hour to provide service at a rate μ. Suppose also that we incur a gross profit of A dollars for each customer served. If the system has a capacity N, what service rate μ maximizes our total profit?

Solution: To solve this, suppose that we use rate μ. Let us determine the amount of money coming in per hour and subtract from this the amount going out each hour. This will give us our profit per hour, and we can choose μ so as to maximize this.

Now, potential customers arrive at a rate λ. However, a certain proportion of them do not join the system; namely, those who arrive when there are N customers already in the system. Hence, since P_N is the proportion of time that the system is full, it follows that entering customers arrive at a rate of $\lambda(1 - P_N)$. Since each customer pays A, it follows that money comes in at an hourly rate of $\lambda(1 - P_N)A$ and since it goes out at an hourly rate of $c\mu$, it follows that our total profit per hour is given by

$$\text{Profit per hour} = \lambda(1 - P_N)A - c\mu$$

$$= \lambda A\left[1 - \frac{(\lambda/\mu)^N(1 - \lambda/\mu)}{1 - (\lambda/\mu)^{N+1}}\right] - c\mu$$

$$= \frac{\lambda A[1 - (\lambda/\mu)^N]}{1 - (\lambda/\mu)^{N+1}} - c\mu$$

For instance if $N = 2$, $\lambda = 1$, $A = 10$, $c = 1$, then

$$\text{Profit per hour} = \frac{10[1 - (1/\mu)^2]}{1 - (1/\mu)^3} - \mu$$

$$= \frac{10(\mu^3 - \mu)}{\mu^3 - 1} - \mu$$

In order to maximize profit we differentiate to obtain

$$\frac{d}{d\mu}[\text{Profit per hour}] = 10\frac{(2\mu^3 - 3\mu^2 + 1)}{(\mu^3 - 1)^2} - 1$$

The value of μ that maximizes our profit now can be obtained by equating to zero and solving numerically.   ◊

In the previous two models, it has been quite easy to define the state of the system. Namely, it was defined as the number of people in the system. Now we shall consider some examples where a more detailed state space is necessary.

3.3. A Shoeshine Shop

Consider a shoeshine shop consisting of two chairs. Suppose that an entering customer first will go to chair 1. When his work is completed in chair 1, he will go either to chair 2 if that chair is empty or else wait in chair 1 until chair 2 becomes empty. Suppose that a potential customer will enter this shop as long as chair 1 is empty. (Thus, for instance, a potential customer might enter even if there is a customer in chair 2).

If we suppose that potential customers arrive in accordance with a Poisson process at rate λ, and that the service times for the two chairs are independent and have respective exponential rates of μ_1 and μ_2, then

(a) what proportion of potential customers enter the system?

(b) what is the mean number of customers in the system?

(c) what is the average amount of time that an entering customer spends in the system?

To begin we must first decide upon an appropriate state space. It is clear that the state of the system must include more information than merely the number of customers in the system. For instance, it would not be enough to specify that there is one customer in the system as we would also have to know which chair he was in. Further, if we only know that there are two customers in the system, then we would not know if the man in chair 1 is still being served or if he is just waiting for the person in chair 2 to finish. To account for these points, the following state space, consisting of the five states, $(0, 0)$, $(1, 0)$, $(0, 1)$, $(1, 1)$, and $(b, 1)$, will be used. The states have the following interpretation:

State	Interpretation
$(0, 0)$	There are no customers in the system.
$(1, 0)$	There is one customer in the system, and he is in chair 1.
$(0, 1)$	There is one customer in the system, and he is in chair 2.
$(1, 1)$	There are two customers in the system, and both are presently being served.
$(b, 1)$	There are two customers in the system, but the customer in the first chair has completed his work in that chair and is waiting for the second chair to become free.

It should be noted that when the system is in state $(b, 1)$, the person in chair 1, though not being served, is nevertheless "blocking" potential arrivals from entering the system.

As a prelude to writing down the balance equations, it is usually worthwhile to make a transition diagram. This is done by first drawing a circle for each state and then drawing an arrow labeled by the rate at which the process goes from one state to another. The transition diagram for this model is shown in Figure 8.1. The explanation for the diagram is as follows:

The arrow from state $(0, 0)$ to state $(1, 0)$ which is labeled λ means that when the process is in state $(0, 0)$, that is when the system is empty, then it goes to state $(1, 0)$ at a rate λ, that is via an arrival. The arrow from $(0, 1)$ to $(1, 1)$ is similarly explained.

When the process is in state $(1, 0)$, it will go to state $(0, 1)$ when the customer in chair 1 is finished and this occurs at a rate μ_1; hence the arrow from $(1, 0)$ to $(0, 1)$ labeled μ_1. The arrow from $(1, 1)$ to $(b, 1)$ is similarly explained.

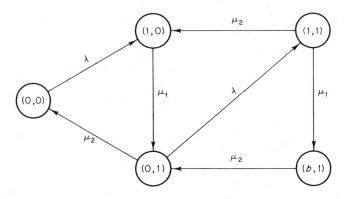

Figure 8.1.   A transition diagram

When in state $(b, 1)$ the process will go to state $(0, 1)$ when the customer in chair 2 completes his service (which occurs at rate μ_2); hence the arrow from $(b, 1)$ to $(0, 1)$ labeled μ_2. Also when in state $(1, 1)$ the process will go to state $(1, 0)$ when the man in chair 2 finishes and hence the arrow from $(1, 1)$ to $(1, 0)$ labeled μ_2. Finally, if the process is in state $(0, 1)$, then it will go to state $(0, 0)$ when the man in chair 2 completes his service, hence the arrow from $(0, 1)$ to $(0, 0)$ labeled μ_2.

As there are no other possible transitions, this completes the transition diagram.

To write the balance equations we equate the sum of the arrows (multiplied by the probability of the states where they originate) coming into a state with the sum of the arrows (multiplied by the probability of the state) going out of that state. This gives

State	*Rate that the process leaves = rate that it enters*
$(0, 0)$	$\lambda P_{00} = \mu_2 P_{01}$
$(1, 0)$	$\mu_1 P_{10} = \lambda P_{00} + \mu_2 P_{11}$
$(0, 1)$	$(\lambda + \mu_2)P_{01} = \mu_1 P_{10} + \mu_2 P_{b1}$
$(1, 1)$	$(\mu_1 + \mu_2)P_{11} = \lambda P_{01}$
$(b, 1)$	$\mu_2 P_{b1} = \mu_1 P_{11}$

These along with the equation

$$P_{00} + P_{10} + P_{01} + P_{11} + P_{b1} = 1$$

may be solved to determine the limiting probabilities. Though it is easy to solve the above equations, the resultant solutions are quite involved and hence will not be explicitly presented. However, it is easy to answer our questions in terms of these limiting probabilities. First, since a potential customer will enter the system when the state is either $(0, 0)$ or $(0, 1)$, it follows that the proportion of customers entering the system is $P_{00} + P_{01}$. Secondly, since there is one customer in the system whenever the state is $(0, 1)$ or $(1, 0)$ and two customers in the system whenever the state is $(1, 1)$ or $(b, 1)$, it follows that L, the average number in the system, is given by

$$L = P_{01} + P_{10} + 2(P_{11} + P_{b1})$$

To derive the average amount of time that an entering customer spends in the system, we use the relationship $W = L/\lambda_a$. Since a potential customer will enter the system when in state $(0, 0)$ or $(0, 1)$, it follows that $\lambda_a = \lambda(P_{00} + P_{01})$ and hence

$$W = \frac{P_{01} + P_{10} + 2(P_{11} + P_{b1})}{\lambda(P_{00} + P_{01})}$$

Example 3c

(a) If $\lambda = 1$, $\mu_1 = 1$, $\mu_2 = 2$, then calculate the above quantities of interest.

(b) If $\lambda = 1$, $\mu_1 = 2$, $\mu_2 = 1$, then calculate the above.

Solution:

(a) Solving the balance equations, yields that

$$P_{00} = \tfrac{12}{37}, \qquad P_{10} = \tfrac{16}{37}, \qquad P_{11} = \tfrac{2}{37}$$
$$P_{01} = \tfrac{6}{37}, \qquad P_{b1} = \tfrac{1}{37}$$

Hence,

$$L = \tfrac{28}{37}, \qquad W = \tfrac{28}{18}$$

(b) Solving the balance equations yields

$$P_{00} = \tfrac{3}{11}, \qquad P_{10} = \tfrac{2}{11}, \qquad P_{11} = \tfrac{1}{11}$$
$$P_{b1} = \tfrac{2}{11}, \qquad P_{01} = \tfrac{3}{11}$$

Hence,

$$L = 1, \qquad W = \tfrac{11}{6} \quad \Diamond$$

3.4. A Queueing System with Bulk Service

In this model, we consider a single-server exponential queueing system in which the server is able to serve two customers at the same time. Whenever the server completes a service, he then serves the next two customers at the same time. However, if there is only one customer in line, then he serves that customer by himself. We shall assume that his service time is exponential at rate μ whether he is serving one or two customers. As usual, we suppose that customers arrive at an exponential rate λ. One example of such a system might be an elevator or a cable car which can take at most two passengers at any time.

It would seem that the state of the system would have to tell us not only how many customers there are in the system, but also whether one or two are presently being served. However, it turns out that we can solve the problem easier not by concentrating on the number of customers in the system, but rather on the number in *queue*. So let us define the state as the number of customers waiting in queue, with two states when there is no one in queue. That is, let us have as a state space $0'$, $0, 1, 2, \ldots, \ldots$ with the interpretation

State	Interpretation
$0'$	No one in service
0	Server busy; no one waiting
$n, \quad n > 0$	n customers waiting

The transition diagram is shown in Figure 8.2 and the balance equations are

State	Rate at which the process leaves = rate at which it enters
$0'$	$\lambda P_{0'} = \mu P_0$
0	$(\lambda + \mu)P_0 = \lambda P_{0'} + \mu P_1 + \mu P_2$
$n, \quad n \geq 1$	$(\lambda + \mu)P_n = \lambda P_{n-1} + \mu P_{n+2}$

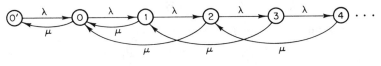

Figure 8.2.

Now the set of equations

$$(\lambda + \mu)P_n = \lambda P_{n-1} + \mu P_{n+2} \qquad n = 1, 2, \ldots \tag{3.9}$$

have a solution of the form

$$P_n = \alpha^n P_0$$

To see this, substitute the above in Equation (3.9) to obtain

$$(\lambda + \mu)\alpha^n P_0 = \lambda\alpha^{n-1}P_0 + \mu\alpha^{n+2}P_0$$

or

$$(\lambda + \mu)\alpha = \lambda + \mu\alpha^3$$

Solving this for α yields the three roots:

$$\alpha = 1, \qquad \alpha = \frac{-1 - \sqrt{1 + 4\lambda/\mu}}{2}, \quad \text{and} \quad \alpha = \frac{-1 + \sqrt{1 + 4\lambda/\mu}}{2}$$

As the first two are clearly not possible, it follows that

$$\alpha = \frac{\sqrt{1 + 4\lambda/\mu} - 1}{2}$$

Hence,

$$P_n = \alpha^n P_0$$

$$P_{0'} = \frac{\mu}{\lambda} P_0$$

where the bottom equation follows from the first balance equation. (We can ignore the second balance equation as one of these equations is always redundant.) To obtain P_0, we use

$$P_0 + P_{0'} + \sum_{n=1}^{\infty} P_n = 1$$

or

$$P_0 \left[1 + \frac{\mu}{\lambda} + \sum_{n=1}^{\infty} \alpha^n \right] = 1$$

or

$$P_0 \left[\frac{1}{1-\alpha} + \frac{\mu}{\lambda} \right] = 1$$

or

$$P_0 = \frac{\lambda(1-\alpha)}{\lambda + \mu(1-\alpha)}$$

and thus

$$P_n = \frac{\alpha^n \lambda(1-\alpha)}{\lambda + \mu(1-\alpha)}, \qquad n \geq 0 \tag{3.10}$$

$$P_{0'} = \frac{\mu(1 - \alpha)}{\lambda + \mu(1 - \alpha)}$$

where

$$\alpha = \frac{\sqrt{1 + 4\lambda/\mu} - 1}{2}$$

It should be noted that for the above to be valid we need $\alpha < 1$ or equivalently $\lambda/\mu < 2$ which is intuitive since the maximum service rate is 2μ which must be larger than the arrival rate λ to avoid overloading the system.

All the relevant quantities of interest now can be determined. For instance, to determine the proportion of customers that are served alone, we first note that the rate at which customers are served alone is $\lambda P_{0'} + \mu P_1$, since when the system is empty a customer will be served alone upon the next arrival and when there is one customer in queue he will be served alone upon a departure. As the rate at which customers are served is λ, it follows that

$$\text{Proportion of customers that are served alone} = \frac{\lambda P_{0'} + \mu P_1}{\lambda}$$

$$= P_{0'} + \frac{\mu}{\lambda} P_1$$

Also,

$$L_Q = \sum_{n=1}^{\infty} n P_n$$

$$= \frac{\lambda(1 - \alpha)}{\lambda + \mu(1 - \alpha)} \sum_{n=1}^{\infty} n\alpha^n \quad \text{from Equation (3.10)}$$

$$= \frac{\lambda\alpha}{(1 - \alpha)[\lambda + \mu(1 - \alpha)]} \quad \text{by algebraic identity} \sum_{1}^{\infty} n\alpha^n = \frac{\alpha}{(1 - \alpha)^2},$$

and

$$W_Q = \frac{L_Q}{\lambda}$$

$$W = W_Q + \frac{1}{\mu}$$

$$L = \lambda W$$

4. Network of Queues

4.1. Open Systems

Consider a two-server system in which customers arrive at a Poisson rate λ at server 1. After being served by server 1 they then join the queue in front of server 2. We suppose there is infinite waiting space at both servers. Each server serves one customer at a time with server i taking an exponential time with rate μ_i for a service, $i = 1, 2$. Such a system is called a *tandem* or *sequential* system (see Figure 8.3).

To analyze this system we need keep track of the number of customers at server 1 and the number at server 2. So let us define the state by the pair (n, m)—meaning that there are n customers at server 1 and m at server 2. The balance equations are

State	*Rate that the process leaves = rate that it enters*
$0, 0$	$\lambda P_{0,0} = \mu_2 P_{0,1}$
$n, 0; n > 0$	$(\lambda + \mu_1)P_{n,0} = \mu_2 P_{n,1} + \lambda P_{n-1,0}$
$0, m; m > 0$	$(\lambda + \mu_2)P_{0,m} = \mu_2 P_{0,m+1} + \mu_1 P_{1,m-1}$
$n, m; nm > 0$	$(\lambda + \mu_1 + \mu_2)P_{n,m} = \mu_2 P_{n,m+1} + \mu_1 P_{n+1,m-1}$

$$+ \lambda P_{n-1,m} \qquad (4.1)$$

Rather than directly attempting to solve the above (along with the equation $\Sigma_{n,m} P_{n,m} = 1$) we shall guess at a solution and then verify that it indeed satisfies the above. We first note that the situation at server 1 is just as in an $M/M/1$ model. Similarly, as it was shown in Section 6 of Chapter 6 that the departure process of an $M/M/1$ queue is a Poisson process with rate λ, it follows that what server 2 faces is also an $M/M/1$ queue. Hence, the probability that there are n customers at server 1 is

$$P\{n \text{ at server } 1\} = \left(\frac{\lambda}{\mu_1}\right)^n \left(1 - \frac{\lambda}{\mu_1}\right)$$

and, similarly,

Figure 8.3. A tandem queue.

$$P\{m \text{ at server } 2\} = \left(\frac{\lambda}{\mu_2}\right)^m \left(1 - \frac{\lambda}{\mu_2}\right)$$

Now if the numbers of customers at servers 1 and 2 were independent random variables, then it would follow that

$$P_{n,m} = \left(\frac{\lambda}{\mu_1}\right)^n \left(1 - \frac{\lambda}{\mu_1}\right)\left(\frac{\lambda}{\mu_2}\right)^m \left(1 - \frac{\lambda}{\mu_2}\right) \tag{4.2}$$

To verify that $P_{n,m}$ is indeed equal to the above (and thus that the number of customers at server 1 is independent of the number at server 2), all we need do is verify that the above satisfies the set of equations (4.1)—this suffices since we know that the $P_{n,m}$ are the unique solution of equations (4.1). Now, for instance, if we consider the first equation of (4.1), we need to show that

$$\lambda\left(1 - \frac{\lambda}{\mu_1}\right)\left(1 - \frac{\lambda}{\mu_2}\right) = \mu_2\left(1 - \frac{\lambda}{\mu_1}\right)\left(\frac{\lambda}{\mu_2}\right)\left(1 - \frac{\lambda}{\mu_2}\right)$$

which is easily verified. We leave it as an exercise to show that the $P_{n,m}$, as given by Equation (4.2), satisfy all of the equations (4.1), and are thus the limiting probabilities.

From the above we see that L, the average number of customers in the system, is given by

$$L = \sum_{n,m} (n + m)P_{n,m}$$

$$= \sum_{n} n\left(\frac{\lambda}{\mu_1}\right)^n \left(1 - \frac{\lambda}{\mu_1}\right) + \sum_{m} m\left(\frac{\lambda}{\mu_2}\right)^m \left(1 - \frac{\lambda}{\mu_2}\right)$$

$$= \frac{\lambda}{\mu_1 - \lambda} + \frac{\lambda}{\mu_2 - \lambda}$$

and from this we see that the average time a customer spends in the system is

$$W = \frac{L}{\lambda} = \frac{1}{\mu_1 - \lambda} + \frac{1}{\mu_2 - \lambda}$$

Remarks

(i) The result (Equation (4.2)) could have been obtained as a direct consequence of the time reversibility of an $M/M/1$ (see Section 6 of Chapter 6). For not only does time reversibility imply that the output from server 1 is a Poisson process, but it also implies (Problem 16 of Chapter 6) that the number of customers at server 1 is independent of the past departure times from server 1. As these past departure times constitute the arrival process to server 2, the independence of the numbers of customers in the two systems follows.

(ii) Since a Poisson arrival sees time averages, it follows that in a tandem queue the numbers of customers an arrival (to server 1) sees at the two servers are independent random variables. However, it should be noted that this does not imply that the waiting times of a given customer at the two servers are independent. For a counter example suppose that λ is very small with respect to $\mu_1 = \mu_2$; and thus almost all customers have zero wait in queue at both servers. However, given that the wait in queue of a customer at server 1 is positive, his wait in queue at server 2 also will be positive with probability at least as large as $\frac{1}{2}$ (why?). Hence, the waiting times in queue are not independent. Remarkably enough, however, it turns out that the total times (that is, service time plus wait in queue) that an arrival spends at the two servers are indeed independent random variables. $\diamond$

The above result can be substantially generalized. To do so, consider a system of k servers. Customers arrive from outside the system to each server $i, i = 1, \ldots k$, in accordance with a Poisson process at rate r_i; they then join the queue at i until their turn at service comes. Once a customer is served by server i, he then joins the queue in front of server $j, j = 1, \ldots, k$, with probability P_{ij}. Hence, $\Sigma_{j=1}^{k} P_{ij} \le 1$, and $1 - \Sigma_{j=1}^{k} P_{ij}$ represents the probability that a customer departs the system after being served by server i.

If we let λ_j denote the total arrival rate of customers to server j, then the λ_j can be obtained as the solution of

$$\lambda_j = r_j + \sum_{i=1}^{k} \lambda_i P_{ij}, \qquad i = 1, \ldots, k \qquad (4.3)$$

Equation (4.3) follows since r_j is the arrival rate of customers to j coming from outside the system and, as λ_i is the rate at which customers depart server i (rate in must equal rate out), $\lambda_i P_{ij}$ is the arrival rate to j of those coming from server i.

It turns out that the number of customers at each of the servers is independent and of the form

$$P\{n \text{ customers at server } j\} = \left(\frac{\lambda_j}{\mu_j}\right)^n \left(1 - \frac{\lambda_j}{\mu_j}\right), \qquad n \ge 1$$

where μ_j is the exponential service rate at server j and the λ_j are the solution of Equation (4.3). Of course, it is necessary that $\lambda_j/\mu_j < 1$ for all j. In order to prove this, we first note that it is equivalent to asserting that the limiting probabilities $P(n_1, n_2, \ldots, n_k) = P\{n_j$ at server j, $j = 1, \ldots, k\}$ are given by

$$P(n_1, n_2, \ldots, n_k) = \prod_{j=1}^{k} \left(\frac{\lambda_j}{\mu_j}\right)^{n_j} \left(1 - \frac{\lambda_j}{\mu_j}\right) \qquad (4.4)$$

which can be verified by showing that it satisfies the balance equations for this model.

The average number of customers in the system is

$$L = \sum_{j=1}^{k} \text{average number at server } j$$

$$= \sum_{j=1}^{k} \frac{\lambda_j}{\mu_j - \lambda_j}$$

The average time a customer spends in the system can be obtained from $L = \lambda W$ with $\lambda = \sum_{j=1}^{k} r_j$ (Why not $\lambda = \sum_{j=1}^{k} \lambda_j$?). This yields

$$W = \frac{\displaystyle\sum_{j=1}^{k} \frac{\lambda_j}{\mu_j - \lambda_j}}{\displaystyle\sum_{j=1}^{k} r_j}$$

Remarks   The result embodied in Equation (4.4) is rather remarkable in that it says that the distribution of the number of customers at server i is the same as in an $M/M/1$ system with rates λ_i and μ_i. What is remarkable is that in the network model the arrival process at node i need *not* be a Poisson process. For if there is a possibility that a customer may visit a server more than once (a situation called *feedback*), then the arrival process will not be Poisson. An easy example illustrating this is to suppose that there is a single server whose service rate is very large with respect to the arrival rate from outside. Suppose also that with probability $p = .9$ a customer upon completion of service is fed back into the system. Hence, at an arrival time epoch there is a large probability of another arrival in a short time (namely, the feedback arrival); whereas at an arbitrary time point there will be only a very slight chance of an arrival occurring shortly (since λ is so very small). Hence, the arrival process does not possess independent increments and so cannot be Poisson. In face even though it is straightforward to verify Equation (4.4) there does not appear to be, at present, any simple explanation as to why it is, in fact, true.

Thus, we see that when feedback is allowed the steady-state probabilities of the number of customers at any given station has the same distribution as in an $M/M/1$ model even though the model is not $M/M/1$. (Presumably such quantities as the joint distribution of the number at the station at two different time points will not be the same as for an $M/M/1$.) ◊

Example 4a Consider a system of two servers where customers from outside the system arrive at server 1 at a Poisson rate 4 and at server 2 at a Poisson rate 5. The service rates of 1 and 2 are respectively 8 and 10. A customer upon completion of service at server 1 is equally likely either to go to server 2 or to leave the system (i.e., $P_{11} = 0$, $P_{12} = \frac{1}{2}$); whereas a departure from server 2 will go 25 percent of the time to server 1 and will depart the system otherwise (i.e., $P_{21} = \frac{1}{4}$, $P_{22} = 0$). Determine the limiting probabilities, L, and W.

Solution: The total arrival rates to servers 1 and 2—call them λ_1 and λ_2—can be obtained from Equation (4.3). That is, we have

$$\lambda_1 = 4 + \tfrac{1}{4}\lambda_2$$
$$\lambda_2 = 5 + \tfrac{1}{2}\lambda_1$$

implying that

$$\lambda_1 = 6, \qquad \lambda_2 = 8$$

Hence,

$$P\{n \text{ at server 1, } m \text{ at server 2}\} = \left(\tfrac{3}{4}\right)^n \tfrac{1}{4} \left(\tfrac{4}{5}\right)^m \tfrac{1}{5}$$
$$= \tfrac{1}{20} \left(\tfrac{3}{4}\right)^n \left(\tfrac{4}{5}\right)^m$$

and

$$L = \frac{6}{8 - 6} + \frac{8}{10 - 8} = 7$$
$$W = \frac{L}{9} = \frac{7}{9} \quad ◊$$

4.2. Closed Systems

The queueing systems described in Section 4.1 are called *open systems* since customers are able to enter and depart the system. A system in which new customers never enter and existing ones never depart is called a *closed system*.

Let us suppose that we have m customers moving among a system of k servers. When a customer completes service at server i, she then joins the queue in front of server j, $j = 1, \ldots, k$, with probability P_{ij}, where we now suppose that $\Sigma_{j=1}^{k} P_{ij} = 1$ for all $i = 1, \ldots, k$. That is, $\mathbf{P} = [P_{ij}]$ is Markov transition probability matrix, which we shall assume is irreducible. Let $\boldsymbol{\pi} = (\pi_1, \ldots, \pi_k)$ denote the stationary probabilities for this Markov chain; that is, $\boldsymbol{\pi}$ is the unique positive solution of

$$\pi_j = \sum_{i=1}^{k} \pi_i P_{ij}$$

$$\sum_{j=1}^{k} \pi_j = 1 \tag{4.5}$$

If we denote the average arrival rate (or equivalently the average service completion rate) at server j by $\lambda_m(j)$, $j = 1, \ldots, k$ then, analogous to Equation (4.3), the $\lambda_m(j)$ satisfy

$$\lambda_m(j) = \sum_{i=1}^{k} \lambda_m(i) P_{ij}$$

Hence, from (4.5) we can conclude that

$$\lambda_m(j) = \lambda_m \pi_j, \qquad j = 1, 2, \ldots, k \tag{4.6}$$

where

$$\lambda_m = \sum_{j=1}^{k} \lambda_m(j) \tag{4.7}$$

From Equation (4.7), we see that λ_m is the average service completion rate of the entire system, that is, it is the system *throughput* rate.*

If we let $P_m(n_1, n_2, \ldots, n_k)$ denote the limiting probabilities

$$P_m(n_1, n_2, \ldots, n_k) = P\{n_j \text{ customers at server } j, j = 1, \ldots, k\}$$

then, by verifying that they satisfy the balance equation, it can be shown that

---

*We are using the notation of $\lambda_m(j)$ and λ_m to indicate the dependence on the number of customers in the closed system. This will be used in recursive relations we will develop.

$$P_m(n_1, n_2, \ldots, n_k) = \begin{cases} K_m \prod_{j=1}^{k} (\lambda_m(j)/\mu_j)^{n_j} & \text{if } \sum_{j=1}^{k} n_j = m \\ 0 & \text{otherwise} \end{cases}$$

But from (4.6) we thus obtain that

$$P_m(n_1, n_2, \ldots, n_k) = \begin{cases} C_m \prod_{j=1}^{k} (\pi_j/\mu_j)^{n_j} & \text{if } \sum_{j=1}^{k} n_j = m \\ 0 & \text{otherwise} \end{cases} \qquad (4.8)$$

where

$$C_m = \left[\sum_{\substack{n_1, \ldots, n_k: \\ \Sigma n_j = m}} \prod_{j=1}^{k} (\pi_j/\mu_j)^{n_j} \right]^{-1} \qquad (4.9)$$

The above formula (4.8) is not as useful as one might suppose for in order to utilize it we must determine the normalizing constant C_m given by (4.9) which requires summing the products $\Pi_{j=1}^{k} (\pi_j/\mu_j)^{n_j}$ over all the feasible vectors $(n_1, \ldots, n_k): \Sigma_{j=1}^{k} n_j = m$. Hence, since there are $\binom{m+k-1}{m}$ vectors this is only computationally feasible for relatively small values of m and k.

We will now present an approach that will enable us to recursively determine many of the quantities of interest in this model without first computing the normalizing constants. To begin, consider a customer who has just left server i and is headed to server j, and let us determine the probability of the system as seen by this customer. In particular, let us determine the probability that this customer observes, at that moment, n_l customers at server l, $l = 1, \ldots, k$, $\Sigma_{l=1}^{k} n_l = m - 1$. This is done as follows:

$P\{$customer observes n_l at server l, l

$= 1, \ldots, k|$customer goes from i to $j\}$

$= \dfrac{P\{\text{state is } (n_1, \ldots, n_i + 1, \ldots, n_j, \ldots, n_k), \text{ customer goes from } i \text{ to } j\}}{P\{\text{customer goes from } i \text{ to } j\}}$

$= \dfrac{P_m(n_1, \ldots, n_i + 1, \ldots, n_j, \ldots, n_k)\mu_i P_{ij}}{\displaystyle\sum_{\mathbf{n}:\Sigma n_j = m-1} P_m(n_1, \ldots, n_i + 1, \ldots, n_k)\mu_i P_{ij}}$

$$= \frac{\dfrac{\pi_i}{\mu_i} \displaystyle\prod_{j=1}^{k} (\pi_j/\mu_j)^{n_j}}{K} \quad \text{from (4.8)}$$

$$= C \prod_{j=1}^{k} (\pi_j/\mu_j)^{n_j}$$

where C does not depend on $n_1, \ldots, n_k$. But as the above is a probability density on the set of vectors $(n_1, \ldots, n_k)$, $\Sigma_{j=1}^{k} n_j = m - 1$ it follows from (4.8) that it must equal $P_{m-1}(n_1, \ldots, n_k)$. Hence

$P\{\text{customer observes } n_l \text{ at server } l, l$

$$= 1, \ldots, k | \text{customer goes from } i \text{ to } j\}$$

$$= P_{m-1}(n_1, \ldots, n_k), \quad \sum_{i=1}^{k} n_i = m - 1 \tag{4.10}$$

As (4.10) is true for all i, we thus have proven the following proposition, known as the Arrival Theorem.

Proposition 4.1.   (The Arrival Theorem):   In the closed network system with m customers, the system as seen by arrivals to server j, is distributed as the stationary distribution in the same network system when there are only $m - 1$ customers.

Denote by $L_m(j)$ and $W_m(j)$ the average number of customers and the average time a customer spends at server j when there are m customers in the network. Upon conditioning on the number of customers found at server j by an arrival to that server, it follows that

$$W_m(j) = \frac{1 + E_m[\text{number at server } j \text{ as seen by an arrival}]}{\mu_j}$$

$$= \frac{1 + L_{m-1}(j)}{\mu_j} \tag{4.11}$$

where the last equality follows from the Arrival Theorem. Now when there are $m - 1$ customers in the system, then from (4.6) $\lambda_{m-1}(j)$, the average arrival rate to server j, satisfies

$$\lambda_{m-1}(j) = \lambda_{m-1}\pi_j$$

Now applying the basic cost identity Equation (2.1) with the cost rule being that each customer in the network system of $m - 1$ customers pays one per unit time while at server j we obtain

$$L_{m-1}(j) = \lambda_{m-1}\pi_j W_{m-1}(j) \tag{4.12}$$

Using (4.11) this yields,

$$W_m(j) = \frac{1 + \lambda_{m-1} \pi_j W_{m-1}(j)}{\mu_j} \qquad (4.13)$$

Also using the fact that $\Sigma_{j=1}^k L_{m-1}(j) = m - 1$ (why?) we obtain, from (4.12), that

$$m - 1 = \lambda_{m-1} \sum_{j=1}^k \pi_j W_{m-1}(j)$$

or

$$\lambda_{m-1} = \frac{m - 1}{\displaystyle\sum_{i=1}^k \pi_i W_{m-1}(i)} \qquad (4.14)$$

Hence, from (4.13), we obtain the recursion

$$W_m(j) = \frac{1}{\mu_j} + \frac{(m - 1)\pi_j W_{m-1}(j)}{\mu_j \displaystyle\sum_{i=1}^k \pi_i W_{m-1}(i)} \qquad (4.15)$$

Starting with the stationary probabilities π_j, $j = 1, \ldots, k$ and $W_1(j) = 1/\mu_j$ we can now use (4.15) to recursively determine $W_2(j)$, $W_3(j), \ldots, W_m(j)$. We can then determine the throughput rate λ_m by using (4.14), and this will determine $L_m(j)$ by (4.12). This recursive approach is called *mean value analysis.*

Example 4b Consider a k server network in which the customers move in a cyclic permutation. That is,

$$P_{i, i+1} = 1, \quad i = 1, 2, \ldots, k - 1, \quad P_{k, 1} = 1$$

Let us determine the average number of customers at server j when there are two customers in the system. Now, for this network

$$\pi_i = 1/k, \quad i = 1, \ldots, k$$

and as

$$W_1(j) = \frac{1}{\mu_j}$$

we obtain from (4.15) that

$$W_2(j) = \frac{1}{\mu_j} + \frac{\dfrac{1}{k}\dfrac{1}{\mu_j}}{\mu_j \sum_{i=1}^{k} \dfrac{1}{k}\dfrac{1}{\mu_i}}$$

$$= \frac{1}{\mu_j} + \frac{1}{\mu_j^2 \sum_{i=1}^{k} 1/\mu_i}$$

Hence, from (4.14)

$$\lambda_2 = \frac{2}{\sum_{l=1}^{k} \dfrac{1}{k} W_2(l)} = \frac{2k}{\sum_{l=1}^{k}\left(\dfrac{1}{\mu_l} + \dfrac{1}{\mu_l^2 \sum_{i=1}^{k} 1/\mu_i}\right)}$$

and finally, using (4.12),

$$L_2(j) = \lambda_2 \frac{1}{k} W_2(j)$$

$$= \frac{2\left(\dfrac{1}{\mu_j} + \dfrac{1}{\mu_j^2 \sum_{i=1}^{k} 1/\mu_i}\right)}{\sum_{l=1}^{k}\left(\dfrac{1}{\mu_l} + \dfrac{1}{\mu_l^2 \sum_{i=1}^{k} 1/\mu_i}\right)} \qquad \Diamond$$

5. The System $M/G/1$

5.1. Preliminaries: Work and Another Cost Identity

For an arbitrary queueing system, let us define the work in the system at any time t to be the sum of the remaining service times of all customers in the system at time t. For instance, suppose there are three customers in the system—the one in service having been there for three of his required five units of service time, and both people in queue having service times of six units. Then the work at that time is $2 + 6 +$

6 = 14. Let V denote the (time) average work in the system.

Now recall the fundamental cost Equation (2.1) which states that the

Average rate at which the system earns

$= \lambda_a \times$ average amount a customer pays,

and consider the following cost rule: *Each customer pays at a rate of y/unit time when his remaining service time is y, whether he is in queue or in service.* Thus, the rate at which the system earns is just the work in the system; so the basic identity yields that

$$V = \lambda_a E[\text{amount paid by a customer}]$$

Now, let S and W_Q^* denote respectively the service time and the amount of time the customer spends waiting in queue. Then, since the customer pays at a constant rate of S per unit time while he waits in queue and at a rate of $S - x$ after spending an amount of time x in service, we have

$$E[\text{amount paid by a customer}] = E\left[SW_Q^* + \int_0^S (S - x)dx\right]$$

and thus

$$V = \lambda_a E[SW_Q^*] + \frac{\lambda_a E[S^2]}{2} \tag{5.1}$$

It should be noted that the above is a basic queueing identity (like Equations (2.2)–(2.4)) and as such valid in almost all models. In addition, if a customer's service time is independent of his wait in queue (as is usually, but not always the case),[*] then we have from Equation (5.1) that

$$V = \lambda_a E[S]W_Q + \frac{\lambda_a E[S^2]}{2} \tag{5.2}$$

5.2. Application of Work to M/G/1

The $M/G/1$ model assumes (i) Poisson arrivals at rate λ; (ii) a general service distribution; and (iii) a single server. In addition, we will suppose that customers are served in the order of their arrival.

---

[*] For an example where it is not true, see Section 6.2 of this chapter.

Now for an arbitrary customer in an $M/G/1$ system,

customer's wait in queue = work in the system when he arrives   (5.3)

This follows since there is only a single server (think about it!). Taking expectations of both sides of Equation (5.3) yields

$$W_Q = \text{average work as seen by an arrival.}$$

But, due to Poisson arrivals, the average work as seen by an arrival will equal V, the time average work in the system. Hence, for the model $M/G/1$,

$$W_Q = V$$

The above in conjunction with the identity

$$V = \lambda E[S]W_Q + \frac{\lambda E[S^2]}{2}$$

yields the so-called Pollaczek–Khintchine formula,

$$W_Q = \frac{\lambda E[S^2]}{2(1 - \lambda E[S])} \tag{5.4}$$

where $E[S]$ and $E[S^2]$ are the first two moments of the service distribution.

The quantities L, L_Q, and W can be obtained from Equation (5.4) as

$$L_Q = \lambda W_Q = \frac{\lambda^2 E[S^2]}{2(1 - \lambda E[S])}$$

$$W = W_Q + E[S] = \frac{\lambda E[S^2]}{2(1 - \lambda E[S])} + E[S]$$

$$L = \lambda W = \frac{\lambda^2 E[S^2]}{2(1 - \lambda E[S])} + \lambda E[S] \tag{5.5}$$

Remarks

(i) For the above quantities to be finite, we need $\lambda E[S] < 1$. This condition is intuitive since we know from renewal theory that if the server was always busy, then the departure rate would be $1/E[S]$ (see Section 3 of Chapter 7), which must be larger than the arrival rate λ to keep things finite.

(ii) Since $E[S^2] = \text{Var}(S) + (E[S])^2$, we see from Equations (5.4) and (5.5) that, for fixed mean service time, L, L_Q, W, and W_Q all increase as the variance of the service distribution increases.

(iii) Another approach to obtain W_Q is presented in Problem 22.

5.3. Busy Periods

The system alternates between idle periods (when there are no customers in the system, and so the server is idle) and busy periods (when there is at least one customer in the system, and so the server is busy).

Let us denote by I_n and B_n, respectively, the lengths of the nth idle and the nth busy period, $n \geq 1$. Hence, in the first $\sum_{j=1}^{n}(I_j + B_j)$ time units the server will be idle for a time $\sum_{j=1}^{n} I_j$, and so the proportion of time that the server will be idle, which of course is just P_0, can be expressed as

$$P_0 = \text{proportion of idle time}$$

$$= \lim_{n \to \infty} \frac{I_1 + \cdots + I_n}{I_1 + \cdots + I_n + B_1 + \cdots + B_n}$$

Now it is easy to see that the $I_1, I_2, \ldots$ are independent and identically distributed as the $B_1, B_2, \ldots$ are. Hence, by dividing the numerator and the denominator of the right side of the above by n, and then applying the strong law of large numbers, we obtain

$$P_0 = \lim_{n \to \infty} \frac{(I_1 + \cdots + I_n)/n}{(I_1 + \cdots + I_n)/n + (B_1 + \cdots + B_n)/n}$$

$$= \frac{E[I]}{E[I] + E[B]} \tag{5.6}$$

where I and B represent idle and busy time random variables.

Now I represents the time from when a customer departs and leaves the system empty until the next arrival. Hence, from Poisson arrivals, it follows that I is exponential with rate λ, and so

$$E[I] = \frac{1}{\lambda} \tag{5.7}$$

To compute P_0, we note from Equation (2.4) (obtained from the fundamental cost equation by supposing that a customer pays at a rate of one per unit time while in service) that

Average number of busy servers = $\lambda E[S]$

However, as the left-hand side of the above equals $1 - P_0$ (why?), we have

$$P_0 = 1 - \lambda E[S] \qquad (5.8)$$

and from Equations (5.6)–(5.8)

$$1 - \lambda E[S] = \frac{1/\lambda}{1/\lambda + E[B]}$$

or

$$E[B] = \frac{E[S]}{1 - \lambda E[S]}$$

Another quantity of interest is C, the number of customers served in a busy period. The mean of C can be computed by noting that, on the average, for every $E[C]$ arrivals exactly one will find the system empty (namely, the first customer in the busy period). Hence,

$$a_0 = \frac{1}{E[C]}$$

and, as $a_0 = P_0 = 1 - \lambda E[S]$ because of Poisson arrivals, we see that

$$E[C] = \frac{1}{1 - \lambda E[S]}$$

6. Variations on the $M/G/1$

6.1. The $M/G/1$ with Random-Sized Batch Arrivals

Suppose that, as in the $M/G/1$, arrivals occur in accordance with a Poisson process having rate λ. But now suppose that each arrival consists not of a single customer but of a random number of customers. As before there is a single server whose service times have distribution G.

Let us denote by $\alpha_j, j \geq 1$, the probability that an arbitrary batch consists of j customers; and let N denote a random variable representing the size of a batch and so $P\{N = j\} = \alpha_j$. Since $\lambda_a = \lambda E(N)$, the basic formula for work (Equation 5.2) becomes

$$V = \lambda E[N] \left[E(S) W_Q + \frac{E(S^2)}{2} \right] \qquad (6.1)$$

To obtain a second equation relating V to W_Q, consider an average customer. We have that

His wait in queue = work in system when he arrives

+ his waiting time due to those in his batch

Taking expectations and using the fact that Poisson arrivals see time averages yields

$$W_Q = V + E[\text{waiting time due to those in his batch}]$$

$$= V + E[W_B] \tag{6.2}$$

Now $E(W_B)$ can be computed by conditioning on the number in the batch, but we must be careful. For the probability that our average customer comes from a batch of size j is *not* α_j. For α_j is the proportion of batches which are of size j, and if we pick a customer at random, it is more likely that he comes from a larger rather than a smaller batch. (For instance, suppose $\alpha_1 = \alpha_{100} = \frac{1}{2}$, then half the batches are of size 1 but 100/101 of the customers will come from a batch of size 100!)

To determine the probability that our average customer came from a batch of size j we reason as follows: Let M be a large number. Then of the first M batches approximately $M\alpha_j$ will be of size j, $j \geq 1$, and thus there would have been approximately $jM\alpha_j$ customers that arrived in a batch of size j. Hence, the proportion of arrivals in the first M batches that were from batches of size j is approximately $jM\alpha_j/\Sigma_j jM\alpha_j$. This proportion becomes exact as $M \to \infty$, and so we see that

$$\text{Proportion of customers from batches of size } j = \frac{j\alpha_j}{\displaystyle\sum_j j\alpha_j}$$

$$= \frac{j\alpha_j}{E[N]}$$

We are now ready to compute $E(W_B)$, the expected wait in queue due to others in the batch.

$$E[W_B] = \sum_j E[W_B|\text{batch of size } j] \frac{j\alpha_j}{E[N]} \tag{6.3}$$

Now if there are j customers in his batch, then our customer would have to wait for $i - 1$ of them to be served if he was ith in line among his batch members. As he is equally likely to be either 1st, 2nd, . . . or jth in line we see that

$$E[W_B|\text{batch is of size } j] = \sum_{i=1}^{j} (i - 1)E(S) \frac{1}{j}$$

$$= \frac{j-1}{2} E[S]$$

Substituting this in Equation (6.3) yields

$$E[W_B] = \frac{E[S]}{2E[N]} \sum_{j} (j - 1)j\alpha_j$$

$$= \frac{E[S](E[N^2] - E[N])}{2E[N]}$$

and from Equations (6.1) and (6.2) we obtain

$$W_Q = \frac{E[S](E[N^2] - E[N])/2E[N] + \lambda E[N]E[S^2]/2}{1 - \lambda E[N]E[S]}$$

Remarks

(i) Note that the condition for W_Q to be finite is that

$$\lambda E(N) < \frac{1}{E[S]}$$

which again says that the arrival rate must be less than the service rate (when the server is busy).

(ii) For fixed value of $E[N]$, W_Q is increasing in $\text{Var}[N]$, again indicating that "single-server queues do not like variation."

(iii) The other quantities L, L_Q, and W can be obtained by using

$$W = W_Q + E[S]$$

$$L = \lambda_a W = \lambda E[N]W$$

$$L_Q = \lambda E[N]W_Q$$

6.2. Priority Queues

Priority queueing systems are ones in which customers are classified into types and then given service priority according to their type. Consider the situation where there are two types of customers, which arrive according to independent Poisson processes with respective rates λ_1 and λ_2, and have service distributions G_1 and G_2. We suppose that type 1 customers are given service priority, in that service will never begin on a type 2 customer if a type 1 is waiting. However, if a type 2 is being served and a type 1 arrives, we assume that the service of the type 2 is continued until completion. That is, there is no preemption once service has begun.

Let W_Q^i denote the average wait in queue of a type i customer, $i = 1, 2$. Our objective is to compute the W_Q^i.

First, note that the total work in the system at any time would be exactly the same no matter what priority rule was employed (as long as the server is always busy whenever there are customers in the system). This is so since the work will always decrease at a rate of one per unit time when the server is busy (no matter who is in service) and will always jump by the service time of an arrival. Hence, the work in the system is exactly as it would be if there was no priority rule but rather a first-come, first-served (called FIFO) ordering. However, under FIFO the above model is just $M/G/1$ with

$$\lambda = \lambda_1 + \lambda_2$$

$$G(x) = \frac{\lambda_1}{\lambda} G_1(x) + \frac{\lambda_2}{\lambda} G_2(x) \tag{6.4}$$

which follows since the combination of two independent Poisson processes is itself a Poisson process whose rate is the sum of the rates of the component processes. The service distribution G can be obtained by conditioning on which priority class the arrival is from—as is done in Equation (6.4).

Hence, from the results of Section 5, it follows that V, the average work in the priority queueing system, is given by

$$V = \frac{\lambda E[S^2]}{2(1 - \lambda E[S])}$$

$$= \frac{\lambda \left(\frac{\lambda_1}{\lambda} E[S_1^2] + \frac{\lambda_2}{\lambda} E[S_2^2] \right)}{2 \left[1 - \lambda \left(\frac{\lambda_1}{\lambda} E[S_1] + \frac{\lambda_2}{\lambda} E[S_2] \right) \right]}$$

$$= \frac{\lambda_1 E[S_1^2] + \lambda_2 E[S_2^2]}{2(1 - \lambda_1 E[S_1] - \lambda_2 E[S_2])} \tag{6.5}$$

where S_i has distribution G_i, $i = 1, 2$.

Continuing in our quest for W_Q^i, let us note that S and W_Q^*, the service and wait in queue of an arbitrary customer, are not independent in the priority model since knowledge about S gives us information as to the type of customer which in turn gives us information about W_Q^*. To get around this we will compute separately the average amount of type 1 and type 2 work in the system. Denoting V^i as the average amount of type i work we have, exactly as in Section 5.1,

$$V^i = \lambda_i E[S_i] W_Q^i + \frac{\lambda_i E[S_i^2]}{2}, \qquad i = 1, 2 \tag{6.6}$$

If we define

$$V_Q^i \equiv \lambda_i E[S_i] W_Q^i$$

$$V_S^i \equiv \frac{\lambda_i E[S_i^2]}{2}$$

then we may interpret V_Q^i as the average amount of type i work in queue, and V_S^i as the average amount of type i work in service (why?).

Now we are ready to compute W_Q^1. To do so, consider an arbitrary type 1 arrival. Then

His delay = amount of type 1 work in the system when he arrives

+ amount of type 2 work in service when he arrives.

Taking expectations and using the fact that Poisson arrivals see time averages yields

$$W_Q^1 = V^1 + V_S^2$$

$$= \lambda_1 E[S_1] W_Q^1 + \frac{\lambda_1 E[S_1^2]}{2} + \frac{\lambda_2 E[S_2^2]}{2} \tag{6.7}$$

or

$$W_Q^1 = \frac{\lambda_1 E[S_1^2] + \lambda_2 E[S_2^2]}{2(1 - \lambda_1 E[S_1])} \tag{6.8}$$

To obtain W_Q^2 we first note that since $V = V^1 + V^2$, we have from Equations (6.5) and (6.6) that

$$\frac{\lambda_1 E[S_1^2] + \lambda_2 E[S_2^2]}{2(1 - \lambda_1 E[S_1] - \lambda_2 E[S_2])} = \lambda_1 E[S_1] W_Q^1 + \lambda_2 E[S_2] W_Q^2$$

$$+ \frac{\lambda_1 E[S_1^2]}{2} + \frac{\lambda_2 E[S_2^2]}{2}$$

$$= W_Q^1 + \lambda_2 E[S_2] W_Q^2 \quad \text{(from equation (6.7))}$$

Now, using Equation (6.8), we obtain

$$\lambda_2 E[S_2] W_Q^2 =$$

$$\frac{\lambda_1 E[S_1^2] + \lambda_2 E[S_2^2]}{2} \left[\frac{1}{1 - \lambda_1 E[S_1] - \lambda_2 E[S_2]} - \frac{1}{1 - \lambda_1 E[S_1]} \right]$$

or

$$W_Q^2 = \frac{\lambda_1 E[S_1^2] + \lambda_2 E[S_2^2]}{2(1 - \lambda_1 E[S_1] - \lambda_2 E[S_2])(1 - \lambda_1 E[S_1])} \tag{6.9}$$

Remarks

(i) Note that from Equation (6.8), the condition for W_Q^1 to be finite is that $\lambda_1 E[S_1] < 1$, which is independent of the type 2 parameters. (Is this intuitive?) For W_Q^2 to be finite, we need, from Equation (6.9), that

$$\lambda_1 E[S_1] + \lambda_2 E[S_2] < 1$$

Since the arrival rate of all customers is $\lambda = \lambda_1 + \lambda_2$, and the average service time of a customer is $(\lambda_1/\lambda)E[S_1] + (\lambda_2/\lambda)E[S_2]$, the above condition is just that the average arrival rate be less than the average service rate.

(ii) If there are n types of customers, we can solve for V^j, $j = 1, \ldots, n$; in a similar fashion. First, note that the total amount of work in the system of customers of types $1, \ldots, j$ is independent of the internal priority rule concerning types $1, \ldots, j$ and only depends on the fact that each of them is given priority over any customers of types $j + 1, \ldots, n$. (Why is this? Reason it out!) Hence, $V^1 + \cdots + V^j$ is the same as it would be if types $1, \ldots, j$ were considered as a single type I priority class and types $j + 1, \ldots, n$ as a single type II priority class. Now, from Equations (6.6) and (6.8),

$$V^I = \frac{\lambda_I E[S_I^2] + \lambda_I \lambda_{II} E[S_I] E[S_{II}^2]}{2(1 - \lambda_I E[S_I])}$$

where

$$\lambda_I = \lambda_1 + \cdots + \lambda_j$$

$$\lambda_{II} = \lambda_{j+1} + \cdots + \lambda_n$$

$$E[S_I] = \sum_{i=1}^{j} \frac{\lambda_i}{\lambda_I} E[S_i]$$

$$E[S_I^2] = \sum_{i=1}^{j} \frac{\lambda_i}{\lambda_I} E[S_i^2]$$

$$E[S_{II}^2] = \sum_{i=j+1}^{n} \frac{\lambda_i}{\lambda_{II}} E[S_i^2]$$

Hence, as $V^I = V^1 + \cdots + V^j$, we have an expression for $V^1 + \cdots + V^j$, for each $j = 1, \ldots, n$, which then can be solved for the individual $V^1, V^2, \ldots, V^n$. We now can obtain W_Q^i from Equation (6.6). The result of all this (which we leave for an exercise) is that

$$W_Q^i = \frac{\lambda_1 E[S_1^2] + \cdots + \lambda_n E[S_n^2]}{2 \prod_{j=1}^{i} (1 - \lambda_1 E[S_1] - \cdots - \lambda_j E[S_j])}, \quad i = 1, \ldots, n \quad (6.10)$$

7. The Model $G/M/1$

The model $G/M/1$ assumes that the times between successive arrivals have an arbitrary distribution G. The service times are exponentially distributed with rate μ and there is a single server.

The immediate difficulty in analyzing this model stems from the fact that the number of customers in the system is not informative enough to serve as a state space. For in summarizing what has occurred up to the present we would need to know not only the number in the system, but also the amount of time that has elapsed since the last arrival (since G is not memoryless). (Why need we not be concerned with the amount of time the person being served has already spent in service?) To get around this problem we shall only look at the system when a customer arrives; and so let us define X_n, $n \geq 1$, by

$X_n \equiv$ the number in the system as seen by the nth arrival

It is easy to see that the process $\{X_n, n \geq 1\}$ is a Markov chain. To compute the transition probabilities P_{ij} for this Markov chain let us first note that, as long as there are customers to be served, the number of services in any length of time t is a Poisson random variable with mean μt. This is true since the time between successive services is exponential and, as we know, this implies that the number of services thus constitutes a Poisson process. Hence,

$$P_{i,\,i+1-j} = \int_0^\infty e^{-\mu t}\,\frac{(\mu t)^j}{j!}\,dG(t), \qquad j = 0, 1, \ldots, i$$

which follows since if an arrival finds i in the system, then the next arrival will find $i + 1$ minus the number served, and the probability that j will be served is easily seen to equal the right side of the above (by conditioning on the time between the successive arrivals).

The formula for P_{i0} is a little different (it is the probability that *at least* $i + 1$ Poisson events occur in a random length of time having distribution G) and can be obtained from

$$P_{i0} = 1 - \sum_{j=0}^{i} P_{i,\,i+1-j}$$

The limiting probabilities π_k, $k = 0, 1, \ldots$, can be obtained as the unique solution of

$$\pi_k = \sum_i \pi_i P_{ik}, \qquad k \geq 0$$

$$\sum_k \pi_k = 1$$

which, in this case, reduce to

$$\pi_k = \sum_{i=k-1}^{\infty} \pi_i \int_0^\infty e^{-\mu t}\,\frac{(\mu t)^{i+1-k}}{(i+1-k)!}\,dG(t), \qquad k \geq 1$$

$$\sum_0^\infty \pi_k = 1 \tag{7.1}$$

(We have not included the equation $\pi_0 = \Sigma \pi_i P_{i0}$ since one of the equations is always redundant.)

To solve the above, let us try a solution of the form $\pi_k = c\beta^k$. Substitution into Equation (7.1) leads to

$$c\beta^k = c \sum_{i=k-1}^{\infty} \beta^i \int_0^\infty e^{-\mu t}\,\frac{(\mu t)^{i+1-k}}{(i+1-k)!}\,dG(t)$$

$$= c \int_0^\infty e^{-\mu t}\beta^{k-1} \sum_{i=k-1}^{\infty} \frac{(\beta\mu t)^{i+1-k}}{(i+1-k)!}\,dG(t) \tag{7.2}$$

However,

$$\sum_{i=k-1}^{\infty} \frac{(\beta\mu t)^{i+1-k}}{(i+1-k)!} = \sum_{j=0}^{\infty} \frac{(\beta\mu t)^j}{j!}$$

$$= e^{\beta\mu t}$$

and thus Equation (7.2) reduces to

$$\beta^k = \beta^{k-1} \int_0^\infty e^{-\mu t(1-\beta)} dG(t)$$

or

$$\beta = \int_0^\infty e^{-\mu t(1-\beta)} dG(t) \qquad (7.3)$$

The constant c can be obtained from $\Sigma_k \pi_k = 1$ which implies that

$$c \sum_0^\infty \beta^k = 1$$

or

$$c = 1 - \beta$$

As the π_k is the *unique* solution to Equation (7.1), and $\pi_k = (1 - \beta)\beta^k$ satisfies, it follows that

$$\pi_k = (1 - \beta)\beta^k, \qquad k = 0, 1, \ldots$$

where β is the solution of Equation (7.3). (It can be shown that if the mean of G is greater than the mean service time $1/\mu$, then there is a unique value of β satisfying Equation (7.3) which is between 0 and 1.) The exact value of β usually can only be obtained by numerical methods.

As π_k is the limiting probability that an arrival sees k customers, it is just the a_k as defined in Section 2. Hence,

$$a_k = (1 - \beta)\beta^k, \qquad k \geq 0 \qquad (7.4)$$

We can obtain W by conditioning on the number in the system when a customer arrives. This yields

$$W = \sum_k E[\text{time in system}|\text{arrival sees } k](1 - \beta)\beta^k$$

$$= \sum_k \frac{k+1}{\mu}(1 - \beta)\beta^k \qquad \begin{array}{l}\text{(Since if an arrival sees } k\text{, then he} \\ \text{spends } k+1 \text{ service periods in} \\ \text{the system.)}\end{array}$$

$$= \frac{1}{\mu(1 - \beta)} \qquad \left(\text{by using } \sum_0^\infty kx^k = \frac{x}{(1-x)^2}\right)$$

and

$$W_Q = W - \frac{1}{\mu} = \frac{\beta}{\mu(1 - \beta)}$$

$$L = \lambda W = \frac{\lambda}{\mu(1 - \beta)}$$

$$L_Q = \lambda W_Q = \frac{\lambda\beta}{\mu(1 - \beta)} \qquad (7.5)$$

where λ is the reciprocal of the mean interarrival time. That is, $1/\lambda = \int_0^\infty x \, dG(x)$.

In fact, in exactly the same manner as shown for the $M/M/1$ in Section 3.1 and Problem 4 we can show that

W^* is exponential with rate $\mu(1 - \beta)$

$$W_Q^* = \begin{cases} 0 \text{ with probability } 1 - \beta \\ \text{exponential with rate } \mu(1 - \beta) \text{ with probability } \beta \end{cases}$$

where W^* and W_Q^* are the amounts of time that a customer spends in system and queue, respectively (their means are W and W_Q).

Whereas $a_k = (1 - \beta)\beta^k$ is the probability that an arrival sees k in the system, it is not equal to the proportion of time during which there are k in the system (since the arrival process is not Poisson). To obtain the P_k we first note that the rate at which the number in the system changes from $k - 1$ to k must equal the rate at which it changes from k to $k - 1$ (why?). Now the rate at which it changes from $k - 1$ to k is equal to the arrival rate λ multiplied by the proportion of arrivals finding $k - 1$ in the system. That is,

Rate number in system goes from $k - 1$ to $k = \lambda a_{k-1}$

Similarly, the rate at which the number in the system changes from k to $k - 1$ is equal to the proportion of time during which there are k in the system multiplied by the (constant) service rate. That is,

Rate number in system goes from k to $k - 1 = P_k \mu$

Equating these rates yields

$$P_k = \frac{\lambda}{\mu} a_{k-1}, \qquad k \geq 1$$

and so from Equation (7.4)

$$P_k = \frac{\lambda}{\mu}(1 - \beta)\beta^{k-1}, \qquad k \geq 1$$

and, as $P_0 = 1 - \Sigma_{k=1}^\infty P_k$, we obtain

$$P_0 = 1 - \frac{\lambda}{\mu}$$

Remarks   In the above analysis we guessed at a solution of the stationary probabilities of the Markov chain of the form $\pi_k = c\beta^k$, then verified such a solution by substituting in the stationary equation (7.1). However, it could have been argued directly that the stationary probabilities of the Markov chain are of this form. To do so, define β_i to be the expected number of times that state $i + 1$ is visited in the Markov chain between two successive visits to state i, $i \geq 0$. Now it is not difficult to see (and we will let the reader argue it out for him or herself) that

$$\beta_0 = \beta_1 = \beta_2 = \cdots = \beta$$

Now it can be shown by using renewal reward processes that

$$\pi_{i+1} = \frac{E[\text{number of visits to state } i + 1 \text{ in an } i - i \text{ cycle}]}{E[\text{number of transitions in an } i - i \text{ cycle}]}$$

$$= \frac{\beta_i}{1/\pi_i}$$

and so,

$$\pi_{i+1} = \beta_i \pi_i = \beta \pi_i, \qquad i \geq 0$$

implying, since $\Sigma_0^\infty \pi_i = 1$, that

$$\pi_i = \beta^i(1 - \beta), \qquad i \geq 0 \quad \diamond$$

7.1. The $G/M/1$ Busy and Idle Periods

Suppose that an arrival has just found the system empty—and so initiates a busy period—and let N denote the number of customers served in that busy period. Since the Nth arrival (after the initiator of the busy period) will also find the system empty, it follows that N is the number of transitions for the Markov chain (of Section 7) to go from state 0 to state 0. Hence, $1/E[N]$ is the proportion of transitions that take the Markov chain into state 0; or equivalently, it is the proportion of arrivals that find the system empty. Therefore,

$$E[N] = \frac{1}{a_0} = \frac{1}{1 - \beta}$$

Also, as the next busy period begins after the Nth interarrival, it follows that the cycle time (that is, the sum of a busy and idle period) is equal to the time until the Nth interarrival. In other words, the sum of a busy and idle period can be expressed as the sum of N interarrival times. Thus, if T_i is the ith interarrival time after the busy period begins, then

$$E[\text{Busy}] + E[\text{Idle}] = E\left[\sum_{i=1}^{N} T_i\right]$$

$$= E[N]\ E[T] \qquad \text{(by Wald's Equation)}$$

$$= \frac{1}{\lambda(1 - \beta)} \tag{7.6}$$

For a second relation between $E[\text{Busy}]$ and $E[\text{Idle}]$, we can use the same argument as in Section 5.3 to conclude that

$$1 - P_0 = \frac{E[\text{Busy}]}{E[\text{Idle}] + E[\text{Busy}]}$$

and since $P_0 = 1 - \lambda/\mu$, we obtain, upon combining this with (7.6), that

$$E[\text{Busy}] = \frac{1}{\mu(1 - \beta)}$$

$$E[\text{Idle}] = \frac{\mu - \lambda}{\lambda\mu(1 - \beta)}$$

8. Multiserver Queues

By and large, systems which have more than one server are much more difficult to analyze than those with a single server. In Section 8.1 we start first with a Poisson arrival system in which no queue is allowed, and then consider in Section 8.2 the infinite capacity $M/M/k$ system. For both of these models we are able to present the limiting probabilities. In Section 8.3 we consider the model $G/M/k$. The analysis here is similar to that of the $G/M/1$ (Section 7) except that in the place of a single quantity β given as the solution of an integral equation, we have k such quantities. We end in Section 8.4 with the model $M/G/k$ for which unfortunately our previous technique (used in $M/G/1$) no longer enables us to derive W_Q, and we content ourselves with an approximation.

8.1. Erlang's Loss System

A loss system is a queueing system in which arrivals that find all servers busy do not enter but rather are lost to the system. The simplest such system is the $M/M/k$ loss system in which customers arrive according to a Poisson process having rate λ, enter the system if at least one of the k servers is free, and then spend an exponential amount of time with rate μ being served. The balance equations for this system are

State	Rate leave = rate enter
0	$\lambda P_0 = \mu P_1$
1	$(\lambda + \mu)P_1 = 2\mu P_2 + \lambda P_0$
2	$(\lambda + 2\mu)P_2 = 3\mu P_3 + \lambda P_1$
$i, 0 < i < k$	$(\lambda + i\mu)P_i = (i + 1)\mu P_{i+1} + \lambda P_{i-1}$
k	$k\mu P_k = \lambda P_{k-1}$

Rewriting gives

$$\lambda P_0 = \mu P_1$$

$$\lambda P_1 = 2\mu P_2$$

$$\lambda P_2 = 3\mu P_3$$

$$\vdots$$

$$\lambda P_{k-1} = k\mu P_k$$

or

$$P_1 = \frac{\lambda}{\mu} P_0$$

$$P_2 = \frac{\lambda}{2\mu} P_1 = \frac{(\lambda/\mu)^2}{2} P_0$$

$$P_3 = \frac{\lambda}{3\mu} P_2 = \frac{(\lambda/\mu)^3}{3!} P_0$$

$$\vdots$$

$$P_k = \frac{\lambda}{k\mu} P_{k-1} = \frac{(\lambda/\mu)^k}{k!} P_0$$

and, using $\Sigma_0^k P_i = 1$, we obtain

$$P_i = \frac{(\lambda/\mu)^i/i!}{\displaystyle\sum_{j=0}^{k}(\lambda/\mu)^j/j!}, \qquad i = 0, 1, \ldots, k$$

Since $E[S] = 1/\mu$, where $E[S]$ is the mean service time, the above can be written as

$$P_i = \frac{(\lambda E[S])^i/i!}{\sum\limits_{j=0}^{k} (\lambda E[S])^j/j!}, \qquad i = 0, 1, \ldots, k \qquad (8.1)$$

Consider now the same system except that the service distribution is general—that is, consider the $M/G/k$ with no queue allowed. This model is sometimes called the *Erlang loss system*. It can be shown (though the proof is advanced) that Equation (8.1) (which is called Erlang's loss formula) remains valid for this more general system.

8.2. The $M/M/k$ Queue

The $M/M/k$ infinite capacity queue can be analyzed by the balance equation technique. We leave it for the reader to verify that

$$P_i = \begin{cases} \dfrac{\dfrac{(\lambda/\mu)^i}{i!}}{\sum\limits_{i=0}^{k-1} \dfrac{(\lambda/\mu)^i}{i!} + \dfrac{(\lambda/\mu)^k}{k!} \dfrac{k\mu}{k\mu - \lambda}}, & i \leq k \\[4em] \dfrac{\left(\dfrac{\lambda}{k\mu}\right)^i k^k}{k!} P_0 & i > k \end{cases}$$

We see from the above that we need to impose the condition $\lambda < k\mu$.

8.3. The $G/M/k$ Queue

In this model we again suppose that there are k servers, each of whom serves at an exponential rate μ. However, we now allow the time between successive arrivals to have an arbitrary distribution G. In order to ensure that a steady-state (or limiting) distribution exists, we assume the condition $1/\mu_G < k\mu$ where μ_G is the mean of G.*

The analysis for this model is similar to that presented in Section 6 for the case $k = 1$. Namely, to avoid having to keep track of the time since the last arrival, we look at the system only at arrival epochs. Once

---

*It follows from renewal theory (Proposition 3.1 of Chapter 7) that customers arrive at rate $1/\mu_G$, and as the maximum service rate is $k\mu$, we clearly need that $1/\mu_G < k\mu$ for limiting probabilities to exist.

again, if we define X_n as the number in the system at the moment of the nth arrival, then $\{X_n, n \geq 0\}$ is a Markov chain.

To derive the transition probabilities of the Markov chain, it helps to first note the relationship

$$X_{n+1} = X_n + 1 - Y_n, \qquad n \geq 0$$

where Y_n denotes the number of departures during the interarrival time between the nth and $(n + 1)$st arrival. The transition probabilities can now be calculated as

Case (i): $j > i + 1$.

In this case it easily follows that $P_{ij} = 0$.

Case (ii): $j \leq i + 1 \leq k$.

In this case if an arrival finds i in the system, then as $i < k$ the new arrival will also immediately enter service. Hence, the next arrival will find j if of the $i + 1$ services exactly $i + 1 - j$ are completed during the interarrival time. Conditioning on the length of this interarrival time yields

$$P_{ij} = P\{i + 1 - j \text{ of } i + 1 \text{ services are completed in an interarrival time}\}$$

$$= \int_0^\infty P\{i + 1 - j \text{ of } i + 1 \text{ are completed} | \text{interarrival time is } t\} \, dG(t)$$

$$= \int_0^\infty \binom{i+1}{j}(1 - e^{-\mu t})^{i+1-j}(e^{-\mu t})^j \, dG(t)$$

where the last equality follows since the number of service completions in a time t will have a binomial distribution.

Case (iii): $i + 1 \geq j \geq k$

To evaluate P_{ij} in this case we first note that when all servers are busy, the departure process is a Poisson process with rate $k\mu$ (why?). Hence, again conditioning on the interarrival time we have

$$P_{ij} = P\{i + 1 - j \text{ departures}\}$$

$$= \int_0^\infty P\{i + 1 - j \text{ departures in time } t\} \, dG(t)$$

$$= \int_0^\infty e^{-k\mu t} \frac{(k\mu t)^{i+1-j}}{(i + 1 - j)!} \, dG(t)$$

Case (iv): $i + 1 \geq k > j$

In this case since when all servers are busy the departure process is a Poisson process, it follows that the length of time until there will only be k in the system will have a gamma distribution with parameters $i + 1 - k$, $k\mu$ (the time until $i + 1 - k$ events of a Poisson process with rate $k\mu$ occur is gamma distributed with parameters $i + 1 - k$, $k\mu$). Conditioning first on the interarrival time and then on the time until there are only k in the system (call this latter random variable T_k) yields

$$P_{ij} = \int_0^\infty P\{i + 1 - j \text{ departures in time } t\}dG(t)$$

$$= \int_0^\infty \int_0^t P\{i + 1 - j \text{ departures in } t \mid T_k = s\} k\mu e^{-k\mu s} \frac{(k\mu s)^{i-k}}{(i - k)!} ds\, dG(t)$$

$$= \int_0^\infty \int_0^t \binom{k}{j}(1 - e^{-\mu(t-s)})^{k-j}(e^{-\mu(t-s)})^j k\mu e^{-k\mu s} \frac{(k\mu s)^{i-k}}{(i - k)!} ds\, dG(t)$$

where the last equality follows since of the k people in service at time s the number whose service will end by time t is binomial with parameters k and $1 - e^{-\mu(t-s)}$.

We now can verify either by a direct substitution into the equations $\pi_j = \Sigma_i \pi_i P_{ij}$, or by the same argument as presented in the remark at the end of Section 6, that the limiting probabilities of this Markov chain are of the form

$$\pi_{k-1+j} = c\beta^j, \qquad j = 0, 1, \ldots .$$

Substitution into any of the equations $\pi_j = \Sigma_i \pi_i P_{ij}$ when $j > k$ yields that β is given as the solution of

$$\beta = \int_0^\infty e^{-k\mu t(1-\beta)}dG(t)$$

The values $\pi_0, \pi_1, \ldots , \pi_{k-2}$, can be obtained by recursively solving the first $k - 1$ of the steady-state equations, and c can then be computed by using $\Sigma_0^\infty \pi_i = 1$.

If we let W_Q^* denote the amount of time that a customer spends in queue, then in exactly the same manner as in $G/M/1$ we can show that

$$W_Q^* = \begin{cases} 0 & \text{with probability } \displaystyle\sum_0^{k-1} \pi_i = 1 - \frac{c\beta}{1 - \beta} \\[2em] \text{Exp}(k\mu(1 - \beta)) & \text{with probability } \displaystyle\sum_k^\infty \pi_i = \frac{c\beta}{1 - \beta} \end{cases}$$

where $\mathrm{Exp}(k\mu(1 - \beta))$ is an exponential random variable with rate $k\mu(1 - \beta)$.

8.4. The $M/G/k$ Queue

In this section we consider the $M/G/k$ system in which customers arrive at a Poisson rate λ and are served by any of k servers, each of whom has the service distribution G. If we attempt to mimic the analysis presented in Section 5 for the $M/G/1$ system, then we would start with the basic identity

$$V = \lambda E[S]W_Q + \lambda E[S^2]/2 \tag{8.2}$$

and then attempt to derive a second equation relating V and W_Q.

Now if we consider an arbitrary arrival, then we have the following identity:

Work in system when customer arrives

$$= k \times \text{time customer spends in queue} + R \tag{8.3}$$

where R is the sum of the remaining service times of all other customers in service at the moment when our arrival enters service.

The foregoing follows since while the arrival is waiting in queue, work is being processed at a rate k per unit time (since all servers are busy). Thus, an amount of work $k \times$ time in queue is processed while he waits in queue. Now all of this work was present when he arrived and in addition the remaining work on those still being served when he enters service was also present when he arrived—so we obtain Equation (8.3). For an illustration, suppose that there are three servers all of whom are busy when the customer arrives. Suppose, in addition, that there are no other customers in the system and also that the remaining service times of the three people in service are 3, 6, and 7. Hence, the work seen by the arrival is $3 + 6 + 7 = 16$. Now the arrival will spend 3 time units in queue, and at the moment he enters service, the remaining times of the other two customers are $6 - 3 = 3$ and $7 - 3 = 4$. Hence, $R = 3 + 4 = 7$ and as a check of Equation (8.3) we see that $16 = 3 \times 3 + 7$.

Taking expectations of Equation (8.2) and using the fact that Poisson arrivals see time averages, we obtain that

$$V = kW_Q + E[R]$$

which, along with Equation (8.2), would enable us to solve for W_Q if we could compute $E[R]$. However there is no known method for computing $E[R]$ and in fact, there is no known exact formula for W_Q. The following approximation for W_Q was obtained in Reference 4 by using the above approach and then approximating $E[R]$.

$$W_Q \approx \frac{\lambda^k E[S^2](E[S])^{k-1}}{2(k-1)!(k-\lambda E[S])^2 \left[\displaystyle\sum_{n=0}^{k-1} \frac{(\lambda E[S])^n}{n!} + \frac{(\lambda E[S])^k}{(k-1)!(k-\lambda E[S])} \right]}$$

(8.4)

The above approximation has been shown to be quite close to the W_Q when the service distribution is gamma. It is also exact when G is exponential.

Problems

1. For the $M/M/1$ queue, compute
   (a) the expected number of arrivals during a service period; and
   (b) the probability that no customers arrive during a service period.
   Hint: "Condition."

2. Machines in a factory break down at an exponential rate of six per hour. There is a single repairman who fixes machines at an exponential rate of eight per hour. The cost incurred in lost production when machines are out of service is \$10 per hour per machine. What is the average cost rate incurred due to failed machines?

3. The manager of a market can hire either Mary or Alice. Mary, who gives service at an exponential rate of 20 customers per hour, can be hired at a rate of \$3 per hour. Alice, who gives service at an exponential rate of 30 customers per hour, can be hired at a rate of \$$C$ per hour. The manager estimates that, on the average, each customer's time is worth \$1 per hour and should be accounted for in the model. If customers arrive at a Poisson rate of 10 per hour, then
   (a) what is the average cost per hour if Mary is hired? if Alice is hired?
   (b) find C if the average cost per hour is the same for Mary and Alice.

4. For the $M/M/1$ queue, show that the probability that a customer spends an amount of time x or less in queue is given by

$$1 - \frac{\lambda}{\mu}, \qquad \text{if } x = 0$$

$$1 - \frac{\lambda}{\mu} + \frac{\lambda}{\mu}(1 - e^{-(\mu-\lambda)x}), \qquad \text{if } x > 0$$

5. Two customers move about among three servers. Upon completion of service at server i, the customer leaves that server and enters service at whichever of the other two servers is free. (Therefore, there are always two busy servers.) If the service times at server i are exponential with rate μ_i, $i = 1,2,3$, what proportion of time is server i idle?

6. Show that W is smaller in an $M/M/1$ model having arrivals at rate λ and service at rate 2μ than it is in a two-server $M/M/2$ model with arrivals at rate λ and with each server at rate μ. Can you give an intuitive explanation for this result? Would it also be true for W_Q?

7. Consider a single-server queue with Poisson arrivals and exponential service times having the following variation: Whenever a service is completed a departure occurs only with probability α. With probability $1 - \alpha$ the customer, instead of leaving, joins the end of the queue. Note that a customer may be serviced more than once.
   (a) Set up the balance equations and solve for the steady-state probabilities, stating conditions for it to exist.
   (b) Find the expected waiting time of a customer from the time he arrives until he enters service for the first time.
   (c) What is the probability that a customer enters service exactly n times, for $n = 1, 2, \ldots$?
   (d) What is the expected amount of time that a customer spends in service (which does not include the time he spends waiting in line)?
   Hint: Use (c).
   (e) What is the distribution of the total length of time a customer spends being served?
   Hint: Is it memoryless?

8. A supermarket has two exponential checkout counters, each operating at rate μ. Arrivals are Poisson at rate λ. The counters operate in the following way:
   (i) One queue feeds both counters.
   (ii) One counter is operated by a permanent checker and the other by a stock clerk who instantaneously begins checking whenever there are two or more customers in the system. The clerk returns to stocking whenever he completes a service, and there are less than two customers in the system.
   (a) Let P_n = proportion of time there are n in the system. Set up equations for P_n and solve.

(b) At what rate does the number in the system go from 0 to 1? from 2 to 1?

(c) What proportion of time is the stock clerk checking? *Hint:* Be a little careful when there is one in the system.

9. Customers arrive at a single-service facility at a Poisson rate of 40 per hour. When two or fewer customers are present, a single attendant operates the facility, and the service time for each customer is exponentially distributed with a mean value of two minutes. However, when there are three or more customers at the facility, the attendant is joined by an assistant and, working together, they reduce the mean service time to one minute. Assuming a system capacity of four customers,

(a) what proportion of time are both servers free?

(b) each man is to receive a salary proportional to the amount of time he is actually at work servicing customers, the rate being the same for both. If together they earn $100 per day, how should this money be split?

10. Consider a sequential-service system consisting of two servers, A and B. Arriving customers will enter this system only if server A is free. If a customer does enter, then he is immediately served by server A. When his service by A is completed, he then goes to B if B is free, or if B is busy, he leaves the system. Assuming that the (Poisson) arrival rate is two customers an hour, and that A and B serve at respective (exponential) rates of four and two customers an hour,

(a) what proportion of customers enter the system?

(b) what proportion of entering customers receive service from B?

(c) what is the average number of customers in the system?

(d) what is the average amount of time that an entering customer spends in the system?

11. Poisson (λ) arrivals join a queue in front of two parallel servers A and B, having exponential service rates μ_A and μ_B. When the system is empty, arrivals go into server A with probability α and into B with probability $1 - \alpha$. Otherwise, the head of the queue takes the first free server.

Figure 8.4.

(a) Define states 0, 2, . . . and set up the balance equations. Do not solve.

(b) In terms of the probabilities in part (a), what is the average number in the system? Average number of servers idle?
(c) In terms of the probabilities in part (a), what is the probability that an arbitrary arrival will get serviced in A?

12. In a queue with unlimited waiting space, arrivals are Poisson (parameter λ) and service times are exponentially distributed (parameter μ). However, the server waits until K people are present before beginning service on the first customer; thereafter, he services one at a time until all K units, and all subsequent arrivals, are serviced. The server is then "idle" until K new arrivals have occurred.
(a) Define an appropriate state space, draw the transition diagram, and set up the balance equations.
(b) In terms of the limiting probabilities, what is the average time a customer spends in queue?
(c) What conditions on λ and μ are necessary?

13. Consider a single-server exponential system in which ordinary customers arrive at a rate λ and have service rate μ. In addition, there is a special customer who has a service rate μ_1. Whenever this special customer arrives, it goes directly into service (if anyone else is in service, then this person is bumped back into queue). When the special customer is not being serviced, the customer spends an exponential amount of time (with mean $1/\Theta$) out of the system.
(a) What is the average arrival rate of the special customer?
(b) Define an appropriate state space and set up balance equations.
(c) Find the probability that an ordinary customer is bumped n times.

14. Let D denote the time between successive departures in a stationary $M/M/1$ queue with $\lambda < \mu$. Show, by conditioning on whether or not a departure has left the system empty, that D is exponential with rate λ.

Hint: By conditioning on whether or not the departure has left the system empty we see that

$$D = \begin{cases} \text{Exponential } (\mu) & \text{with probability } \lambda/\mu \\ \text{Exponential } (\lambda) * \text{Exponential } (\mu) & \text{with probability } 1 - \lambda/\mu \end{cases}$$

where Exponential $(\lambda) *$ Exponential (μ) represents the sum of two independent exponential random variables having rates μ and λ.

Now use moment-generating functions to show that D has the required distribution.

Note that the above does not prove that the departure process is Poisson. To prove this we need show not only that the interdeparture times are all exponential with rate λ, but also that they are independent.

15. For the tandem queue model verify that $P_{n,m} = (\lambda/\mu_1)^n (1 - \lambda/\mu_1)(\lambda/\mu_2)^m (1 - \lambda/\mu_2)$ satisfies the balance equation (4.1).

16. Verify Equation (4.4) for a system of two servers by showing that it satisfies the balance equations for this model.

17. Consider a network of three stations. Customers arrive at stations 1, 2, 3 in accordance with Poisson processes having respective rates 5, 10, 15. The service times at the three stations are exponential with respective rates 10, 50, 100. A customer completing service at station 1 is equally likely to either (a) go to station 2, (b) go to station 3, or (c) leave the system. A customer departing service at station 2 always goes to station 3. A departure from service at station 3 is equally likely to either go to station 2 or leave the system.

   (i) What is the average number of customers in the system (consisting of all three stations)?

   (ii) What is the average time a customer spends in the system?

18. Consider a closed queueing network consisting of two customers moving among two servers, and suppose that after each service completion the customer is equally likely to go to either server—that is, $P_{1,2} = P_{2,1} = \frac{1}{2}$. Let μ_i denote the exponential service rate at server i, $i = 1, 2$.

   (a) Determine the average number of customers at each server.

   (b) Determine the service completion rate for each server.

19. State and prove the equivalent of the Arrival Theorem for open queueing networks.

20. Compare the $M/G/1$ system for first-come, first-served queue discipline with one of last-come, first-served (for instance, in which units for service are taken from the top of a stack). Would you think that the queue size, waiting time, and busy-period distributions differ? What about their means? What if the queue discipline was always to choose at random among those waiting? Intuitively which discipline would result in the smallest variance in the waiting time distribution?

21. In an $M/G/1$ queue

(a) what proportion of departures leave behind 0 work?
(b) what is the average work in the system as seen by a departure?

22. For the $M/G/1$ queue, let X_n denote the number in the system left behind by the nth departure.

(a) If

$$X_{n+1} = \begin{cases} X_n - 1 + Y_n, & \text{if} \quad X_n \geq 1 \\ Y_n, & \text{if} \quad X_n = 0 \end{cases}$$

what does Y_n represent?

(b) Rewrite the above as

$$X_{n+1} = X_n - 1 + Y_n + \delta_n \tag{9.1}$$

where

$$\delta_n = \begin{array}{l} 0, \quad \text{if} \quad X_n \geq 1 \\ 1, \quad \text{if} \quad X_n = 0 \end{array}$$

Take expectations and let $n \to \infty$ in Equation (9.1) to obtain

$$E[\delta_\infty] = 1 - \lambda E[S]$$

(c) Square both sides of Equation (9.1), take expectations, and then let $n \to \infty$ to obtain

$$E[X_\infty] = \frac{\lambda^2 E[S^2]}{2(1 - \lambda E[S])} + \lambda E[S]$$

(d) Argue that $E[X_\infty]$, the average number as seen by a departure, is equal to L.

23. Consider an $M/G/1$ system in which the first customer in a busy period has service distribution G_1 and all others have distribution G_2. Let C denote the number of customers in a busy period, and let S denote the service time of a customer chosen at random.

Argue that

(a) $a_0 = P_0 = 1 - \lambda E[S]$
(b) $E[S] = a_0 E[S_1] + (1 - a_0) E[S_2]$ where S_i has distribution G_i.
(c) Use (a) and (b) to show that $E[B]$, the expected length of a busy period, is given by

$$E[B] = \frac{E[S_1]}{1 - \lambda E[S_2]}$$

24. Consider a $M/G/1$ system with $\lambda E[S] < 1$.

(a) Suppose that service is about to begin at a moment when there are n customers in the system.

    (i) Argue that the additional time until there are only $n - 1$ customers in the system has the same distribution as a busy period.

    (ii) What is the expected additional time until the system is empty?

(b) Suppose that the work in the system at some moment is A. We are interested in the expected additional time until the system is empty—call it $E[T]$. Let N denote the number of arrivals during the first A units of time.

    (i) Compute $E[T|N]$.

    (ii) Compute $E[T]$.

25. Carloads of customers arrive at a single-server station in accordance to a Poisson process with rate 4 per hour. The service times are exponentially distributed with rate 20 per hour. If each carload contains either 1, 2, or 3 customers with respective probabilities $\frac{1}{4}, \frac{1}{2}, \frac{1}{4}$, compute the average customer delay in queue.

26. In the two-class priority queueing model of Section 6.2, what is W_Q? Show that W_Q is less than it would be under FIFO if $E[S_1] < E[S_2]$ and greater than under FIFO if $E[S_1] > E[S_2]$.

27. In a two-class priority queueing model suppose that a cost of C_i per unit time is incurred for each type i customer that waits in queue, $i = 1, 2$. Show that type 1 customers should be given priority over type 2 (as opposed to the reverse) if

$$\frac{E[S_1]}{C_1} < \frac{E[S_2]}{C_2}$$

28. Consider the priority queueing model of Section 6.2 but now suppose that if a type 2 customer is being served when a type 1 arrives then the type 2 customer is bumped out of service. This is called the preemptive case. Suppose that when a bumped type 2 customer goes back in service his service begins at the point where it left off when he was bumped.

(a) Argue that the work in the system at any time is the same as in the nonpreemptive case.

(b) Derive W_Q^1.

Hint: How do type 2 customers affect type 1's?

(c) Why is it not true that

$$V_Q^2 = \lambda_2 E[S_2] W_Q^2$$

(d) Argue that the work seen by a type 2 arrival is the same as in the nonpreemptive case, and so

$$W_Q^2 = W_Q^2(\text{nonpreemptive}) + E[\text{extra time}]$$

where the extra time is due to the fact that he may be bumped.

(e) Let N denote the number of times a type 2 customer is bumped. Why is

$$E[\text{extra time}|N] = \frac{NE[S_1]}{1 - \lambda_1 E[S_1]}$$

Hint: When a type 2 is bumped, relate the time until he gets back in service to a "busy period."

(f) Let S_2 denote the service time of a type 2. What is $E[N|S_2]$?

(g) Combine the above to obtain

$$W_Q^2 = W_Q^2(\text{nonpreemptive}) + \frac{\lambda_1 E[S_1]E[S_2]}{1 - \lambda_1 E[S_1]}$$

29. Verify Equation (6.10).

30. In the $G/M/1$ model if G is exponential with rate λ show that $\beta = \lambda/\mu$.

31. Verify Erlang's loss formula, Equation (8.1), when $k = 1$.

32. Verify the formula given for the P_i of the $M/M/k$.

33. In the Erlang loss system suppose the Poisson arrival rate is $\lambda = 2$, and suppose there are three servers each of whom has a service distribution that is uniformly distributed over $(0, 2)$. What proportion of potential customers is lost?

34. In the $M/M/k$ system,

(i) what is the probability that a customer will have to wait in queue?

(ii) determine L and W.

35. Verify the formula for the distribution of W_Q^* given for the $G/M/k$ model.

36. Consider a system where the interarrival times have an arbitrary distribution F, and there is a single server whose service distribution is G. Let D_n denote the amount of time the nth customer spends waiting in queue. Interpret S_n, T_n so that

$$D_{n+1} = \begin{cases} D_n + S_n - T_n, & \text{if } D_n + S_n - T_n \geq 0 \\ 0, & \text{if } D_n + S_n - T_n < 0 \end{cases}$$

37. Consider a model in which the interarrival times have an arbitrary distribution F, and there are k servers each having service distribution G. What condition on F and G do you think would be necessary for there to exist limiting probabilities?

References

1. J. Cohen, "The Single Server Queue," North-Holland Publ., Amsterdam, 1969.

2. D. R. Cox and W. L. Smith, "Queues," Wiley, New York, 1961.

3. L. Kleinrock, "Queueing Systems," Vol. 1., Wiley, New York, 1975.

4. S. Nozaki and S. Ross, "Approximations in Finite Capacity Multiserver Queues with Poisson Arrivals," *J. Appl. Prob.* 15, 826–834 (1978).

5. N. U. Prabhu, "Queues and Inventories," Wiley, New York, 1965.

6. L. Takacs, "Introduction to the Theory of Queues," Oxford University Press, London and New York, 1962.

Chapter 9

Reliability Theory

1. Introduction

Reliability theory is concerned with determining the probability that a system, possibly consisting of many components, will function. We shall suppose that whether or not the system functions is determined solely from a knowledge of which components are functioning. For instance, a *series* system will function if and only if all of its components are functioning, while a *parallel* system will function if and only if at least one of its components is functioning. In Section 2, we explore the possible ways in which the functioning of the system may depend upon the functioning of its components. In Section 3, we suppose that each component will function with some known probability (independently of each other) and show how to obtain the probability that the system will function. As this probability often is difficult to explicitly compute, we also present useful upper and lower bounds in Section 4. In Section 5 we look at a system dynamically over time by supposing that each component initially functions and does so for a random length of time at which it fails. We then discuss the relationship between the distribution of the amount of time that a system functions and the distributions of the component lifetimes. In particular, it turns out that if the amount of time that a component functions has an *increasing failure rate on the average* (IFRA) distribution, then so does the distribution of system lifetime. In section 6 we consider the problem of obtaining the mean lifetime of a system. In the final section we analyze the system when failed components are subjected to repair.

2. Structure Functions

Consider a system consisting of n components, and suppose that each component is either functioning or has failed. To indicate whether or not the ith component is functioning, we define the indicator variable x_i by

$$x_i = \begin{cases} 1, & \text{if the } i\text{th component is functioning} \\ 0, & \text{if the } i\text{th component has failed} \end{cases}$$

The vector $\mathbf{x} = (x_1, \ldots, x_n)$ is called the state vector. It indicates which of the components are functioning and which have failed.

We further suppose that whether or not the system as a whole is functioning is completely determined by the state vector $\mathbf{x}$. Specifically, it is supposed that there exists a function $\phi(\mathbf{x})$ such that

$$\phi(\mathbf{x}) = \begin{cases} 1, & \text{if the system is functioning when the state vector is } \mathbf{x} \\ 0, & \text{if the system has failed when the state vector is } \mathbf{x} \end{cases}$$

The function $\phi(\mathbf{x})$ is called the *structure function* of the system.

Example 2a (The Series Structure): A series system functions if and only if all of its components are functioning. Hence, its structure function is given by

$$\phi(\mathbf{x}) = \min(x_1, \ldots, x_n) = \prod_{i=1}^{n} x_i$$

We shall find it useful to represent the structure of a system in terms of a diagram. The relevant diagram for the series structure is shown in Figure 9.1.

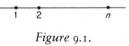

Figure 9.1.

The idea is that if a signal is initiated at the left end of the diagram then in order for it to successfully reach the right end, it must pass through all of the components; hence, they must all be functioning. ◊

Example 2b (The Parallel Structure): A parallel system functions if and only if at least one of its components is functioning. Hence, its structure function is given by

$$\phi(\mathbf{x}) = \max(x_1, \ldots, x_n)$$

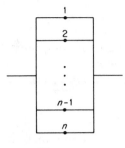

Figure 9.2

A parallel structure may be pictorially illustrated by Figure 9.2. This follows since a signal at the left end can successfully reach the right end as long as at least one component is functioning.   ◇

Example 2c   (The k-Out-of-n Structure): The series and parallel systems are both special cases of a k-out-of-n system. Such a system functions if and only if at least k of the n components are functioning. As $\sum_{i=1}^{n} x_i$ equals the number of functioning components, the structure function of a k-out-of-n system is given by

$$\phi(\mathbf{x}) = \begin{cases} 1, & \text{if } \sum_{i=1}^{n} x_i \geq k \\ 0, & \text{if } \sum_{i=1}^{n} x_i < k \end{cases}$$

series and parallel systems are respectively n-out-of-n and 1-out-of-n system.

The two-out-of-three system may be diagramed as shown in Figure 9.3.

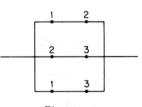

Figure 9.3.

Example 2d   (A Four-Component Structure): Consider a system consisting of four components, and suppose that the system functions

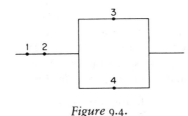

Figure 9.4.

if and only if components 1 and 2 both function and at least one of components 3 and 4 function. Its structure function is given by

$$\phi(\mathbf{x}) = x_1 x_2 \max(x_3, x_4)$$

Pictorially, the system is as shown in Figure 9.4. A useful identity, easily checked, is that for binary variables,* x_i, $i = 1, \ldots, n$,

$$\max(x_1, \ldots, x_n) = 1 - \prod_{i=1}^{n} (1 - x_i)$$

When $n = 2$, this yields

$$\max(x_1, x_2) = 1 - (1 - x_1)(1 - x_2) = x_1 + x_2 - x_1 x_2$$

Hence, the structure function in the above example may be written as

$$\phi(\mathbf{x}) = x_1 x_2 (x_3 + x_4 - x_3 x_4) \quad \Diamond$$

It is natural to assume that replacing a failed component by a functioning one will never lead to a deterioration of the system. In other words, it is natural to assume that the structure function $\phi(\mathbf{x})$ is an increasing function of $\mathbf{x}$, that is, if $x_i \leq y_i$, $i = 1, \ldots, n$, then $\phi(\mathbf{x}) \leq \phi(\mathbf{y})$. Such an assumption shall be made in this chapter and the system will be called *monotone*.

2.1. Minimal Path and Minimal Cut Sets

In this section, we show how any system can be represented both as a series arrangement of parallel structures and as a parallel arrangement of series structures. As a preliminary, we need the following concepts.

---

*A binary variable is one which assumes either the value 0 or 1.

A state vector **x** is called a *path vector* if $\phi(\mathbf{x}) = 1$. If, in addition, $\phi(\mathbf{y}) = 0$ for all $\mathbf{y} < \mathbf{x}$, then **x** is said to be a *minimal path vector*.** If **x** is a minimal path vector, then the set $A = \{i : x_i = 1\}$ is called a *minimal path set*. In other words, a minimal path set is a minimal set of components whose functioning ensures the functioning of the system.

Example 2.1a   Consider a five-component system whose structure is illustrated by Figure 9.5. Its structure function equals

$$\phi(\mathbf{x}) = \max(x_1, x_2)\max(x_3 x_4, x_5)$$

$$= (x_1 + x_2 - x_1 x_2)(x_3 x_4 + x_5 - x_3 x_4 x_5)$$

There are four minimal path sets, namely, $\{1, 3, 4\}$, $\{2, 3, 4\}$, $\{1, 5\}$, $\{2, 5\}$.   ◇

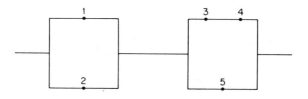

Figure 9.5.

Example 2.1b   In a k-out-of-n system, there are $\binom{n}{k}$ minimal path sets, namely, all of the sets consisting of exactly k components.   ◇

Let $A_1, \ldots, A_s$ denote the minimal path sets of a given system. We define $\alpha_j(\mathbf{x})$, the indicator function of the jth minimal path set, by

$$\alpha_j(\mathbf{x}) = \begin{cases} 1, & \text{if all the components of } A_j \text{ are functioning} \\ 0, & \text{otherwise} \end{cases}$$

$$= \prod_{i \in A_j} x_i$$

By definition, it follows that the system will function if all the components of at least one minimal path set are functioning. That is, if $\alpha_j(\mathbf{x}) = 1$ for some j. On the other hand, if the system functions, then the set of functioning components must include a minimal path set. Therefore, *a system will function if and only if all the components of at least one minimal path set are functioning.* Hence,

---

**We say that $\mathbf{y} < \mathbf{x}$ if $y_i \le x_i$, $i = 1, \ldots, n$, with $y_i < x_i$ for some i.

$$\phi(\mathbf{x}) = \begin{cases} 1, & \text{if } \alpha_j(\mathbf{x}) = 1 \quad \text{for some } j \\ 0, & \text{if } \alpha_j(\mathbf{x}) = 0 \quad \text{for all } j \end{cases}$$

or equivalently

$$\phi(\mathbf{x}) = \max_j \alpha_j(\mathbf{x})$$

$$= \max_j \prod_{i \in A_j} x_i$$

(2.1.1)

Since $\alpha_j(\mathbf{x})$ is a series structure function of the components of the jth minimal path set, Equation (2.1.1) expresses an arbitrary system as a parallel arrangement of series systems.

Example 2.1c Consider the system of Example 2.1a. As its minimal path sets are $A_1 = \{1, 3, 4\}$, $A_2 = \{2, 3, 4\}$, $A_3 = \{1, 5\}$, and $A_4 = \{2, 5\}$, we have by Equation (2.1.1) that

$$\phi(\mathbf{x}) = \max\{x_1 x_3 x_4, x_2 x_3 x_4, x_1 x_5, x_2 x_5\}$$
$$= 1 - (1 - x_1 x_3 x_4)(1 - x_2 x_3 x_4)(1 - x_1 x_5)(1 - x_2 x_5)$$

The reader should verify that this equals the value of $\phi(\mathbf{x})$ given in Example 2.1a. (Make use of the fact that, since x_i equals 0 or 1, $x_i^2 = x_i$.) This representation may be pictured as shown in Figure 9.6. ◊

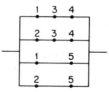

Figure 9.6.

Example 2.1d The system whose structure is as pictured in Figure 9.7 is called the bridge system.

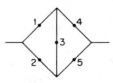

Figure 9.7.

Its minimal path sets are {1, 4}, {1, 3, 5}, {2, 5}, and {2, 3, 4}. Hence, by Equation (2.1.1), its structure function may be expressed as

$$\phi(\mathbf{x}) = \max\{x_1 x_4, x_1 x_3 x_5, x_2 x_5, x_2 x_3 x_4\}$$
$$= 1 - (1 - x_1 x_4)(1 - x_1 x_3 x_5)(1 - x_2 x_5)(1 - x_2 x_3 x_4)$$

This representation of $\phi(\mathbf{x})$ is diagramed as shown in Figure 9.8.   ◇

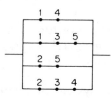

Figure 9.8.

    A state vector $\mathbf{x}$ is called a *cut vector* if $\phi(\mathbf{x}) = 0$. If, in addition, $\phi(\mathbf{y}) = 1$ for all $\mathbf{y} > \mathbf{x}$, then $\mathbf{x}$ is said to be a *minimal cut vector*. If $\mathbf{x}$ is a minimal cut vector, then the set $C = \{i : x_i = 0\}$ is called a *minimal cut set*. In other words, a minimal cut set is a minimal set of components whose failure ensures the failure of the system.

    Let $C_1, \ldots, C_k$ denote the minimal cut sets of a given system. We define $\beta_j(\mathbf{x})$, the indicator function of the jth minimal cut set, by

$$\beta_j(\mathbf{x}) = \begin{cases} 1, & \text{if at least one component of the } j\text{th minimal} \\ & \text{cut set is functioning} \\ 0, & \text{if all of the components of the } j\text{th minimal} \\ & \text{cut set are not functioning} \end{cases}$$
$$= \max_{i \in C_j} x_i$$

Since a system is not functioning if and only if all the components of at least one minimal cut set are not functioning, it follows that

$$\phi(\mathbf{x}) = \prod_{j=1}^{k} \beta_j(\mathbf{x})$$
$$= \prod_{j=1}^{k} \max_{i \in C_j} x_i \qquad (2.1.2)$$

    Since $\beta_j(\mathbf{x})$ is a parallel structure function of the components of the jth minimal cut set, Equation (2.1.2) represents an arbitrary system as a series arrangement of parallel systems.

Example 2.1e The minimal cut sets of the bridge structure shown in Figure 9.9 are {1, 2}, {1, 3, 5}, {2, 3, 4}, and {4, 5}.

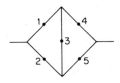

Figure 9.9.

Hence, from Equation (2.1.2), we may express $\phi(\mathbf{x})$ by

$$\phi(\mathbf{x}) = \max(x_1, x_2)\max(x_1, x_3, x_5)\max(x_2, x_3, x_4)\max(x_4, x_5)$$
$$= [1 - (1 - x_1)(1 - x_2)][1 - (1 - x_1)(1 - x_3)(1 - x_5)]$$
$$\times [1 - (1 - x_2)(1 - x_3)(1 - x_4)][1 - (1 - x_4)(1 - x_5)]$$

This representation of $\phi(\mathbf{x})$ is pictorially expressed as Figure 9.10.

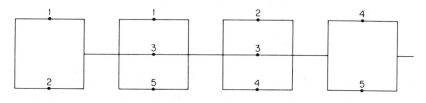

Figure 9.10.

3. Reliability of Systems of Independent Components

In this section, we suppose that X_i, the state of the ith component, is a random variable such that

$$P\{X_i = 1\} = p_i = 1 - P\{X_i = 0\}$$

The value p_i, which equals the probability that the ith component is functioning, is called the *reliability* of the ith component. If we define r by

$$r = P\{\phi(\mathbf{X}) = 1\} \quad \text{where} \quad \mathbf{X} = (X_1, \ldots, X_n)$$

then r is called the reliability of the system. When the components, that is, the random variables X_i, $i = 1, \ldots, n$, are independent, we may express r as a function of the component reliabilities. That is,

$$r = r(\mathbf{p}) \quad \text{where} \quad \mathbf{p} = (p_1, \ldots, p_n)$$

The function $r(\mathbf{p})$ is called the *reliability function*. We shall assume throughout the remainder of this chapter that the components are independent.

Example 3a (The Series System): The reliability function of the series system of n independent components is given by

$$\begin{aligned} r(\mathbf{p}) &= P\{\phi(\mathbf{X}) = 1\} \\ &= P\{X_i = 1 \quad \text{for all} \quad i = 1, \ldots, n\} \\ &= \prod_{i=1}^{n} p_i \quad \diamond \end{aligned}$$

Example 3b (The Parallel System): The reliability function of the parallel system of n independent components is given by

$$\begin{aligned} r(\mathbf{p}) &= P\{\phi(\mathbf{X}) = 1\} \\ &= P\{X_i = 1 \quad \text{for some} \quad i = 1, \ldots, n\} \\ &= 1 - P\{X_i = 0 \quad \text{for all} \quad i = 1, \ldots, n\} \\ &= 1 - \prod_{i=1}^{n} (1 - p_i) \quad \diamond \end{aligned}$$

Example 3c (The k-Out-of-n System with Equal Probabilities): Consider a k-out-of-n system. If $p_i = p$ for all $i = 1, \ldots, n$, then the reliability function is given by

$$\begin{aligned} r(p, \ldots, p) &= P\{\phi(\mathbf{X}) = 1\} \\ &= P\left\{ \sum_{i=1}^{n} X_i \geq k \right\} \\ &= \sum_{i=k}^{n} \binom{n}{i} p^i (1 - p)^{n-i} \quad \diamond \end{aligned}$$

Example 3d (The Two-Out-of-Three System): The reliability function of a two-out-of-three system is given by

$$\begin{aligned} r(\mathbf{p}) &= P\{\phi(\mathbf{X}) = 1\} \\ &= P\{\mathbf{X} = (1, 1, 1)\} + P\{\mathbf{X} = (1, 1, 0)\} + P\{\mathbf{X} = (1, 0, 1)\} \\ &\quad + P\{\mathbf{X} = (0, 1, 1)\} \\ &= p_1 p_2 p_3 + p_1 p_2 (1 - p_3) + p_1 (1 - p_2) p_3 + (1 - p_1) p_2 p_3 \\ &= p_1 p_2 + p_1 p_3 + p_2 p_3 - 2 p_1 p_2 p_3 \quad \diamond \end{aligned}$$

Example 3e (The Three-Out-of-Four System): The reliability function of a three-out-of-four system is given by

$$
\begin{aligned}
r(\mathbf{p}) &= P\{\mathbf{X} = (1, 1, 1, 1)\} + P\{\mathbf{X} = (1, 1, 1, 0)\} + P\{\mathbf{X} = (1, 1, 0, 1)\} \\
&\quad + P\{\mathbf{X} = (1, 0, 1, 1)\} + P\{\mathbf{X} = (0, 1, 1, 1)\} \\
&= p_1 p_2 p_3 p_4 + p_1 p_2 p_3 (1 - p_4) + p_1 p_2 (1 - p_3) p_4 \\
&\quad + p_1 (1 - p_2) p_3 p_4 + (1 - p_1) p_2 p_3 p_4 \\
&= p_1 p_2 p_3 + p_1 p_2 p_4 + p_1 p_3 p_4 + p_2 p_3 p_4 - 3 p_1 p_2 p_3 p_4 \quad \Diamond
\end{aligned}
$$

Example 3f (A Five-Component System): Consider a five-component system that functions if and only if component 1, component 2, and at least one of the remaining components function. Its reliability function is given by

$$
\begin{aligned}
r(\mathbf{p}) &= P\{X_1 = 1, X_2 = 1, \max(X_3, X_4, X_5) = 1\} \\
&= P\{X_1 = 1\} P\{X_2 = 1\} P\{\max(X_3, X_4, X_5) = 1\} \\
&= p_1 p_2 [1 - (1 - p_3)(1 - p_4)(1 - p_5)] \quad \Diamond
\end{aligned}
$$

Since $\phi(\mathbf{X})$ is a $0 - 1$ (that is, a Bernoulli) random variable, we may also compute $r(\mathbf{p})$ by taking its expectation. That is,

$$
\begin{aligned}
r(\mathbf{p}) &= P\{\phi(\mathbf{X}) = 1\} \\
&= E[\phi(\mathbf{X})]
\end{aligned}
$$

Example 3g (A Four-Component System): A four-component system that functions when both components 1, 4, and at least one of the other components function has its structure function given by

$$
\phi(\mathbf{x}) = x_1 x_4 \max(x_2, x_3)
$$

Hence,

$$
\begin{aligned}
r(\mathbf{p}) &= E[\phi(\mathbf{X})] \\
&= E[X_1 X_4 (1 - (1 - X_2)(1 - X_3))] \\
&= p_1 p_4 [1 - (1 - p_2)(1 - p_3)] \quad \Diamond
\end{aligned}
$$

An important and intuitive property of the reliability function $r(\mathbf{p})$ is given by the following proposition.

Proposition 3.1. If $r(\mathbf{p})$ is the reliability function of a system of independent components, then $r(\mathbf{p})$ is an increasing function of $\mathbf{p}$.

Proof: By conditioning on X_i and using the independence of the components, we obtain

$$r(\mathbf{p}) = E[\phi(\mathbf{X})]$$
$$= p_i E[\phi(\mathbf{X})|X_i = 1] + (1 - p_i)E[\phi(\mathbf{X})|X_i = 0]$$
$$= p_i E[\phi(1_i, \mathbf{X})] + (1 - p_i)E[\phi(0_i, \mathbf{X})]$$

where

$$(1_i, \mathbf{X}) = (X_1, \ldots, X_{i-1}, 1, X_{i+1}, \ldots, X_n)$$
$$(0_i, \mathbf{X}) = (X_1, \ldots, X_{i-1}, 0, X_{i+1}, \ldots, X_n)$$

Subtracting and adding $p_i E[\phi(0_i, \mathbf{X})]$ to the above yields

$$r(\mathbf{p}) = p_i E[\phi(1_i, \mathbf{X}) - \phi(0_i, \mathbf{X})] + E[\phi(0_i, \mathbf{X})]$$

However, since ϕ is an increasing function, it follows that

$$E[\phi(1_i, \mathbf{X}) - \phi(0_i, \mathbf{X})] \geq 0$$

and so the above is increasing in p_i for all i. Hence the result is proven. $\diamond$

Let us now consider the following situation: A system consisting of n different components is to be built from a stockpile containing exactly two of each type of component. How should we use the stockpile so as to maximize our probability of attaining a functioning system? In particular, should we build two separate systems, in which case the probability of attaining a functioning one would be

$$P\{\text{at least one of the two systems function}\}$$
$$= 1 - P\{\text{neither of the systems function}\}$$
$$= 1 - [(1 - r(\mathbf{p}))(1 - r(\mathbf{p}'))]$$

where $p_i(p_i')$ is the probability that the first (second) number i component functions; or should we build a single system whose ith component functions if at least one of the number i components function? In this latter case, the probability that the system will function equals

$$r[1 - (1 - \mathbf{p})(1 - \mathbf{p}')]$$

since $1 - (1 - p_i)(1 - p_i')$ equals the probability that the ith component in the single system will function.* We now show that replication at the component level is more effective than replication at the system level.

---

*Notation: If $\mathbf{x} = (x_1, \ldots, x_n)$, $\mathbf{y} = (y_1, \ldots, y_n)$, then $\mathbf{xy} = (x_1 y_1, \ldots, x_n y_n)$.

Theorem 3.1.   For any reliability function r and vectors $\mathbf{p}, \mathbf{p}'$

$$r[1 - (1 - \mathbf{p})(1 - \mathbf{p}')] \geq 1 - [1 - r(\mathbf{p})][1 - r(\mathbf{p}')]$$

Proof: Let $X_1, \ldots, X_n, X_1', \ldots, X_n'$ be mutually independent $0 - 1$ random variables with

$$p_i = P\{X_i = 1\}, \qquad p_i' = P\{X_i' = 1\}$$

Since $P\{\max(X_i, X_i') = 1\} = 1 - (1 - p_i)(1 - p_i')$, it follows that

$$r[1 - (1 - \mathbf{p})(1 - \mathbf{p}')] = E(\phi[\max(\mathbf{X}, \mathbf{X}')])$$

However, by the monotonicity of ϕ, we have that $\phi[\max(\mathbf{X}, \mathbf{X}')]$ is greater than or equal to both $\phi(\mathbf{X})$ and $\phi(\mathbf{X}')$ and hence is at least as large as $\max[\phi(\mathbf{X}), \phi(\mathbf{X}')]$. Hence, from the above, we have that

$$
\begin{aligned}
r[1 - (1 - \mathbf{p})(1 - \mathbf{p}')] &\geq E[\max(\phi(\mathbf{X}), \phi(\mathbf{X}'))] \\
&= P\{\max[\phi(\mathbf{X}), \phi(\mathbf{X}')] = 1\} \\
&= 1 - P\{\phi(\mathbf{X}) = 0, \phi(\mathbf{X}') = 0\} \\
&= 1 - [1 - r(\mathbf{p})][1 - r(\mathbf{p}')]
\end{aligned}
$$

where the first equality follows from the fact that $\max[\phi(\mathbf{X}), \phi(\mathbf{X}')]$ is a $0 - 1$ random variable and hence its expectation equals the probability of its equaling 1. $\Diamond$

As an illustration of the above theorem, suppose that we want to build a series system of two different types of components from a stockpile consisting of two of each of the kinds of components. Suppose that the reliability of each component is $\frac{1}{2}$. If we use the stockpile to build two separate systems, then the probability of attaining a working system is

$$1 - (\tfrac{3}{4})^2 = \tfrac{7}{16}$$

while if we build a single system, replicating components, then the probability of attaining a working system is

$$(\tfrac{3}{4})^2 = \tfrac{9}{16}$$

Hence, replicating components leads to a higher reliability than replicating systems (as, of course, it must by Theorem 3.1).

4. Bounds on the Reliability Function

Consider the bridge system of Example 2.1d, which is represented by Figure 9.11.

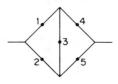

Figure 9.11.

Using the minimal path representation, we have that

$$\phi(\mathbf{x}) = 1 - (1 - x_1 x_4)(1 - x_1 x_3 x_5)(1 - x_2 x_5)(1 - x_2 x_3 x_4)$$

Hence,

$$r(\mathbf{p}) = 1 - E[(1 - X_1 X_4)(1 - X_1 X_3 X_5)(1 - X_2 X_5)(1 - X_2 X_3 X_4)]$$

However, since the minimal path sets overlap (that is, they have components in common), the random variables $(1 - X_1 X_4)$, $(1 - X_1 X_3 X_5)$, $(1 - X_2 X_5)$, and $(1 - X_2 X_3 X_4)$ are not independent, and thus the expected value of their product is not equal to the product of their expected values. Therefore, in order to compute $r(\mathbf{p})$, we must first multiply the four random variables and take the expected value. Doing so, we obtain

$$
\begin{aligned}
r(\mathbf{p}) &= E[X_1 X_4 + X_2 X_5 + X_1 X_3 X_5 + X_2 X_3 X_4 - X_1 X_2 X_3 X_4 \\
&\quad - X_1 X_2 X_3 X_5 - X_1 X_2 X_4 X_5 - X_1 X_3 X_4 X_5 - X_2 X_3 X_4 X_5 \\
&\quad + 2 X_1 X_2 X_3 X_4 X_5] \\
&= p_1 p_4 + p_2 p_5 + p_1 p_3 p_5 + p_2 p_3 p_4 - p_1 p_2 p_3 p_4 - p_1 p_2 p_3 p_5 \\
&\quad - p_1 p_2 p_4 p_5 - p_1 p_3 p_4 p_5 - p_2 p_3 p_4 p_5 + 2 p_1 p_2 p_3 p_4 p_5
\end{aligned}
$$

As can be seen by the above example, it is often quite tedious to evaluate $r(\mathbf{p})$, and thus it would be useful if we had a simple way of obtaining bounds. We now consider two methods for this.

4.1. Method of Inclusion and Exclusion

The following is a well-known formula for the probability of the union of the events $E_1, E_2, \ldots, E_n$.

$$P\left(\bigcup_{i=1}^{n} E_i\right) = \sum_{i=1}^{n} P(E_i) - \sum\sum_{i<j} P(E_i E_j) + \sum\sum\sum_{i<j<k} P(E_i E_j E_k)$$

$$- \cdots + (-1)^{n+1} P(E_1 E_2 \ldots E_n) \tag{4.1}$$

A result, not as well known, is the following set of inequalities:

$$P\left(\bigcup_{1}^{n} E_i\right) \le \sum_{i=1}^{n} P(E_i)$$

$$P\left(\bigcup_{1}^{n} E_i\right) \ge \sum_{i} P(E_i) - \sum_{i<j} P(E_i E_j)$$

$$P\left(\bigcup_{1}^{n} E_i\right) \le \sum_{i} P(E_i) - \sum\sum_{i<j} P(E_i E_j) + \sum\sum\sum_{i<j<k} P(E_i E_j E_k)$$

$$\ge \cdots$$

$$\le \cdots \tag{4.2}$$

where the inequality always changes direction as we add an additional term of the expansion of $P(\cup_{i=1}^{n} E_i)$.

The equality (4.1) is usually proven by induction on the number of events. However, let us now present another approach that will not only prove (4.1) but also establish the inequalities (4.2).

To begin, define the indicator variables $I_j, j = 1, \ldots, n$ by

$$I_j = \begin{cases} 1 & \text{if } E_j \text{ occurs} \\ 0 & \text{otherwise} \end{cases}$$

Letting

$$N = \sum_{j=1}^{n} I_j$$

then N denotes the number of the E_j, $1 \le j \le n$, that occur. Also, let

$$I = \begin{cases} 1 & \text{if } N > 0 \\ 0 & \text{if } N = 0 \end{cases}$$

Then, as

$$1 - I = (1 - 1)^N$$

we obtain, upon application of the binomial theorem, that

$$1 - I = \sum_{i=0}^{N} \binom{N}{i}(-1)^i$$

or

$$I = N - \binom{N}{2} + \binom{N}{3} - \cdots \pm \binom{N}{N} \tag{4.3}$$

We now make use of the following combinatorial identity (which is easily established by induction on i):

$$\binom{n}{i} - \binom{n}{i+1} + \cdots \pm \binom{n}{n} = \binom{n-1}{i-1} \geq 0, \qquad i \leq n$$

The above thus implies that

$$\binom{N}{i} - \binom{N}{i+1} + \cdots \pm \binom{N}{N} \geq 0 \tag{4.4}$$

From (4.3) and (4.4) we obtain that

$$I \leq N \qquad\qquad\qquad \text{by letting } i = 2 \text{ in (4.4)}$$

$$I \geq N - \binom{N}{2} \qquad\qquad \text{by letting } i = 3 \text{ in (4.4)}$$

$$I \leq N - \binom{N}{2} + \binom{N}{3} \tag{4.5}$$

$$\vdots$$

and so on. Now since $N \leq n$ and $\binom{m}{i} = 0$ whenever $i > m$ we can rewrite (4.3) as

$$I = \sum_{i=1}^{n} \binom{N}{i} (-1)^{i+1} \tag{4.6}$$

The equality (4.1) and inequalities (4.2) now follow upon taking expectations of (4.5) and (4.6). This is the case since

$$E[I] = P\{N > 0\}$$
$$= P\{\text{at least one of the } E_j \text{ occurs}\}$$
$$= P\left(\bigcup_{1}^{n} E_j \right)$$
$$E[N] = E\left[\sum_{j=1}^{n} I_j \right]$$
$$= \sum_{j=1}^{n} P(E_j)$$

Also,

$$E\left[\binom{N}{2}\right] = E[\text{number of pairs of the } E_j \text{ that occur}]$$

$$= E\left[\sum_{i<j}\sum I_i I_j\right]$$

$$= \sum_{i<j}\sum P(E_i E_j)$$

and, in general,

$$E\left[\binom{N}{i}\right] = E[\text{number of sets of size } i \text{ that occur}]$$

$$= E\left[\sum_{j_1<j_2<\ \cdots\ <j_i}\sum \cdots \sum I_{j_1} I_{j_2} \cdots I_{j_i}\right]$$

$$= \sum_{j_1<j_2<\ \cdots\ <j_i}\sum \cdots \sum P(E_{j_1} E_{j_2} \cdots E_{j_i})$$

The bounds expressed in Equation (4.2) are commonly called the *inclusion-exclusion bounds*. To apply them in order to obtain bounds on the reliability function, let $A_1, A_2, \ldots, A_s$ denote the minimal path sets of a given structure ϕ, and define the events $E_1, E_2, \ldots, E_s$ by

$$E_i = \{\text{all components in } A_i \text{ function}\}$$

Now, since the system functions if and only if at least one of the events E_i occurs, we have

$$r(\mathbf{p}) = P\left(\bigcup_1^s E_i\right)$$

Applying (4.2) yields the desired bounds on $r(\mathbf{p})$. The terms in the summation are computed thusly:

$$P(E_i) = \prod_{\ell \in A_i} p_\ell$$

$$P(E_i E_j) = \prod_{\ell \in A_i \cup A_j} p_\ell$$

$$P(E_i E_j E_k) = \prod_{\ell \in A_i \cup A_j \cup A_k} p_\ell$$

and so forth for intersections of more than three of the events. (The above follows since, for instance, in order for the event $E_i E_j$ to occur, all of the components in A_i and all of them in A_j must function; or, in other words, all components in $A_i \cup A_j$ must function.)

When the p_i's are small the probabilities of the intersection of many of the events E_i should be quite small and the convergence should be relatively rapid.

Example 4a Consider the bridge structure with identical component probabilities. That is, take p_i to equal p for all i. Letting $A_1 = \{1, 4\}$, $A_2 = \{1, 3, 5\}$, $A_3 = \{2, 5\}$, and $A_4 = \{2, 3, 4\}$ denote the minimal path sets, we have that

$$P(E_1) = P(E_3) = p^2$$
$$P(E_2) = P(E_4) = p^3$$

Also, as exactly five of the six $= \binom{4}{2}$ unions of A_i and A_j contain four components (the exception being $A_2 \cup A_4$ which contains all five components) we have

$$P(E_1 E_2) = P(E_1 E_3) = P(E_1 E_4) = P(E_2 E_3) = P(E_3 E_4) = p^4$$
$$P(E_2 E_4) = p^5$$

Hence, the first two inclusion-exclusion bounds yield

$$2(p^2 + p^3) - 5p^4 - p^5 \le r(p) \le 2(p^2 + p^3)$$

where $r(p) = r(p, p, p, p, p)$. For instance when $p = .4$, we have

$$.3098 \le r(.4) \le .448$$
$$\text{and when } p = .3$$
$$.191 \le r(.3) \le .234 \quad \diamond$$

Just as we can define events in terms of the minimal path sets whose union is the event that the system functions, so can we define events in terms of the minimal cut sets whose union is the event that the system fails. Let $C_1, C_2, \ldots, C_r$ denote the minimal cut sets and define the events $F_1, \ldots, F_r$ by

$$F_i = \{\text{all components in } C_i \text{ are failed}\}$$

Now, as the system is failed if and only if all of the components of at least one minimal cut set are failed we have that

$$1 - r(\mathbf{p}) = P\left(\bigcup_1^r F_i\right)$$

$$1 - r(\mathbf{p}) \le \sum_i P(F_i)$$

$$1 - r(\mathbf{p}) \ge \sum_i P(F_i) - \sum_{i<j}\sum P(F_iF_j)$$

$$1 - r(\mathbf{p}) \le \sum_i P(F_i) - \sum_{i<j}\sum P(F_iF_j) + \sum_{i<j<k}\sum\sum P(F_iF_jF_k)$$

and so on. As

$$P(F_i) = \prod_{l\in C_i} (1 - p_l)$$

$$P(F_iF_j) = \prod_{l\in C_i\cup C_j} (1 - p_l)$$

$$P(F_iF_jF_k) = \prod_{l\in C_i\cup C_j\cup C_k} (1 - p_l)$$

The convergence should be relatively rapid when the p_i's are large.

Example 4b (A Random Graph): Let us recall from Section 6.2 of Chapter 3 that a graph consists of a set N of nodes and a set A of pairs of nodes, called arcs. For any two nodes i and j we say that the sequence of arcs $(i, i_1)(i_1, i_2), \ldots , (i_k, j)$ constitutes an $i - j$ path. If there is an $i - j$ path between all the $\binom{n}{2}$ pairs of nodes i and j, $i \ne j$, then the graph is said to be connected. If we think of the nodes of a graph as representing geographical locations and the arcs as representing direct communication links between the nodes, then the graph will be connected if any two nodes can communicate with each other—if not directly, then at least through the use of intermediary nodes.

A graph can always be subdivided into nonoverlapping connected subgraphs called components. For instance, the graph in Figure 9.12 with nodes $N = \{1, 2, 3, 4, 5, 6\}$ and arcs $A\{(1, 2)$,

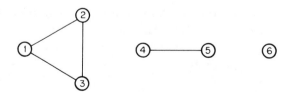

Figure 9.12.

(1, 3), (2, 3), (4, 5)} consists of three components (A graph consisting of a single node is considered to be connected).

Consider now the random graph having nodes $1, 2, \ldots, n$ which is such that there is an arc from node i to node j with probability P_{ij}. Assume in addition that the occurrence of these arcs constitute independent events. That is, assume that the $\binom{n}{2}$ random variables X_{ij}, $i \neq j$, are independent where

$$X_{ij} = \begin{cases} 1, & \text{if } (i, j) \text{ is an arc} \\ 0, & \text{otherwise} \end{cases}$$

We are interested in the probability that this graph will be connected.

We can think of the above as being a reliability system of $\binom{n}{2}$ components—each component corresponding to a potential arc. The component is said to work if the corresponding arc is indeed an arc of the network, and the system is said to work if the corresponding graph is connected. As the addition of an arc to a connected graph cannot disconnect the graph, it follows that the structure so defined is monotone.

Let us start by determining the minimal path and minimal cut sets. It is easy to see that a graph will not be connected if and only if the set of arcs can be partitioned into two nonempty subsets X and X^c in such a way that there is no arc connecting a node from X with one from X^c. For instance, if there are six nodes and if there are no arcs connecting any of the nodes 1, 2, 3, 4 with either 5 or 6, then clearly the graph will not be connected. Thus we see that any partition of the nodes into two nonempty subsets X and X^c corresponds to the minimal cut set defined by

$$\{(i, j): \quad i \in X, \quad j \in X^c\}$$

As there are $2^{n-1} - 1$ such partitions (there are $2^n - 2$ ways of choosing a nonempty proper subset X and, as the partition X, X^c is the same as X^c, X, we must divide by 2) there are therefore this number of minimal cut sets.

To determine the minimal path sets, we must characterize a minimal set of arcs which result in a connected graph. Now the graph in Figure 9.13 is connected but it would remain connected if any one of the arcs from the cycle shown in Figure 9.14 were removed. In fact it is not difficult to see that the minimal path sets are exactly those sets of arcs which result in a graph being connected but not having any cycles (a cycle being a path

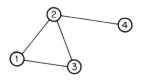

Figure 9.13.   *Figure 9.14.*

from a node to itself). Such sets of arcs are called spanning trees (Figure 9.15).

Figure 9.15. Two spanning trees (minimal path sets) when $n = 4$.

It is easily verified that any spanning tree contains exactly $n - 1$ arcs, and it is a famous result in graph theory (due to Cayley) that there are exactly n^{n-2} of these minimal path sets.

Because of the large number of minimal path and minimal cut sets (n^{n-2} and $2^{n-1} - 1$, respectively), it is difficult to obtain any useful bounds without making further restrictions. So, let us assume that all the P_{ij} equal the common value p. That is, we suppose that each of the ( ) possible arcs exists, independently, with the same probability p. We shall start by deriving a recursive formula for the probability that the graph is connected, which is computationally useful when n is not too large, and then we shall present an asymptotic formula for this probability when n is large.

Let us denote by P_n the probability that the random graph having n nodes is connected. To derive a recursive formula for P_n we first concentrate attention on a single node—say node 1—and try to determine the probability that node 1 will be part of a component of size k in the resultant graph. Now for a given set of $k - 1$ other nodes these nodes along with node 1 will form a component if

(i) there are no arcs connecting any of these k nodes with any of the remaining $n - k$ nodes;

(ii) the random graph, restricted to these k nodes (and $\binom{k}{2}$ potential arcs—each independently appearing with probability p) is connected.

The probability that (i) and (ii) both occur is

$$q^{k(n-k)}P_k$$

where $q = 1 - p$. As there are $\binom{n-1}{k-1}$ ways of choosing $k - 1$ other nodes (to form along with node 1 a component of size k) we see that

$$P\{\text{node 1 is part of a component of size } k\}$$

$$= \binom{n-1}{k-1} q^{k(n-k)}P_k, \qquad k = 1, 2, \ldots, n$$

Now since the sum of the above probabilities as k ranges from 1 thru n clearly must equal 1, and as the graph is connected if and only if node 1 is part of a component of size n, we see that

$$P_n = 1 - \sum_{k=1}^{n-1} \binom{n-1}{k-1} q^{k(n-k)}P_k, \qquad n = 2, 3, \ldots \qquad (4.7)$$

Starting with $P_1 = 1$, $P_2 = p$, Equation (4.7) can be used to recursively determine P_n when n is not too large. It is particularly suited for numerical computation.

To determine an asymptotic formula for P_n when n is large, first note from Equation (4.7) that since $P_k \leq 1$, we have

$$1 - P_n \leq \sum_{k=1}^{n-1} \binom{n-1}{k-1} q^{k(n-k)}$$

As it can be shown that for $q < 1$ and n sufficiently large

$$\sum_{k=1}^{n-1} \binom{n-1}{k-1} q^{k(n-k)} \leq (n + 1)q^{n-1}$$

we have that for n large

$$1 - P_n \leq (n + 1)q^{n-1} \qquad (4.8)$$

To obtain a bound in the other direction, we concentrate our attention on a particular type of minimal cut set—namely, those that separate one node from all others in the graph. Specifically, define the minimal cut set C_i as

$$C_i = \{(i, j): j \neq i\}$$

and define F_i to be the event that all arcs in C_i are not working (and thus, node i is isolated from the other nodes). Now,

$$1 - P_n = P(\text{graph is not connected}) \geq P\left(\bigcup_i F_i\right)$$

since, if any of the events F_i occur, then the graph will be disconnected. By the inclusion-exclusion bounds, we have that

$$P\left(\bigcup_i F_i\right) \geq \sum_i P(F_i) - \sum\sum_{i<j} P(F_i F_j)$$

As $P(F_i)$ and $P(F_i F_j)$ are just the respective probabilities that a given set of $n - 1$ arcs and that a given set of $2n - 3$ arcs are not in the graph (why?), it follows that

$$P(F_i) = q^{n-1},$$
$$P(F_i F_j) = q^{2n-3}, \qquad i \neq j$$

and so,

$$1 - P_n \geq nq^{n-1} - \binom{n}{2} q^{2n-3}$$

Combining this with Equation (4.8) yields that for n sufficiently large

$$nq^{n-1} - \binom{n}{2} q^{2n-3} \leq 1 - P_n \leq (n + 1)q^{n-1}$$

and as $\binom{n}{2}q^{2n-3}/nq^{n-1} \to 0$, as $n \to \infty$, we see that for large n

$$1 - P_n \approx nq^{n-1}$$

Thus, for instance, when $n = 20$ and $p = \frac{1}{2}$, the probability that the random graph will be connected is approximately given by

$$P_{20} \approx 1 - 20(\tfrac{1}{2})^{19} = .99998 \quad \Diamond$$

4.2. Second Method for Obtaining Bounds on $r(\mathbf{p})$

Our second approach to obtaining bounds on $r(\mathbf{p})$ is based on expressing the desired probability as the probability of the intersection of events. To do so, let $A_1, A_2, \ldots, A_s$ denote the minimal path sets as before, and define the events, $D_i, i = 1, \ldots$ by

$$D_i = \{\text{at least one component in } A_i \text{ has failed}\}$$

Now since the system will have failed if and only if at least one component in each of the minimal path sets has failed we have that

$$1 - r(\mathbf{p}) = P(D_1 D_2 \ldots D_s)$$
$$= P(D_1)P(D_2|D_1) \ldots P(D_s|D_1 D_2 \ldots D_{s-1}) \quad (4.9)$$

Now it is quite intuitive that the information that at least one component of A_1 is down can only increase the probability that at least one component of A_2 is down (or else leave the probability unchanged if A_1 and A_2 do not overlap). Hence, intuitively

$$P(D_2|D_1) \geq P(D_2)$$

To prove this inequality, we write

$$P(D_2) = P(D_2|D_1)P(D_1) + P(D_2|D_1^c)(1 - P(D_1)) \qquad (4.10)$$

and note that

$$P(D_2|D_1^c) = P\{\text{at least one failed in } A_2|\text{all functioning in } A_1\}$$

$$= 1 - \prod_{\substack{j \in A_2 \\ j \notin A_1}} p_j$$

$$\leq 1 - \prod_{j \in A_2} p_j$$

$$= P(D_2)$$

Hence, from Equation (4.10) we see that

$$P(D_2) \leq P(D_2|D_1)P(D_1) + P(D_2)(1 - P(D_1))$$

or

$$P(D_2|D_1) \geq P(D_2)$$

By the same reasoning it also follows that

$$P(D_i|D_1 \ldots D_{i-1}) \geq P(D_i)$$

and so from Equation (4.9) we have

$$1 - r(\mathbf{p}) \geq \prod_i P(D_i)$$

or, equivalently,

$$r(\mathbf{p}) \leq 1 - \prod_i \left(1 - \prod_{j \in A_i} p_j\right)$$

To obtain a bound in the other direction, let $C_1, \ldots, C_r$ denote the minimal cut sets and define the events $U_1, \ldots, U_r$ by

$$U_i = \{\text{at least one component in } C_i \text{ is functioning}\}$$

Then, since the system will function if and only if all of the events U_i occur, we have

$$r(\mathbf{p}) = P(U_1 U_2 \ldots U_r)$$
$$= P(U_1)P(U_2|U_1) \ldots P(U_r|U_1 \ldots U_{r-1})$$
$$\geq \prod_i P(U_i)$$

where the last inequality is established in exactly the same manner as for the D_i. Hence,

$$r(\mathbf{p}) \geq \prod_i \left[1 - \prod_{j \in C_i} (1 - p_j) \right]$$

and we thus have the following bounds for the reliability function

$$\prod_i \left[1 - \prod_{j \in C_i} (1 - p_j) \right] \leq r(\mathbf{p}) \leq 1 - \prod_i \left(1 - \prod_{j \in A_i} p_j \right) \quad (4.11)$$

It is to be expected that the upper bound should be close to the actual $r(\mathbf{p})$ if there is not too much overlap in the minimal path sets, and the lower bound to be close if there is not too much overlap in the minimal cut sets.

Example 4c   For the three-out-of-four system the minimal path sets are $A_1 = \{1, 2, 3\}$, $A_2 = \{1, 2, 4\}$, $A_3 = \{1, 3, 4\}$, and $A_4 = \{2, 3, 4\}$; and the minimal cut sets are $C_1 = \{1, 2\}$, $C_2 = \{1, 3\}$, $C_3 = \{1, 4\}$, $C_4 = \{2, 3\}$, $C_5 = \{2, 4\}$, and $C_6 = \{3, 4\}$. Hence, by Equation (4.11) we have

$$(1 - q_1 q_2)(1 - q_1 q_3)(1 - q_1 q_4)(1 - q_2 q_3)(1 - q_2 q_4)(1 - q_3 q_4)$$
$$\leq r(\mathbf{p}) \leq 1 - (1 - p_1 p_2 p_3)(1 - p_1 p_2 p_4)(1 - p_1 p_3 p_4)(1 - p_2 p_3 p_4)$$

where $q_i \equiv 1 - p_i$. For instance, if $p_i = \frac{1}{2}$ for all i, then the above yields that

$$.18 \leq r(\tfrac{1}{2}, \ldots, \tfrac{1}{2}) \leq .59$$

The exact value for this structure is easily computed to be

$$r(\tfrac{1}{2}, \ldots, \tfrac{1}{2}) = \tfrac{5}{16} = .31 \quad \Diamond$$

5. System Life as a Function of Component Lives

For a random variable having distribution function G we define $\bar{G}(a) \equiv 1 - G(a)$ to be the probability that the random variable is greater than a.

Consider *a* system in which the *i*th component functions for a random length of time having distribution F_i and then fails. Once failed it remains in that state forever. Assuming that the individual component lifetimes are independent, how can we express the distribution of system lifetime as a function of the system reliability function $r(\mathbf{p})$ and the individual component lifetime distributions F_i, $i = 1, \ldots, n$?

To answer the above we first note that the system will function for a length of time t or greater if and only if it is still functioning at time t. That is, letting F denote the distribution of system lifetime, we have

$$\overline{F}(t) = P\{\text{system life} > t\}$$
$$= P\{\text{system is functioning at time } t\}$$

But, by the definition of $r(\mathbf{p})$ we have that

$$P\{\text{system is functioning at time } t\} = r(P_1(t), \ldots, P_n(t))$$

where

$$P_i(t) = P\{\text{component } i \text{ is functioning at } t\}$$
$$= P\{\text{lifetime of } i > t\}$$
$$= \overline{F}_i(t)$$

Hence we see that

$$\overline{F}(t) = r(\overline{F}_1(t), \ldots, \overline{F}_n(t)) \tag{5.1}$$

Example 5a   In a series system, $r(\mathbf{p}) = \prod_1^n p_i$ and so from Equation (5.1)

$$\overline{F}(t) = \prod_1^n \overline{F}_i(t)$$

which is, of course, quite obvious since for a series system the system life is equal to the minimum of the component lives and so will be greater than t if and only if all component lives are greater than t.   ◇

Example 5b   In a parallel system $r(\mathbf{p}) = 1 - \prod_1^n (1 - p_i)$ and so

$$\overline{F}(t) = 1 - \prod_1^n F_i(t)$$

The above is also easily derived by noting that, in the case of a parallel system, the system life is equal to the maximum of the component lives.   ◇

For a continuous distribution G, we define $\lambda(t)$, the *failure rate function* of G, by

$$\lambda(t) = \frac{g(t)}{\bar{G}(t)}$$

where $g(t) = d/dt\, G(t)$. In Section 2.2 of Chapter 5, it is shown that if G is the distribution of the lifetime of an item, then $\lambda(t)$ represents the probability intensity that a t-year-old item will fail. We say that G is an *increasing failure rate* (IFR) distribution if $\lambda(t)$ is an increasing function of t. Similarly, we say that G is a *decreasing failure rate* (DFR) distribution if $\lambda(t)$ is a decreasing function of t.

Example 5c (The Weibull Distribution): A random variable is said to have the *Weibull* distribution if its distribution is given, for some $\lambda > 0$, $\alpha > 0$, by

$$G(t) = 1 - e^{-(\lambda t)^{\alpha}}, \qquad t \geq 0$$

The failure rate function for a Weibull distribution equals

$$\lambda(t) = \frac{e^{-(\lambda t)^{\alpha}} \alpha (\lambda t)^{\alpha - 1} \lambda}{e^{-(\lambda t)^{\alpha}}} = \alpha \lambda (\lambda t)^{\alpha - 1}$$

Thus, the Weibull distribution is IFR when $\alpha \geq 1$, and DFR when $0 < \alpha \leq 1$; when $\alpha = 1$, $G(t) = 1 - e^{-\lambda t}$, the exponential distribution, which is both IFR and DFR. $\diamond$

Example 5d (The Gamma Distribution): A random variable is said to have a *gamma* distribution if its density is given, for some $\lambda > 0$, $\alpha > 0$, by

$$g(t) = \frac{\lambda e^{-\lambda t}(\lambda t)^{\alpha - 1}}{\Gamma(\alpha)} \quad \text{for} \quad t \geq 0$$

where $\Gamma(\alpha) \equiv \int_0^{\infty} e^{-t} t^{\alpha - 1}\, dt$.

For the gamma distribution,

$$\frac{1}{\lambda(t)} = \frac{\bar{G}(t)}{g(t)} = \frac{\displaystyle\int_t^{\infty} \lambda e^{-\lambda x}(\lambda x)^{\alpha - 1}\, dx}{\lambda e^{-\lambda t}(\lambda t)^{\alpha - 1}}$$

$$= \int_t^{\infty} e^{-\lambda(x - t)}\left(\frac{x}{t}\right)^{\alpha - 1} dx$$

With the change of variables $u = x - t$, we obtain that

$$\frac{1}{\lambda(t)} = \int_0^\infty e^{-\lambda u} \left(1 + \frac{u}{t}\right)^{\alpha - 1} du$$

Hence, G is IFR when $\alpha \geq 1$ and is DFR when $0 < \alpha \leq 1$. ◇

Suppose that the lifetime distribution of each component in a monotone system is IFR. Does this imply that the system lifetime is also IFR? To answer this, let us at first suppose that each component has the same lifetime distribution, which we denote by G. That is, $F_i(t) = G(t)$, $i = 1, \ldots, n$. To determine whether the system lifetime is IFR, we must compute $\lambda_F(t)$, the failure rate function of F. Now, by definition,

$$\lambda_F(t) = \frac{\dfrac{d}{dt} F(t)}{\overline{F}(t)}$$

$$= \frac{\dfrac{d}{dt} [1 - r(\overline{G}(t))]}{r(\overline{G}(t))}$$

where

$$r(\overline{G}(t)) \equiv r(\overline{G}(t), \ldots, \overline{G}(t))$$

Hence,

$$\lambda_F(t) = \frac{r'(\overline{G}(t))}{r(\overline{G}(t))} G'(t)$$

$$= \frac{\overline{G}(t) r'(\overline{G}(t))}{r(\overline{G}(t))} \frac{G'(t)}{\overline{G}(t)}$$

$$= \lambda_G(t) \left. \frac{p r'(p)}{r(p)} \right|_{p = \overline{G}(t)} \tag{5.2}$$

Since $\overline{G}(t)$ is a decreasing function of t, it follows from Equation (5.2) that *if each component of a coherent system has the same IFR lifetime distribution, then the distribution of system lifetime will be IFR if $p r'(p)/r(p)$ is a decreasing function of p.*

Example 5e (The k-Out-of-n System with Identical Components): Consider the k-out-of-n system which will function if and only if k or more components function. When each component has the same probability p of functioning, the number of functioning components will have a binomial distribution with parameters n and p. Hence,

$$r(p) = \sum_{i=k}^{n} \binom{n}{i} p^{i}(1-p)^{n-i}$$

which, by continual integration by parts, can be shown to be equal to

$$r(p) = \frac{n!}{(k-1)!(n-k)!} \int_{0}^{p} x^{k-1}(1-x)^{n-k}dx$$

Upon differentiation, we obtain

$$r'(p) = \frac{n!}{(k-1)!(n-k)!} p^{k-1}(1-p)^{n-k}$$

Therefore,

$$\frac{pr'(p)}{r(p)} = \left[\frac{r(p)}{pr'(p)}\right]^{-1}$$

$$= \left[\frac{1}{p} \int_{0}^{p} \left(\frac{x}{p}\right)^{k-1} \left(\frac{1-x}{1-p}\right)^{n-k} dx\right]^{-1}$$

Letting $y = x/p$, yields

$$\frac{pr'(p)}{r(p)} = \left[\int_{0}^{1} y^{k-1} \left(\frac{1-yp}{1-p}\right)^{n-k} dy\right]^{-1}$$

Since $(1-yp)/(1-p)$ is increasing in p, it follows that $pr'(p)/r(p)$ is decreasing in p. Thus, if a k-out-of-n system is composed of independent, like components having an increasing failure rate, the system itself has an increasing failure rate.   ◊

It turns out, however, that for a k-out-of-n system, in which the independent components have different IFR lifetime distributions, the system lifetime need not be IFR. Consider the following example of a two-out-of-two (that is, a parallel) system.

Example 5f (A Parallel System That Is Not IFR): The life distribution of a parallel system of two independent components, the ith component having an exponential distribution with mean $1/i$, $i = 1$, 2, is given by

$$\bar{F}(t) = 1 - (1 - e^{-t})(1 - e^{-2t})$$
$$= e^{-2t} + e^{-t} - e^{-3t}$$

Therefore,

$$\lambda(t) = \frac{f(t)}{\bar{F}(t)}$$

$$= \frac{2e^{-2t} + e^{-t} - 3e^{-3t}}{e^{-2t} + e^{-t} - e^{-3t}}$$

It easily follows, upon differentiation, that the sign of $\lambda'(t)$ is determined by $e^{-5t} - e^{-3t} + 3e^{-4t}$ which is positive for small values and negative for large values of t. Therefore, $\lambda(t)$ is initially strictly increasing and then strictly decreasing. Hence, F is not IFR.   ◇

Remarks   The result of the above example is quite surprising at first glance. To obtain a better feel for it we need the concept of a mixture of distribution function. The distribution function G is said to be a *mixture of* the distributions G_1 and G_2 if for some p, $0 < p < 1$,

$$G(x) = pG_1(x) + (1 - p)G_2(x) \tag{5.3}$$

Mixtures occur when we sample from a population made up of two distinct groups. For example, suppose we have a stockpile of items of which the fraction p are type 1 and the fraction $1 - p$ are type 2. Suppose that the lifetime distribution of type 1 items is G_1 and of type 2 items is G_2. If we choose an item at random from the stockpile, then its life distribution is as given by Equation (5.3).

Consider now a mixture of two exponential distributions having rates λ_1 and λ_2 where $\lambda_1 < \lambda_2$. We are interested in determining whether or not this mixture distribution is IFR. To do so, we note that if the item selected has survived up to time t, then its distribution of remaining life is still a mixture of the two exponential distributions. This is so since its remaining life will still be exponential with rate λ_1 if it is type 1 or with rate λ_2 if it is a type 2 item. However, the probability that it is a type 1 item is no longer the (prior) probability p but is now a conditional probability given that it has survived to time t. In fact, its probability of being a type 1 is

$$P\{\text{type 1}|\text{life} > t\} = \frac{P\{\text{type 1, life} > t\}}{P\{\text{life} > t\}}$$

$$= \frac{pe^{-\lambda_1 t}}{pe^{-\lambda_1 t} + (1 - p)e^{-\lambda_2 t}}$$

As the above is increasing in t, it follows that the larger t is, the more likely it is that the item in use is a type 1 (the better one, since $\lambda_1 < \lambda_2$). Hence, the older the item is, the less likely it is to fail, and thus the mixture of exponentials far from being IFR is, in fact, DFR.

Now let us return to the parallel system of two exponential components having respective rates λ_1 and λ_2. The lifetime of such a system can be expressed as the sum of two independent random variables—namely

$$\text{System life} = \text{Exp}(\lambda_1 + \lambda_2) + \begin{cases} \text{Exp}(\lambda_1) \text{ with probability } \dfrac{\lambda_2}{\lambda_1 + \lambda_2} \\[2ex] \text{Exp}(\lambda_2) \text{ with probability } \dfrac{\lambda_1}{\lambda_1 + \lambda_2} \end{cases}$$

The first random variable whose distribution is exponential with rate $\lambda_1 + \lambda_2$ represents the time until one of the components fails, and the second, which is a mixture of exponentials, is the additional time until the other component fails. (Why are these two random variables independent?)

Now given that the system has survived a time t, it is very unlikely when t is large that both components are still functioning, but instead it is far more likely that one of the components has failed. Hence, for large t, the distribution of remaining life is basically a mixture of two exponentials—and so as t becomes even larger its failure rate should decrease (as indeed occurs). ◊

Recall that the failure rate function of a distribution $F(t)$ having density $f(t) = F'(t)$ is defined by

$$\lambda(t) = \frac{f(t)}{1 - F(t)}$$

By integrating both sides of the above, we obtain

$$\int_0^t \lambda(s)\,ds = \int_0^t \frac{f(s)}{1 - F(s)}\,ds$$
$$= -\log \bar{F}(t)$$

Hence,

$$\bar{F}(t) = e^{-\Lambda(t)} \tag{5.4}$$

where

$$\Lambda(t) = \int_0^t \lambda(s)\,ds$$

The function $\Lambda(t)$ is called the *hazard function* of the distribution F.

Definition 5.1. A distribution F is said to have *increasing failure on the average* (IFRA) if

$$\frac{\Lambda(t)}{t} = \frac{\displaystyle\int_0^t \lambda(s)\,ds}{t} \tag{5.5}$$

increases in t for $t \geq 0$.

In other words, Equation (5.5) states that the average failure rate up to time t increases as t increases. It is not difficult to show that if F is IFR, then F is IFRA; but the reverse need not be true.

Note that F is IFRA if $\Lambda(s)/s \leq \Lambda(t)/t$ whenever $0 \leq s \leq t$, which is equivalent to

$$\frac{\Lambda(\alpha t)}{\alpha t} \leq \frac{\Lambda(t)}{t} \quad \text{for} \quad 0 \leq \alpha \leq 1, \quad \text{all } t \geq 0$$

But by Equation (5.4) we see that $\Lambda(t) = -\log \bar{F}(t)$, and so the above is equivalent to

$$-\log \bar{F}(\alpha t) \leq -\alpha \log \bar{F}(t)$$

or equivalently

$$\log \bar{F}(\alpha t) \geq \log \bar{F}^\alpha(t)$$

which, since $\log x$ is a monotone function of x, shows that *F is IFRA if and only if*

$$\bar{F}(\alpha t) \geq \bar{F}^\alpha(t) \quad \text{for} \quad 0 \leq \alpha \leq 1, \quad \text{all } t \geq 0 \tag{5.6}$$

For a vector $\mathbf{p} = (p_1, \ldots, p_n)$, we define $\mathbf{p}^\alpha = (p_1^\alpha, \ldots, p_n^\alpha)$. We shall need the following proposition.

Proposition 5.1. Any reliability function $r(\mathbf{p})$ satisfies

$$r(\mathbf{p}^\alpha) \geq [r(\mathbf{p})]^\alpha, \quad 0 \leq \alpha \leq 1$$

Proof: We prove this by induction on n, the number of components in the system. If $n = 1$, then either $r(p) \equiv 0$, $r(p) \equiv 1$, or $r(p) \equiv p$. Hence, the proposition follows in this case.

So assume that Proposition 5.1 is valid for all monotone systems of $n - 1$ components and consider a system of n components having structure function ϕ. By conditioning upon whether or not the nth component is functioning, we obtain that

$$r(\mathbf{p}^\alpha) = p_n^\alpha r(1_n, \mathbf{p}^\alpha) + (1 - p_n^\alpha) r(0_n, \mathbf{p}^\alpha) \tag{5.7}$$

Now consider a system of components 1 through $n - 1$ having a structure function $\phi_1(\mathbf{x}) = \phi(1_n, \mathbf{x})$. The reliability function for this system is given by $r_1(\mathbf{p}) = r(1_n, \mathbf{p})$; hence, from the induction assumption (valid for all monotone systems of $n - 1$ components), we have that

$$r(1_n, \mathbf{p}^\alpha) \geq [r(1_n, \mathbf{p})]^\alpha$$

Similarly, by considering the system of components 1 through $n - 1$ and structure function $\phi_0(\mathbf{x}) = \phi(0_n, \mathbf{x})$, we obtain

$$r(0_n, \mathbf{p}^\alpha) \geq [r(0_n, \mathbf{p})]^\alpha$$

Thus, from Equation (5.7), we obtain

$$r(\mathbf{p}^\alpha) \geq p_n^\alpha [r(1_n, \mathbf{p})]^\alpha + (1 - p_n^\alpha)[r(0_n, \mathbf{p})]^\alpha$$

which, by using the lemma to follow [with $\lambda = p_n$, $x = r(1_n, \mathbf{p})$, $y = r(0_n, \mathbf{p})$], implies that

$$r(\mathbf{p}^\alpha) \geq [p_n r(1_n, \mathbf{p}) + (1 - p_n) r(0_n, \mathbf{p})]^\alpha$$
$$= [r(\mathbf{p})]^\alpha$$

which proves the result.

Lemma 5.1. If $0 \leq \alpha \leq 1$, $0 \leq \lambda \leq 1$, then

$$h(y) = \lambda^\alpha x^\alpha + (1 - \lambda^\alpha) y^\alpha - (\lambda x + (1 - \lambda) y)^\alpha \geq 0$$

for all $0 \leq y \leq x$.

Proof: The proof is left as an exercise. ◇

We are now ready to prove the following important theorem.

Theorem 5.1. For a monotone system of independent components, if each component has an IFRA lifetime distribution, then the distribution of system lifetime is itself IFRA.

Proof: The distribution of system lifetime F is given by

$$\bar{F}(\alpha t) = r(\bar{F}_1(\alpha t), \ldots, \bar{F}_n(\alpha t))$$

Hence, since r is a monotone function, and since each of the component distributions $\bar{F}_i$ are IFRA, we obtain from Equation (5.6) that

$$\bar{F}(\alpha t) \geq r(\bar{F}_1^\alpha(t), \ldots, \bar{F}_n^\alpha(t))$$
$$\geq [r(\bar{F}_1(t), \ldots, \bar{F}_n(t)]^\alpha$$
$$= \bar{F}^\alpha(t)$$

which by Equation (5.6) proves the theorem. The last inequality followed, of course, from Proposition 5.1.   ◇

6. Expected System Lifetime

In this section, we show how the mean lifetime of a system can be determined, at least in theory, from a knowledge of the reliability function $r(\mathbf{p})$ and the component lifetime distributions F_i, $i = 1, \ldots, n$.

Since the system's lifetime will be t or larger if and only if the system is still functioning at time t, we have that

$$P\{\text{system life} \geq t\} = r(\bar{\mathbf{F}}(t))$$

where $\bar{\mathbf{F}}(t) = (\bar{F}_1(t), \ldots, \bar{F}_n(t))$. Hence, by a well-known formula which states that for any nonnegative random variable X, $E[X] = \int_0^\infty P\{X \geq x\}dx$, we see that[*]

$$E[\text{system life}] = \int_0^\infty r(\bar{\mathbf{F}}(t))dt \qquad (6.1)$$

Example 6a   (A Series System of Uniformly Distributed Components): Consider a series system of three independent components each of which functions for an amount of time (in hours) uniformly distributed over (0, 10). Hence, $r(\mathbf{p}) = p_1 p_2 p_3$ and

$$F_i(t) = \begin{cases} t/10, & 0 \leq t \leq 10, \\ 1, & t > 10 \end{cases} \qquad i = 1, 2, 3$$

---

[*]That $E[X] = \int_0^\infty P\{X \geq x\}dx$ can be shown as follows when X has density f.

$$\int_0^\infty P\{X \geq x\}dx = \int_0^\infty \int_x^\infty f(y)dydx = \int_0^\infty \int_0^y f(y)dxdy = \int_0^\infty yf(y)dy = E[X]$$

Therefore,

$$r(\bar{\mathbf{F}}(t)) = \begin{cases} \left(\dfrac{10 - t}{10}\right)^3, & 0 \le t \le 10 \\ 0, & t > 10 \end{cases}$$

and so from Equation (6.1) we obtain

$$E[\text{system life}] = \int_0^{10} \left(\frac{10 - t}{10}\right)^3 dt$$

$$= 10 \int_0^1 y^3 \, dy$$

$$= \tfrac{5}{2} \quad \lozenge$$

Example 6b (A Two-Out-of-Three System): Consider a two-out-of-three system of independent components, in which each component's lifetime is (in months) uniformly distributed over $(0, 1)$. As was shown in Example 3d, the reliability of such a system is given by

$$r(\mathbf{p}) = p_1 p_2 + p_1 p_3 + p_2 p_3 - 2 p_1 p_2 p_3$$

Since

$$F_i(t) = \begin{cases} t, & 0 \le t \le 1 \\ 1, & t > 1 \end{cases}$$

we see from Equation (6.1) that

$$E[\text{system life}] = \int_0^1 [3(1 - t)^2 - 2(1 - t)^3] dt$$

$$= \int_0^1 (3y^2 - 2y^3) dy$$

$$= 1 - \tfrac{1}{2}$$

$$= \tfrac{1}{2} \quad \lozenge$$

Example 6c (A Four-Component System): Consider the four-component system that functions when components 1 and 2 and at least one of components 3 and 4 function. Its structure function is given by

$$\phi(\mathbf{x}) = x_1 x_2 (x_3 + x_4 - x_3 x_4)$$

and thus its reliability function equals

$$r(\mathbf{p}) = p_1 p_2 (p_3 + p_4 - p_3 p_4)$$

Let us compute the mean system lifetime when the ith component is uniformly distributed over $(0, i)$, $= 1, 2, 3, 4$.

Now

$$\bar{F}_1(t) = \begin{cases} 1 - t, & 0 \le t \le 1 \\ 0, & t > 1 \end{cases}$$

$$\bar{F}_2(t) = \begin{cases} 1 - t/2, & 0 \le t \le 2 \\ 0, & t > 2 \end{cases}$$

$$\bar{F}_3(t) = \begin{cases} 1 - t/3, & 0 \le t \le 3 \\ 0, & t > 3 \end{cases}$$

$$\bar{F}_4(t) = \begin{cases} 1 - t/4, & 0 \le t \le 4 \\ 0, & t > 4 \end{cases}$$

Hence,

$$r(\bar{\mathbf{F}}(t)) = \begin{cases} (1 - t)\left(\dfrac{2 - t}{2}\right)\left[\dfrac{3 - t}{3} + \dfrac{4 - t}{4} - \dfrac{(3 - t)(4 - t)}{12}\right], & 0 \le t \le 1 \\ 0 & t > 1 \end{cases}$$

Therefore,

$$\begin{aligned} E[\text{system life}] &= \frac{1}{24} \int_0^1 (1 - t)(2 - t)(12 - t^2)\,dt \\ &= \frac{593}{(24)(60)} \\ &\approx .41 \quad \Diamond \end{aligned}$$

We end this section by obtaining the mean lifetime of a k-out-of-n system of independent identically distributed exponential components. If θ is the mean lifetime of each component, then

$$\bar{F}_i(t) = e^{-t/\theta}$$

Hence, since for a k-out-of-n system

$$r(p, p, \ldots, p) = \sum_{i=k}^{n} \binom{n}{i} p^i (1 - p)^{n-i}$$

we obtain from Equation (6.1) that

$$E[\text{system life}] = \int_0^\infty \sum_{i=k}^n \binom{n}{i}(e^{-t/\theta})^i(1 - e^{-t/\theta})^{n-i}dt$$

Making the substitution

$$y = e^{-t/\theta}, \quad dy = -\frac{1}{\theta}e^{-t/\theta}\,dt = -\frac{y}{\theta}\,dt$$

yields

$$E[\text{system life}] = \theta \sum_{i=k}^n \binom{n}{i} \int_0^1 y^{i-1}(1 - y)^{n-i}dy$$

Now it is not difficult to show that*

$$\int_0^1 y^n(1 - y)^m dy = \frac{m!n!}{(m + n + 1)!} \tag{6.2}$$

Thus, the above equals

$$E[\text{system life}] = \theta \sum_{i=k}^n \frac{n!}{(n - i)!i!} \frac{(i - 1)!(n - i)!}{n!}$$

$$= \theta \sum_{i=k}^n 1/i \tag{6.3}$$

Remarks Equation (6.3) could have been proven directly by making use of special properties of the exponential distribution. First note that the lifetime of a k-out-of-n system can be written as $T_1 + \cdots + T_{n-k+1}$, where T_i represents the time between the $(i - 1)$st and ith failure. This is true since $T_1 + \cdots + T_{n-k+1}$ equals the time at which the $(n - k + 1)$st component fails, which is also the first time that the number of functioning components is less than k. Now, when all n components are functioning, the rate at which failures occur is n/θ. That is, T_1 is exponentially distributed with mean θ/n. Similarly, since T_i represents the time until the next failure when there are $n - (i - 1)$ functioning components, it follows that T_i is exponentially distributed with mean $\theta/(n - i + 1)$. Hence, the mean system lifetime equals

---

*Let $C(n, m) = \int_0^1 y^n(1 - y)^m dy$. Integration by parts yields that $C(n, m) = [m/(n + 1)]C(n + 1, m - 1)$. Starting with $C(n, 0) = 1/(n + 1)$, Equation (6.2) follows by mathematical induction.

$$E[T_1 + \cdots + T_{n-k+1}] = \theta \left[\frac{1}{n} + \cdots + \frac{1}{k} \right]$$

Note also that it follows, from the lack of memory of the exponential, that the T_i, $i = 1, \ldots, n - k + 1$, are independent random variables.   ◇

7. Systems with Repair

Consider an n component system having reliability function $r(\mathbf{p})$. Suppose that component i functions for an exponentially distributed time with rate λ_i and then fails; once failed it takes an exponential time with rate μ_i to be repaired, $i = 1, \ldots, n$. All components act independently. Let us suppose that all components are initially working, and let

$$A(t) = P\{\text{system is working at } t\}$$

$A(t)$ is called the *availability* at time t. Since the components act independently, $A(t)$ can be expressed in terms of the reliability function as follows:

$$A(t) = r(A_1(t), \ldots, A_n(t)) \tag{7.1}$$

where

$$A_i(t) = P\{\text{component } i \text{ is functioning at } t\}$$

Now the state of component i—either on or off—changes in accordance with a two-state continuous time Markov chain. Hence, from the results of Example 4c of Chapter 6 we have

$$A_i(t) = P_{00}(t) = \frac{\mu_i}{\mu_i + \lambda_i} + \frac{\lambda_i}{\mu_i + \lambda_i} e^{-(\lambda_i + \mu_i)t}$$

Thus we obtain that

$$A(t) = r\left(\frac{\mu}{\mu + \lambda} + \frac{\lambda}{\mu + \lambda} e^{-(\lambda + \mu)t} \right)$$

If we let t approach ∞, then we obtain that the limiting availability—call it A—is given by

$$A = \lim_{t \to \infty} A(t) = r\left(\frac{\mu}{\lambda + \mu} \right)$$

Remarks

(i) If the on and off distributions for component i are arbitrary continuous distributions with respective means $1/\lambda_i$ and $1/\mu_i$, $i = 1, \ldots, n$, then it follows from the theory of alternating renewal processes (see Section 5.1 of Chapter 7) that

$$A_i(t) \to \frac{1/\lambda_i}{1/\lambda_i + 1/\mu_i} = \frac{\mu_i}{\mu_i + \lambda_i}$$

and thus, using the continuity of the reliability function, it follows from (7.1) that the limiting availability is

$$A = \lim_{t \to \infty} A(t) = r\left(\frac{\mu}{\mu + \lambda}\right)$$

Hence, A depends only on the on and off distributions through their means.

(ii) It can be shown (using the theory of regenerative processes as presented in Section 5 of Chapter 7) that A will also equal the long run proportion of time that the system will be functioning.

Example 7a For a series system $r(\mathbf{p}) = \prod_{i=1}^n p_i$ and so

$$A(t) = \prod_{i=1}^n \left[\frac{\mu_i}{\mu_i + \lambda_i} + \frac{\lambda_i}{\mu_i + \lambda_i} e^{-(\lambda_i + \mu_i)t}\right]$$

and

$$A = \prod_{i=1}^n \frac{\mu_i}{\mu_i + \lambda_i} \quad \Diamond$$

Example 7b For a parallel system $r(\mathbf{p}) = 1 - \prod_{i=1}^n (1 - p_i)$ and thus

$$A(t) = 1 - \prod_{i=1}^n \left[\frac{\lambda_i}{\mu_i + \lambda_i}(1 - e^{-(\lambda_i + \mu_i)t})\right]$$

and

$$A = 1 - \prod_{i=1}^n \frac{\lambda_i}{\mu_i + \lambda_i} \quad \Diamond$$

The above system will alternate between periods when it is up and periods when it is down. Let us denote by U_i and D_i, $i \geq 1$, the lengths of the ith up and down period respectively. For instance in a two-of-three system, U_1 will be the time until two components are down; D_1, the additional time until two are up; U_2 the additional time until two are down, and so on. Let

$$\overline{U} = \lim_{n \to \infty} \frac{U_1 + \cdots + U_n}{n}$$

$$\overline{D} = \lim_{n \to \infty} \frac{D_1 + \cdots + D_n}{n}$$

denote the average length of an up and down period respectively.*

To determine $\overline{U}$ and $\overline{D}$, note first that in the first n up-down cycles—that is in time $\Sigma_{i=1}^{n}(U_i + D_i)$—the system will be up for a time $\Sigma_{i=1}^{n}U_i$. Hence, the proportion of time the system will be up in the first n up-down cycles is

$$\frac{U_1 + \cdots + U_n}{U_1 + \cdots + U_n + D_1 + \cdots + D_n} = \frac{\displaystyle\sum_{i=1}^{n} U_i/n}{\displaystyle\sum_{i=1}^{n} U_i/n + \sum_{i=1}^{n} D_i/n}$$

As $n \to \infty$ this must converge to A, the long run proportion of time the system is up. Hence,

$$\frac{\overline{U}}{\overline{U} + \overline{D}} = A = r\left(\frac{\mu}{\lambda + \mu}\right) \tag{7.2}$$

However, to solve for $\overline{U}$ and $\overline{D}$ we need a second equation. To obtain one consider the rate at which the system fails. As there will be n failures in time $\Sigma_{i=1}^{n}(U_i + D_i)$, it follows that the rate at which the system fails is

$$\text{rate at which system fails} = \lim_{n \to \infty} \frac{n}{\displaystyle\sum_{1}^{n} U_i + \sum_{1}^{n} D_1}$$

$$= \lim_{n \to \infty} \frac{1}{\displaystyle\sum_{1}^{n} U_i/n + \sum_{1}^{n} D_i/n} = \frac{1}{\overline{U} + \overline{D}} \tag{7.3}$$

---

*It can be shown using the theory of regenerative processes that, with Probability 1, the above limits will exist and will be constants.

That is, the above yields the intuitive result that, on average, there is one failure every $\overline{U} + \overline{D}$ time units. To utilize this let us determine the rate at which a failure of component i causes the system to go from up to down. Now the system will go from up to down when component i fails if the states of the other components $x_1, \ldots, x_{i-1}, x_{i+1}, \ldots, x_n$ are such that $\phi(1_i, \mathbf{x}) = 1$, $\phi(0_i, \mathbf{x}) = 0$. That is the states of the other components must be such that

$$\phi(1_i, \mathbf{x}) - \phi(0_i, \mathbf{x}) = 1 \tag{7.4}$$

Since component i will, on average, have one failure every $1/\lambda_i + 1/\mu_i$ time units, it follows that the rate at which component i fails is equal to $(1/\lambda_i + 1/\mu_i)^{-1} = \lambda_i \mu_i / (\lambda_i + \mu_i)$. In addition, the states of the other components will be such that (7.4) holds with probability

$P\{\phi(1_i, X(\infty)) - \phi(0_i, X(\infty)) = 1\}$
$= E[\phi[1_i, X(\infty)] - \phi(0_i, X(\infty))]$ since $\phi(1_i, X(\infty)) - \phi(0_i, X(\infty))$ is a Bernoulli random variable

$$= r\left(1_i, \frac{\mu}{\lambda + \mu}\right) - r\left(0_i, \frac{\mu}{\lambda + \mu}\right)$$

Hence, putting the above together we see that

$$\begin{array}{l} \text{rate at which component} \\ i \text{ causes the system to fail} \end{array} = \frac{\lambda_i \mu_i}{\lambda_i + \mu_i} \left[r\left(1_i, \frac{\mu}{\lambda + \mu}\right) - r\left(0_i, \frac{\mu}{\lambda + \mu}\right) \right]$$

Summing the above over all components i thus gives

$$\text{rate at which system fails} = \sum_i \frac{\lambda_i \mu_i}{\lambda_i + \mu_i} \left[r\left(1_i, \frac{\mu}{\lambda + \mu}\right) - r\left(0_i, \frac{\mu}{\lambda + \mu}\right) \right]$$

Finally equating the above with (7.3) yields

$$\frac{1}{\overline{U} + \overline{D}} = \sum_i \frac{\lambda_i \mu_i}{\lambda_i + \mu_i} \left[r\left(1_i, \frac{\mu}{\lambda + \mu}\right) - r\left(0_i, \frac{\mu}{\lambda + \mu}\right) \right] \tag{7.5}$$

Solving (7.2) and (7.5) we obtain

$$\overline{U} = \frac{r\left(\dfrac{\mu}{\lambda + \mu}\right)}{\displaystyle\sum_{i=1}^{n} \frac{\lambda_i \mu_i}{\lambda_i + \mu_i} \left[r\left(1_i, \dfrac{\mu}{\lambda + \mu}\right) - r\left(0_i, \dfrac{\mu}{\lambda + \mu}\right) \right]} \tag{7.6}$$

$$D = \frac{\left[1 - r\left(\frac{\mu}{\lambda + \mu}\right)\right]\overline{U}}{r\left(\frac{\mu}{\lambda + \mu}\right)} \tag{7.7}$$

Also (7.5) yields the rate at which the system fails.

Remarks   In establishing the formulas for $\overline{U}$ and $\overline{D}$, we did not make use of the assumption of exponential on and off times, and, in fact, our derivation is valid and (7.6) and (7.7) hold whenever $\overline{U}$ and $\overline{D}$ are well-defined (a sufficient condition is that all on and off distributions are continuous). The quantities λ_i, μ_i, $i = 1, \ldots,$ n will represent, respectively, the reciprocals of the mean lifetimes and mean repair times. $\diamond$

Example 7c   For a series system

$$\overline{U} = \frac{\prod_i \frac{\mu_i}{\mu_i + \lambda_i}}{\sum_i \frac{\lambda_i \mu_i}{\lambda_i + \mu_i} \prod_{j \neq i} \frac{\mu_j}{\mu_j + \lambda_j}} = \frac{1}{\sum_i \lambda_i}$$

$$\overline{D} = \frac{1 - \prod_i \frac{\mu_i}{\mu_i + \lambda_i}}{\prod_i \frac{\mu_i}{\mu_i + \lambda_i}} \frac{1}{\sum_i \lambda_i}$$

whereas for a parallel system

$$\overline{U} = \frac{1 - \prod_i \frac{\lambda_i}{\mu_i + \lambda_i}}{\sum_i \frac{\lambda_i \mu_i}{\lambda_i + \mu_i} \prod_{j \neq i} \frac{\lambda_j}{\lambda_j + \mu_j}}$$

$$\overline{D} = \frac{\left[\prod_i \frac{\lambda_i}{\mu_i + \lambda_i}\right]}{1 - \prod_i \frac{\lambda_i}{\mu_i + \lambda_i}} \overline{U} = \frac{1}{\sum_i \mu_i}$$

The above formulas hold for arbitrary continuous up and down distributions with $1/\lambda_i$ and $1/\mu_i$ denoting respectively the mean up and down times of component i, $i = 1, \ldots, n$. ◇

Remarks The model of this section would arise when the components are separately maintained with each having its own repair facility. For models having a common repair facility, the interested reader should see Examples 5d and 6a of Chapter 6. ◇

Problems

1. Prove that for any structure function φ,

$$\phi(\mathbf{x}) = x_i\phi(1_i, \mathbf{x}) + (1 - x_i)\phi(0_i, \mathbf{x})$$

where

$$(1_i, \mathbf{x}) = (x_1, \ldots, x_{i-1}, 1, x_{i+1}, \ldots, x_n)$$
$$(0_i, \mathbf{x}) = (x_1, \ldots, x_{i-1}, 0, x_{i+1}, \ldots, x_n)$$

2. Show that
(a) If $\phi(0, 0, \ldots, 0) = 0$ and $\phi(1, 1, \ldots, 1) = 1$, then

$$\min x_i \le \phi(\mathbf{x}) \le \max x_i$$

(b) $\phi(\max(\mathbf{x}, \mathbf{y})) \ge \max(\phi(\mathbf{x}), \phi(\mathbf{y}))$
(c) $\phi(\min(\mathbf{x}, \mathbf{y})) \le \min(\phi(\mathbf{x}), \phi(\mathbf{y}))$.

3. For any structure function φ, we define the dual structure ϕ^D by

$$\phi^D(\mathbf{x}) = 1 - \phi(1 - \mathbf{x})$$

(a) Show that the dual of a parallel (series) system is a series (parallel) system.
(b) Show that the dual of a dual structure is the original structure.
(c) What is the dual of a k-out-of-n structure?
(d) Show that a minimal path (cut) set of the dual system is a minimal cut (path) set of the original structure.
4. Write the structure function corresponding to the following:
(a)

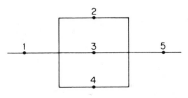

Figure 9.16.

(b)

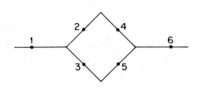

Figure 9.17.

(c)

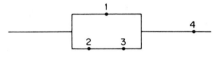

Figure 9.18.

5. Find the minimal path and minimal cut sets for:
   (a)

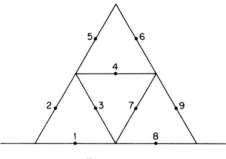

Figure 9.19

(b)

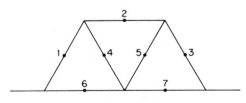

Figure 9.20.

6. Component i is said to be *relevant* to the system if for some state vector $\mathbf{x}$,

$$\phi(1_i, \mathbf{x}) = 1, \ \phi(0_i, \mathbf{x}) = 0$$

Otherwise, it is said to be *irrelevant*.

  (a) Explain in words what it means for a component to be irrelevant.

  (b) Let $A_1, \ldots, A_s$ be the minimal path sets of a system, and let S denote the set of components. Show that $S = \bigcup_{i=1}^{s} A_i$ if and only if all components are relevant.

  (c) Let $C_1, \ldots, C_k$ denote the minimal cut sets. Show that $S = \bigcup_{i=1}^{k} C_i$ if and only if all components are relevant.

7. Let t_i denote the time of failure of the ith component; let $\tau_\phi(t)$ denote the time to failure of the system ϕ as a function of the vector $\mathbf{t} = (t_1, \ldots, t_n)$. Show that

$$\max_{1 \leq j \leq s} \min_{i \in A_j} t_i = \tau_\phi(\mathbf{t}) = \min_{1 \leq j \leq k} \max_{i \in C_j} t_i$$

where $C_1, \ldots, C_k$ are the minimal cut sets, and $A_1, \ldots, A_s$ the minimal path sets.

8. Let $r(\mathbf{p})$ be the reliability function. Show that

$$r(\mathbf{p}) = p_i r(1_i, \mathbf{p}) + (1 - p_i) r(0_i, \mathbf{p})$$

9. Compute the reliability function of the bridge system (see Figure 9.11) by conditioning upon whether or not component 3 is working.

10. Compute upper and lower bounds of the reliability function (using Method 2) for the systems given in Problem 4, and compare them with the exact values when $p_i \equiv \frac{1}{2}$.

11. Compute the upper and lower bounds of $r(\mathbf{p})$ using both methods for the

  (a) two-out-of-three system and

  (b) two-out-of-four-system.

  (c) Compare these bounds with the exact reliability when

  (i) $p_i \equiv .5$

  (ii) $p_i \equiv .8$

  (iii) $p_i \equiv .2$.

12. Let N be a nonnegative, integer-valued random variable. Show that

$$P\{N > 0\} \geq \frac{(E[N])^2}{E[N^2]}$$

and explain how this inequality can be used to derive additional bounds on a reliability function.

Hint: $E[N^2] = E[N^2|N > 0]P\{N > 0\}$ (Why?)
$$\geq (E[N|N > 0])^2 P\{N > 0\}. \text{ (Why?)}$$

Now multiply both sides by $P\{N > 0\}$.

13. Consider a structure in which the minimal path sets are $\{1, 2, 3\}$ and $\{3, 4, 5\}$.
    (a) What are the minimal cut sets?
    (b) If the component lifetimes are independent uniform $(0, 1)$ random variables, determine the probability that the system life will be less than $\frac{1}{2}$.

14. Let $X_1, X_2, \ldots, X_n$ denote independent and identically distributed random variables and define the so-called order statistics $X_{(1)}, \ldots, X_{(n)}$ by

$$X_{(i)} \equiv i\text{th smallest of } X_1, \ldots, X_n$$

Show that if the distribution of X_j is IFR, then so is the distribution of $X_{(i)}$.
    Hint: Relate this to one of the examples of this chapter.

15. Consider the following four structures:

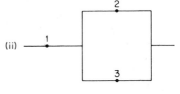

(ii)

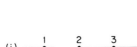

(i)

Figure 9.21.

Figure 9.22.

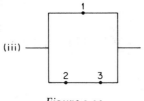

(iii)

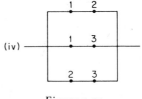

(iv)

Figure 9.23.

Figure 9.24

Let F_1, F_2, and F_3 be the corresponding component failure distributions; each of which is assumed to be IFR (increasing failure rate). Let F be the system failure distribution. All components are independent.

(a) For which structures is F necessarily IFR if $F_1 = F_2 = F_3$? Give reasons.

(b) For which structures is F necessarily IFR if $F_2 = F_3$? Give reasons.

(c) For which structures is F necessarily IFR if $F_1 \neq F_2 \neq F_3$? Give reasons.

16. Let X denote the lifetime of an item. Suppose the item has reached the age of t. Let X_t denote its remaining life and define

$$\bar{F}_t(a) = P\{X_t > a\}$$

In words, $\bar{F}_t(a)$ is the probability that a t-year-old item survives an additional time a. Show that

(a) $\bar{F}_t(a) = \bar{F}(t + a)/\bar{F}(t)$ where F is the distribution function of X.

(b) Another definition of IFR is to say that F is IFR if $\bar{F}_t(a)$ decreases in t, for all a. Show that this definition is equivalent to the one given in the text when F has a density.

17. Show that if each (independent) component of a series system has an IFR distribution, then the system lifetime is itself IFR by

(a) showing that

$$\lambda_F(t) = \sum_i \lambda_i(t)$$

where $\lambda_F(t)$ is the failure rate function of the system; and $\lambda_i(t)$ the failure rate function of the lifetime of component i.

(b) using the definition of IFR given in problem 16.

18. Show that if F is IFR, then it is also IFRA, and show by counterexample that the reverse is not true.

19. We say that ζ is a p-percentile of the distribution F if $F(\zeta) = p$. Show that if ζ is a p-percentile of the IFRA distribution F, then

$$1 - F(x) \leq e^{-\theta x}, \qquad x \geq \zeta$$
$$1 - F(x) \geq e^{-\theta x}, \qquad x \leq \zeta$$

where

$$\theta = \frac{-\log(1 - p)}{\zeta}$$

20. Prove Lemma 5.1.

Hint: Let $x = y + \delta$. Note that $f(t) = t^\alpha$ is a concave function when $0 \le \alpha \le 1$, and use the fact that for a concave function $f(t + h) - f(t)$ is decreasing in t.

21. Let $r(p) = r(p, p, \ldots, p)$. Show that if $r(p_0) = p_0$, then

$$r(p) \ge p \quad \text{for} \quad p \ge p_0$$
$$r(p) \le p \quad \text{for} \quad p \le p_0$$

Hint: Use Proposition 5.1.

22. Find the mean lifetime of a series system of two components when the component lifetimes are respectively uniform on $(0, 1)$ and uniform on $(0, 2)$. Repeat for a parallel system.

23. Show that the mean lifetime of a parallel system of two components is

$$\frac{1}{\mu_1 + \mu_2} + \frac{\mu_1}{(\mu_1 + \mu_2)\mu_2} + \frac{\mu_2}{(\mu_1 + \mu_2)\mu_1}$$

when the first component is exponentially distributed with mean $1/\mu_1$ and the second is exponential with mean $1/\mu_2$.

24. Compute the expected system lifetime of a three-out-of-four system when the first two component lifetimes are uniform on $(0, 1)$ and the second two are uniform on $(0, 2)$.

25. Show that the variance of the lifetime of a k-out-of-n system of components, each of whose lifetimes is exponential with mean θ, is given by

$$\theta^2 \sum_{i=k}^{n} 1/i^2$$

26. For the model of Section 7, compute (i) the average up time, (ii) the average down time, and (iii) the system failure rate for a k-out-of-n structure.

27. Prove the combinatorial identity

$$\binom{n-1}{i-1} = \binom{n}{i} - \binom{n}{i+1} + \cdots \pm \binom{n}{n}, \quad i \le n$$

(a) by induction on i
(b) by a backwards induction argument on i—that is, prove it first for $i = n$, then assume it for $i = k$ and show that this implies that it is true for $i = k - 1$.

References

1. R. E. Barlow and F. Proschan, "Statistical Theory of Reliability and Life Testing," Holt, New York, 1975.

2. H. Frank and I. Frisch, "Communication, Transmission, and Transportation Network," Addison-Wesley, Reading, Massachusetts, 1971.

Chapter 10

Brownian Motion and Stationary Processes

1. Brownian Motion

Let us start by considering the symmetric random walk which in each time unit is equally likely to take a unit step either to the left or to the right. That is, it is a Markov chain with $P_{i,i+1} = \frac{1}{2} = P_{i,i-1}$, $i = 0, \pm 1, \ldots$. Now suppose that we speed up this process by taking smaller and smaller steps in smaller and smaller time intervals. If we now go to the limit in the right manner what we obtain is Brownian motion.

More precisely, suppose that each Δt time unit we take a step of size Δx either to the left or the right with equal probabilities. If we let $X(t)$ denote the position at time t then

$$X(t) = \Delta x(X_1 + \cdots + X_{[t/\Delta t]}) \tag{1.1}$$

where

$$X_i = \begin{cases} +1 & \text{if the } i\text{th step of length } \Delta x \text{ is to the right} \\ -1 & \text{if it is to the left} \end{cases}$$

and $[t/\Delta t]$ is the largest integer less than or equal to $t/\Delta t$, and where the X_i are assumed independent with

$$P\{X_i = 1\} = P\{X_i = -1\} = \frac{1}{2}$$

As $E[X_i] = 0$, $\text{Var}(X_i) = E[X_i^2] = 1$, we see from (1.1) that

$$E[X(t)] = 0$$

$$\text{Var}(X(t)) = (\Delta x)^2 \left[\frac{t}{\Delta t}\right] \qquad (1.2)$$

We shall now let Δx and Δt go to 0. However, we must do it in a way to keep the resulting limiting process nontrivial (for instance, if we let $\Delta x = \Delta t$ and let $\Delta t \to 0$, then from the above we see that $E[X(t)]$ and $\text{Var}(X(t))$ would both converge to 0 and thus $X(t)$ would equal 0 with probability 1). If we let $\Delta x = c\sqrt{\Delta t}$ for some positive constant c then from (1.2) we see that as $\Delta t \to 0$

$$E[X(t)] = 0$$
$$\text{Var}(X(t)) \to c^2 t$$

We now list some intuitive properties of this limiting process obtained by taking $\Delta x = c\sqrt{\Delta t}$ and then letting $\Delta t \to 0$. From (1.1) and the Central Limit Theorem it seems reasonable that

(i) $X(t)$ is normal with mean 0 and variance $c^2 t$. In addition, as the changes of value of the random walk in nonoverlapping time intervals are independent, we have

(ii) $\{X(t), t \geq 0\}$ has independent increments, in that for all $t_1 < t_2 < \cdots < t_n$

$$X(t_n) - X(t_{n-1}), X(t_{n-1}) - X(t_{n-2}), \ldots, X(t_2) - X(t_1), X(t_1)$$

are independent. Finally, as the distribution of the change in position of the random walk over any time interval depends only on the length of that interval, it would appear that

(iii) $\{X(t), t \geq 0\}$ has stationary increments, in that the distribution of $X(t + s) - X(t)$ does not depend on t. We are now ready for the following formal definition.

Definition 1.1. A stochastic process $[X(t), t \geq 0]$ is said to be a *Brownian motion* process if

(i) $X(0) = 0$;
(ii) $\{X(t), t \geq 0\}$ has stationary and independent increments;
(iii) for every $t > 0$, $X(t)$ is normally distributed with mean 0 and variance $c^2 t$.

The Brownian motion process, sometimes called the Wiener process is one of the most useful stochastic processes in applied probability theory. It originated in physics as a description of Brownian motion. This phenomenon, named after the English botanist Robert Brown who discovered it, is the motion exhibited by a small particle which is totally immersed in a liquid or gas. Since then, the process has been used beneficially in such areas as statistical testing of goodness of fit, analyzing the price levels on the stock market, and quantum mechanics.

The first explanation of the phenomenon of Brownian motion was given by Einstein in 1905. He showed that Brownian motion could be explained by assuming that the immersed particle was continually being subjected to bombardment by the molecules of the surrounding medium. However, the above concise definition of this stochastic process underlying Brownian motion was given by Wiener in a series of papers originating in 1918.

When $c = 1$ the process is often called Standard Brownian Motion. As any Brownian motion can always be converted to the standard process by looking at $X(t)/c$ we shall suppose throughout that $c = 1$.

The interpretation of Brownian motion as the limit of the random walks (1.1) suggests that $X(t)$ should be a continuous function of t. This turns out to be the case, and it may be proven that, with probability 1, $X(t)$ is indeed a continuous function of t. This fact is quite deep, and no proof shall be attempted.

As $X(t)$ is normal with mean 0 and variance t, its density function is given by

$$f_t(x) = \frac{1}{\sqrt{2\pi t}} \, e^{-x^2/2t}$$

To obtain the joint density function of $X(t_1)$, $X(t_2)$, . . . , $X(t_n)$ for $t_1 < \cdot \cdot \cdot < t_n$, note first that the set of equalities

$$X(t_1) = x_1$$
$$X(t_2) = x_2$$
$$\vdots$$
$$X(t_n) = x_n$$

is equivalent to

$$X(t_1) = x_1$$
$$X(t_2) - X(t_1) = x_2 - x_1$$
$$\vdots$$
$$X(t_n) - X(t_{n-1}) = x_n - x_{n-1}$$

However, by the independent increment assumption it follows that $X(t_1)$, $X(t_2) - X(t_1)$, . . . , $X(t_n) - X(t_{n-1})$, are independent and, by the stationary increment assumption, that $X(t_k) - X(t_{k-1})$ is normal with mean 0 and variance $t_k - t_{k-1}$. Hence, the joint density of $X(t_1)$, . . . , $X(t_n)$ is given by

$$f(x_1, x_2, \ldots, x_n) = f_{t_1}(x_1) f_{t_2-t_1}(x_2 - x_1) \ldots f_{t_n-t_{n-1}}(x_n - x_{n-1})$$

$$= \frac{\exp\left\{-\frac{1}{2}\left[\frac{x_1^2}{t_1} + \frac{(x_2 - x_1)^2}{t_2 - t_1} + \cdots + \frac{(x_n - x_{n-1})^2}{t_n - t_{n-1}}\right]\right\}}{(2\pi)^{n/2}[t_1(t_2 - t_1) \ldots (t_n - t_{n-1})]^{1/2}} \tag{1.3}$$

By using (1.3), we may compute in principle any desired probabilities. For instance, suppose we require the conditional distribution of $X(s)$ given that $X(t) = B$ where $s < t$. The conditional density is

$$f_{s|t}(x \mid B) = \frac{f_s(x) f_{t-s}(B - x)}{f_t(B)}$$

$$= K_1 \exp\left\{-x^2/2s - (B - x)^2/2(t - s)\right\}$$

$$= K_2 \exp\left\{-x^2\left(\frac{1}{2s} + \frac{1}{2(t - s)}\right) + \frac{Bx}{t - s}\right\}$$

$$= K_2 \exp\left\{-\frac{t}{2s(t - s)}\left(x^2 - 2\frac{sB}{t}x\right)\right\}$$

$$= K_3 \exp\left\{-\frac{t(x - Bs/t)^2}{2s(t - s)}\right\}$$

where K_1, K_2, and K_3 do not depend on x. Hence, we see from the above that the conditional distribution of $X(s)$ given that $X(t) = B$ is, for $s < t$, normal with mean and variance given by

$$E[X(s)|X(t) = B] = \frac{s}{t} B$$

$$\mathrm{Var}[X(s)|X(t) = B] = \frac{s}{t}(t - s) \tag{1.4}$$

2. Hitting Times, Maximum Variable, and the Gambler's Ruin Problem

Let us denote by T_a the first time the Brownian motion process hits a. When $a > 0$ we will compute $P\{T_a \le t\}$ by considering $P\{X(t) \ge a\}$ and conditioning on whether or not $T_a \le t$. This gives

$$P\{X(t) \ge a\} =$$
$$P\{X(t) \ge a|T_a \le t\}P\{T_a \le t\} + P\{X(t) \ge a|T_a > t\}P\{T_a > t\} \tag{2.1}$$

Now if $T_a \le t$, then the process hits a at some point in $[0, t]$ and, by symmetry, it is just as likely to be above a or below a at time t. That is.

$$P\{X(t) \geq a \mid T_a \leq t\} = \tfrac{1}{2}$$

As the second right hand term of (2.1) is clearly equal to 0 (since, by continuity, the process value cannot be greater than a without having yet hit a), we see that

$$P\{T_a \leq t\} = 2P\{X(t) \geq a\}$$

$$= \frac{2}{\sqrt{2\pi t}} \int_a^\infty e^{-x^2/2t} \, dx$$

$$= \frac{2}{\sqrt{2\pi}} \int_{a/\sqrt{t}}^\infty e^{-y^2/2} \, dy, \quad a > 0 \tag{2.2}$$

For $a < 0$, the distribution of T_a is, by symmetry, the same as that of T_{-a}. Hence, from (2.2) we obtain

$$P\{T_a \leq t\} = \frac{2}{\sqrt{2\pi}} \int_{|a|/\sqrt{t}}^\infty e^{-y^2/2} \, dy \tag{2.3}$$

Another random variable of interest is the maximum value the process attains in $[0, t]$. Its distribution is obtained as follows: For $a > 0$

$$P\{\max_{0 \leq s \leq t} X(s) \geq a\} = P\{T_a \leq t\} \qquad \text{by continuity}$$

$$= 2P\{X(t) \geq a\} \qquad \text{from (2.2)}$$

$$= \frac{2}{\sqrt{2\pi}} \int_{a/\sqrt{t}}^\infty e^{-y^2/2} \, dy$$

Let us now consider the probability that Brownian motion hits A before $-B$ where $A > 0$, $B > 0$. To compute this we shall make use of the interpretation of Brownian motion as being a limit of the symmetric random walk. To start let us recall from the results of the gambler's ruin problem (see Section 5 of Chapter 4) that the probability that the symmetric random walk goes up A before going down B when each step is equally likely to be either up or down a distance Δx is (by Equation (5.1) of Section 5 of Chapter 4 with $N = (A + B)/\Delta x$, $i = B/\Delta x$) equal to $B\Delta x/(A + B)\Delta x = B/(A + B)$.

Hence, upon letting $\Delta x \to 0$ we see that

$$P\{\text{up } A \text{ before down } B\} = \frac{B}{A + B}$$

Example 2a Some people tend to believe that if $X(t)$ represents the market value of a given stock at time t then $\log X(t)$ is approximately a Brownian motion process. If this is so the probability that a stock whose price is presently equal to x_0 will reach the value αx_0 before declining to x_0/β, where $\alpha > 1$, $\beta > 1$, is equal to $\log \beta/(\log \alpha + \log \beta)$.

3. White Noise

Let $\{X(t), t \geq 0\}$ denote a standard Brownian motion process and let f be a function having a continuous derivative in the region $[a, b]$. The stochastic integral $\int_a^b f(t)dX(t)$ is defined as follows:

$$\int_a^b f(t)dX(t) \equiv \lim_{\substack{n \to \infty \\ \max(t_i - t_i - 1) \to 0}} \sum_{i=1}^n f(t_{i-1})[X(t_i) - X(t_{i-1})] \tag{3.1}$$

where $a = t_0 < t_1 < \ldots < t_n = b$ is a partition of the region $[a, b]$. Using the identity (the integration by parts formula applied to sums)

$$\sum_{i=1}^n f(t_{i-1})[X(t_i) - X(t_{i-1})]$$

$$= f(b)X(b) - f(a)X(a) - \sum_{i=1}^n X(t_i)[f(t_i) - f(t_{i-1})]$$

we see that

$$\int_a^b f(t)dX(t) = f(b)X(b) - f(a)X(a) - \int_a^b X(t)df(t) \tag{3.2}$$

The above Equation (3.2) is usually taken as the definition of $\int_a^b f(t)dX(t)$.

By using the right side of (3.2) we obtain, upon assuming the interchangeability of expectation and limit, that

$$E\left[\int_a^b f(t)dX(t)\right] = 0$$

Also,

$$\text{Var}\left(\sum_{i=1}^n f(t_{i-1})[X(t_i) - X(t_{i-1})]\right) = \sum_{i=1}^n f^2(t_{i-1})\text{Var}[X(t_i) - X(t_{i-1})]$$

$$= \sum_{i=1}^n f^2(t_{i-1})(t_i - t_{i-1})$$

where the top equality follows from the independent increments of Brownian motion. Hence, we obtain from (3.1) upon taking limits of the above that

$$\mathrm{Var}\left[\int_a^b f(t)dX(t)\right] = \int_a^b f^2(t)dt$$

Remarks   The above gives operational meaning to the family of quantities $\{dX(t), 0 \le t < \infty\}$ by viewing it as an operator that carries functions f into the values $\int_a^b f(t)dX(t)$. This is called a white noise transformation, or more loosely $\{dX(t), 0 \le t < \infty\}$ is called white noise since it can be imagined that a time varying function f travels through a white noise medium to yield the output (at time b) $\int_a^b f(t)dX(t)$.

Example 3a   Consider a particle of unit mass that is suspended in a liquid and suppose that, due to the liquid, there is a viscous force that retards the velocity of the particle at a rate proportional to its present velocity. In addition, let us suppose that the velocity instantaneously changes according to a constant multiple of white noise. That is, if $V(t)$ denotes the particle's velocity at t suppose that

$$V'(t) = -\beta V(t) + \alpha X'(t)$$

where $\{X(t), t \ge 0\}$ is standard Brownian motion. The above can be written as follows:

$$e^{\beta t}[V'(t) + \beta V(t)] = \alpha e^{\beta t} X'(t)$$

or

$$\frac{d}{dt}[e^{\beta t} V(t)] = \alpha e^{\beta t} X'(t)$$

Hence, upon integration, we obtain

$$e^{\beta t} V(t) = V(0) + \alpha \int_0^t e^{\beta s} X'(s)ds$$

or,

$$V(t) = V(0)e^{-\beta t} + \alpha \int_0^t e^{-\beta(t-s)}dX(s)$$

Hence, from (3.2)

$$V(t) = V(0)\,e^{-\beta t} + \alpha\left[X(t) - \int_0^t X(s)\,\beta e^{-\beta(t-s)}\,ds\right] \quad \diamond$$

4. Gaussian Processes

We start with the following definition.

Definition 4.1. A Stochastic Process $X(t)$, $t \geq 0$ is called a *Gaussian* or a *normal* process if $X(t_1)$, . . . , $X(t_n)$ has a multivariate normal distribution for all t_1, . . . , t_n.

If $\{X(t), t \geq 0\}$ is a Brownian motion process, then as each of $X(t_1)$, $X(t_2)$, . . . , $X(t_n)$ can be expressed as a linear combination of the independent normal random variables $X(t_1)$, $X(t_2) - X(t_1)$, $X(t_3) - X(t_2)$, . . . , $X(t_n) - X(t_{n-1})$ it follows that Brownian motion is a Gaussian process.

As a multivariate normal distribution is completely determined by the marginal mean values and the covariance values (see Section 6 of Chapter 2) it follows that Brownian motion could also be defined as a Gaussian process having $E[X(t)] = 0$ and, for $s \leq t$,

$$
\begin{aligned}
\text{Cov}\,(X(s), X(t)) &= \text{Cov}\,(X(s), X(s) + X(t) - X(s)) \\
&= \text{Cov}\,(X(s), X(s)) + \text{Cov}\,(X(s), X(t) - X(s)) \\
&= \text{Cov}\,(X(s), X(s))\ \text{by independent increments} \\
&= s\ \text{since Var}\,(X(s)) = s
\end{aligned}
\tag{4.1}
$$

Let $\{X(t), t \geq 0\}$ be a Brownian motion process and consider the process values between 0 and 1 conditional on $X(1) = 0$. That is, consider the conditional stochastic process $\{X(t), 0 \leq t \leq 1 | X(1) = 0\}$. Since the conditional distribution of $X(t_1)$, . . . , $X(t_n)$ is multivariate normal it follows that this conditional process, known as the *Brownian Bridge* (as it is tied down both at 0 and at 1), is a Gaussian process. Let us compute its covariance function. As, from (1.4)

$$
E[X(s) | X(1) = 0] = 0, \quad \text{for} \quad s < 1
$$

we have that, for $s < t < 1$,

$$
\begin{aligned}
\text{Cov}\,[(X(s), X(t)) | X(1) = 0] \\
&= E[X(s)X(t) | X(1) = 0] \\
&= E[E[X(s)X(t) | X(t), X(1) = 0] | X(1) = 0] \\
&= E[X(t)E[X(s) | X(t)] | X(1) = 0] \\
&= E\left[X(t)\frac{s}{t}X(t) | X(1) = 0\right] \quad \text{by (1.4)}
\end{aligned}
$$

$$= \frac{s}{t} E[X^2(t)|X(1) = 0]$$

$$= \frac{s}{t} t(1 - t) \quad \text{by (1.4)}$$

$$= s(1 - t)$$

Thus the Brownian Bridge can be defined as a Gaussian process with mean value 0 and covariance function $s(1 - t)$, $s \leq t$. This leads to an alternative approach to obtaining such a process.

Proposition 4.1. If $\{X(t), t \geq 0\}$ is Brownian motion, then $\{Z(t), 0 \leq t \leq 1\}$ is a Brownian Bridge process when $Z(t) = X(t) - tX(1)$.

Proof: As it is immediate that $\{Z(t), t \geq 0\}$ is a Gaussian process, all we need verify is that $E[Z(t)] = 0$ and Cov $(Z(s), Z(t)) = s(1 - t)$, when $s \leq t$. The former is immediate and the latter follows from

$$\begin{aligned}
\text{Cov}\ (Z(s), Z(t)) &= \text{Cov}\ (X(s) - sX(1), X(t) - tX(1)) \\
&= \text{Cov}\ (X(s), X(t)) - t\ \text{Cov}\ (X(s), X(1)) \\
&= -s\ \text{Cov}\ (X(1), X(t)) + st\ \text{Cov}\ (X(1), X(1)) \\
&= s - st - st + st \\
&= s(1 - t)
\end{aligned}$$

and the proof is complete.   ◇

If $\{X(t), t \geq 0\}$ is Brownian motion, then the process $\{Z(t), t \geq 0\}$ defined by

$$Z(t) = \int_0^t X(s)\ ds \tag{4.2}$$

is called *Integrated Brownian Motion*. As an illustration of how such a process may arise in practice, suppose we are interested in modelling the price of a commodity throughout time. Letting $Z(t)$ denote the price at t then, rather than assuming that $\{Z(t)\}$ is Brownian motion (or that log $Z(t)$ is Brownian motion), we might want to assume that the rate of change of $Z(t)$ follows a Brownian motion. For instance we might suppose that the rate of change of the commodity's price is the current inflation rate which is imagined to vary as Brownian motion. Hence,

$$\frac{d}{dt} Z(t) = X(t)$$

or

$$Z(t) = Z(0) + \int_0^t X(s) \, ds$$

It follows from the fact that Brownian motion is a Gaussian process that $\{Z(t), t \geq 0\}$ is also Gaussian. To prove this, first recall that W_1, . . . , W_n is said to have a multivariate normal distribution if they can be represented as

$$W_i = \sum_{j=1}^m a_{ij} U_j, \quad i = 1, \ldots n$$

where $U_j, j = 1, \ldots, m$ are independent normal random variables. From this it follows that any set of partial sums of $W_1, \ldots, W_n$ are also jointly normal. The fact that $Z(t_1), \ldots, Z(t_n)$ is multivariate normal can now be shown by writing the integral in (4.2) as a limit of approximating sums.

As $\{Z(t), t \geq 0\}$ is Gaussian it follows that its distribution is characterized by its mean value and covariance function. We now compute these

$$E[Z(t)] = E\left[\int_0^t X(s) \, ds\right]$$

$$= \int_0^t E[X(s)] \, ds$$

$$= 0$$

For $s \leq t$,

$$\text{Cov}[Z(s), Z(t)] = E[Z(s) Z(t)]$$

$$= E\left[\int_0^s X(y) \, dy \int_0^t X(u) \, du\right]$$

$$= E\left[\int_0^s \int_0^t X(y) X(u) \, dy \, du\right]$$

$$= \int_0^s \int_0^t E[X(y) X(u)] \, dy \, du$$

$$= \int_0^s \int_0^t \min(y, u) \, dy \, du \qquad \text{by (4.1)}$$

$$= \int_0^s \left(\int_0^u y\,dy + \int_u^t u\,dy \right) du$$

$$= s^2 \left(\frac{t}{2} - \frac{s}{6} \right) \quad \Diamond$$

5. Stationary and Weakly Stationary Processes

A stochastic process $\{X(t), t \geq 0\}$ is said to be a *stationary process* if for all n, s, $t_1, \ldots, t_n$ the random vectors $X(t_1), \ldots, X(t_n)$ and $X(t_1 + s), \ldots, X(t_n + s)$ have the same joint distribution. In other words, a process is stationary if, in choosing any fixed point s as the origin, the ensuing process has the same probability law. Two examples of stationary processes are:

(i) An ergodic continuous time Markov chain $\{X(t), t \geq 0\}$ when

$$P\{X(0) = j\} = P_j, \quad j \geq 0$$

where $\{P_j, j \geq 0\}$ are the limiting probabilities.

(ii) $\{X(t), t \geq 0\}$ when $X(t) = N(t + L) - N(t)$, $t \geq 0$, where $L > 0$ is a fixed constant and $\{N(t), t \geq 0\}$ is a Poisson process having rate λ.

The first one of the above processes is stationary for it is a Markov chain whose initial state is chosen according to the limiting probabilities, and it can thus be regarded as an ergodic Markov chain that one starts observing at time ∞. Hence the continuation of this process a time s after observation begins is just the continuation of the chain starting at time $\infty + s$, which clearly has the same probability law for all s. That the second example—where $X(t)$ represents the number of events of a Poisson process that occur between t and $t + L$—is stationary follows from the stationary and independent increment assumption of the Poisson process which implies that the continuation of a Poisson process at any time s remains a Poisson process.

Example 5a. (The Random Telegraph Signal Process):   Let $\{N(t), t \geq 0\}$ denote a Poisson process and let X_0 be independent of this process and be such that $P\{X_0 = 1\} = P\{X_0 = -1\} = \frac{1}{2}$. Defining $X(t) = X_0(-1)^{N(t)}$ then $\{X(t), t \geq 0\}$ is called *random telegraph signal* process. To see that it is stationary, note first that starting at any time t, no matter what the value of $N(t)$, as X_0 is equally likely to be either plus or minus 1, it follows that $X(t)$ is equally likely to be either plus or minus 1. Hence, as the continuation of a Poisson process beyond any time remains a Poisson process it follows that $\{X(t), t \geq 0\}$ is a stationary process.

Let us compute the mean and covariance function of the random telegraph signal

$$
\begin{aligned}
E[X(t)] &= E[X_0(-1)^{N(t)}] \\
&= E[X_0]E[(-1)^{N(t)}] \quad \text{by independence} \\
&= 0 \quad \text{since} \quad E[X_0] = 0
\end{aligned}
$$

$$
\begin{aligned}
\text{Cov}\,[X(t), X(t+s)] &= E[X(t)X(t+s)] \\
&= E[X_0^2(-1)^{N(t)+N(t+s)}] \\
&= E[(-1)^{2N(t)}(-1)^{N(t+s)-N(t)}] \\
&= E[(-1)^{N(t+s)-N(t)}] \\
&= E[(-1)^{N(s)}] \\
&= \sum_{i=0}^{\infty} (-1)^i e^{-\lambda s} \frac{(\lambda s)^i}{i!} \\
&= e^{-2\lambda s} \tag{5.1}
\end{aligned}
$$

For an application of the random telegraph signal consider a particle moving at a constant unit velocity along a straight line and suppose that collisions involving this particle occur at a Poisson rate λ. Also suppose that each time the particle suffers a collision it reverses direction. Therefore, if X_0 represents the initial velocity of the particle, then its velocity at time t—call it $X(t)$— is given by $X(t) = X_0(-1)^{N(t)}$ where $N(t)$ denotes the number of collisions involving the particle by time t. Hence, if X_0 is equally likely to be plus or minus 1, and is independent of $\{N(t), t \geq 0\}$ then $\{X(t), t \geq 0\}$ is a random telegraph signal process. If we now let

$$
D(t) = \int_0^t X(s)\,ds
$$

then $D(t)$ represents the displacement of the particle at time t from its position at time 0. The mean and variance of $D(t)$ are obtained as follows:

$$
E[D(t)] = \int_0^t E[X(s)]\,ds = 0
$$

$$
\text{Var}\,[D(t)] = E[D^2(t)]
$$

$$= E\left[\int_0^t X(y)dy \int_0^t X(u)du\right]$$

$$= \int_0^t \int_0^t E[X(y)X(u)]dy\,du$$

$$= 2 \iint_{0<y<u<t} E[X(y)X(u)]dy\,du$$

$$= 2 \int_0^t \int_0^u e^{-2\lambda(u-y)}dy\,du \qquad \text{by (5.1)}$$

$$= \frac{1}{\lambda}\left(t - \frac{1}{2\lambda} + \frac{1}{2\lambda}e^{-2\lambda t}\right) \qquad \diamond$$

The condition for a process to be stationary is rather stringent and so we define the process $\{X(t), t \geq 0\}$ to be a *second order stationary* or a *weakly stationary* process if $E[X(t)] = c$ and $\text{Cov}\,[X(t), X(t+s)]$ does not depend on t. That is, a process is second order stationary if the first two moments of $X(t)$ are the same for all t and the covariance between $X(s)$ and $X(t)$ depends only on $|t - s|$. For a second order stationary process, let

$$R(s) = \text{Cov}\,[X(t), X(t+s)]$$

As the finite dimensional distributions of a Gaussian process (being multivariate normal) are determined by their means and covariance, it follows that a second order stationary Gaussian process is stationary.

Example 5b (The Ornstein-Uhlenbeck Process): Let $\{X(t), t \geq 0\}$ be a Brownian motion process, and define, for $\alpha > 0$,

$$V(t) = e^{-\alpha t/2}X(e^{\alpha t})$$

The process $\{V(t), t \geq 0\}$ is called the Ornstein-Uhlenbeck process. It has been proposed as a model for describing the velocity of a particle immersed in a liquid or gas, and as such is useful in statistical mechanics. Let us compute its mean and covariance function.

$$E[V(t)] = 0$$
$$\text{Cov}\,[V(t), V(t+s)] = e^{-\alpha t/2}e^{-\alpha(t+s)/2}\text{Cov}\,[X(e^{\alpha t}), X(e^{\alpha(t+s)})]$$
$$= e^{-\alpha t}e^{-\alpha s/2}e^{\alpha t} \qquad \text{by Equation (4.1)}$$
$$= e^{-\alpha s/2}$$

Hence, $\{V(t), t \geq 0\}$ is weakly stationary and as it is clearly a Gaussian process (since Brownian motion is Gaussian) we can conclude that it is stationary. It is interesting to note that (with $\alpha = 4\lambda$) it has the same mean and covariance function as the Random Telegraph Signal process, thus illustrating that two quite different processes can have the same second order properties. (Of course if two Gaussian processes have the same mean and covariance functions then they are identically distributed.) ◊

As the following examples show, there are many types of second order stationary processes that are not stationary.

Example 5c (An Auto Regressive Process): Let $Z_0, Z_1, Z_2, \ldots,$ be uncorrelated random variables with $E[Z_n] = 0, n \geq 0$ and

$$\text{Var}(Z_n) = \begin{cases} \sigma^2/(1 - \lambda^2), & n = 0 \\ \sigma^2, & n \geq 1 \end{cases}$$

where $\lambda^2 < 1$. Define

$$X_0 = Z_0 \tag{5.2}$$
$$X_n = \lambda X_{n-1} + Z_n, \qquad n \geq 1$$

The process $\{X_n, n \geq 0\}$ is called a *first-order auto regressive process*. It says that the state at time n (that is, X_n) is a constant multiple of the state at time $n - 1$ plus a random error term Z_n.
Iterating (5.2) yields

$$X_n = \lambda(\lambda X_{n-2} + Z_{n-1}) + Z_n$$
$$= \lambda^2 X_{n-2} + \lambda Z_{n-1} + Z_n$$
$$\vdots$$
$$= \sum_{i=0}^{n} \lambda^{n-i} Z_i$$

and so

$$\text{Cov}(X_n, X_{n+m}) = \text{Cov}\left(\sum_{i=0}^{n} \lambda^{n-i} Z_i, \sum_{i=0}^{n+m} \lambda^{n+m-i} Z_i\right)$$
$$= \sum_{i=0}^{n} \lambda^{n-i} \lambda^{n+m-i} \text{Cov}(Z_i, Z_i)$$
$$= \sigma^2 \lambda^{2n+m}\left(\frac{1}{1 - \lambda^2} + \sum_{i=1}^{n} \lambda^{-2i}\right)$$

$$= \frac{\sigma^2 \lambda^m}{1 - \lambda^2}$$

where the above uses the fact that Z_i and Z_j are uncorrelated when $i \neq j$. As $E[X_n] = 0$, we see that $\{X_n, n \geq 0\}$ is weakly stationary (the definition for a discrete time process is the obvious analog of that given for continuous time processes).   ◇

Example 5d If, in the random telegraph signal process, we drop the requirement that $P\{X_0 = 1\} = P\{X_0 = -1\} = \frac{1}{2}$ and only require that $E[X_0] = 0$, then the process $\{X(t), t \geq 0\}$ need no longer be stationary. (It will remain stationary if X_0 has a symmetric distribution in the sense that $-X_0$ has the same distribution as X_0.) However, the process will be weakly stationary since

$$E[X(t)] = E[X_0]E[(-1)^{N(t)}] = 0$$
$$\text{Cov}\,[X(t), X(t + s)] = E[X(t)X(t + s)]$$
$$= E[X_0^2]E[(-1)^{N(t)+N(t+s)}]$$
$$= E[X_0^2]e^{-2\lambda s} \qquad \text{from (5.1)}   ◇$$

Example 5e Let $W_0, W_1, W_2, \ldots$ be uncorrelated with $E[W_n] = \mu$ and $\text{Var}(W_n) = \sigma^2$, $n \geq 0$, and for some positive integer k define

$$X_n = \frac{W_n + W_{n-1} + \cdots + W_{n-k}}{k + 1}, \qquad n \geq k$$

The process $\{X_n, n \geq k\}$, which at each time keeps track of the arithmetic average of the most recent $k + 1$ values of the W's, is called a moving average process. Using the fact that the $W_n, n \geq 0$ are uncorrelated, we see that

$$\text{Cov}\,(X_n, X_{n+m}) = \begin{cases} \dfrac{(k + 1 - m)\sigma^2}{(k + 1)^2} & \text{if}\quad 0 \leq m \leq k \\[2mm] 0 & \text{if}\quad m > k \end{cases}$$

Hence, $\{X_n, n \geq k\}$ is a second order stationary process.   ◇

Let $\{X_n, n \geq 1\}$ be a second order stationary process with $E[X_n] = \mu$. An important question is when, if ever, does $\overline{X}_n \equiv \sum_{i=1}^{n} X_i / n$ converge to μ? The following proposition, which we state without proof, shows that $E[(\overline{X}_n - \mu)^2] \to 0$ if and only if $\sum_{i=1}^{n} R(i)/n \to 0$. That is, the expected square of the difference between $\overline{X}_n$ and μ will converge to 0 if and only if the limiting average value of $R(i)$ converges to 0.

Proposition 5.1. Let $\{X_n, n \geq 1\}$ be a second order stationary process having mean μ and covariance function $R(i) = \text{Cov}(X_n, X_{n+i})$, and let $\overline{X}_n \equiv \Sigma_{i=1}^n X_i/n$. Then $\lim_{n \to \infty} E[(\overline{X}_n - \mu)^2] = 0$ if and only if $\lim_{n \to \infty} \Sigma_{i=1}^n R(i)/n = 0$.

6. Harmonic Analysis of Weakly Stationary Processes

Suppose that the stochastic processes $\{X(t), -\infty < t < \infty\}$ and $\{Y(t), -\infty < t < \infty\}$ are related as follows:

$$Y(t) = \int_{-\infty}^{\infty} X(t - s)h(s)ds \qquad (6.1)$$

We can imagine that a signal, whose value at time t is $X(t)$, is passed through a physical system that distorts its value so that $Y(t)$, the received value at t, is given by (6.1). The processes $\{X(t)\}$ and $\{Y(t)\}$ are called respectively the input and output processes. The function h is called the *impulse response* function. If $h(s) = 0$ whenever $s < 0$, then h is also called a weighting function since (6.1) expresses the output at t as a weighted integral of all the inputs prior to t with $h(s)$ representing the weight given the input s time units ago.

The relationship expressed by (6.1) is a special case of a time invariant linear filter. It is called a filter because we can imagine that the input process $\{X(t)\}$ is passed through a medium and then filtered to yield the output process $\{Y(t)\}$. It is a linear filter because if the input processes $\{X_i(t)\}$, $i = 1, 2$, result in the output processes $\{Y_i(t)\}$—that is, if $Y_i(t) = \int_0^\infty X_i(t - s)h(s)ds$—then the output process corresponding to the input process $\{aX_1(t) + bX_2(t)\}$ is just $\{aY_1(t) + bY_2(t)\}$. It is called time invariant since lagging the input process by a time τ—that is, considering the new input process $\overline{X}(t) = X(t + \tau)$—results in a lag of τ in the output process since $\int_0^\infty \overline{X}(t - s)h(s)ds = \int_0^\infty X(t + \tau - s) h(s)ds = Y(t + \tau)$.

Let us now suppose that the input process $\{X(t), -\infty < t < \infty\}$ is weakly stationary with $E[X(t)] = 0$ and covariance function $R_X(s) = \text{Cov}[X(t), X(t + s)]$. Let us compute the mean value and covariance function of the output process $\{Y(t)\}$.

Assuming that we can interchange the expectation and integration operations (a sufficient condition being that $\int |h(s)|ds < \infty^*$ and, for some $M < \infty$, $E|X(t)| < M$ for all t) we obtain that

---

*The range of all integrals in this section is from $-\infty$ to $+\infty$.

$$E[Y(t)] = \int E[X(t - s)]h(s)ds$$
$$= 0$$

Similarly,

$$\text{Cov}\,[Y(t_1), Y(t_2)] = \text{Cov}\left[\int X(t_1 - s_1)h(s_1)ds_1, \int X(t_2 - s_2)h(s_2)ds_2\right]$$

$$= \int\int \text{Cov}\,[X(t_1 - s_1), X(t_2 - s_2)]h(s_1)h(s_2)ds_1ds_2$$

$$= \int\int R_X(t_2 - s_2 - t_1 + s_1)h(s_1)h(s_2)ds_1ds_2 \qquad (6.2)$$

Hence, Cov $[Y(t_1), Y(t_2)]$ depends on t_1, t_2 only through $t_2 - t_1$; thus showing that $\{Y(t)\}$ is also weakly stationary.

The above expression for $R_Y(t_2 - t_1) = \text{Cov}\,[Y(t_1), Y(t_2)]$ is, however, more compactly and usefully expressed in terms of Fourier transforms of R_X and R_Y. Let, for $i = \sqrt{-1}$,

$$\tilde{R}_X(w) = \int e^{-iws}R_X(s)ds$$

and

$$\tilde{R}_Y(w) = \int e^{-iws}R_Y(s)ds$$

denote the Fourier transforms respectively of R_X and R_Y. The function $\tilde{R}_X(w)$ is also called the *power spectral density* of the process $\{X(t)\}$. Also let

$$\tilde{h}(w) = \int e^{-iws}h(s)ds$$

denote the Fourier transform of the function h. Then, from (6.2)

$$\tilde{R}_Y(w) = \int\int\int e^{-iws}R_X(s - s_2 + s_1)h(s_1)h(s_2)ds_1ds_2ds$$

$$= \int\int\int e^{-iw(s-s_2+s_1)}R_X(s - s_2 + s_1)ds\,e^{-iws_2}h(s_2)ds_2e^{iws_1}h(s_1)ds_1$$

$$= \tilde{R}_X(w)\tilde{h}(w)\tilde{h}(-w) \qquad (6.3)$$

Now, using the representation

$$e^{ix} = \cos x + i \sin x$$
$$e^{-ix} = \cos(-x) + i \sin(-x) = \cos x - i \sin x$$

we obtain that

$$\tilde{h}(w)\tilde{h}(-w) = \left[\int h(s) \cos(ws)ds - i \int h(s) \sin(ws)ds \right]$$

$$\times \left[\int h(s) \cos(ws)ds + i \int h(s) \sin(ws)ds \right]$$

$$= \left[\int h(s) \cos(ws)ds \right]^2 + \left[\int h(s) \sin(ws)ds \right]^2$$

$$= \left| \int h(s)e^{-iws}ds \right|^2 = |\tilde{h}(w)|^2$$

Hence, from (6.3) we obtain

$$\tilde{R}_Y(w) = \tilde{R}_X(w)|\tilde{h}(w)|^2$$

In words, the Fourier transform of the covariance function of the output process is equal to the square of the amplitude of the Fourier transform of the impulse function multiplied by the Fourier transform of the covariance function of the input process.

Problems

In the following problems, unless otherwise specified, $\{X(t), t \geq 0\}$ is a standard Brownian motion process and T_a denotes the time it takes this process to hit a.

1. What is the distribution of $X(s) + X(t)$, $s \leq t$?

2. Compute the conditional distribution of $X(s)$ given that $X(t_1) = A$, $X(t_2) = B$ where $0 < t_1 < s < t_2$.

3. Let $Y(t) = X(t) + \mu t$ for some constant μ. The process $\{Y(t), 0 < t < \infty\}$ is called a Brownian motion process with drift rate μ. Compute the joint density of $Y(s)$ and $Y(t)$, $s < t$.

4. Consider the random walk which in each Δt time unit either goes up or down the amount $\sqrt{\Delta t}$ with respective probabilities p and $1 - p$ where $p = 1/2 (1 + \mu\sqrt{\Delta t})$.
   (a) Argue that as $\Delta t \to 0$ the resulting limiting process is a Brownian motion process with drift rate μ.

(b) Using (a) and the results of the gambler's ruin problem, compute the probability that a Brownian motion process with drift rate μ goes up A before going down B, $A > 0$, $B > 0$.

5. Compute $E[X(t_1)X(t_2)X(t_3)]$ for $t_1 < t_2 < t_3$.

6. Show that

$$P\{T_a < \infty\} = 1$$
$$E[T_a] = \infty, \, a \neq 0$$

7. What is $P\{T_1 < T_{-1} < T_2\}$?

8. Suppose you own one share of a stock whose price changes according to a standard Brownian motion process. Suppose that you purchased the stock at a price $b + c$, $c > 0$, and the present price is b. You have decided to sell the stock either when it reaches the price $b + c$ or when an additional time t goes by (whichever occurs first). What is the probability that you do not recover your purchase price?

9. Compute an expression for

$$P\{\max_{t_1 \leq s \leq t_2} X(s) > x\}$$

10. Compute the mean and variance of
(a) $\int_0^1 t \, dX(t)$.
(b) $\int_0^1 t^2 \, dX(t)$.

11. Let $Y(t) = tX(1/t)$, $t > 0$ and $Y(0) = 0$.
(a) What is the distribution of $Y(t)$?
(b) Compute $\text{Cov}(Y(s), Y(t))$.
(c) Argue that $\{Y(t), t \geq 0\}$ is a standard Brownian motion process.

12. Let $Y(t) = X(a^2t)/a$ for $a > 0$. Argue that $\{Y(t)\}$ is a standard Brownian motion process.

13. Let $\{Z(t), t \geq 0\}$ denote a Brownian Bridge process. Show that if

$$Y(t) = (t + 1)Z(t/(t + 1))$$

then $\{Y(t), t \geq 0\}$ is a standard Brownian motion process.

14. Let $X(t) = N(t + 1) - N(t)$ where $\{N(t), t \geq 0\}$ is a Poisson process with rate λ. Compute

$$\text{Cov}\,[X(t), X(t + s)]$$

15. Let $\{N(t), t \geq 0\}$ denote a Poisson process with rate λ and define $Y(t)$ to be the time from t until the next Poisson event.

(a) Argue that $\{Y(t), t \geq 0\}$ is a stationary process.

(b) Compute Cov $[Y(t), Y(t + s)]$.

16. Let $\{X(t), -\infty < t < \infty\}$ be a weakly stationary process having covariance function $R_X(s) = \text{Cov}[X(t), X(t + s)]$.

(a) Show that

$$\text{Var}(X(t + s) - X(t)) = 2R_X(0) - 2R_X(t)$$

(b) If $Y(t) = X(t + 1) - X(t)$ show that $\{Y(t), -\infty < t < \infty\}$ is also weakly stationary having a covariance function $R_Y(s) = \text{Cov}[Y(t), Y(t + s)]$ that satisfies

$$R_Y(s) = 2R_X(s) - R_X(s - 1) - R_X(s + 1)$$

17. Let Y_1 and Y_2 be independent unit normal random variables and for some constant w set

$$X(t) = Y_1 \cos wt + Y_2 \sin wt, \qquad -\infty < t < \infty$$

(a) Show that $\{X(t)\}$ is a weakly stationary process.

(b) Argue that $\{X(t)\}$ is a stationary process.

18. Let $\{X(t), -\infty < t < \infty\}$ be weakly stationary with covariance function $R(s) = \text{Cov}(X(t), X(t + s))$ and let $\tilde{R}(w)$ denote the power spectral density of the process.

(i) Show that $\tilde{R}(w) = \tilde{R}(-w)$. It can be shown that

$$R(s) = \frac{1}{2\pi} \int_{-\infty}^{\infty} \tilde{R}(w)e^{iws}\,dw$$

(ii) Use the above to show that

$$\int_{-\infty}^{\infty} \tilde{R}(w)\,dw = 2\pi E[X^2(t)]$$

References

1. M. S. Bartlett, "An Introduction to Stochastic Processes," Cambridge Univ. Press, London, 1954.

2. U. Grenander and M. Rosenblatt, "Statistical Analysis of Stationary Time Series," John Wiley, New York, 1957.

3. S. Karlin and H. Taylor, "A Second Course in Stochastic Processes," Academic Press, Orlando, Fla, 1981.

4. L. H. Koopmans, "The Spectral Analysis of Time Series," Academic Press, Orlando, Fla, 1974.

5. S. Ross, "Stochastic Processes," John Wiley, New York, 1983.

Chapter 11

Simulation

1. Introduction

Let $\mathbf{X} = (X_1, \ldots, X_n)$ denote a random vector having a given density function $f(x_1, \ldots, x_n)$ and suppose we are interested in computing

$$E[g(\mathbf{X})] = \int\int \cdots \int g(x_1, \ldots, x_n) f(x_1, \ldots, x_n) dx_1 dx_2 \ldots dx_n$$

for some n dimensional function g. For instance g could represent the total delay in queue of the first $[\frac{n}{2}]$ customers when the X values represent the first $[\frac{n}{2}]$ interarrival and service times.* In many situations, it is not analytically possible either to compute the above multiple intergral exactly or even to numerically approximate it within a given accuracy. One possibility that remains is to approximate $E[g(\mathbf{X})]$ by means of simulation.

To approximate $E[g(\mathbf{X})]$, start by generating a random vector $\mathbf{X}^{(1)} = (X_1^{(1)}, \ldots, X_n^{(1)})$ having the joint density $f(x_1, \ldots, x_n)$ and then compute $Y^{(1)} = g(\mathbf{X}^{(1)})$. Now generate a second random vector (independent of the first) $\mathbf{X}^{(2)}$ and compute $Y^{(2)} = g(\mathbf{X}^{(2)})$. Keep on doing this until r, a fixed number, of independent and identically distributed random variables $Y^{(i)} = g(\mathbf{X}^{(i)})$, $i = 1, \ldots, r$ have been generated. Now by the strong law of large numbers, we know that

---

*We are using the notation [a] to represent the largest integer less than or equal to a.

$$\lim_{r \to \infty} \frac{Y^{(1)} + \cdots + Y^{(r)}}{r} = E[Y^{(i)}] = E[g(\mathbf{X})]$$

and so we can use the average of the generated Y's as an estimate of $E[g(\mathbf{X})]$. This approach to estimating $E[g(\mathbf{X})]$ is called the *Monte-Carlo Simulation* approach.

Clearly there remains the problem of how to generate, or *simulate,* random vectors having a specified joint distribution. The first step in doing this is to be able to generate random variables from a uniform distribution on $(0, 1)$. One way to do this would be to take 10 identical slips of paper, numbered $0, 1, \ldots, 9$, place them in a hat and then successively select n slips, with replacement, from the hat. The sequence of digits obtained (with a decimal point in front) can be regarded as the value of a uniform $(0, 1)$ random variable rounded off to the nearest $(\frac{1}{10})^n$. For instance, if the sequence of digits selected is 3, 8, 7, 2, 1, then the value of the uniform $(0, 1)$ random variable is .38721 (to the nearest .00001). Tables of the values of uniform $(0, 1)$ random variables, known as random number tables, have been extensively published [for instance, see The RAND Corporation, *A Million Random Digits with 100,000 Normal Deviates* (New York: The Free Press, 1955)].

However, the above is not the way in which digital computers simulate uniform $(0, 1)$ random variables. In practice, they use pseudo random numbers instead of truly random ones. Most random number generators start with an initial value X_0, called the seed, and then recursively compute values by specifying positive integers a, c, and m, and then letting

$$X_{n+1} = (aX_n + c) \text{ modulo } m, \, n \geq 0$$

where the above means that $aX_n + c$ is divided by m and the remainder is taken as the value of X_{n+1}. Thus each X_n is either $0, 1, \ldots, m - 1$ and the quantity X_n/m is taken as an approximation to a uniform $(0, 1)$ random variable. It can be shown that subject to suitable choices for a, c, m, the above gives rise to a sequence of numbers that looks as if it was generated from independent uniform $(0, 1)$ random variables.

As our starting point in the simulation of random variables from an arbitrary distribution, we shall suppose that we can simulate from the uniform $(0, 1)$ distribution, and we shall use the term "random numbers" to mean independent random variables from this distribution. In Sections 2 and 3 we present both general and special techniques for simulating continuous random variables; and in Section 4 we do the same for discrete random variables. In Section 5 we discuss the simulation both of jointly distributed random variables and stochastic processes. Particular attention is given to the simulation of nonhomoge-

Table 11.1   **A Random Number Table**

04839	96423	24878	82651	66566	14778	76797	14780	13300	87074
68086	26432	46901	20849	89768	81536	86645	12659	92259	57102
39064	66432	84673	40027	32832	61362	98947	96067	64760	64584
25669	26422	44407	44048	37937	63904	45766	66134	75470	66520
64117	94305	26766	25940	39972	22209	71500	64568	91402	42416
87917	77341	42206	35126	74087	99547	81817	42607	43808	76655
62797	56170	86324	88072	76222	36086	84637	93161	76038	65855
95876	55293	18988	27354	26575	08625	40801	59920	29841	80150
29888	88604	67917	48708	18912	82271	65424	69774	33611	54262
73577	12908	30883	18317	28290	35797	05998	41688	34952	37888
27958	30134	04024	86385	29880	99730	55536	84855	29080	09250
90999	49127	20044	59931	06115	20542	18059	02008	73708	83517
18845	49618	02304	51038	20655	58727	28168	15475	56942	53389
94824	78171	84610	82834	09922	25417	44137	48413	25555	21246
35605	81263	39667	47358	56873	56307	61607	49518	89356	20103
33362	64270	01638	92477	66969	98420	04880	45585	46565	04102
88720	82765	34476	17032	87589	40836	32427	70002	70663	88863
39475	46473	23219	53416	94970	25832	69975	94884	19661	72828
06990	67245	68350	82948	11398	42878	80287	88267	47363	46634
40980	07391	58745	25774	22987	80059	39911	96189	41151	14222
83974	29992	65381	38857	50490	83765	55657	14361	31720	57375
33339	31926	14883	24413	59744	92351	97473	89286	35931	04110
31662	25388	61642	34072	81249	35648	56891	69352	48373	45578
93526	70765	10592	04542	76463	54328	02349	17247	28865	14777
20492	38391	91132	21999	59516	81652	27195	48223	46751	22923
04153	53381	79401	21438	83035	92350	36693	31238	59649	91754
05520	91962	04739	13092	97662	24822	94730	06496	35090	04822
47498	87637	99016	71060	88824	71013	18735	20286	23153	72924
23167	49323	45021	33132	12544	41035	80780	45393	44812	12515
23792	14422	15059	45799	22716	19792	09983	74353	68668	30429
85900	98275	32388	52390	16815	69298	82732	38480	73817	32523
42559	78985	05300	22164	24369	54224	35083	19687	11062	91491
14349	82674	66523	44133	00697	35552	35970	19124	63318	29686
17403	53363	44167	64486	64758	75366	76554	31601	12614	33072
23632	27889	47914	02584	37680	20801	72152	39339	34806	08930

neous Poisson processes, and in fact three different approaches for this are discussed. Simulation of two-dimensional Poisson processes is discussed in Section 5.2. In Section 6 we discuss various methods for increasing the precision of the simulation estimates by reducing their variance; and in Section 7 we consider the problem of choosing the number of simulation runs needed to attain a desired level of precision. Before beginning this program, however, let us consider two applications of simulation to combinatorial problems.

Example 1a   (Generating a Random Permutation): Suppose we are interested in generating a permutation of the numbers $1, 2, \ldots, n$ that is such that all $n!$ possible orderings are equally likely. The following algorithm will accomplish this by first choosing one of the numbers 1, . . . , n at random and then putting that number in position n; it then chooses at random one of the remaining $n - 1$ numbers and puts that number in position $n - 1$; it then chooses at random one of the remaining $n - 2$ numbers and puts it in position $n - 2$, and so on (where choosing a number at random means that each of the remaining numbers is equally likely to be chosen). However, so that we do not have to consider exactly which of the numbers remain to be positioned, it is convenient and efficient to keep the numbers in an ordered list and then randomly choose the position of the number rather than the number itself. That is, starting with any initial ordering $p_1, p_2, \ldots,$ p_n, we pick one of the positions 1, . . . , n at random and then interchange the number in that position with the one in position n. Now we randomly choose one of the positions 1, . . . , $n - 1$ and interchange the number in this position with the one in position $n - 1$, and so on.

In order to implement the above we need to be able to generate a random variable that is equally likely to take on any of the values 1, 2, . . . , k. To accomplish this, let U denote a random number—that is, U is uniformly distributed over $(0,1)$—and note that kU is uniform on $(0,k)$ and so

$$P\{i - 1 < kU < i\} = \frac{1}{k}, \qquad i = 1, \ldots, k$$

Hence, if we let Int(kU) denote the largest integer less than or equal to kU, then the random variable $I = \text{Int}(kU) + 1$ will be such that

$$P\{I = i\} = P\{\text{Int}(kU) = i - 1\} = P\{i - 1 < kU < i\} = \frac{1}{k}$$

The above algorithm for generating a random permutation can now be written as follows:

Step 1: Let $p_1, p_2, \ldots, p_n$ be any permutation of 1, 2, . . . , n (for instance, we can choose $p_j = j, j = 1, \ldots, n$).
Step 2: Set $k = n$.
Step 3: Generate a random number U and let $I = \text{Int}(kU) + 1$.
Step 4: Interchange the values of p_I and p_k.
Step 5: Let $k = k - 1$ and if $k > 1$ go to Step 3.
Step 6: $p_1, \ldots, p_n$ is the desired random permutation.

For instance, suppose $n = 4$ and the initial permutation is 1, 2, 3, 4. If the first value of I (which is equally likely to be either

1,2,3,4) is $I = 3$, then the new permutation is 1,2,4,3. If the next value of I is $I = 2$ then the new permutation is 1,4,2,3. If the final value of I is $I = 2$, then the final permutation is 1,4,2,3, and this is the value of the random permutation.

One very important property of the above algorithm is that it can also be used to generate a random subset, say of size r, of the integers $1, \ldots, n$. Namely, just follow the algorithm until the positions $n, n - 1, \ldots, n - r + 1$ are filled. The elements in these positions constitute the random subset. $\diamond$

Example 1b (Estimating the Number of Distinct Entries in a Large List): Consider a list of n entries where n is very large, and suppose we are interested in estimating d, the number of distinct elements in the list. If we let m_i denote the number of times that the element in position i appears on the list, then we can express d by

$$d = \sum_{i=1}^{n} 1/m_i$$

To estimate d, suppose that we generate a random value X equally likely to be either $1, 2, \ldots, n$ (that is, we take $X = [nU] + 1$) and then let $m(X)$ denote the number of times the element in position X appears on the list. Then

$$E[1/m(X)] = \sum_{i=1}^{n} \frac{1}{m_i} \frac{1}{n} = d/n$$

Hence, if we generate k such random variables $X_1, \ldots, X_k$ we can estimate d by

$$d \approx \frac{n \sum_{i=1}^{k} 1/m(X_i)}{k}$$

Suppose now that each item in the list has a value attached to it—$v(i)$ being the value of the ith element. The sum of the values of the distinct items—call it v—can be expressed as

$$v = \sum_{i=1}^{n} v(i)/m(i)$$

Now if $X = [nU] + 1$, where U is a random number, then

$$E\left[\frac{v(X)}{m(X)}\right] = \sum_{i=1}^{n} \frac{v(i)}{m(i)} \frac{1}{n} = v/n$$

Hence, we can estimate v by generating $X_1, \ldots, X_k$ and then estimating v by

$$v \approx \frac{n}{k} \sum_{i=1}^{k} \frac{v(X_i)}{m(X_i)}$$

For an important application of the above, let $A_i = \{a_{i,1}, \ldots, a_{i,n_i}\}$, $i = 1, \ldots, s$ denote events, and suppose we are interested in estimating $P(\cup_{i=1}^{s} A_i)$. Since

$$P\left(\bigcup_{i=1}^{s} A_i\right) = \sum_{a \in \cup A_i} P(a) = \sum_{i=1}^{s} \sum_{j=1}^{n_i} P(a_{i,j})/m(a_{i,j})$$

where $m(a_{i,j})$ is the number of events to which the point $a_{i,j}$ belongs, the above method can be used to estimate $P(\cup_1^s A_i)$.

It should be noted that the above procedure for estimating v can be effected without a prior knowledge of the set of values $\{v_1, \ldots, v_n\}$. That is, it suffices that we can determine the value of an element in a specific place and the number of times that element appears on the list. When the set of values is a priori known, there is a more efficient approach available as will be shown in Example 4e. ◇

2. General Techniques for Simulating Continuous Random Variables

In this section we present three methods for simulating continuous random variables.

2.1 The Inverse Transformation Method

A general method for simulating a random variable having a continuous distribution—called the *Inverse Transformation Method*—is based on the following proposition.

Proposition 2.1. Let U be a uniform $(0, 1)$ random variable. For any continuous distribution function F if we define the random variable X by

$$X = F^{-1}(U)$$

then the random variable X has distribution function F. [$F^{-1}(u)$ is defined to equal that value x for which $F(x) = u$.]

Proof:

$$F_X(a) = P\{X \le a\}$$
$$= P\{F^{-1}(U) \le a\} \qquad (2.1)$$

Now, since $F(x)$ is a monotone function, it follows that $F^{-1}(U) \le a$ if and only if $U \le F(a)$. Hence, from Equation (2.1), we see that

$$F_X(a) = P\{U \le F(a)\}$$
$$= F(a) \quad \diamond$$

Hence we can simulate a random variable X from the continuous distribution F, when F^{-1} is computable, by simulating a random number U and then setting $X = F^{-1}(U)$.

Example 2a (Simulating an Exponential Random Variable): If $F(x) = 1 - e^{-x}$, then $F^{-1}(u)$ is that value of x such that

$$1 - e^{-x} = u$$

or

$$x = -\log(1 - u)$$

Hence, if U is a uniform $(0, 1)$ variable, then

$$F^{-1}(U) = -\log(1 - U)$$

is exponentially distributed with mean 1. Since $1 - U$ is also uniformly distributed on $(0, 1)$ it follows that $-\log U$ is exponential with mean 1. Since cX is exponential with mean c when X is exponential with mean 1, it follows that $-c \log U$ is exponential with mean c. $\quad \diamond$

2.2 The Rejection Method

Suppose that we have a method for simulating a random variable having density function $g(x)$. We can use this as the basis for simulating from the continuous distribution having density $f(x)$ by simulating Y from g and then accepting this simulated value with a probability proportional to $f(Y)/g(Y)$.

Specifically let c be a constant such that

$$\frac{f(y)}{g(y)} \le c \quad \text{for all } y$$

We then have the following technique for simulating a random variable having density f.

Rejection Method

Step 1: Simulate Y having density g and simulate a random number U.

Step 2: If $U \leq f(Y)/cg(Y)$ set $X = Y$. Otherwise return to Step 1.

Proposition 2.2. The random variable X generated by the Rejection Method has density function f.

Proof: Let X be the value obtained, and let N denote the number of necessary iterations. Then

$$P\{X \leq x\} = P\{Y_N \leq x\}$$

$$= P\{Y \leq x | U \leq f(Y)/cg(Y)\}$$

$$= \frac{P\left\{Y \leq x, U \leq \dfrac{f(Y)}{cg(Y)}\right\}}{K}$$

$$= \frac{\int P\left\{Y \leq x, U \leq \dfrac{f(Y)}{cg(Y)} \,\middle|\, Y = y\right\} g(y)dy}{K}$$

$$= \frac{\int_{-\infty}^{x} \dfrac{f(y)}{cg(y)} g(y)dy}{K}$$

$$= \frac{\int_{-\infty}^{x} f(y)dy}{Kc}$$

where $K = P\{U \leq f(Y)/cg(Y)\}$. Letting $x \to \infty$ shows that $K = 1/c$ and the proof is complete. ◊

Remarks

　　(1) The above method was originally presented by Von-Neumann in the special case where g was positive only in some finite interval (a, b), and Y was chosen to be uniform over (a, b). [That is, $Y = a + (b - a)U$].

　　(2) Note that the way in which we "accept the value Y with

probability $f(Y)/cg(Y)$" is by generating a uniform $(0, 1)$ random variable U and then accepting Y if $U \le f(Y)/cg(Y)$.

(3) Since each iteration of the method will, independently, result in an accepted value with probability $P\{U \le f(Y)/cg(Y)\} = 1/c$ it follows that the number of iterations is geometric with mean c.

(4) Actually, it is not necessary to generate a new uniform random number when deciding whether or not to accept, since at a cost of some additional computation, a single random number, suitably modified at each iteration, can be used throughout. To see how, note that the actual value of U is not used—only whether or not $U < f(X)/cg(X)$. Hence, if X is rejected—that is, if $U > f(X)/cg(X)$—we can use the fact that, given X,

$$\frac{U - \dfrac{f(X)}{cg(X)}}{1 - \dfrac{f(X)}{cg(X)}} = \frac{cUg(X) - f(X)}{cg(X) - f(X)}$$

is uniform on $(0, 1)$. Hence, this may be used as a uniform random number in the next iteration. As this saves the generation of a random number at the cost of the computation above, whether it is a net savings depends greatly upon the method being used to generate random numbers.   ◇

Example 2b   Let us use the rejection method to generate a random variable having density function

$$f(x) = 20\, x(1 - x)^3, \qquad 0 < x < 1$$

Since this random variable (which is beta with parameters 2,4) is concentrated in the interval $(0,1)$, let us consider the rejection method with

$$g(x) = 1, \qquad 0 < x < 1$$

To determine the constant c such that $f(x)/g(x) \le c$, we use calculus to determine the maximum value of

$$\frac{f(x)}{g(x)} = 20 \times (1 - x)^3$$

Differentiation of this quantity yields

$$\frac{d}{dx}\left[\frac{f(x)}{g(x)}\right] = 20\ [(1 - x)^3 - 3x(1 - x)^2]$$

Setting this equal to 0 shows that the maximal value is attained when $x = \frac{1}{4}$, and thus

$$\frac{f(x)}{g(x)} \leq 20\left(\frac{1}{4}\right)\left(\frac{3}{4}\right)^3 = \frac{135}{64} \equiv c$$

Hence,

$$\frac{f(x)}{cg(x)} = \frac{256}{27}x(1 - x)^3$$

and thus the rejection procedure is as follows:

Step 1: Generate random numbers U_1 and U_2.
Step 2: If $U_2 \leq \frac{256}{27} U_1 (1 - U_1)^3$, stop and set $X = U_1$. Otherwise return to Step 1.

The average number of times that Step 1 will be performed is $c = \frac{135}{64}$. ◇

Example 2c (Simulating a Normal Random Variable): To simulate a unit normal random variable Z (that is, one with mean 0 and variance 1) note first that the absolute value of Z has density function

$$f(x) = \frac{2}{\sqrt{2\pi}}\ e^{-x^2/2}, \qquad 0 < x < \infty \tag{2.1}$$

We will start by simulating from the above density by using the rejection method with

$$g(x) = e^{-x}, \qquad 0 < x < \infty$$

Now, note that

$$\frac{f(x)}{g(x)} = \sqrt{2e/\pi}\ \exp\{-(x - 1)^2/2\} \leq \sqrt{2e/\pi}$$

Hence, using the rejection method we can simulate from (2.1) as follows:

(a) Generate independent random variables Y and U, Y being exponential with rate 1 and U being uniform on $(0, 1)$.
(b) If $U \leq \exp\{-(Y - 1)^2/2\}$, or equivalently, if

$$-\log U \geq (Y - 1)^2/2$$

set $X = Y$. Otherwise return to (a).

Once we have simulated a random variable X having density function (2.1) we can then generate a unit normal random variable Z by letting Z be equally likely to be either X or $-X$.

To improve upon the above, note first that from Example 2a it follows that $-\log U$ will also be exponential with rate 1. Hence, steps (a) and (b) above are equivalent to

(a′) generate independent exponentials with rate 1, Y_1, and Y_2.
(b′) set $X = Y_1$ if $Y_2 \geq (Y_1 - 1)^2/2$. Otherwise return to (a′).

Now suppose that we accept step (b′). It then follows by the lack of memory property of the exponential that the amount by which Y_2 exceeds $(Y_1 - 1)^2/2$ will also be exponential with rate 1.

Hence, summing up, we have the following algorithm which generates an exponential with rate 1 and an independent unit normal random variable.

Step 1: Generate Y_1, an exponential random variable with rate 1.

Step 2: Generate Y_2, an exponential with rate 1.

Step 3: If $Y_2 - (Y_1 - 1)^2/2 > 0$, set $Y = Y_2 - (Y_1 - 1)^2/2$ and go to Step 4. Otherwise go to Step 1.

Step 4: Generate a random number U and set

$$Z = \begin{cases} Y_1 & \text{if } U \leq \frac{1}{2} \\ -Y_1 & \text{if } U > \frac{1}{2} \end{cases}$$

The random variables Z and Y generated by the above are independent with Z being normal with mean 0 and variance 1 and Y being exponential with rate 1. (If we want the normal random variable to have mean μ and variance σ^2, just take $\mu + \sigma Z$).   ◊

Remarks

(1) Since $c = \sqrt{2e/\pi} \approx 1.32$, the above requires a geometric distributed number of iterations of Step 2 with mean 1.32.

(2) The final random number of Step 4 need not be separately simulated but rather can be obtained from the first digit of any random number used earlier. That is, suppose we generate a random number to simulate an exponential; then we can strip off the initial digit of this random number and just use the remaining digits (with the decimal point moved one step to the right) as the random number. If this initial digit is 0, 1, 2, 3, or 4 (or 0 if the computer is generating binary digits), then we take the sign of Z to be positive and take it negative otherwise.

(3) If we are generating a sequence of unit normal random variables, then we can use the exponential obtained in Step 4 as the initial exponential needed in Step 1 for the next normal to be generated. Hence, on the average, we can simulate a unit normal by generating 1.64 exponentials and computing 1.32 squares.

2.3 Hazard Rate Method

Let F be a continuous distribution function with $\bar{F}(0) = 1$. Recall that $\lambda(t)$, the hazard rate function of F, is defined by

$$\lambda(t) = \frac{f(t)}{\bar{F}(t)}, \qquad t \geq 0$$

(where $f(t) = F'(t)$ is the density function). Recall also that $\lambda(t)$ represents the instantaneous probability intensity that an item having life distribution F will fail at time t given it has survived to that time.

Suppose now that we are given a bounded function $\lambda(t)$, such that $\int_0^\infty \lambda(t)dt = \infty$, and we desire to simulate a random variable S having $\lambda(t)$ as its hazard rate function.

To do so let λ be such that

$$\lambda(t) \leq \lambda \quad \text{for all} \quad t \geq 0$$

To simulate from $\lambda(t)$, $t \geq 0$ we will

(a) simulate a Poisson process having rate λ. We will then only "accept" or "count" certain of these Poisson events. Specifically we will

(b) count an event that occurs at time t, independently of all else, with probability $\lambda(t)/\lambda$.

We now have the following proposition.

Proposition 2.3.   The time of the first counted event—call it S—is a random variable whose distribution has hazard rate function $\lambda(t)$, $t \geq 0$.

Proof:

$P\{t < S < t + dt | S > t\}$

$= P\{\text{first counted event in } (t, t + dt) \,|\, \text{no counted events prior to } t\}$

$= P\{\text{poisson event in } (t, t + dt), \text{ it is counted}|$

  $\text{no counted events prior to } t\}$

$= P\{\text{poisson event in } (t, t + dt), \text{ it is counted}\}$

$= [\lambda dt + o(dt)]\, \dfrac{\lambda(t)}{\lambda} = \lambda(t)dt + o(dt)$

which completes the proof. It should be noted that the next to last equality follows from the independent increment property of Poisson processes. $\Diamond$

As the interarrival times of a Poisson process having rate λ are exponential with rate λ it thus follows from Example 2a and the previous proposition that the following algorithm will generate a random variable having hazard rate function $\lambda(t)$, $t \geq 0$.

Hazard Rate Method for Generating S: $\lambda_S(t) = \lambda(t)$

Let λ be such that $\lambda(t) \leq \lambda$ for all $t \geq 0$. Generate pairs of random variables U_i, X_i, $i \geq 1$, with X_i being exponential with rate λ and U_i being uniform $(0, 1)$, stopping at

$$N = \min\left\{ n : U_n \leq \lambda \left(\sum_{i=1}^{n} X_i \right) \Big/ \lambda \right\}$$

Set

$$S = \sum_{i=1}^{N} X_i \quad \Diamond$$

To compute $E[N]$ we need the result, known as Wald's Equation, which states that if $X_1, X_2, \ldots$ are independent and identically distributed random variables that are observed in sequence up to some random time N then

$$E\left[\sum_{i=1}^{N} X_i \right] = E[N]E[X]$$

More precisely let X_1, X_1, . . . denote a sequence of independent random variables and consider the following definition.

Definition 2.3.   An integer-valued random variable N is said to be a *stopping time* for the sequence X_1, X_2, . . . if the event $\{N = n\}$ is independent of X_{n+1}, X_{n+2}, . . . for all $n = 1, 2, \ldots$.

Intuitively, we observe the X_n's in sequential order and N denotes the number observed before stopping. If $N = n$, then we have stopped after observing $X_1, \ldots, X_n$ and before observing X_{n+1}, X_{n+2}, . . . for all $n = 1, 2, \ldots$.

Example.   Let X_n, $n = 1, 2, \ldots$ be independent and such that

$$P\{X_n = 0\} = P\{X_n = 1\} = \tfrac{1}{2}, \qquad n = 1, 2, \ldots.$$

If we let

$$N = \min\{n: X_1 + \cdots + X_n = 10\}$$

then N is a stopping time. We may regard N as being the stopping time of an experiment that successively flips a fair coin and then stops when the number of heads reaches 10.   ◇

Proposition 2.4.   (Wald's Equation): If X_1, X_2, . . . are independent and identically distributed random variables having finite expectations, and if N is a stopping time for X_1, X_2, . . . such that $E[N] < \infty$, then

$$E\left[\sum_1^N X_n\right] = E[N]E[X]$$

Proof: Letting

$$I_n = \begin{cases} 1 & \text{if} \quad N \geq n \\ 0 & \text{if} \quad N < n \end{cases}$$

we have that

$$\sum_{n=1}^N X_n = \sum_{n=1}^\infty X_n I_n$$

Hence,

$$E\left[\sum_{n=1}^N X_n\right] = E\left[\sum_{n=1}^\infty X_n I_n\right] = \sum_{n=1}^\infty E[X_n I_n] \qquad (2.2)$$

However, $I_n = 1$ if only if we have not stopped after successively observing $X_1, \ldots, X_{n-1}$. Therefore, I_n is determined by X_1,

. . . , X_{n-1} and is thus independent of X_n. From (2.2) we thus obtain

$$E\left[\sum_{n=1}^{N} X_n\right] = \sum_{n=1}^{\infty} E[X_n]E[I_n]$$

$$= E[X]\sum_{n=1}^{\infty} E[I_n]$$

$$= E[X]E\left[\sum_{n=1}^{\infty} I_n\right]$$

$$= E[X]E[N]   \diamondsuit$$

Returning to the hazard rate method, we have that

$$S = \sum_{i=1}^{N} X_i$$

As $N = \min\{n : U_n \le \lambda(\Sigma_1^n X_i)/\lambda\}$ it follows that the event that $N = n$ is independent of $X_{n+1}, X_{n+2}, ...$. Hence, by Wald's Equation

$$E[S] = E[N]E[X_i]$$

$$= \frac{E[N]}{\lambda}$$

or

$$E[N] = \lambda E[S]$$

where $E[S]$ is the mean of the desired random variable.

3. Special Techniques for Simulating Continuous Random Variables

Special techniques have been devised to simulate from most of the common continuous distribution. We now present certain of these.

3.1 The Normal Distribution

Let X and Y denote independent unit normal random variables and thus have the joint density function

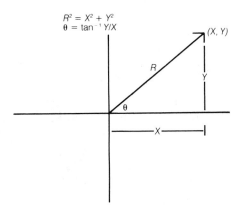

Figure 11.1

$$f(x, y) = \frac{1}{2\pi} e^{-(x^2+y^2)/2} \qquad -\infty < x < \infty, \ -\infty < y < \infty$$

Consider now the polar coordinates of the point (X, Y). As shown in Figure 11.1, that is

$$R^2 = X^2 + Y^2$$
$$\Theta = \tan^{-1} Y/X$$

To obtain the joint density of R^2 and Θ consider the transformation

$$d = x^2 + y^2, \qquad \theta = \tan^{-1} y/x$$

The Jacobian of this transformation is

$$J = \begin{vmatrix} \dfrac{\partial d}{\partial x} & \dfrac{\partial d}{\partial y} \\[2ex] \dfrac{\partial \theta}{\partial x} & \dfrac{\partial \theta}{\partial y} \end{vmatrix}$$

$$= \begin{vmatrix} 2x & 2y \\[2ex] \dfrac{1}{1 + \dfrac{y^2}{x^2}} \left(\dfrac{-y}{x^2} \right) & \dfrac{1}{1 + \dfrac{y^2}{x^2}} \left(\dfrac{1}{x} \right) \end{vmatrix} = 2 \begin{vmatrix} x & y \\[2ex] -\dfrac{y}{x^2 + y^2} & \dfrac{x}{x^2 + y^2} \end{vmatrix} = 2$$

Hence, from Section 5.3 of Chapter 2 the joint density of R^2 and Θ is given by

$$f_{R^2,\Theta}(d, \theta) = \frac{1}{2\pi} e^{-d/2} \frac{1}{2}$$

$$= \frac{1}{2} e^{-d/2} \frac{1}{2\pi} \qquad 0 < d < \infty, 0 < \theta < 2\pi$$

Thus, we can conclude that R^2 and Θ are independent with R^2 having an exponential distribution with rate $\frac{1}{2}$ and Θ being uniform on $(0, 2\pi)$.

Let us now go in reverse from the polar to the rectangular coordinates. From the above if we start with W, an exponential random variable with rate $\frac{1}{2}$ (W plays the role of R^2) and with V, independent of W and uniformly distributed over $(0, 2\pi)$ (V plays the role of Θ) then $X = \sqrt{W} \cos V$, $Y = \sqrt{W} \sin V$ will be independent unit normals. Hence using the results of Example 2a we see that if U_1 and U_2 are independent uniform $(0, 1)$ random numbers then

$$X = (-2 \log U_1)^{1/2} \cos(2 \pi U_2)$$

and   (3.1)

$$Y = (-2 \log U_1)^{1/2} \sin(2\pi U_2)$$

are independent unit normal random variables.

Remarks   The fact that $X^2 + Y^2$ has an exponential distribution with rate $\frac{1}{2}$ is quite interesting for, by the definition of the chi-square distribution, $X^2 + Y^2$ has a chi-square distribution with 2 degrees of freedom. Hence, these two distributions are identical. $\diamond$

The above approach to generating unit normal random variables is called the Box–Muller approach. Its efficiency suffers somewhat from its need to compute the above sine and cosine values. There is, however, a way to get around this potentially time-consuming difficulty. To begin, note that if U is uniform on $(0, 1)$, then $2U$ is uniform on $(0, 2)$, and so $2U - 1$ is uniform on $(-1, 1)$. Thus, if we generate random numbers U_1 and U_2 and set

$$V_1 = 2U_1 - 1$$
$$V_2 = 2U_2 - 1$$

then (V_1, V_2) is uniformly distributed in the square of area 4 centered at $(0, 0)$ (see Figure 11.2).

Suppose now that we continually generate such pairs (V_1, V_2) until we obtain one that is contained in the circle of radius 1 centered at $(0, 0)$—that is, until (V_1, V_2) is such that $V_1^2 + V_2^2 \leq 1$. It now follows that such a pair (V_1, V_2) is uniformly distributed in the circle. If we let $\overline{R}, \overline{\Theta}$

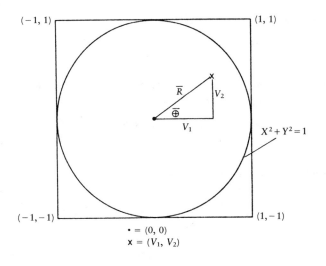

$\bullet = (0, 0)$
$\mathbf{x} = (V_1, V_2)$

Figure 11.2

denote the polar coordinates of this pair, then it is easy to verify that $\overline{R}$ and $\overline{\Theta}$ are independent, with $\overline{R}^2$ being uniformly distributed on $(0, 1)$, and $\overline{\Theta}$ uniformly distributed on $(0, 2\pi)$.

Since

$$\sin \overline{\Theta} = V_2/\overline{R} = -\frac{V_2}{\sqrt{V_1^2 + V_2^2}}$$

$$\cos \overline{\Theta} = V_1/\overline{R} = \frac{V_1}{\sqrt{V_1^2 + V_2^2}}$$

it follows from Equation (3.1) that we can generate independent unit normals X and Y by generating another random number U and setting

$$X = (-2 \log U)^{1/2} \, V_1/\overline{R}$$

$$Y = (-2 \log U)^{1/2} \, V_2/\overline{R}$$

In fact, since (conditional on $V_1^2 + V_2^2 \le 1$) $\overline{R}^2$ is uniform on $(0, 1)$ and is independent of $\overline{\Theta}$, we can use it instead of generating a new random number U; thus showing that

$$X = (-2 \log \overline{R}^2)^{1/2} \, V_1/\overline{R} = \sqrt{\frac{-2 \log S}{S}} \, V_1$$

$$Y = (-2 \log \overline{R}^2)^{1/2} \, V_2/\overline{R} = \sqrt{\frac{-2 \log S}{S}} \, V_2$$

are independent unit normals, where

$$S = \overline{R}^2 = V_1^2 + V_2^2$$

Summing up, we thus have the following approach to generating a pair of independent unit normals:

Step 1: Generate random numbers U_1 and U_2.
Step 2: Set $V_1 = 2U_1 - 1$, $V_2 = 2U_2 - 1$, $S = V_1^2 + V_2^2$.
Step 3: If $S > 1$, return to Step 1.
Step 4: Return the independent unit normals

$$X = \sqrt{\frac{-2 \log S}{S}}\, V_1, \qquad Y = \sqrt{\frac{-2 \log S}{S}}\, V_2$$

The above is called the polar method. Since the probability that a random point in the square will fall within the circle is equal to $\pi/4$ (the area of the circle divided by the area of the square), it follows that, on average, the polar method will require $4/\pi = 1.273$ iterations of Step 1. Hence, it will, on average, require 2.546 random numbers, 1 logarithm, 1 square root, 1 division, and 4.546 multiplications to generate 2 independent unit normals.   ◊

3.2 The Gamma Distribution

To simulate from a gamma distribution with parameters (n, λ), when n is an integer, we use the fact that the sum of n independent exponential random variables each having rate λ has this distribution. Hence, if $U_1, \ldots, U_n$ are independent uniform $(0, 1)$ random variables,

$$X = -\sum_{i=1}^{n} \frac{1}{\lambda} \log U_i = -\frac{1}{\lambda} \log\left(\prod_{i=1}^{n} U_i\right)$$

has the desired distribution.

When n is large, there are other techniques available that do not require so many random numbers. One possibility is to use the rejection procedure with $g(x)$ being taken as the density of an exponential random variable with mean n/λ (as this is the mean of the gamma). It can be shown that for large n the average number of iterations needed by the rejection algorithm is $e[(n - 1)/2\pi]^{1/2}$. In addition, if we wanted to generate a series of gammas, then, just as in Example 2b, we can arrange things so that upon acceptance we obtain not only a gamma random variable but also, for free, an exponential random variable which can then be used in obtaining the next gamma (see Problem 8).

3.3 The Chi-Squared Distribution

The chi-squared distribution with n degrees of freedom is the distribution of $\chi_n^2 = Z_1^2 + \cdots + Z_n^2$ where Z_i, $i = 1, \ldots, n$ are independent unit normals. Using the fact noted in the remark at the end of Section 3.1 we see that $Z_1^2 + Z_2^2$ has an exponential distribution with rate $\frac{1}{2}$. Hence, when n is even—say $n = 2k$—χ_{2k}^2 has a gamma distribution with parameters $(k, \frac{1}{2})$. Hence, $-2 \log (\Pi_{i=1}^k U_i)$ has a chi-squared distribution with $2k$ degrees of freedom. We can simulate a chi-squared random variable with $2k + 1$ degrees of freedom by first simulating a unit normal random variable Z and then adding Z^2 to the above. That is,

$$\chi_{2k+1}^2 = Z^2 - 2 \log\left(\prod_{i=1}^k U_i\right)$$

where $Z, U_1, \ldots, U_n$ are independent with Z being a unit normal and the others being uniform $(0, 1)$ random variables.

3.4 The Beta (n,m) Distribution

The random variable X is said to have a beta distribution with parameters n, m if its density is given by

$$f(x) = \frac{(n + m - 1)!}{(n - 1)!(m - 1)!} x^{n-1} (1 - x)^{m-1}, \qquad 0 < x < 1$$

One approach to simulating from the above distribution is to let $U_1, \ldots, U_{n+m-1}$ be independent uniform $(0, 1)$ random variables and consider the nth smallest value of this set—call it $U_{(n)}$. Now $U_{(n)}$ will equal x if of the $n + m - 1$ variables

    (i) $n-1$ are smaller than x
    (ii) one equals x
    (iii) $m-1$ are greater than x

Hence, if the $n + m - 1$ uniform random variables are partitioned into three subsets of sizes $n - 1$, 1, and $m - 1$ the probability (density) that each of the variables in the first set is less than x, the variable in the second set equals x, and all the variables in the third set are greater than x is given by

$$(P\{U < x\})^{n-1} f_u(x)(P\{U > x\})^{m-1} = x^{n-1}(1 - x)^{m-1}$$

Hence as there are $(n + m - 1)!/(n - 1)!(m - 1)!$ possible partitions it follows that $U_{(n)}$ is beta with parameters (n, m).

Thus one way to simulate from the beta distribution is to find the nth smallest of a set of $n + m - 1$ random numbers. However, when n and m are large, this procedure is not particularly efficient.

For another approach consider a Poisson process with rate 1, and recall that given S_{n+m}, the time of the $(n + m)$th event, the set of the first $n + m - 1$ event times is distributed independently and uniformly on $(0, S_{n+m})$. Hence, given S_{n+m}, the nth smallest of the first $n + m - 1$ event times—that is S_n—is distributed as the nth smallest of a set of $n + m - 1$ uniform $(0, S_{n+m})$ random variables. But from the above we can thus conclude that S_n/S_{n+m} has a Beta distribution with parameters (n, m). Therefore, if $U_1, \ldots, U_{n+m}$ are random numbers,

$$\frac{-\log \prod_{i=1}^{n} U_i}{-\log \prod_{i=1}^{m+n} U_i} \text{ is Beta with parameters } (n, m)$$

By writing the above as

$$\frac{-\log \prod_{i=1}^{n} U_i}{-\log \prod_{1}^{n} U_i - \log \prod_{n+1}^{n+m} U_i}$$

we see that it has the same distribution as $X/(X + Y)$ where X and Y are independent gamma random variables with respective parameters $(n, 1)$ and $(m, 1)$. Hence, when n and m are large, we can efficiently simulate a beta by first simulating two gamma random variables.

3.5 The Exponential Distribution—The Von-Neumann Algorithm

As we have seen, an exponential random variable with rate 1 can be simulated by computing the negative of the logarithm of a random number. Most computer programs for computing a logarithm, however, involve a power series expansion, and so it might be useful to have at hand a second method that is computationally easier. We now present such a method due to Von-Neumann.

To begin let U_1, U_2, . . . be independent uniform $(0, 1)$ random variables and define N, $N \geq 2$, by

$$N = \min\{n : U_1 \geq U_2 \geq \cdots \geq U_{n-1} < U_n\}$$

That is, N is the first random number that is greater than its predecessor. Let us now compute the joint distribution of N and U_1.

$$P\{N > n, U_1 \leq y\} = \int_0^1 P\{N > n, U_1 \leq y | U_1 = x\} dx$$

$$= \int_0^y P\{N > n | U_1 = x\} dx$$

Now, given that $U_1 = x$, N will be greater than n if $x \geq U_2 \geq \cdots \geq U_n$ or, equivalently, if

(a) $U_i \leq x$, $i = 2, \ldots, n$

and

(b) $U_2 \geq \cdots \geq U_n$

Now (a) has probability x^{n-1} of occurring and given (a), as all the $(n - 1)!$ possible rankings of $U_2, \ldots, U_n$ are equally likely, (b) has probability $1/(n - 1)!$ of occurring. Hence,

$$P\{N > n | U_1 = x\} = \frac{x^{n-1}}{(n - 1)!}$$

and so

$$P\{N > n, U_1 \leq y\} = \int_0^y \frac{x^{n-1}}{(n - 1)!} dx = \frac{y^n}{n!}$$

which yields that

$$P\{N = n, U_1 \leq y\} = P\{N > n - 1, U_1 \leq y\} - P\{N > n, U_1 \leq y\}$$

$$= \frac{y^{n-1}}{(n - 1)!} - \frac{y^n}{n!}$$

Upon summing over all the even integers, we see that

$$P\{N \text{ is even}, U_1 \leq y\} = y - \frac{y^2}{2!} + \frac{y^3}{3!} - \frac{y^4}{4!} - \cdots$$

$$= 1 - e^{-y} \tag{3.2}$$

We are now ready for the following algorithm for generating an exponential random variable with rate 1.

Step 1: Generate uniform random numbers $U_1, U_2, \ldots$ stopping at $N = \min\{n : U_1 \geq \cdots \geq U_{n-1} < U_n\}$.

Step 2: If N is even accept that run, and go to Step 3. If N is odd reject the run, and return to Step 1.

Step 3: Set X equal to the number of failed runs plus the first random number in the successful run.

To show that X is exponential with rate 1, first note that the probability of a successful run is, from (3.2) with $y = 1$,

$$P\{N \text{ is even}\} = 1 - e^{-1}$$

Now in order for X to exceed x, the first $[x]$ runs must all be unsuccessful and the next run must either be unsuccessful or be successful but have $U_1 > x - [x]$ (where $[x]$ is the largest integer not exceeding x). As

$$P\{N \text{ even}, U_1 > y\} = P\{N \text{ even}\} - P\{N \text{ even}, U_1 \leq y\}$$
$$= 1 - e^{-1} - (1 - e^{-y}) = e^{-y} - e^{-1}$$

we see that

$$P\{X > x\} = e^{-[x]}[e^{-1} + e^{-(x-[x])} - e^{-1}] = e^{-x}$$

which yields the result.

Let T denote the number of trials needed to generate a successful run. As each trial is a success with probability $1 - e^{-1}$ it follows that T is geometric with mean $1/(1 - e^{-1})$. If we let N_i denote the number of uniform random variables used on the ith run, $i \geq 1$, then T (being the first run i for which N_i is even) is a stopping time for this sequence. Hence, by Wald's equation, the mean number of uniform random variables needed by this algorithm is given by

$$E\left[\sum_{i=1}^{T} N_i\right] = E[N]E[T]$$

Now,

$$E[N] = \sum_{n=0}^{\infty} P\{N > n\}$$

$$= 1 + \sum_{n=1}^{\infty} P\{U_1 \geq \cdots \geq U_n\}$$

$$= 1 + \sum_{n=1}^{\infty} 1/n!$$

$$= e$$

and so

$$E\left[\sum_{i=1}^{T} N_i\right] = \frac{e}{1 - e^{-1}} \approx 4.3$$

Hence, this algorithm, which computationally speaking, is quite easy to perform, requires on the average about 4.3 random numbers to execute.

Remarks Since in any run the actual value of U_N is not utilized but only the fact that $U_N > U_{N-1}$ we can, for any failed run, use $(U_N - U_{N-1})/(1 - U_{N-1})$ as a uniform random number in the next run. This will result in a saving of one random number on each run, after the initial one, and thus will reduce the expected number of random numbers that need be generated by the amount

$$\frac{1}{1 - e^{-1}} - 1 = \frac{e^{-1}}{1 - e^{-1}} \approx 0.58 \quad \lozenge$$

4. Simulating from Discrete Distributions

All of the general methods for simulating from continuous distributions have analogs in the discrete case. For instance, if we want to simulate a random variable X having probability mass function

$$P\{X = x_j\} = P_j, \qquad j = 0, 1, \ldots, \qquad \sum_j P_j = 1$$

We can use the following discrete time analog of the inverse transform technique.

To simulate X for which $P\{X = x_j\} = P_j$
let U be uniformly distributed over $(0, 1)$, and set

$$X = \begin{cases} x_1 & \text{if} \quad U < P_1 \\ x_2 & \text{if} \quad P_1 < U < P_1 + P_2 \\ \vdots \\ x_j & \text{if} \quad \sum_1^{j-1} P_i < U < \sum_i^j P_i \\ \vdots \end{cases}$$

As,

$$P\{X = x_j\} = P\left\{ \sum_1^{j-1} P_i < U < \sum_1^j P_i \right\} = P_j$$

we see that X has the desired distribution.

Example 4a (The Geometric Distribution):   Suppose we want to simulate X such that

$$P\{X = i\} = p(1 - p)^{i-1}, \qquad i \geq 1$$

As

$$\sum_{i=1}^{j-1} P\{X = i\} = 1 - P\{X > j - 1\} = 1 - (1 - p)^{j-1}$$

we can simulate such a random variable by generating a random number U and then setting X equal to that value j for which

$$1 - (1 - p)^{j-1} < U < 1 - (1 - p)^j$$

or, equivalently, for which

$$(1 - p)^j < 1 - U < (1 - p)^{j-1}$$

As $1 - U$ has the same distribution as U we can thus define X by

$$X = \min\{j : (1 - p)^j < U\} = \min\left\{ j : j > \frac{\log U}{\log(1 - p)} \right\}$$

$$= 1 + \left[\frac{\log U}{\log(1 - p)} \right] \qquad \diamond$$

As in the continuous case, special simulation techniques have been developed for the more common discrete distributions. We now present certain of these.

Example 4b (Simulating a Binomial Random Variable): A binomial (n, p) random variable can be most easily simulated by recalling that it can be expressed as the sum of n independent Bernoulli random variables. That is, if $U_1, \ldots, U_n$ are independent uniform $(0, 1)$ variables, then letting

$$X_i = \begin{cases} 1 & \text{if } U_i < p \\ 0 & \text{otherwise} \end{cases}$$

it follows that $X \equiv \Sigma_{i=1}^{n} X_i$ is a binomial random variable with parameters n and p.

One difficulty with the above procedure is that it requires the generation of n random numbers. To show how to reduce the number of random numbers needed, note first that the above procedure does not use the actual value of a random number U but only whether or not it exceeds p. Using this and the result that the conditional distribution of U given that $U < p$ is uniform on $(0, p)$ and the conditional distribution of U given that $U > p$ is uniform on $(p, 1)$, we now show how we can simulate a binomial (n, p) random variable using only a single random number:

Step 1: Let $\alpha = 1/p$, $\beta = 1/(1 - p)$.

Step 2: Set $k = 0$.

Step 3: Generate a uniform random number U.

Step 4: If $k = n$ stop. Otherwise reset k to equal $k + 1$.

Step 5: If $U \leq p$ set $X_k = 1$ and reset U to equal αU. If $U > p$ set $X_k = 0$ and reset U to equal $\beta(U - p)$. Return to step 4.

The above procedure generates $X_1, \ldots, X_n$ and $X = \Sigma_{i=1}^{n} X_i$ is the desired random variable. It works by noting whether $U_k \leq p$ or $U_k > p$; in the former case it takes U_{k+1} to equal U_k/p, and in the latter case it takes U_{k+1} to equal $(U_k - p)/(1 - p)$.* ◊

Example 4c (Simulating a Poisson Random Variable): To simulate a Poisson random variable with mean λ, generate independent uniform $(0, 1)$ random variables $U_1, U_2, \ldots$ stopping at

---

*Because of computer roundoff errors, a single random number should not be continuously used when n is large.

$$N + 1 = \min\left\{n : \prod_{i=1}^{n} U_i < e^{-\lambda}\right\}$$

The random variable N has the desired distribution, which can be seen by noting that

$$N = \max\left\{n : \sum_{i=1}^{n} -\log U_i < \lambda\right\}$$

But $-\log U_i$ is exponential with rate 1, and so if we interpret $-\log U_i$, $i \geq 1$, as the interarrival times of a Poisson process having rate 1, we see that $N = N(\lambda)$ would equal the number of events by time λ. Hence N is Poisson with mean λ.

When λ is large we can reduce the amount of computation in the above simulation of $N(\lambda)$, the number of events by time λ of a Poisson process having rate 1, by first choosing an integer m and simulating S_m, the time of the mth event of the Poisson process and then simulating $N(\lambda)$ according to the conditional distribution of $N(\lambda)$ given S_m. Now the conditional distribution of $N(\lambda)$ given S_m is as follows:

$$N(\lambda)|S_m = s \sim m + \text{Poisson}(\lambda - s) \quad \text{if} \quad s < \lambda$$

$$N(\lambda)|S_m = s \sim \text{Binomial}\left(m - 1, \frac{\lambda}{s}\right) \quad \text{if} \quad s > \lambda$$

where $\sim$ means "has the distribution of." The above follows since if the mth event occurs at time s, where $s < \lambda$, then the number of events by time λ is m plus the number of events in (s, λ). On the other hand given that $S_m = s$ the set of times at which the first $m - 1$ events occur has the same distribution as a set of $m - 1$ uniform $(0, s)$ random variables (see Section 3.5 of Chapter 5). Hence, when $\lambda < s$, the number of these which occur by time λ is binomial with parameters $m - 1$ and λ/s. Hence, we can simulate $N(\lambda)$ by first simulating S_m and then simulate either $P(\lambda - S_m)$, a Poisson random variable with mean $\lambda - S_m$ when $S_m < \lambda$, or simulate $\text{Bin}(m - 1, \lambda/S_m)$, a binomial random variable with parameters $m - 1$, and λ/S_m, when $S_m > \lambda$; and then setting

$$N(\lambda) = \begin{cases} m + P(\lambda - S_m) & \text{if} \quad S_m < \lambda \\ \text{Bin}(m - 1, \lambda/S_m) & \text{if} \quad S_m > \lambda \end{cases}$$

In the above it has been found computationally effective to let m be approximately $\frac{7}{8} \lambda$. Of course S_m is simulated by simulating from a gamma (m, λ) distribution via an approach that is computationally fast when m is large (see Section 3.3).   ◊

There are also rejection and hazard rate methods for discrete distributions but we leave their development as exercises. However, there is a technique available for simulating finite discrete random variables—called the alias method—which, though requiring some set up time, is very fast to implement.

4.1 The Alias Method

In what follows, the quantities $\mathbf{P}, \mathbf{P}^{(k)}, \mathbf{Q}^{(k)}, k \le n - 1$ will represent probability mass functions on the integers $1, 2, \ldots, n$—that is they will be n-vectors of nonnegative numbers summing to 1. In addition, the vector $\mathbf{P}^{(k)}$ will have at most k nonzero components, and each of the $\mathbf{Q}^{(k)}$ will have at most two nonzero components. We will show that any probability mass function $\mathbf{P}$ can be represented as an equally weighted mixture of $n - 1$ probability mass functions $\mathbf{Q}$ (each having at most two nonzero components). That is, we show then for suitably defined $\mathbf{Q}^{(1)}, \ldots, \mathbf{Q}^{(n-1)}$, $\mathbf{P}$ can be expressed as

$$\mathbf{P} = \frac{1}{n-1} \sum_{k=1}^{n-1} \mathbf{Q}^{(k)} \qquad (4.1)$$

As a prelude to presenting the method for obtaining this representation, we will need the following simple lemma whose proof is left as an exercise.

Lemma 4.1. Let $\mathbf{P} = \{P_i, i = 1, \ldots, n\}$ denote a probability mass function, then

(a) there exists an i, $1 \le i \le n$, such that $P_i < 1/(n - 1)$, and
(b) for this i there exists a j, $j \ne i$, such that $P_i + P_j \ge 1/(n - 1)$.

Before presenting the general technique for obtaining the representation (4.1), let us illustrate it by an example.

Example 4d Consider the three point distribution $\mathbf{P}$ with $P_1 = \frac{7}{16}$, $P_2 = \frac{1}{2}$, $P_3 = \frac{1}{16}$. We start by choosing i and j satisfying the conditions of Lemma 4.1. As $P_3 < \frac{1}{2}$ and $P_3 + P_2 > \frac{1}{2}$, we can work with $i = 3$ and $j = 2$. We will now define a 2-point mass function $Q^{(1)}$ putting all of its weight on 3 and 2 and such that $\mathbf{P}$ will be expressible as an equally weighted mixture between $Q^{(1)}$ and a second 2-point mass function $Q^{(2)}$. Secondly, all of the mass of point 3 will be contained in $Q^{(1)}$. As we will have

$$P_j = \tfrac{1}{2}(Q_j^{(1)} + Q_j^{(2)}), j = 1, 2, 3 \qquad (4.2)$$

and, by the above, $Q_3^{(2)}$ is supposed to equal 0, we must therefore take

$$Q_3^{(1)} = 2P_3 = \tfrac{1}{8}, \qquad Q_2^{(1)} = 1 - Q_3^{(1)} = \tfrac{7}{8}, \qquad Q_1^{(1)} = 0$$

To satisfy (4.2), we must then set

$$Q_3^{(2)} = 0, \qquad Q_2^{(2)} = 2P_2 - \tfrac{7}{8} = \tfrac{1}{8}, \qquad Q_1^{(2)} = 2P_1 = \tfrac{7}{8}$$

Hence, we have the desired representation in this case. Suppose now that the original distribution was the following 4-point mass function.

$$P_1 = \tfrac{7}{16}, \qquad P_2 = \tfrac{1}{4}, \qquad P_3 = \tfrac{1}{8}, \qquad P_4 = \tfrac{3}{16}$$

Now, $P_3 < \tfrac{1}{3}$ and $P_3 + P_1 > \tfrac{1}{3}$. Hence our initial 2-point mass function—$Q^{(1)}$—will concentrate on points 3 and 1 (giving no weight to 2 and 4). As the final representation will give weight $\tfrac{1}{3}$ to $Q^{(1)}$ and in addition the other $Q^{(j)}$, $j = 2, 3$, will not give any mass to the value 3, we must have that

$$\tfrac{1}{3} Q_3^{(1)} = P_3 = \tfrac{1}{8}$$

Hence,

$$Q_3^{(1)} = \tfrac{3}{8}, \qquad Q_1^{(1)} = 1 - \tfrac{3}{8} = \tfrac{5}{8}$$

Also, we can write

$$\mathbf{P} = \tfrac{1}{3} Q^{(1)} + \tfrac{2}{3} \mathbf{P}^{(3)}$$

where $\mathbf{P}^{(3)}$, to satisfy the above, must be the vector

$$\mathbf{P}_1^{(3)} = \tfrac{3}{2}(P_1 - \tfrac{1}{3} Q_1^{(1)}) = \tfrac{11}{32}$$
$$\mathbf{P}_2^{(3)} = \tfrac{3}{2} P_2 = \tfrac{3}{8}$$
$$\mathbf{P}_3^{(3)} = 0$$
$$\mathbf{P}_4^{(3)} = \tfrac{3}{2} P_4 = \tfrac{9}{32}$$

Note that $\mathbf{P}^{(3)}$ gives no mass to the value 3. We can now express the mass function $\mathbf{P}^{(3)}$ as an equally weighted mixture of two point mass functions $Q^{(2)}$ and $Q^{(3)}$, and we will end up with

$$\mathbf{P} = \tfrac{1}{3} Q^{(1)} + \tfrac{2}{3}(\tfrac{1}{2} Q^{(2)} + \tfrac{1}{2} Q^{(3)})$$
$$= \tfrac{1}{3}(Q^{(1)} + Q^{(2)} + Q^{(3)})$$

(We leave it as an exercise for the reader to fill in the details.)   ◊

The above example outlines the following general procedure for writing the n point mass function $\mathbf{P}$ in the form (4.1) where each of the $\mathbf{Q}^{(i)}$ are mass functions giving all their mass to at most 2-points. To start, we choose i and j satisfying the conditions of Lemma 4.1. We now define the mass function $\mathbf{Q}^{(1)}$ concentrating on the points i and j and which will contain all of the mass for point i by noting that in the representation (4.1) $Q_i^{(k)} = 0$ for $k = 2, \ldots, n - 1$, implying that

$$Q_i^{(1)} = (n - 1)P_i \quad \text{and so} \quad Q_j^{(1)} = 1 - (n - 1)P_i$$

Writing

$$\mathbf{P} = \frac{1}{n - 1} \mathbf{Q}^{(1)} + \frac{n - 2}{n - 1} \mathbf{P}^{(n-1)} \tag{4.3}$$

where $\mathbf{P}^{(n-1)}$ represents the remaining mass, we see that

$$P_i^{(n-1)} = 0$$

$$P_j^{(n-1)} = \frac{n-1}{n-2} \left(P_j - \frac{1}{n-1} Q_j^{(1)} \right) = \frac{n-1}{n-2} \left(P_i + P_j - \frac{1}{n-1} \right)$$

$$P_k^{(n-1)} = \frac{n-1}{n-2} P_k, \qquad k \neq i \text{ or } j$$

That the above is indeed a probability mass function is easily checked— for instance, the nonnegativity of $P_j^{(n-1)}$ follows from the fact that j was chosen so that $P_i + P_j \geq 1/(n-1)$.

We may now repeat the above procedure in the $(n-1)$ point probability mass function $\mathbf{P}^{(n-1)}$ to obtain

$$\mathbf{P}^{(n-1)} = \frac{1}{n-2} \mathbf{Q}^{(2)} + \frac{n-3}{n-2} \mathbf{P}^{(n-2)}$$

and thus from (4.3) we have

$$\mathbf{P} = \frac{1}{n-1} \mathbf{Q}^{(1)} + \frac{1}{n-1} \mathbf{Q}^{(2)} + \frac{n-3}{n-1} \mathbf{P}^{(n-2)}$$

We now repeat the procedure on $\mathbf{P}^{(n-2)}$ and so on until we finally obtain

$$\mathbf{P} = \frac{1}{n-1} (\mathbf{Q}^{(1)} + \cdots + \mathbf{Q}^{(n-1)})$$

In this way we are able to represent $\mathbf{P}$ as an equally weighted mixture of $n - 1$ 2-point mass functions. We can now easily simulate from $\mathbf{P}$ by first generating a random integer N equally likely to be either 1, 2,

. . . , $n - 1$. If the resulting value N is such that $\mathbf{Q}^{(N)}$ puts positive weight only on the points i_N and j_N, then we can set X equal to i_N if a second random number is less than $Q_{i_N}^{(N)}$ and equal to j_N otherwise. The random variable X will have probability mass function $\mathbf{P}$. That is, we have the following procedure for simulating from $\mathbf{P}$.

Step 1: Generate U_1 and set $N = 1 + [(n - 1)U_1]$.

Step 2: Generate U_2 and set

$$X = \begin{cases} i_N & \text{if} \quad U_2 < Q_{i_N}^{(N)} \\ j_N & \text{otherwise} \end{cases}$$

Remarks

(1) The above is called the alias method because by a renumbering of the $\mathbf{Q}$'s we can always arrange things so that for each k, $Q_k^{(k)} > 0$. (That is, we can arrange things so that the kth 2-point mass function gives positive weight to the value k.) Hence, the procedure calls for simulating N, equally likely to be $1, 2, \ldots,$ $n - 1$, and then if $N = k$ it either accepts k as the value of X, or it accepts for the value of X the "alias" of k (namely, the other value that $Q^{(k)}$ gives positive weight).

(2) Actually, it is not necessary to generate a new random number in Step 2. As $N - 1$ is the integer part of $(n - 1)U_1$, it follows that the remainder $(n - 1)U_1 - (N - 1)$ is independent of U_1 and is uniformly distributed in $(0, 1)$. Hence, rather than generating a new random number U_2 in Step 2, we can use $(n - 1)$ $U_1 - (N - 1) = (n - 1)U_1 - [(n - 1)U_1]$.

Example 4e Let us return to the problem of Example 1a which considers a list of n, not necessarily distinct, items. Each item has a value—$v(i)$ being the value of the item in position i—and we are interested in estimating

$$v = \sum_{i=1}^{n} v(i)/m(i)$$

where $m(i)$ is the number of times the item in position i appears on the list. In words, v is the sum of the values of the (distinct) items on the list.

To estimate v, note that if X is a random variable such that

$$P\{X = i\} = v(i) \bigg/ \sum_{1}^{n} v(j), \qquad i = 1, \ldots, n$$

then

$$E[1/m(X)] = \frac{\sum\limits_i v(i)/m(i)}{\sum\limits_j v(j)} = v \bigg/ \sum_{j=1}^n v(j)$$

Hence, we can estimate v by using the alias (or any other) method to generate independent random variables $X_1, \ldots, X_k$ having the same distribution as X and then estimating v by

$$v \approx \sum_{j=1}^n v(j) \frac{\sum\limits_{i=1}^k 1/m(X_i)}{k} \qquad \diamond$$

5. Stochastic Processes

One can usually simulate a stochastic process by simulating a sequence of random variables. For instance, to simulate the first t time units of a renewal process having interarrival distribution F we can simulate independent random variables $X_1, X_2, \ldots$ having distribution F stopping at

$$N = \min\{n : X_1 + \cdots + X_n > t\}$$

The X_i, $i \geq 1$, represent the interarrival times of the renewal process and so the above simulation yields $N - 1$ events by time t—the events occurring at times $X_1, X_1 + X_2, \ldots, X_1 + \ldots + X_{N-1}$.

Actually there is another approach for simulating a Poisson process that is quite efficient. Suppose we want to simulate the first t time units of a Poisson process having rate λ. To do so we can first simulate $N(t)$, the number of events by t, and then use the result that given the value of $N(t)$, the set of $N(t)$ event times are distributed as a set of n independent uniform $(0, t)$ random variables. Hence we start by simulating $N(t)$, a Poisson random variable with mean λt (by one of the methods given in Example 4c). Then, if $N(t) = n$, generate a new set of n random numbers—call them $U_1, \ldots, U_n$, and $\{tU_1, \ldots, tU_n\}$ will represent the set of $N(t)$ event times. If we could stop here this would be much more efficient than simulating the exponentially distributed interarrival times. However, we usually desire the event times in increasing order—for instance for $s < t$,

$$N(s) = \text{number of } U_i : tU_i \leq s$$

and so to compute the function $N(s)$, $s \leq t$, it is best to first order the values U_i, $i = 1, \ldots, n$ before multiplying by t. However, in doing so one should not use an all-purpose sorting algorithm, such as quicksort (see Example 4e of Chapter 3), but rather one that takes into account that the elements to be sorted come from a uniform $(0, 1)$ population. Such a sorting algorithm of n uniform $(0, 1)$ variables is as follows: Rather than a single list to be sorted of length n we will consider n ordered, or linked, lists of random size. The value U will be put in list i if its value is between $(i - 1)/n$ and i/n—that is, U is put in list $[nU] + 1$. The individual lists are then ordered and the total linkage of all the lists is the desired ordering. As almost all of the n lists will be of relatively small size [for instance, if $n = 1000$ the mean number of lists of size greater than 4 is (using the Poisson approximation to the binomial) approximately equal to $1000(1 - \frac{65}{24} e^{-1}) \simeq 4$] the sorting of individual lists will be quite quick, and so the running time of such an algorithm will be proportional to n (rather than to $n \log n$ as in the best all-purpose sorting algorithms).

An extremely important counting process for modeling purposes is the nonhomogeneous Poisson process, which relaxes the Poisson process assumption of stationary increments. Thus it allows for the possibility that the arrival rate need not be constant but can vary with time. However, there are few analytical studies that assume a nonhomogeneous Poisson arrival process for the simple reason that such models are not usually mathematically tractable. (For example, there is no known expression for the average customer delay in the single server exponential service distribution queueing model which assumes a nonhomogeneous arrival process).[*] Clearly such models are strong candidates for simulation studies.   ◊

5.1 Simulating a Nonhomogeneous Poisson Process

We now present three methods for simulating a nonhomogeneous Poisson process having intensity function $\lambda(t)$, $0 \leq t < \infty$.

---

[*]One queueing model that assumes a nonhomogeneous Poisson arrival process and is mathematically tractable is the infinite server model.

Method 1. Sampling a Poisson Process

To simulate the first T time units of a nonhomogeneous Poisson process with intensity function $\lambda(t)$, let λ be such that

$$\lambda(t) \leq \lambda \quad \text{for all} \quad t \leq T$$

Now as shown in Chapter 5, such a nonhomogeneous Poisson process can be generated by a random selection of the event times of a Poisson process having rate λ. That is, if an event of a Poisson process with rate λ that occurs at time t is counted (independently of what has transpired previously) with probability $\lambda(t)/\lambda$ then the process of counted events is a nonhomogeneous Poisson process with intensity function $\lambda(t)$, $0 \leq t \leq T$. Hence, by simulating a Poisson process and then randomly counting its events, we can generate the desired nonhomogeneous Poisson process. We thus have the following procedure:

Generate independent random variables $X_1, U_1, X_2, U_2, \ldots$ where the X_i is exponential with rate λ and the U_i are random numbers, stopping at

$$N = \min\left\{ n : \sum_{i=1}^{n} X_i > T \right\}$$

Now let, for $j = 1, \ldots, N - 1$

$$I_j = \begin{cases} 1 & \text{if} \quad U_j \leq \lambda \left(\sum_{i=1}^{j} X_i \right) \Big/ \lambda \\ 0 & \text{otherwise} \end{cases}$$

and set

$$J = \{ j : I_j = 1 \}$$

Thus the counting process having events at the set of times $\{\sum_{i=1}^{j} X_i : j \in J\}$ constitutes the desired process.

The above procedure, referred to as the thinning algorithm (because it "thins" the homogeneous Poisson points) will clearly be most efficient, in the sense of having the fewest number of rejected event times, when $\lambda(t)$ is near λ throughout the interval. Thus, an obvious improvement is to break up the interval into subintervals and then use the procedure over each subinterval. That is, determine appropriate values k, $0 < t_1 < t_2 < \cdots < t_k < T$, $\lambda_1, \ldots, \lambda_{k+1}$, such that

$$\lambda(s) \leq \lambda_i \quad \text{when} \quad t_{i-1} \leq s < t_i, \quad i = 1, \ldots, k + 1 \quad \text{(where}$$
$$t_0 = 0, \, t_{k+1} = T) \tag{5.1}$$

Now simulate the nonhomogeneous Poisson process over the interval (t_{i-1}, t_i) by generating exponential random variables with rate λ_i and accepting the generated event occurring at time s, $s \in (t_{i-1}, t_i)$, with probability $\lambda(s)/\lambda_i$. Because of the memoryless property of the exponential and the fact that the rate of an exponential can be changed upon multiplication by a constant, it follows that there is no loss of efficiency in going from one subinterval to the next. In order words, if we are at $t \in [t_{i-1}, t_i)$ and generate X, an exponential with rate λ_i, which is such that $t + X > t_i$ then we can use $\lambda_i[X - (t_i - t)]/\lambda_{i+1}$ as the next exponential with rate λ_{i+1}. Thus, we have the following algorithm for generating the first t time units of a nonhomogeneous Poisson process with intensity function $\lambda(s)$ when the relations (5.1) are satisfied. In the algorithm, t will present the present time and I the present interval (that is, $I = i$ when $t_{i-1} \leq t < t_i$).

Step 1: $t = 0$, $I = 1$.

Step 2: Generate an exponential random variable X having rate λ_I.

Step 3: If $t + X < t_I$, reset $t = t + X$, generate a random number U, and accept the event time t if $U \leq \lambda(t)/\lambda_I$. Return to Step 2.

Step 4: Step reached if $t + X \geq t_I$. Stop if $I = k + 1$. Otherwise, reset $X = (X - t_I + t)\lambda_I/\lambda_{I+1}$. Also reset $t = t_I$ and $I = I + 1$, and go to Step 3.

Suppose now that over some subinterval (t_{i-1}, t_i) it follows that $\underline{\lambda}_i > 0$ where

$$\underline{\lambda}_i \equiv \text{infimum}\{\lambda(s) : t_{i-1} \leq s < t_i\}.$$

In such a situation, we should not use the thinning algorithm directly but rather should first simulate a Poisson process with rate $\underline{\lambda}_i$ over the desired interval and then simulate a nonhomogeneous Poisson process with the intensity function $\lambda(s) = \lambda(s) - \underline{\lambda}_i$ when $s \in (t_{i-1}, t_i)$. (The final exponential generated for the Poisson process, which carries one beyond the desired boundary, need not be wasted but can be suitably transformed so as to be reusable.) The superposition (or, merging) of the two processes yields the desired process over the interval. The reason for doing it this way is that it saves the need to generate uniform random variables for a Poisson distributed number, with mean $\underline{\lambda}_i(t_i - t_{i-1})$ of the event times. For instance, consider the case where

$$\lambda(s) = 10 + s, \qquad 0 < s < 1.$$

Using the thinking method with $\lambda = 11$ would generate an expected number of 11 events each of which would require a random number to determine whether or not to accept it. On the other hand, to generate a Poisson process with rate 10 and then merge it with a generated nonhomogeneous Poisson process with rate $\lambda(s) = s$, $0 < s < 1$, would yield an equally distributed number of event times but with the expected number needing to be checked to determine acceptance being equal to 1.

Another way to make the simulation of nonhomogeneous Poisson processes more efficient is to make use of superpositions. For instance, consider the process where

$$\lambda(t) = \begin{cases} \exp\{t^2\} & 0 < t < 1.5 \\ \exp\{2.25\} & 1.5 < t < 2.5 \\ \exp\{(4 - t)^2\} & 2.5 < t < 4 \end{cases}$$

A plot of this intensity function is given in Figure 11.3. One way of simulating this process up to time 4 is to first generate a Poisson process with rate 1 over this interval; then generate a Poisson process with rate $e - 1$ over this interval and then accept all events in $(1,3)$ and only accept an event at time t which is not contained in $(1,3)$ with probability $[\lambda(t) - 1]/(e - 1)$; then generate a Poisson process with rate $e^{2.25} - e$ over the interval $(1,3)$, accepting all event times between 1.5

Figure 11.3

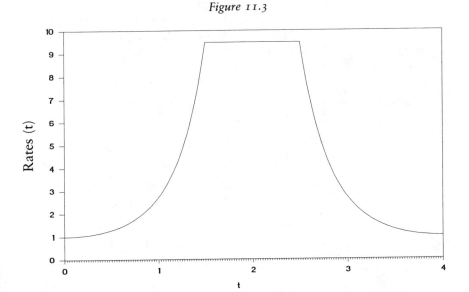

and 2.5 and any event time t outside this interval with probability $[\lambda(t) - e]/(e^{2.25} - e)$. The superposition of these processes is the desired nonhomogeneous Poisson process. In other words, what we have done is to break up $\lambda(t)$ into the following non-negative parts

$$\lambda(t) = \lambda_1(t) + \lambda_2(t) + \lambda_3(t), \qquad 0 < t < 4$$

where

$$\lambda_1(t) \equiv 1$$

$$\lambda_2(t) = \begin{cases} \lambda(t) - 1 & 0 < t < 1 \\ e - 1 & 1 < t < 3 \\ \lambda(t) - 1 & 3 < t < 4 \end{cases}$$

$$\lambda_3(t) = \begin{cases} \lambda(t) - e & 3 < t < 4 \\ 0 & \text{otherwise} \end{cases}$$

and where the thinning algorithm (with a single interval in each case) was used to simulate the constituent nonhomogeneous processes.

Method 2. Conditional Distribution of the Arrival Times

Recall the result for a Poisson process having rate λ that given the number of events by time T the set of event times are independent and identically distributed uniform $(0, T)$ random variables. Now suppose that each of these events is independently counted with a probability that is equal to $\lambda(t)/\lambda$ when the event occurred at time t. Hence, given the number of counted events, it follows that the set of times of these counted events are independent with a common distribution given by $F(s)$ where

$$\begin{aligned} F(s) &= P\{\text{Time} \le s | \text{counted}\} \\ &= \frac{P\{\text{Time} \le s, \text{counted}\}}{P\{\text{counted}\}} \\ &= \frac{\displaystyle\int_0^T P\{\text{Time} \le s, \text{counted} | \text{Time} = x\}\, \frac{dx}{T}}{P\{\text{counted}\}} \\ &= \frac{\displaystyle\int_0^s \lambda(x)\, dx}{\displaystyle\int_0^T \lambda(x)\, dx} \end{aligned}$$

The above (somewhat heuristic) argument thus shows that given n events of a nonhomogeneous Poisson process by time T the n event times are independent with a common density function

$$f(s) = \frac{\lambda(s)}{m(T)}, \qquad 0 < s < T, \qquad m(T) = \int_0^T \lambda(s)\,ds \qquad (5.2)$$

Since $N(T)$, the number of events by time T, is Poisson distributed with mean $m(T)$, we can simulate the nonhomogeneous Poisson process by first simulating $N(T)$ and then simulating $N(T)$ random variables from the density (5.2).

Example 5c If $\lambda(s) = cs$, then we can simulate the first T time units of the nonhomogeneous Poisson process by first simulating $N(T)$, a Poisson random variable having mean $m(T) = \int_0^T cs\,ds = CT^2/2$, and then simulating $N(T)$ random variables having distribution

$$F(s) = \frac{s^2}{T^2}, \qquad 0 < s < T$$

Random variables having the above distribution either can be simulated by use of the inverse transform method (since $F^{-1}(U) = T\sqrt{U}$) or by noting that F is the distribution function of $\max(TU_1, TU_2)$ when U_1 and U_2 are independent random numbers. ◇

If the distribution function specified by (5.2) is not easily invertible, we can always simulate from (5.2) by using the rejection method where we either accept or reject simulated values of uniform $(0, T)$ random variables. That is, let $h(s) = 1/T$, $0 < s < T$. Then

$$\frac{f(s)}{h(s)} = \frac{T\lambda(s)}{m(T)} \le \frac{\lambda T}{m(T)} \equiv C$$

where λ is a bound on $\lambda(s)$, $0 \le s \le T$. Hence, the rejection method is to generate random numbers U_1 and U_2 and then accept TU_1 if

$$U_2 \le \frac{f(TU_1)}{Ch(TU_1)}$$

or, equivalently, if

$$U_2 \le \frac{\lambda(TU_1)}{\lambda}$$

Method 3. Simulating the Event Times

The third method we shall present for simulating a nonhomogeneous Poisson process having intensity function $\lambda(t)$, $t \geq 0$ is probably the most basic approach—namely to simulate the successive event times. So let X_1, X_2, . . . denote the event times of such a process. As these random variables are dependent we will use the conditional distribution approach to simulation. Hence, we need the conditional distribution of X_i given X_1, . . . , X_{i-1}.

To start, note that if an event occurs at time x then, independent of what has occurred prior to x, the time until the next event has the distribution F_x given by

$$\overline{F}_x(t) = P\{0 \text{ events in } (x, x + t)|\text{event at } x\}$$

$$= P\{0 \text{ events in } (x, x + t)\} \text{ by independent increments}$$

$$= \exp\left\{-\int_0^t \lambda(x + y)\,dy\right\}$$

Differentiation yields that the density corresponding to F_x is

$$f_x(t) = \lambda(x + t) \exp\left\{-\int_0^t \lambda(x + y)\,dy\right\}$$

implying that the hazard rate function of F_x is

$$r_x(t) = \frac{f_x(t)}{\overline{F}_x(t)} = \lambda(x + t)$$

We can now simulate the event times X_1, X_2, . . . by simulating X_1 from F_0; then if the simulated value of X_1 is x_1, simulate X_2 by adding x_1 to a value generated from F_{x_1}, and if this sum is x_2 simulate X_3 by adding x_2 to a value generated from F_{x_2}, and so on. The method used to simulate from these distributions should depend, of course, on the form of these distributions. However, it is interesting to note that if we let λ be such that $\lambda(t) \leq \lambda$ and use the hazard rate method to simulate, then we end up with the approach of Method 1 (we leave the verification of this fact as an exercise). Sometimes, however, the distributions F_x can be easily inverted and so the inverse transform method can be applied.

Example 5d Suppose that $\lambda(x) = 1/(x + a)$, $x \geq 0$. Then

$$\int_0^t \lambda(x + y)\,dy = \log\left(\frac{x + a + t}{x + a}\right)$$

Hence,

$$F_x(t) = 1 - \frac{x + a}{x + a + t} = \frac{t}{x + a + t}$$

and so

$$F_x^{-1}(u) = (x + a)\frac{u}{1 - u}$$

We can, therefore, simulate the successive event times $X_1, X_2, \ldots$ by generating $U_1, U_2, \ldots$ and then setting

$$X_1 = \frac{aU_1}{1 - U_1}$$

$$X_2 = (X_1 + a)\frac{U_2}{1 - U_2} + X_1$$

and, in general,

$$X_j = (X_{j-1} + a)\frac{U_j}{1 - U_j} + X_{j-1}, j \geq 2 \quad \diamond$$

5.2. Simulating a Two-Dimensional Poisson Process

A point process consisting of randomly occurring points in the plane is said to be a two-dimensional Poisson process having rate λ if

(a) the number of points in any given region of area A is Poisson distributed with mean λA; and
(b) the numbers of points in disjoint regions are independent.

For a given fixed point **O** in the plane, we now show how to simulate events occurring according to a two-dimensional Poisson process with rate λ in a circular region of radius r centered about **O**. Let R_i, $i \geq 1$, denote the distance between **O** and its ith nearest Poisson point, and let $C(a)$ denote the circle of radius a centered at **O**. Then

$$P\{\pi R_1^2 > b\} = P\left\{R_1 > \sqrt{\frac{b}{\pi}}\right\} = P\{\text{no points in C }(\sqrt{b/\pi})\} = e^{-\lambda b}$$

Also, with $C(a_2) - C(a_1)$ denoting the region between $C(a_2)$ and $C(a_1)$:

$$P\{\pi R_2^2 - \pi R_1^2 > b | R_1 = r\}$$
$$= P\{R_2 > \sqrt{(b + \pi r^2)/\pi} | R_1 = r\}$$
$$= P\{\text{no points in } C(\sqrt{(b + \pi r^2)/\pi}) - C(r) | R_1 = r)$$
$$= P\{\text{no points in } C(\sqrt{(b + \pi r^2)/\pi} - C(r)) \text{ by } (b)$$
$$= e^{-\lambda b}$$

In fact the same argument can be repeated to obtain

Proposition 5.1.   With $R_0 = 0$

$\pi R_i^2 - \pi R_{i-1}^2, i \geq 1$, are independent exponentials with rate λ.

In other words, the amount of area that need be traversed to en-compass a Poisson point is exponential with rate λ. Since, by symmetry, the respective angles of the Poisson points are independent and uni-formly distributed over $(0, 2\pi)$ we thus have the following algorithm for simulating the Poisson process over a circular region of radius r about **O**.

Step 1:   Generate independent exponentials with rate 1, X_1, X_2, . . . , stopping at

$$N = \min\left\{ n : \frac{X_1 + \cdot \cdot \cdot + X_n}{\lambda \pi} > r^2 \right\}$$

Step 2:   If $N = 1$, stop. There are no points in $C(r)$. Otherwise, for $i = 1, \ldots, N - 1$ set

$$R_i = \sqrt{(X_1 + \cdot \cdot \cdot + X_i)/\lambda \pi}$$

Step 3:   Generate independent uniform $(0, 1)$ random variables $U_1, \ldots, U_{N-1}$.

Step 4:   Return the $N - 1$ Poisson points in $C(r)$ whose polar coordinates are

$$(R_i, 2\pi U_i), \qquad i = 1, \ldots, N - 1$$

The above algorithm requires, on average, $1 + \lambda \pi r^2$ exponentials and an equal number of uniform random numbers. Another approach to simulating points in $C(r)$ is to first simulate N, the number of such points, and then use the fact that, given N, the points are uniformly distributed in $C(r)$. This latter procedure requires the simulation of N,

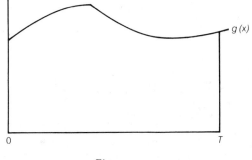

Figure 11.4.

a Poisson random variable with mean $\lambda \pi r^2$ and must then simulate N uniform points on $C(r)$, by simulating R from the distribution $F_R(a) = a^2/r^2$ (see Problem 25) and θ from uniform $(0, 2\pi)$ and must then sort these N uniform values in increasing order of R. The main advantage of the first procedure is that it eliminates the need to sort.

The above algorithm can be thought of as the fanning out of a circle centered at $\mathbf{O}$ with a radius that expands continuously from 0 to r. The successive radii at which Poisson points are encountered is simulated by noting that the additional area necessary to encompass a Poisson point is always, independent of the past, exponential with rate λ. This technique can be used to simulate the process over noncircular regions. For instance, consider a nonnegative function $g(x)$, and suppose we are interested in simulating the Poisson process in the region between the x-axis and g with x going from 0 to T (see Figure 11.2). To do so we can start at the left hand end and fan vertically to the right by considering the successive areas $\int_0^a g(x)\,dx$. Now if $X_1 < X_2 < \ldots$ denote the successive projections of the Poisson points on the x-axis, then analogous to Proposition 5.1, it will follow that (with $X_0 = 0$) $\lambda \int_{X_{i-1}}^{X_i} g(x)\,dx$, $i \geq 1$, will be independent exponentials with rate 1. Hence, we should simulate $\in_1, \in_2, \ldots$, independent exponentials with rate 1, stopping at

$$N = \min\left\{ n : \in_1 + \cdots + \in_n > \lambda \int_0^T g(x)\,dx \right\}$$

and determine $X_1, \ldots, X_{N-1}$ by

$$\lambda \int_0^{X_1} g(x)\,dx = \in_1$$

$$\lambda \int_{X_1}^{X_2} g(x)\,dx \quad = \in_2$$

$$\vdots$$

$$\lambda \int_{X_{N-2}}^{X_{N-1}} g(x)\,dx = \in_{N-1}$$

If we now simulate $U_1, \ldots, U_{N-1}$—independent uniform $(0, 1)$ random numbers—then as the projection on the y-axis of the Poisson point whose x-coordinate is X_i, is uniform on $(0, g(X_i))$, it follows that the simulated Poisson points in the interval are $(X_i, U_i g(X_i))$, $i = 1, \ldots, N - 1$.

Of course, the above technique is most useful when g is regular enough so that the above equations can be solved for the X_i. For instance, if $g(x) = y$ (and so the region of interest is a rectangle), then

$$X_i = \frac{\in_1 + \cdots + \in_i}{\lambda y}, \qquad i = 1, \ldots, N - 1$$

and the Poisson points are

$$(X_i, y U_i), \qquad i = 1, \ldots, N - 1$$

6. Variance Reduction Techniques

Let $X_1, \ldots, X_n$ have a given joint distribution, and suppose we are interested in computing

$$\theta \equiv E[g(X_1, \ldots, X_n)]$$

where g is some specified function. It is often the case that it is not possible to analytically compute the above, and when such is the case we can attempt to use simulation to estimate θ. This is done as follows: Generate $X_1^{(1)}, \ldots, X_n^{(1)}$ having the same joint distribution as $X_1, \ldots, X_n$ and set

$$Y_1 = g(X_1^{(1)}, \ldots, X_n^{(1)})$$

Now simulate a second set of random variables (independent of the first set) $X_1^{(2)}, \ldots, X_n^{(2)}$ having the distribution of $X_1, \ldots, X_n$ and set

$$Y_2 = g(X_1^{(2)}, \ldots, X_n^{(2)})$$

Continue this until you have generated k (some predetermined number) sets, and so have also computed $Y_1, Y_2, \ldots, Y_k$. Now, $Y_1, \ldots, Y_k$ are independent and identically distributed random variables each having the same distribution of $g(X_1, \ldots, X_n)$. Thus, if we let $\overline{Y}$ denote the average of these k random variables—that is

$$\overline{Y} = \sum_{i=1}^{k} Y_i/k$$

then

$$E[\overline{Y}] = \theta$$
$$E[(\overline{Y} - \theta)^2] = \mathrm{Var}(\overline{Y})$$

Hence, we can use $\overline{Y}$ as an estimate of θ. As the expected square of the difference between $\overline{Y}$ and θ is equal to the variance of $\overline{Y}$, we would like this quantity to be as small as possible. (In the above situation, $\mathrm{Var}(\overline{Y}) = \mathrm{Var}(Y_i)/k$, which is usually not known in advance but must be estimated from the generated values $Y_1, \ldots, Y_n$.) We now present three general techniques for reducing the variance of our estimator.

6.1 Use of Antithetic Variables

In the above situation, suppose that we have generated Y_1 and Y_2, identically distributed random variables having mean θ. Now

$$\mathrm{Var}\left(\frac{Y_1 + Y_2}{2}\right) = \tfrac{1}{4}[\mathrm{Var}(Y_1) + \mathrm{Var}(Y_2) + 2\,\mathrm{Cov}(Y_1, Y_2)]$$

$$= \frac{\mathrm{Var}(Y_1)}{2} + \frac{\mathrm{Cov}(Y_1, Y_2)}{2}$$

Hence it would be advantageous (in the sense that the variance would be reduced) if Y_1 and Y_2 rather than being independent were negatively correlated. To see how we could arrange this, let us suppose that the random variables $X_1, \ldots, X_n$ are independent and, in addition, that each is simulated via the inverse transform technique. That is X_i is simulated from $F_i^{-1}(U_i)$ where U_i is a random number and F_i is the distribution of X_i. Hence, Y_1 can be expressed as

$$Y_1 = g(F_1^{-1}(U_1), \ldots, F_n^{-1}(U_n))$$

Now, since $1 - U$ is also uniform over $(0, 1)$ whenever U is a random number (and is negatively correlated with U) it follows that Y_2 defined by

$$Y_2 = g(F_1^{-1}(1 - U_1), \ldots, F_n^{-1}(1 - U_n))$$

will have the same distribution as Y_1. Hence, if Y_1 and Y_2 were negatively correlated, then generating Y_2 by this means would lead to a smaller variance than if it were generated by a new set of random numbers. (In addition, there is a computational savings since rather than having to generate n additional random numbers, we need only subtract each of the previous n from 1.) The following theorem will be the key to showing that this technique—known as the use of antithetic variables—will lead to a reduction in variance whenever g is a monotone function.

Theorem 6.1.   If $X_1, \ldots, X_n$ are independent, then for any increasing functions f and g of n variables

$$E[f(\mathbf{X})g(\mathbf{X})] \geq E[f(\mathbf{X})]E[g(\mathbf{X})] \tag{6.1}$$

where $\mathbf{X} = (X_1, \ldots, X_n)$.

Proof.   The proof is by induction on n. To prove it when $n = 1$ let f and g be increasing functions of a single variable. Then for any x and y

$$(f(x) - f(y))(g(x) - g(y)) \geq 0$$

since if $x \geq y$ $(x \leq y)$ then both factors are nonnegative (nonpositive). Hence, for any random variables X and Y

$$(f(X) - f(Y))(g(X) - g(Y)) \geq 0$$

implying that

$$E[(f(X) - f(Y))(g(X) - g(Y))] \geq 0$$

or, equivalently

$$E[f(X)g(X)] + E[f(Y)g(Y)] \geq E[f(X)g(Y)] + E[f(Y)g(X)]$$

If we now suppose that X and Y are independent and identically distributed then, as in this case,

$$E[f(X)g(X)] = E[f(Y)g(Y)]$$
$$E[f(X)g(Y)] = E[f(Y)g(X)] = E[f(X)]E[g(X)]$$

we obtain the result when $n = 1$.

So assume that (6.1) holds for $n - 1$ variables, and now suppose that $X_1, \ldots, X_n$ are independent and f and g are increasing functions. Then

$E[f(\mathbf{X})g(\mathbf{X})|X_n = x_n]$

$\quad = E[f(X_1, \ldots, X_{n-1}, x_n)g(X_1, \ldots, X_{n-1}, x_n)|X_n = x]$

$\quad = E[f(X_1, \ldots, X_{n-1}, x_n)g(X_1, \ldots, X_{n-1}, x_n)]$

$\quad$ by independence

$\quad \geq E[f(X_1, \ldots, X_{n-1}, x_n)]E[g(X_1, \ldots, X_{n-1}, x_n)]$

$\quad$ by the induction hypothesis

$\quad = E[f(\mathbf{X})|X_n = x_n]E[g(\mathbf{X})|X_n = x_n]$

Hence,

$$E[f(\mathbf{X})g(\mathbf{X})|X_n] \geq E[f(\mathbf{X})|X_n]E[g(\mathbf{X})|X_n]$$

and, upon taking expectations of both sides,

$$E[f(\mathbf{X})g(\mathbf{X})] \geq E[E[f(\mathbf{X})|X_n]E[g(\mathbf{X})|X_n]]$$
$$\geq E[f(\mathbf{X})]E[g(\mathbf{X})]$$

The last inequality follows because $E[f(\mathbf{X})|X_n]$ and $E[g(\mathbf{X})|X_n]$ are both increasing functions of X_n, and so by the result for $n = 1$

$$E[E[f(\mathbf{X})|X_n]E[g(\mathbf{X})|X_n]] \geq E[E[f(\mathbf{X})|X_n]]E[E[g(\mathbf{X})|X_n]]$$
$$= E[f(\mathbf{X})]E[g(\mathbf{X})] \quad \Diamond$$

Corollary 6.2. If $U_1, \ldots, U_n$ are independent, and k is either an increasing or decreasing function, then

$$\text{Cov}(k(U_1, \ldots, U_n), k(1 - U_1, \ldots, 1 - U_n)) \leq 0$$

Proof: Suppose k is increasing. As $-k(1 - U_1, \ldots, 1 - U_n)$ is increasing in $U_1, \ldots, U_n$, then from Theorem 6.1

$$\text{Cov}(k(U_1, \ldots, U_n), \quad -k(1 - U_1, \ldots, 1 - U_n)) \geq 0$$

When k is decreasing just replace k by its negative. $\quad \Diamond$

Since $F_i^{-1}(U_i)$ is increasing in U_i (as F_i, being a distribution function, is increasing) it follows that $g(F_1^{-1}(U_1), \ldots, F_n^{-1}(U_n))$ is a monotone function of $U_1, \ldots, U_n$ whenever g is monotone. Hence, if g is monotone the antithetic variable approach of twice using each set of random numbers $U_1, \ldots, U_n$ by first computing $g(F_1^{-1}(U_1), \ldots, F_n^{-1}(U_n))$ and then $g(F_1^{-1}(1 - U_1), \ldots, F_n^{-1}(1 - U_n))$ will reduce the variance of the estimate of $E[g(X_1, \ldots, X_n)]$. That is, rather than generating k sets of n random numbers, we should generate $k/2$ sets and use each set twice.

Example 6a   (Simulating the Reliability Function): Consider a system of n components in which component i, independently of other components, works with probability p_i, $i = 1, \ldots, n$. Letting

$$X_i = \begin{cases} 1 & \text{if component } i \text{ works} \\ 0 & \text{otherwise} \end{cases}$$

suppose there is a monotone structure function ϕ such that

$$\phi(X_1, \ldots, X_n) = \begin{cases} 1 & \text{if the system works under } X_1, \ldots, X_n \\ 0 & \text{otherwise} \end{cases}$$

We are interested in using simulation to estimate

$$r(p_1, \ldots, p_n) \equiv E[\phi(X_1, \ldots, X_n)] = P\{\phi(X_1, \ldots, X_n) = 1\}$$

Now, we can simulate the X_i by generating uniform random numbers $U_1, \ldots, U_n$ and then setting

$$X_i = \begin{cases} 1 & \text{if} \quad U_i < p_i \\ 0 & \text{otherwise} \end{cases}$$

Hence, we see that

$$\phi(X_1, \ldots, X_n) = k(U_1, \ldots, U_n)$$

where k is a decreasing function of $U_1, \ldots, U_n$. Hence

$$\text{Cov}(k(\mathbf{U}), k(1 - \mathbf{U})) \le 0$$

and so the antithetic variable approach of using $U_1, \ldots, U_n$ to generate both $k(U_1, \ldots, U_n)$ and $k(1 - U_1, \ldots, 1 - U_n)$ results in a smaller variance than if an independent set of random numbers was used to generate the second k.   ◇

Example 6b   (Simulating a Queueing System): Consider a given queueing system, and let D_i denote the delay in queue of the ith arriving customer, and suppose we are interested in simulating the system so as to estimate

$$\theta = E[D_1 + \cdots + D_n]$$

Let $X_1, \ldots, X_n$ denote the first n interarrival times and $S_1, \ldots, S_n$ the first n service times of this system, and suppose these random variables are all independent. Now in most systems $D_1 + \ldots + D_n$ will be a function of $X_1, \ldots, X_n, S_1, \ldots, S_n$—say

$$D_1 + \cdots + D_n = g(X_1, \ldots, X_n, S_1, \ldots, S_n)$$

Also g will usually be increasing in S_i and decreasing in X_i, $i = 1, \ldots, n$. If we use the inverse transform method to simulate X_i, S_i, $i = 1, \ldots, n$—say $X_i = F_i^{-1}(1 - U_i)$, $S_i = G_i^{-1}(\overline{U}_i)$ where $U_1, \ldots, U_n, \overline{U}_1, \ldots, \overline{U}_n$ are independent uniform random numbers—then we may write

$$D_1 + \cdots + D_n = k(U_1, \ldots, U_n, \overline{U}_1, \ldots, \overline{U}_n)$$

where k is increasing in its variates. Hence, the antithetic variable approach will reduce the variance of the estimator of θ. (Thus we would generate U_i, $\overline{U}_i$, $i = 1, \ldots, n$ and set $X_i = F_i^{-1}(1 - U_i)$ and $Y_i = G_i^{-1}(\overline{U}_i)$ for the first run, and $X_i = F_i^{-1}(U_i)$ and $Y_i = G_i^{-1}(1 - \overline{U}_i)$ for the second.) As all the U_i and $\overline{U}_i$ are independent, however, this is equivalent to setting $X_i = F_i^{-1}(U_i)$, $Y_i = G_i^{-1}(\overline{U}_i)$ in the first run and using $1 - U_i$ for U_i and $1 - \overline{U}_i$ for $\overline{U}_i$ in the second. ◇

6.2 Variance Reduction by Conditioning

Let us start by recalling (see Problem 22 of Chapter 3) the conditional variance formula

$$\mathrm{Var}(Y) = E[\mathrm{Var}(Y|Z)] + \mathrm{Var}(E[Y|Z]) \tag{6.2}$$

Now suppose we are interested in estimating $E[g(X_1, \ldots, X_n)]$ by simulating $\mathbf{X} = (X_1, \ldots, X_n)$ and then computing $Y = g(X_1, \ldots, X_n)$. Now, if for some random variable Z we can compute $E[Y|Z]$ then, as $\mathrm{Var}(Y|Z) \geq 0$, it follows from the conditional variance formula that

$$\mathrm{Var}(E[Y|Z]) \leq \mathrm{Var}(Y)$$

implying, since $E[E[Y|Z]] = E[Y]$, that $E[Y|Z]$ is a better estimator of $E[Y]$ than is Y.

In many situations, there are a variety of Z_i that can be conditioned on to obtain an improved estimator. Each of these estimators $E[Y|Z_i]$ will have mean $E[Y]$ and smaller variance than does the raw estimator Y. We now show that for any choice of weights λ_i, $\lambda_i \geq 0$, $\Sigma_i \lambda_i = 1$, $\Sigma_i \lambda_i E[Y|Z_i]$ is also an improvement over Y.

Proposition 6.3. For any $\lambda_i \geq 0$, $\Sigma_{i=1}^{\infty} \lambda_i = 1$

(a) $E\left[\sum_i \lambda_i E[Y|Z_i]\right] = E[Y]$

(b) $\text{Var}\left(\sum_i \lambda_i E[Y|Z_i]\right) \le \text{Var}(Y)$

Proof: The proof of (a) is immediate. To prove (b), let N denote an integer valued random variable independent of all the other random variables under consideration and such that

$$P\{N = i\} = \lambda_i, \qquad i \ge 1$$

From the conditional variance formula

$$\text{Var}(Y) \ge \text{Var}(E[Y|N, Z_N])$$
$$\ge \text{Var}(E[E[Y|N, Z_N]|Z_1, . . .])$$

again by the conditional variance formula

$$= \text{Var}\sum_i \lambda_i E[Y|Z_i] \quad \diamond$$

Example 6c   Consider a queueing system having Poisson arrivals and suppose that any customer arriving when there are already N others in the system is lost. Suppose that we are interested in using simulation to estimate the expected number of lost customers by time t. The raw simulation approach would be to simulate the system up to time t and determine L, the number of lost customers for that run. A better estimate, however, can be obtained by conditioning on the total time in $[0,t]$ that the system is at capacity. Indeed, if we let T denote the time in $[0,t]$ that there are N in the system, then

$$E[L|T] = \lambda T$$

where λ is the Poisson arrival rate. Hence, a better estimate for $E[L]$ than the average value of L over all simulation runs can be obtained by multiplying the average value of T per simulation run by λ. If the arrival process were a nonhomogeneous Poisson process, then we could improve over the raw estimator L by keeping track of those time periods for which the system is at capacity. If we let $I_1, . . . , I_C$ denote the time intervals in $[0,t]$ in which there are N in the system, the..

$$E[L|I_1, \ldots, I_C] = \sum_{i=1}^{C} \int_{I_i} \lambda(s) \, ds$$

where $\lambda(s)$ is the intensity function of the nonhomogeneous Poisson arrival process. The use of the right side of the above would thus lead to a better estimate of $E[L]$ than the raw estimator L. ◇

Example 6d Suppose that we wanted to estimate the expected sum of the times in the system of the first n customers in a queueing system. That is, if W_i is the time that the ith customer spends in the system, then we are interested in estimating

$$\theta = E\left[\sum_{i=1}^{n} W_i \right]$$

Let Y_i denote the "state of the system" at the moment at which the ith customer arrives. It has recently been shown* that for a wide class of models the estimator $\sum_{i=1}^{n} E[W_i|Y_i]$ has (the same mean and) a smaller variance than the estimator $\sum_{i=1}^{n} W_i$. (It should be noted that whereas it is immediate that $E[W_i|Y_i]$ has smaller variance than W_i, because of the covariance terms involved, it is not immediately apparent that $\sum_{i=1}^{n} E[W_i|Y_i]$ has smaller variance than $\sum_{i=1}^{n} W_i$.) For instance, in the model $G/M/1$

$$E[W_i|Y_i] = (N_i + 1)/\mu$$

where N_i is the number in the system encountered by the ith arrival and $1/\mu$ is the mean service time; and the result implies that $\sum_{i=1}^{n} (N_i + 1)/\mu$ is a better estimate of the expected total time in the system of the first n customers than is the raw estimator $\sum_{i=1}^{n} W_i$. ◇

Example 6e (Estimating the Renewal Function by Simulation): Consider a queueing model in which customers arrive daily in accordance with a renewal process having interarrival distribution F. However, suppose that at some fixed time T, for instance 5 P.M.,

---

*S. M. Ross, "Simulating Average Delay—Variance Reduction by Conditioning," *Probability in the Engineering and Informational Sciences* 2(3), (1988), pp. 309–312.

no additional arrivals are permitted and those customers that are still in the system are serviced. At the start of the next, and each succeeding, day customers again begin to arrive in accordance with the renewal process. Suppose we are interested in determining the average time that a customer spends in the system. Upon using the theory of renewal reward processes (with a cycle starting every T time units), it can be shown that

$$\text{average time that a customer spends in the system} = \frac{E[\text{sum of the times in the system of arrivals in } (0,T)]}{m(T)}$$

where $m(T)$ is the expected number of renewals in $(0,T)$.

If we were to use simulation to estimate the above quantity, a run would consist of simulating a single day, and as part of a simulation run, we would observe the quantity $N(T)$, the number of arrivals by time T. Since $E[N(T)] = m(T)$, the natural simulation estimator of $m(T)$ would be the average (over all simulated days) value of $N(T)$ obtained. However, $\text{Var}(N(T))$ is, for large T, proportional to T (its asymptotic form being $T\sigma^2/\mu^3$, where σ^2 is the variance and μ the mean of the interarrival distribution F), and so for large T, the variance of our estimator would be large. A considerable improvement can be obtained by using the analytic formula (see Section 3 of Chapter 7)

$$m(T) = \frac{T}{\mu} - 1 + \frac{E[Y(T)]}{\mu} \qquad (6.3)$$

where $Y(T)$ denotes the time from T until the next renewal—that is, it is the excess life at T. Since the variance of $Y(T)$ does not grow with T (indeed, it converges to a finite value provided the moments of F are finite), it follows that for T large, we would do much better by using the simulation to estimate $E[Y(T)]$ and then use (6.3) to estimate $m(T)$.

However, by employing conditioning, we can improve further on our estimator of $m(T)$. To do so, let $A(T)$ denote the age of the renewal process at time T—that is, it is the time at T since the last renewal. Then, rather than using the value of $Y(T)$, we can reduce the variance by considering $E[Y(T)|A(T)]$. Now knowing that the age at T is equal to x is equivalent to knowing that there was a renewal at time $T - x$ and the next interarrival time

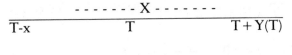

Figure 11.5 $A(T) = x$.

X is greater than x. Since the excess at T will equal $X - x$ (see Figure 11.5), it follows that

$$E[Y(T)|A(T) = x] = E[X - x|X > x]$$

$$= \int_0^\infty \frac{P\{X - x > t\}}{P\{X > x\}} \, dt$$

$$= \int_0^\infty \frac{[1 - F(t + x)]}{1 - F(x)} \, dt$$

which can be numerically evaluated if necessary.

As an illustration of the above, note that if the renewal process is a Poisson process with rate λ, then the raw simulation estimator $N(T)$ will have variance λT, whereas $Y(T)$ will be exponential with rate λ, the estimator based on (6.3) will have variance $\lambda^2 \, \text{Var}(Y(T)) = 1$. On the other hand, since $Y(T)$ will be independent of $A(T)$ (and $E[Y(T)|A(T)] = 1/\lambda$), it follows that the variance of the improved estimator $E[Y(T)|A(T)]$ is 0. That is, conditioning on the age at time T yields, in this case, the exact answer. ◇

Example 6f (Reliability): Suppose as in Example 6a that $X_j, j = 1, \ldots, n$ are independent with $P\{X_j = 1\} = P_j = 1 - P\{X_j = 0\}$, and suppose we are interested in estimating $E[\phi(X_1, \ldots, X_n)]$, where ϕ is a monotone binary function. If we simulate $X_1, \ldots, X_n$, an improvement over the raw estimator, $\phi(X_1, \ldots, X_n)$ is to take its conditional expectation given all the X_j except one. That is, for fixed i, $E[\phi(\mathbf{X})|\in_i(\mathbf{X})]$ is an improved estimator where $\mathbf{X} = (X_1, \ldots, X_n)$ and $\in_i(\mathbf{X}) = (X_1, \ldots, X_{i-1}, X_{i+1}, \ldots, X_n)$. $E[\phi(\mathbf{X})|\in_i(\mathbf{X})]$ will have three possible values—either it will equal 1 (if $\in_i(\mathbf{X})$ is such that the system will function even if $X_i = 0$), or 0 (if $\in_i(\mathbf{X})$ is such that the system will be failed even if $X_i = 1$), or P_i (if $\in_i(\mathbf{X})$ is such that the system will function if $X_i = 1$ and will be failed otherwise.) Also by Proposition 6.3 any estimator of the form

$$\sum_i \lambda_i E[\phi(\mathbf{X})|\in_i(\mathbf{X})], \qquad \sum_i \lambda_i = 1, \qquad \lambda_i \geq 0$$

is an improvement over $\phi(\mathbf{X})$. $\Diamond$

6.3 Control Variates

Again suppose we want to use simulation to estimate $E[g(\mathbf{X})]$ where $\mathbf{X} = (X_1, \ldots, X_n)$. But now suppose that for some function f the expected value of $f(\mathbf{X})$ is known—say $E[f(\mathbf{X})] = \mu$. Then for any constant a we can also use

$$W = g(\mathbf{X}) + a(f(\mathbf{X}) - \mu)$$

as an estimator of $E[g(\mathbf{X})]$. Now

$$\text{Var}(W) = \text{Var}(g(\mathbf{X})) + a^2 \, \text{Var}(f(\mathbf{X})) + 2a \, \text{Cov}(g(\mathbf{X}), f(\mathbf{X}))$$

Simple calculus shows that the above is minimized when

$$a = \frac{-\text{Cov}(f(\mathbf{X}), g(\mathbf{X}))}{\text{Var}(f(\mathbf{X}))}$$

and for this value of a

$$\text{Var}(W) = \text{Var}(g(\mathbf{X})) - \frac{[\text{Cov}(f(\mathbf{X}), g(\mathbf{X}))]^2}{\text{Var}(f(\mathbf{X}))}$$

Unfortunately, neither $\text{Var}(f(\mathbf{X}))$ nor $\text{Cov}(f(\mathbf{X}), g(\mathbf{X}))$ is usually known, so we cannot usually obtain the above reduction in variance. One approach in practice is to guess at these values and hope the resulting W does indeed have smaller variance than does $g(\mathbf{X})$; whereas a second possibility is to use the simulated data to estimate these quantities.

Example 6g (A Queueing System):   Let D_{n+1} denote the delay in queue of the $n + 1$ customer in a queueing system in which the interarrival times are independent and identically distributed (i.i.d.) with distribution F having mean μ_F and are independent of the service times which are i.i.d. with distribution G having mean μ_G. If X_i is the interarrival time between arrival i and $i + 1$, and if S_i is the service time of customer i, $i \geq 1$, we may write

$$D_{n+1} = g(X_1, \ldots, X_n, S_1, \ldots, S_n)$$

To take into account the possibility that the simulated variables X_i, S_i may by chance be quite different from what might be expected we can let

$$f(X_1, \ldots, X_n, S_1, \ldots, S_n) = \sum_{i=1}^{n} (S_i - X_i)$$

As $E[f(\mathbf{X}, \mathbf{S})] = n(\mu_G - \mu_F)$ we could use

$$g(\mathbf{X}, \mathbf{S}) + a[f(\mathbf{X}, \mathbf{S}) - n(\mu_G - \mu_F)]$$

as an estimator of $E[D_{n+1}]$. As D_{n+1} and f are both increasing functions of S_i, $-X_i$, $i = 1, \ldots, n$ it follows from Theorem 6.1 that $f(\mathbf{X}, \mathbf{S})$ and D_{n+1} are positively correlated, and so the simulated estimate of a should turn out to be negative.

If in the above, we wanted to estimate the expected sum of the delays in queue of the first T arrivals (see Example 6e for the motivation), then we could use $\sum_{i=1}^{N(T)} S_i$ as our control variable. Indeed as the arrival process is usually assumed independent of the service times, it follows that

$$E\left[\sum_{i=1}^{N(T)} S_i \right] = E[S] \; E[N(T)]$$

where $E[N(T)]$ can either be computed by the method suggested in Section 8 of Chapter 7 or it can be estimated from the simulation as in Example 6e. This control variable could also be used if the arrival process were a nonhomogeneous Poisson with rate $\lambda(t)$ for, in this case,

$$E[N(T)] = \int_0^T \lambda(t) \, dt \quad \diamond$$

7. Determining the Number of Runs

Suppose that we are going to use simulation to generate r independent and identically distributed random variables $Y^{(1)}, \ldots, Y^{(r)}$ having mean μ and variance σ^2. We are then going to use

$$\overline{Y}_r = \frac{Y^{(1)} + \cdots + Y^{(r)}}{r}$$

as an estimate of μ. The precision of this estimate can be measured by its variance

$$\mathrm{Var}(\overline{Y}_r) = E[(\overline{Y}_r - \mu)^2]$$
$$= \sigma^2/r$$

Hence we would want to choose r, the number of necessary runs, large enough so that σ^2/r is acceptably small. However, the difficulty is that σ^2 is not known in advance. To get around this, one should initially simulate k runs (where $k \geq 30$) and then use the simulated values $Y^{(1)}, \ldots, Y^{(k)}$ to estimate σ^2 by the sample variance

$$\sum_{i=1}^{k} (Y^{(i)} - \bar{Y}_k)^2/(k - 1)$$

Based on this estimate of σ^2 the value of r which attains the desired level of precision can now be determined and an additional $r - k$ runs can be generated.

Problems

1. Suppose it is relatively easy to simulate from the distributions F_i, $i = 1, \ldots, n$. If n is small, how can we simulate from

$$F(x) = \sum_{i=1}^{n} P_i F_i(x), \qquad P_i \geq 0, \sum_i P_i = 1?$$

Give a method for simulating from

$$F(x) = \begin{cases} \dfrac{1 - e^{-2x} + 2x}{3} & 0 < x < 1 \\[2mm] \dfrac{3 - e^{-2x}}{3} & 1 < x < \infty \end{cases}$$

2. Give a method for simulating a negative binomial random variable.

3. Give a method for simulating a hypergeometric random variable.

4. Suppose we want to simulate a point located at random in a circle of radius r centered at the origin. That is, we want to simulate X, Y having joint density

$$f(x, y) = \frac{1}{\pi r^2} \qquad x^2 + y^2 \leq r^2$$

(a) Let $R = \sqrt{X^2 + Y^2}$, $\theta = \tan^{-1} Y/X$ denote the polar coordinates. Compute the joint density of R, θ and use this to give a simulation method. Another method for simulating X, Y is as follows:

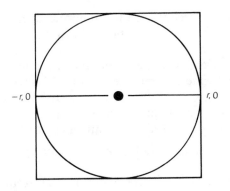

Step 1: Generate independent random numbers U_1, U_2 and set $Z_1 = 2rU_1 - r$, $Z_2 = 2rU_2 - r$. Then Z_1, Z_2 is uniform in the square ϕ whose sides are of length $2r$ and which encloses the circle of radius r.

Step 2: If (Z_1, Z_2) lies in the circle of radius r—that is, if $Z_1^2 + Z_2^2 \le r^2$—set $(X, Y) = (Z_1, Z_2)$. Otherwise return to Step 1.

(b) Prove that this method works, and compute the distribution of the number of random numbers it requires.

5. Suppose it is relatively easy to simulate from F_i for each $i = 1, \ldots, n$. How can we simulate from
(a) $F(x) = \prod_{i=1}^n F_i(x)$.
(b) $F(x) = 1 - \prod_{i=1}^n (1 - F_i(x))$.
(c) Give two methods for simulating from the distribution $F(x) = x^n$, $0 < x < 1$.

6. In Example 2b we simulated the absolute value of a unit normal by using the Von-Neumann rejection procedure on exponential random variables with rate 1. This raises the question of whether we could obtain a more efficient algorithm by using a different exponential density—that is, we could use the density $g(x) = \lambda e^{-\lambda x}$. Show that the mean number of iterations needed in the rejection scheme is minimized when $\lambda = 1$.

7. Give an algorithm for simulating a random variable having density function

$$f(x) = 30(x^2 - 2x^3 + x^4), \qquad 0 < x < 1$$

8. Consider the technique of simulating a gamma (n, λ) random variable by using the rejection method with g being an exponential density with rate λ/n.

(a) Show that the average number of iterations of the algorithm needed to generate a gamma is $n^n e^{1-n}/(n-1)!$.

(b) Use Stirling's approximation to show that for large n the answer to (a) is approximately equal to $e[(n-1)/(2\pi)]^{1/2}$.

(c) Show that the procedure is equivalent to the following:

Step 1: Generate Y_1 and Y_2, independent exponentials with rate 1.

Step 2: If $Y_1 < (n-1)[Y_2 - \log(Y_2) - 1]$, return to Step 1.

Step 3: Set $X = nY_2/\lambda$.

(d) Explain how to obtain an independent exponential along with a gamma from the above algorithm.

9. Set up the alias method for simulating from a binomial random variable with parameters $n = 6$, $p = .4$.

10. Explain how we can number the $\mathbf{Q}^{(k)}$ in the alias method so that k is one of the two points that $\mathbf{Q}^{(k)}$ gives weight.

Hint: Rather than name the initial $\mathbf{Q}$, $\mathbf{Q}^{(1)}$ what else could we call it?

11. Complete the details of Example 4d.

12. Let $X_1, \ldots, X_k$ be independent with

$$P\{X_i = j\} = \frac{1}{n}, \qquad j = 1, \ldots, n, i = 1, \ldots, k$$

If D is the number of distinct values among $X_1, \ldots, X_k$ show that

$$E[D] = n\left[1 - \left(\frac{n-1}{n}\right)^k\right]$$

$$\approx k - \frac{k^2}{2n} \quad \text{when} \quad \frac{k^2}{n} \text{ is small}$$

13. *The Discrete Rejection Method:* Suppose we want to simulate X having probability mass function $P\{X = i\} = P_i$, $i = 1, \ldots, n$ and suppose we can easily simulate from the probability mass function Q_i, $\Sigma_i Q_i = 1$, $Q_i \geq 0$. Let C be such that $P_i \leq CQ_i$, $i = 1, \ldots, n$. Show that the following algorithm generates the desired random variable.

Step 1: Generate Y having mass function $\mathbf{Q}$ and U an independent, random number.

Step 2: If $U \le P_Y/CQ_Y$ set $X = Y$. Otherwise return to Step 1.

14. *The Discrete Hazard Rate Method:* Let X denote a nonnegative integer valued random variable. The function $\lambda(n) = P\{X = n|X \ge n\}$, $n \ge 0$, is called the discrete hazard rate function.

(a) Show that $P\{X = n\} = \lambda(n) \prod_{i=0}^{n-1}(1 - \lambda(i))$.

(b) Show that we can simulate X by generating random numbers $U_1, U_2, \ldots$ stopping at

$$X = \min\{n : U_n \le \lambda(n)\}$$

(c) Apply this method to simulating a geometric random variable. Explain, intuitively, why it works.

(d) Suppose that $\lambda(n) \le p < 1$ for all n. Consider the following algorithm for simulating X and explain why it works: Simulate X_i, U_i, $i \ge 1$ where X_i is geometric with mean $1/p$ and U_i is a random number. Set $S_k = X_1 + \cdots + X_k$ and let

$$X = \min\{S_k : U_k \le \lambda(S_k)/p\}$$

15. Suppose you have just simulated a normal random variable X with mean μ and variance σ^2. Give an easy way to generate a second normal variable with the same mean and variance that is negatively correlated with X.

16. Suppose n balls having weights $w_1, w_2, \ldots, w_n$ are in an urn. These balls are sequentially removed from the urn in the following manner: At each selection, a given ball in the urn is chosen with a probability equal to its weight divided by the sum of the weights of the other balls that are still in the urn. Let $I_1, I_2, \ldots, I_n$ denote the order in which the balls are removed—thus $I_1, \ldots, I_n$ is a random permutation with weights.

(a) Give a method for simulating $I_1, \ldots, I_n$.

(b) Let X_i be independent exponentials with rates w_i, $i = 1, \ldots, n$. Explain how X_i can be utilized to simulate $I_1, \ldots, I_n$.

17. *Order Statistics:* Let $X_1, \ldots, X_n$ be i.i.d. from a continuous distribution F, and let $X_{(i)}$ denote the ith smallest of $X_1, \ldots, X_n$, $i = 1, \ldots, n$. Suppose we want to simulate $X_{(1)} < X_{(2)} < \ldots < X_{(n)}$. One approach is to simulate n values from F, and then order these values. However, this ordering, or *sorting*, can be time-consuming when n is large (the best computer algorithms for sorting n values take on the order of $n \log n$ comparisons).

(a) Suppose that $\lambda(t)$, the hazard rate function of F, is bounded. Show how the hazard rate method can be applied to generate the n variables in such a manner that no sorting is necessary.

Suppose now that F^{-1} is easily computed.

(b) Argue that $X_{(1)}, \ldots, X_{(n)}$ can be generated by simulating $U_{(1)} < U_{(2)} < \ldots < U_{(n)}$—the ordered values of n independent random numbers—and then setting $X_{(i)} = F^{-1}(U_{(i)})$. Explain why this means that $X_{(i)}$ can be generated from $F^{-1}(\beta_i)$ where β_i is beta with parameters $i, n + i + 1$.

(c) Argue that $U_{(1)}, \ldots, U_{(n)}$ can be generated, without any need for sorting, by simulating i.i.d. exponentials $Y_1, \ldots, Y_{n+1}$ and then setting

$$U_{(i)} = \frac{Y_1 + \cdots + Y_i}{Y_1 + \cdots + Y_{n+1}}, i = 1, \ldots, n$$

Hint: Given the time of the $(n + 1)$st event of a Poisson process, what can be said about the set of times of the first n events?

(d) Show that if $U_{(n)} = y$ then $U_{(1)}, \ldots, U_{(n-1)}$ has the same joint distribution as the order statistics of a set of $n - 1$ uniform $(0, y)$ random variables.

(e) Use (d) to show that $U_{(1)}, \ldots, U_{(n)}$ can be generated as follows:

Step 1: Generate random numbers $U_1, \ldots, U_n$

Step 2: Set $U_{(n)} = U_1^{1/n}$, $\quad U_{(n-1)} = U_{(n)}(U_2)^{1/n-1}$,

$$U_{(j-1)} = U_{(j)}(U_{n-j+2})^{1/j-1}, \quad j = 2, \ldots, n - 1$$

18. Let $X_1, \ldots, X_n$ be independent exponential random variables each having rate 1. Set

$$W_1 = X_1/n$$

$$W_i = W_{i-1} + \frac{X_i}{n - i + 1}, \quad i = 2, \ldots, n$$

Explain why $W_1, \ldots, W_n$ has the same joint distribution as the order statistics of a sample of n exponentials each having rate 1.

19. Suppose we want to simulate a large number n of independent exponentials with rate 1—call them $X_1, X_2, \ldots, X_n$. If we were to employ the inverse transform technique we would require one logarithmic computation for each exponential generated. One way to avoid this is to first simulate S_n, a gamma random variable with parameters $(n, 1)$ (say by the method of Section 3.3). Now interpret S_n as the time of the nth event of a Poisson process with rate 1 and use the result that given S_n the set of the first $n - 1$ event times is distributed as the set of $n -$

1 independent uniform $(0, S_n)$ random variables. Based on this, explain why the following algorithm simulates n independent exponentials.

Step 1: Generate S_n, a gamma random variable with parameters $(n, 1)$.

Step 2: Generate $n - 1$ random numbers $U_1, U_2, \ldots, U_{n-1}$.

Step 3: Order the U_i, $i = 1, \ldots, n - 1$ to obtain $U_{(1)} < U_{(2)} < \ldots < U_{(n-1)}$.

Step 4: Let $U_{(0)} = 0$, $U_{(n)} = 1$, and set $X_i = S_n(U_{(i)} - U_{(i-1)})$, $i = 1, \ldots, n$.

When the ordering (Step 3) is performed according to the algorithm described in Section 5, the above is an efficient method for simulating n exponentials when all n are simultaneously required. If memory space is limited, however, and the exponentials can be employed sequentially discarding each exponential from memory once it has been used, then the above may not be appropriate.

20. Consider the following procedure for randomly choosing a subset of size k from the numbers $1, 2, \ldots, n$: Fix p and generate the first n time units of a renewal process whose interarrival distribution is geometric with mean $1/p$—that is $P\{\text{interarrival time} = k\} = p(1 - p)^{k-1}$, $k = 1, 2, \ldots$. Suppose events occur at times $i_1 < i_2 < \ldots < i_m \leq n$. If $m = k$ stop; $i_1, \ldots, i_m$ is the desired set. If $m > k$, then randomly choose (by some method) a subset of size k from $i_1, \ldots, i_m$ and then stop. If $m < k$, take $i_1, \ldots, i_m$ as part of the subset of size k and then select (by some method) a random subset of size $k - m$ from the set $\{1, 2, \ldots, n\} - \{i_1, \ldots, i_m\}$. Explain why this algorithm works. As $E[N(n)] = np$ a reasonable choice of p is to take $p \approx k/n$. (This approach is due to Dieter.)

21. Consider the following algorithm for generating a random permutation of the elements $1, 2, \ldots, n$. In this algorithm, $P(i)$ can be interpreted as the element in position i

Step 1: Set $k = 1$

Step 2: Set $P(1) = 1$

Step 3: If $k = n$, stop. Otherwise, let $k = k + 1$.

Step 4: Generate a random number U and let

$$P(k) = P([kU] + 1)$$
$$P([kU] + 1) = k$$

Go to Step 3.

(a) Explain in words what the algorithm is doing.
(b) Show that at iteration k—that is, when the value of $P(k)$ is initially set—that $P(1), P(2), \ldots, P(k)$ is a random permutation of $1, 2, \ldots, k$.
Hint: Use induction and argue that

$$P_k\{i_1, i_2, \ldots, i_{j-1}, k, i_j, \ldots, i_{k-2}, i\}$$

$$= P_{k-1}\{i_1, i_2, \ldots, i_{j-1}, i, i_j, \ldots, i_{k-2}\}\frac{1}{k}$$

$$= \frac{1}{k!} \text{ by the induction hypothesis}$$

The above algorithm can be used even if n is not initially known.

22. Verify that if we use the hazard rate approach to simulate the event times of a nonhomogeneous Poisson process whose intensity function $\lambda(t)$ is such that $\lambda(t) \leq \lambda$, then we end up with the approach given in Method 1 of Section 5.

23. For a nonhomogeneous Poisson process with intensity function $\lambda(t)$, $t \geq 0$, where $\int_0^\infty \lambda(t)dt = \infty$, let $X_1, X_2, \ldots$ denote the sequence of times at which events occur.
(a) Show that $\int_0^{X_1} \lambda(t)dt$ is exponential with rate 1.
(b) Show that $\int_{X_{i-1}}^{X_i} \lambda(t)dt$, $i \geq 1$, are independent exponentials with rate 1, where $X_0 = 0$.

In words, independent of the past, the additional amount of hazard that must be experienced until an event occurs is exponential with rate 1.

24. Give an efficient method for simulating a nonhomogeneous Poisson process with intensity function

$$\lambda(t) = b + \frac{1}{t + a}, \qquad t \geq 0$$

25. Let (X, Y) be uniformly distributed in a circle of radius r about the origin. That is, their joint density is given by

$$f(x, y) = \frac{1}{\pi r^2}, \qquad 0 \leq x^2 + y^2 \leq r^2$$

Let $R = \sqrt{X^2 + Y^2}$ and $\theta = $ arc tan Y/X denote their polar co-ordinates. Show that R and θ are independent with θ being uniform on $(0, 2\pi)$ and $P\{R < a\} = a^2/r^2$, $0 < a < r$.

26. Let R denote a region in the two-dimensional plane. Show that for a two-dimensional Poisson process, given that there are n points located in R, the points are independently and uniformly distributed in R—that is, their density is $f(x, y) = c$, $(x, y) \in R$ where c is the inverse of the area of R.

27. Let $X_1, \ldots, X_n$ be independent random variables with $E[X_i] = \theta$, $\text{Var}(X_i) = \sigma_i^2$, $i = 1, \ldots, n$, and consider estimates of θ of the form $\sum_{i=1}^{n} \lambda_i X_i$ where $\sum_{i=1}^{n} \lambda_i = 1$. Show that $\text{Var}(\sum_{i=1}^{n} \lambda_i X_i)$ is minimized when $\lambda_i = (1/\sigma_i^2)/(\sum_{j=1}^{n} 1/\sigma_j^2)$, $i = 1, \ldots, n$.

Possible Hint: If you cannot do this for general n, try it first when $n = 2$.

The following three problems are concerned with the estimation of $\int_0^1 g(x)dx = E[g(U)]$ where U is uniform $(0, 1)$.

28. *The Hit-Miss Method:* Suppose g is bounded in $[0, 1]$—for instance, suppose $0 \le g(x) \le b$ for $x \in [0, 1]$. Let U_1, U_2 be independent random numbers and set $X = U_1$, $Y = bU_2$—so the point (X, Y) is uniformly distributed in a rectangle of length 1 and height b. Now set

$$I = \begin{cases} 1 & \text{if} \quad Y < g(X) \\ 0 & \text{otherwise} \end{cases}$$

That is accept (X, Y) if it falls in the shaded area of Figure 11.6.

(a) Show that $E[bI] = \int_0^1 g(x)dx$.

(b) Show that $\text{Var}(bI) \ge \text{Var}(g(U))$, and so hit-miss has larger variance than simply computing g of a random number.

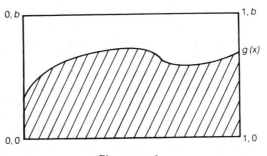

Figure 11.6.

29. *Stratified Sampling:* Let $U_1, \ldots, U_n$ be independent random numbers and set $\overline{U}_i = (U_i + i - 1)/n$, $i = 1, \ldots, n$. Hence, $\overline{U}_i$, $i \ge$

1, is uniform on $((i - 1)/n, i/n)$. $\Sigma_{i=1}^{n} g(\overline{U}_i)/n$ is called the stratified sampling estimator of $\int_0^1 g(x)dx$.

(a) Show that $E[\Sigma_{i=1}^{n} g(\overline{U}_i)/n] = \int_0^1 g(x)dx$.

(b) Show that $\text{Var}[\Sigma_{i=1}^{n} g(\overline{U}_i)/n] \le \text{Var}[\Sigma_{i=1}^{n} g(U_i)/n]$.

Hint: Let U be uniform $(0, 1)$ and define N by $N = i$ if $(i - 1)/n < U < i/n$, $i = 1, \ldots, n$. Now use the conditional variance formula to obtain

$$\text{Var}(g(U)) = E[\text{Var}(g(U)|N)] + \text{Var}(E[g(U)|N])$$
$$\le E[\text{Var}(g(U)|N]$$
$$= \sum_{i=1}^{n} \frac{\text{Var}(g(U)|N = i)}{n} = \sum_{i=1}^{n} \frac{\text{Var}[g(\overline{U}_i)]}{n}$$

30. Let X be a random variable on $[0, 1]$ whose density is $f(x)$. Show that we can estimate $\int_0^1 g(x)dx$ by simulating X and then taking as our estimate $g(X)/f(X)$. This method, called *importance sampling*, tries to choose f similar in shape to g so that $g(X)/f(X)$ has a small variance.

31. Consider a queueing system in which each service time independent of the past has mean μ. Let W_n and D_n denote respectively the amount of time customer n spends in the system and in queue respectively. Hence, $D_n = W_n - S_n$ where S_n is the service time of customer n. Therefore,

$$E[D_n] = E[W_n] - \mu$$

If we use simulation to estimate $E[D_n]$, should we

(a) use the simulated data to determine D_n which is then used as an estimate of $E[D_n]$; or

(b) use the simulated data to determine W_n and then use this quantity minus μ as an estimate of $E[D_n]$.

Repeat if we want to estimate $E[W_n]$.

32. Show that if X and Y have the same distribution then $\text{Var}((X + Y)/2) \le \text{Var}(X)$. Hence, conclude that the use of antithetic variables can never increase variance (though it need not be as efficient as generating an independent set of random numbers).

33. Explain how the use of cyclic permutations, or a variant of such, can be used to reduce variance in a queueing simulation.

References

1. J. Banks and J. Carson, "Discrete Event System Simulation," Prentice-Hall, Englewood Cliffs, New Jersey, 1984.

2. G. Fishman, "Principles of Discrete Event Simulation," John Wiley, New York, 1978.

3. D. Knuth, "Semi Numerical Algorithms," Vol. 2 of *The Art of Computer Programming*, 2nd ed. Reading, Mass.: Addison-Wesley, 1981.

4. A. Law and W. Kelton, "Simulation Modeling and Analysis," McGraw-Hill, New York, 1982.

5. S. Ross and Z. Schechner, "Simulation Uses of the Exponential," Operations Research Center, University of California, Berkeley, 1984.

6. R. Rubenstein, "Simulation and the Monte Carlo Method," John Wiley, New York, 1981.

Answers to Selected Problems

Chapter 1

1. The probability of each point is $\frac{1}{9}$.
2. The sample space consists of 6 points.
3. $\frac{1}{8}$ 5. $\frac{3}{4}$ 13. $\approx .49$

14. $\dfrac{1}{2-p}$ and $\dfrac{1-p}{2-p}$ 17. $\frac{3}{4}, \frac{9}{16}$ 18. $\frac{1}{2}, \frac{1}{3}$

19. $\frac{11}{36}, \frac{1}{3}$ 20. $\frac{5}{12}$ 21. $\frac{20}{21}$

22. $\frac{1}{6}$ 24. 9   26. $\frac{7}{12}$

27. $\frac{3}{5}$ 28. $\frac{4}{5}$ 29. $\frac{1}{2}$

30. $\frac{1}{3}, \frac{1}{5}, 1$ 31. $\frac{4}{9}$ 32. $\frac{1}{11}$

33. $\frac{12}{37}$ 35. Not valid

Chapter 2

1. $\frac{7}{15}$ 3. $\frac{1}{4}, \frac{1}{2}, \frac{1}{4}$

5. $P\{\max = i\} = \frac{11}{36}, \frac{1}{4}, \frac{7}{36}, \frac{5}{36}, \frac{1}{12}, \frac{1}{36}, i = 6, 5, \ldots, 1$

6. p^5 7. .027, .189, .441, .343

8. $p(0) = p(1) = \frac{1}{2}$ 9. $\frac{1}{2}, \frac{1}{10}, \frac{1}{5}, \frac{1}{10}, \frac{1}{10}$

10. $\frac{200}{216}$ 11. $\frac{3}{8}$

12. $\frac{11}{243}$ 20. .054   22. $\frac{1}{32}$

27. .394, .303, .091   28. $c = \frac{3}{4}$ 29. $c = \frac{3}{8}, \frac{11}{16}$

30. $\frac{1}{2}$ 32. $c = 2$ 33. $\frac{31}{6}$

34. $\frac{1}{3}$ 38. $p(1 - p)$ 39. $\frac{n}{n + 1}, 0, 1$

40. $\frac{1}{n + 1}, \frac{1}{2n + 1} - \frac{1}{(n + 1)^2}$ 44. $\frac{e^t - 1}{t}, \frac{1}{2}, \frac{1}{12}$

47. $\frac{pe^t}{1 - (1 - p)e^t}$ 50. $\geq \frac{2}{5}$

51. (i) $\leq \frac{2}{3}$, (ii) $\approx 1 - \Phi(5/\sqrt{10})$ 52. .1498

54. (ii) $\frac{nk}{n + m}$

56. $nP_i, nP_i(1 - P_i), - nP_iP_j, i \neq j, \sum_{i}^{r} (1 - P_i)^n$

Chapter 3

2. $2, \frac{5}{3}, \frac{12}{5}$ 3. No   4. (b) $\frac{5}{3}$

5. $\frac{4}{9}, \frac{8}{27}, \frac{2}{9}, \frac{1}{27}, \frac{5}{3}$ 6. $\frac{9}{5}, 1$ 7. (a) 6, (b) 7

12. $1 + 1/\lambda$ 13. $\frac{1}{4}$ 14. $y^2/3$

17. (a) $\frac{19}{2}$, (b) $\frac{5}{2}$ 18. 21   19. 438

21. 500, 33,333   25. (a) 30, 30, (b) 10, 10

26. (c) No, (d) Yes, since $E[N] = 5$ 32. $\frac{n + 1 - m}{n + 1}$

Chapter 4

2. 8 states   4. Need 6 states   6. .26

7. .6665   12. $\frac{2}{5}$ 14. $\pi_i = \frac{1}{5}$ for all i

15. $\pi_i = \frac{1}{13}$ for all i 18. $\pi_0 = \pi_3 = \frac{1}{20}, \pi_2 = \pi_1 = \frac{9}{20}$

19. $\frac{11}{16}$ 20. $\frac{6}{17}, \frac{7}{17}, \frac{4}{17}$ 22. (a) i, (b) i/m

24. (iii) $\dfrac{pq}{r + q}$, (iv) 0.55   27. (i) No, (ii) Yes

30. $S_{11} = \dfrac{r^2}{4(1 - q)^2}$, $S_{10} = \dfrac{r}{2(1 - q)}$ 34. $\dfrac{n}{1 - \mu}$

36. $\frac{1}{3}, 1, \frac{1}{2}$

37. (d) $(1 - p)^2$, $2p(1 - p)$, p^2 (g) $\displaystyle\sum_{j=1}^{N-i} N/j$

44. 168

Chapter 5

1. e^{-1}, e^{-1} 2. $0, \frac{1}{27}, \frac{1}{4}$

3. e^{-1} 7. $\frac{3}{7}$

8. Z is exponential with rate $\lambda_1 + \lambda_2$ in both cases.

15. $e^{-1/10}$, $e^{-1/4}$, e^{-1} 16. $e^{-3s} + 3se^{-3s}$

18. $\dfrac{\lambda_1}{\lambda_1 + \lambda_2}$ 20. 33/81

21. (a) e^{-2}, (b) 2 P.M., (c) $1 - 5e^{-4}$ 22. (a) e^{-2t}, (b) $2t$

23. (a) $1 - e^{-3}$, (b) 67   24. (a) $\frac{1}{9}$, (b) $\frac{5}{9}$

28. $e^{-11}(11)^n/n!$ 29. Poisson with mean 63

33. 40,000 and 1.6×10^8

Chapter 6

1. $v_{n,m} = \lambda nm$, $P_{(n,m),(n+1,m)} = P_{(n,m),(n,m+1)} = \frac{1}{2}$

2. $v_{n,m} = \alpha n + \beta m$, $P_{(n,m),(n-1,m+1)} = \dfrac{\alpha n}{\alpha n + \beta m} = 1 - P_{(n,m),(n+2,m-1)}$

3. No   4. $\lambda_n = \lambda\alpha_n$, $\mu_n = \mu$

11. (a) $\frac{30}{37}$, (b) $\frac{28}{37}$, (c) would increase business by .45 customers per hour

12. (a) $\frac{245}{272}$, (b) $\frac{125}{272}$ 13. (a) $\frac{116}{143}$, (b) $\frac{148}{175}$

14. $\dfrac{\lambda(1 - \alpha)\mu_2}{\lambda\alpha\mu_1 + \mu_1\mu_2 + \lambda(1 - \alpha)\mu_2}$ 15. (a) $\frac{1068}{761}$, (b) $\frac{336}{761}$

16. (a) 1, (b) $\frac{1}{2}$ 19. Poisson with rate $\min(\lambda, s\mu)$

Chapter 7

1. (a) Yes, (b) No, (c) No
5. One every 5 months
11. (i) Yes, (ii) No

3. $e^{-5/2}$
10. (b) $E[T] = 12$

15. $\dfrac{\mu_G}{\mu + 1/\lambda}$ where μ_G is the mean of G

17. Optimal $T = 8$

24. (i) $\frac{3}{4}$, (ii) $1 - e^{-1}$

19. $\dfrac{cN(N - 1) + 2\lambda KNc + \lambda^2 K^2 c}{2N + 2\lambda K}$

26. (a) $\frac{2}{3}$, (b) $\frac{2}{3}$, $\frac{4}{3}$, $\frac{3}{3}$

Chapter 8

1. (a) λ/μ, (b) $\dfrac{\mu}{\lambda + \mu}$

2. \$30/hr

3. (a) $4, C + \frac{1}{2}$, (b) $\frac{7}{2}$

7. (a) $P_n = (\lambda/\alpha\mu)^n \left(1 - \dfrac{\lambda}{\alpha\mu}\right)$, $n \geq 0$, need $\lambda < \alpha\mu$

(b) $\dfrac{\lambda}{\mu(\alpha\mu - \lambda)}$

(d) $1/\alpha\mu$

8. (a) $P_0 = \dfrac{2\mu - \lambda}{2\mu + \lambda}$, $P_n = \dfrac{\lambda^n}{2^{n-1}\mu^n} P_0$, $n \geq 1$

(c) $\dfrac{\lambda^2(2\mu - \lambda)}{2\mu(\lambda + \mu)(2\mu + \lambda)} + \dfrac{2\lambda^2}{\mu(2\mu - \lambda)}$

9. (a) $\frac{81}{493}$, (b) \$70.72 and \$29.28 per day

10. (a) $\frac{2}{3}$, (b) $\frac{2}{3}$, (c) $\frac{7}{9}$, (d) $\frac{7}{12}$

12. (c) $\lambda < \mu$

13. (a) $(1/\theta + 1/\mu_1)^{-1}$

17. (i) $\frac{49}{6}$, (ii) $\frac{49}{180}$

21. (a) P_0, (b) $\dfrac{\lambda^2 E[S] E[S^2]}{1 - \lambda E[S]} + \lambda(E[S])^2$

24. (b) (ii) $\dfrac{A}{1 - \lambda E[S]}$

25. $\frac{17}{480}$

33. $\frac{8}{38}$

Chapter 9

10. Upper bound is 169/512; lower is 7/32. Exact is 7/32.
22. (a) $\frac{5}{12}$, (b) $\frac{13}{12}$

24. $\frac{31}{60}$

Chapter 10

1. Normal with mean 0 and variance $3s + t$.

2. Normal with mean $A + (B - A)\dfrac{s - t}{t_2 - t_1}$ and variance

$$\dfrac{(t_2 - s)(s - t_1)}{t_2 - t_1}$$

4. (b) $\dfrac{1 - e^{-2\mu B}}{1 - e^{-2\mu(A+B)}}$

5. 0

7. $\frac{1}{6}$

9. *Hint*: Condition on $X(t_1)$

14. $\lambda(1 - s)$ when $s < 1$ and 0 otherwise.

Chapter 11

1. Let U_1, U_2, U_3 be random numbers. Now set

$$X = \begin{cases} \dfrac{-\log U_1}{2} & \text{if} \quad U_3 < \frac{1}{3} \\ U_2 & \text{if} \quad U_3 \geq \frac{1}{3} \end{cases}$$

X has the desired distribution.

5. Let X_i have distribution F_i, $i = 1, \ldots, n$, and let them be independent. For (a) set $X = \max(X_1, \ldots, X_n)$ and for (b) set $X = \min(X_1, \ldots, X_n)$. For c let F_i denote the uniform $(0, 1)$ distribution and apply (a). Another approach to (c) is to use the inverse transform method and thus set $X = U^{1/n}$.

6. With $g(x) = \lambda e^{-\lambda x}$

$$\dfrac{f(x)}{g(x)} = \dfrac{2}{\lambda\sqrt{2\pi}} \dfrac{e^{-x^2/2}}{e^{-\lambda x}} = \dfrac{2}{\lambda\sqrt{2\pi}} \exp\{-\tfrac{1}{2}[(x - \lambda)^2 - \lambda^2]\}$$

$$= \dfrac{2}{\lambda\sqrt{2\pi}} e^{\lambda^2/2} \exp\{-(x - \lambda)^2/2\}$$

Hence, $c = 2e^{\lambda^2/2}/(\lambda\sqrt{2\pi})$ and simple calculus shows that this is minimized when $\lambda = 1$.

16. (b) Order the X_j's and let I_i denote the index of the ith smallest X.

Index